W9-CJP-009

# MECHANISMS OF ANIMAL DISCRIMINATION LEARNING

**N. S. SUTHERLAND**

*Laboratory of Experimental Psychology*
*University of Sussex*
*Brighton, England*

**N. J. MACKINTOSH**

*Department of Psychology*
*Dalhousie University*
*Halifax, Nova Scotia*

**ACADEMIC PRESS • New York and London • 1971**

ACADEMIC PRESS, INC.
111 Fifth Avenue, New York, New York 10003

*United Kingdom Edition published by*
ACADEMIC PRESS, INC. (LONDON) LTD.
Berkeley Square House, London W1X 6BA

LIBRARY OF CONGRESS CATALOG CARD NUMBER: 79-127705

PRINTED IN THE UNITED STATES OF AMERICA

# Contents

## 10. Partial Reinforcement and Extinction



## 11. Partial Reinforcement and Choice Behavior



## 12. Some Comparative Psychology



## 13. Formal Models

# Preface

In this book we attempt a relatively exhaustive review of the field of animal discrimination learning, with excursions into other areas such as generalization, partial reinforcement, and some aspects of comparative psychology. The phenomena discussed are those that in our view may profitably be interpreted in terms of a theory of selective attention: we believe that this includes much of the field traditionally known as "animal learning." Although we are doubtless biased in our interpretation, we have not ignored other approaches, and have attempted to treat many of them with sympathy. Often, of course, there is no conflict between our own approach and that of others: a complete explanation of many of the phenomena discussed will incorporate such concepts as frustration, stimulus aftereffects, and perhaps orienting responses, as well as the notion of selective attention. We believe that one of the problems bedeviling animal psychology is, indeed, that most effects have several different causes.

Two recent books (Lovejoy, 1968; Trabasso and Bower, 1968) have appeared in the same general area as that dealt with in this volume. However, Trabasso and Bower consider primarily human discrimination learning whereas we deal almost exclusively with animal research, and Lovejoy proposes a specific model to cope with a rather more restricted range of phenomena whereas we attempt to cover a wider range and adopt a more eclectic approach.

Most of the ideas in the book have evolved in the course of a close and continuous collaboration between us over the last ten years. N.S.S. accepts primary responsibility for Chapters 1–3, 5, 10, and 13 and N.J.M. for Chapters 4, 6–9, 11, and 12. Having completed writing, we confess to a feeling of disappointment that we have not always been able to explain to our own satisfaction the many phenomena reviewed in terms of selective attention; some of them indeed seem recalcitrant to explanation in any terms. Nonetheless, we feel that although selective attention may not be the whole truth it is at least part of the truth.

# Acknowledgments

Both our work over the last decade and the writing of this book have benefited from discussions with colleagues too numerous to mention. We are particularly indebted to A. R. Wagner who read the entire manuscript with great care, and made a large number of valuable suggestions. We have adopted nearly all of these; some of the errors that remain may be due to our refusal to adopt the remainder. Many others have read individual chapters and clarified our ideas for us: we should particularly like to thank E. Lovejoy, H. M. Jenkins, and C. Turner. Much of the experimental work reported here was conducted by a succession of able research assistants: we are grateful in particular to Lani Andelman, Anne Carr, Valerie Holgate, Lydia Little, and John Wolfe for their conscientious performance of an often tedious task. Finally, we thank Louise Harrison and Margaret Ross for their secretarial assistance in the preparation of the manuscript.

Much of our own work described in the book was supported under contract by the Office of Naval Research, and we should like to acknowledge their generosity in financing research undertaken outside the confines of the United States. Our work has also been supported at different times in England by the Nuffield Foundation, the Science Research Council, and the Medical Research Council, and in Canada by the National Research Council and the Killam Research Fund of Dalhousie University. We are grateful for this financial support.

# Mechanisms of Animal Discrimination Learning

# CHAPTER 1

# The History of Selective Attention

## I. Introduction

William James (1890) began his famous chapter on "Attention" by remarking that it was strange that "so patent a fact as the perpetual presence of selective attention has received hardly any notice from psychologists of the English empirical school [p. 402]." He might have been equally astonished at the dearth of work on this problem had he lived until very recent times.

Elsewhere James (1892) states the problem of selective attention as follows:

> One of the most extraordinary facts of our life is that, although we are besieged at every moment by impressions from our whole sensory surface, we notice so very small a part of them... Yet the physical impressions which do not count are *there* as much as those which do, and affect our sense organs just as energetically. Why they fail to pierce the mind is a mystery [p. 217].

It is towards this mystery or some part of it that this book is directed. We are interested mainly in selective attention in animals. We shall try to show that animals, like men, attend now to one thing, now to another. There are two aspects of the problem. Of the stimuli impinging on peripheral receptors only a few are effective at any one time in determining the animal's response. Secondly, it seems certain that only some features of the stimuli present at any one time are actually stored in memory. We believe that attention defined in this way enters into almost all mammalian behavior, and we shall try to show that many puzzling phenomena connected with learning are to be ascribed to the animal learning to attend to a particular feature of the stimulus rather than to its learning to make some overt response. Unfortunately, work on learning and perception has usually been conducted by different investigators and often in different laboratories. We shall attempt to show that many behavioral findings usually thought of as being "learning" phenomena are really best explained in perceptual terms.

The dichotomy between learning and perception was part of the gulf that divided the two most influential schools of experimental psychology in the twenties and thirties of the present century. For different reasons both the American behaviorists and the Gestaltists largely ignored the problem of selective attention. In this chapter we shall first consider some of the reasons why these two schools neglected the problem with which we are here concerned. We will then review some of the influences that have kept the problem alive since the days of William James and have more recently brought it to the fore.

## II. The Behaviorists

The behaviorists, following Watson, wished to have no truck with unobservables and in particular with mental events. They were guilty of a most extraordinary piece of wishful thinking. Because they wanted to base psychology on observable events, they came to believe that no unobservable events influenced behavior and this led Watson (1913) to revive a proposal made by James himself, namely, that the "so-called higher thought processes go on in terms of faint reinstatements of the original muscular act (including speech here) and these are integrated into systems which respond in serial order." Both James and Watson were anticipated by Sechenov who in 1863 wrote:

> Association is effected through continuous series of reflexes in which the end of a preceding reflex coincides in time with the beginning of a subsequent . . . . Thought in five year olds is mediated through words or whispers, surely through movements of tongue and lips, which is also very frequently (perhaps always, but in different degrees) true of the thinking of adults [pp. 481, 498].[1]

The theory received little or no support from experiments recording from muscles during thinking (Jacobson, 1932; Max, 1937) but the strongest disproof came from experiments showing that human beings under curare can think normally and can remember the results of calculations carried out when the peripheral musculature is blocked by curare (e.g., Smith, Brown, Toman, & Goodman, 1947).

Despite the lack of supporting evidence, Watson's proposal of chain

[1] This quotation and several others in this section were taken from Razran's (1965) excellent article on the history of Russian and American psychology.

reflexes was very influential and was the direct forerunner of Hull's fractional anticipatory goal responses, and of various mediating responses that have been proposed since, for example by Osgood (1953) and by Kendler and Kendler (1962). With his usual acumen in seeing what was wrong with others' theories, Lashley (Lashley and Ball, 1929) put his finger on the crucial point:

> The chief attractiveness of the chain-reflex theories lay in their promise that psychology might progress by simple objective methods and escape the need for indirect inference concerning neural complexities which could not be directly observed. The accumulation of evidence against the existence of the observable muscular contractions which can serve as a basis for the chain-reflex forces us either to seek the completion of the reflex in action currents without observable muscular contraction, or to turn to the central nervous system for explanation of maintenance as well as coordination of activity. There is no direct evidence supporting the excitatory effects of such action currents and it will not be less difficult to disentangle the complex electrical phenomena of tonus and sublimal movement than to measure cerebral changes directly, so that the advantage of greater objectivity cannot be claimed for the motor theory [p. 71].

Although Lashley wrote this in 1929, theorists continued to be confused about the issues. There are at least five different things that can be meant by a mediating response. (1) An actual muscle twitch which gives rise to proprioceptive feedback which in turn serves as a stimulus for a further response. (2) Efferent firing which, although below the threshold required to produce any observable change in muscle tonus, in some unspecified way gives rise to activity in afferents which can serve as a stimulus for the next response. (3) Activity within the central nervous system which does not directly result in a response but which can be conditioned to a stimulus and which can in turn serve as though it were a stimulus for a response. Hebb (1949) proposed mediating processes of this kind. (4) A verbal response mediated at a central level; this is really a subclass of (3) and has been proposed by Kendler and Kendler (1962) among others. (5) A change in consciousness not evinced directly in behavior. If it is assumed that changes in consciousness are mediated by activity in the central nervous system, then (5) is also a subclass of (3). There is of course no reason at all to suppose that central states of the brain cannot be conditioned to stimuli in the same way that responses are, and the only reason why the behaviorists have been so reluctant to accept that this does in fact happen is their desire to limit themselves to easily observable events.

Until recent years, mediating responses were used by the behaviorists only to account for thinking in man and for insightful learning in animals. They were not applied to the problem of why responses should be controlled now by one aspect of the stimulus situation, now by another. It is this latter problem that is our main concern. Hull (1935) applied mediating responses to the problem of insight by suggesting that when an animal eats or drinks, some fraction of the eating or drinking response (the fractional anticipatory goal response) may be conditioned to stimuli which occur in time before the act of eating or drinking. Such a suggestion is not wholly implausible, though how far such a mechanism can explain insightful learning is a debatable point (cf. Deutsch, 1960). Although it is implausible to suppose that overt mediating responses can provide a mechanism to explain all the phenomena of selective attention in animals, it is extremely difficult to prove conclusively from existing evidence that internal mediating processes are necessarily involved in selective attention.

This point can be illustrated by an example drawn from an experiment performed in two stages on rats using a Lashley jumping stand. At the first stage the rat is trained to jump to the left when two horizontal rectangles are presented and to jump right when two vertical rectangles are presented. At the next stage, the animal is trained to approach a white horizontal rectangle and to avoid a black vertical rectangle. The second problem can be solved by learning either to go to the horizontal rectangle and not to the vertical or by learning to approach the white rectangle and avoid the black or by learning simultaneously about both aspects of the rectangles. It turns out that rats pretrained at the first stage to respond in terms of differences in orientation tend to solve the second problem in terms of the orientation of the rectangles rather than in terms of the brightness difference between them. In anthropomorphic terms, the rat has learned at the first stage to pay attention to orientation and hence it solves the second problem in terms of the difference in orientation between the two stimuli rather than in terms of the difference in brightness. If there were some overt response ($R_O$) that must be learned in order for the animal to detect a difference between horizontal and vertical rectangles and if a second response ($R_B$) were needed to detect a difference between black and white, the animal might at the first stage of the problem learn to make the first response. If a difference in orientation can only be detected if response $R_O$ occurs, we would have an explanation of why animals trained at Stage 1 to discriminate an orientation difference should tend to solve the Stage 2 problem in terms of an orientation difference. Since they have already learned the response $R_O$ at Stage 1, they would continue to make this at Stage 2 and hence would learn predominantly in terms of the orientation of the rectangles. Unfortunately the only way of proving that the above phenomenon is not dependent on making external observing responses would be to show that the phenomenon still occurred when stabilized retinal images are used and this experiment has never been conducted.

The experiments that come nearest to demonstrating that selective attention in animals is, at least on some occasions, dependent on the learning of an internal mediating response and not on the learning of an overt orienting response, are those using diffuse light and noise as cues. For example, Kamin (1968) has demonstrated that if rats are trained to make a conditioned emotional response to noise, then they do not subsequently learn to attach the response to a light when at a second stage of the experiment shock onset is signaled by noise plus light. It is hard to believe that animals could learn an external orienting response to the noise that would subsequently prevent them detecting the light.

In what follows, we have adopted the position that much selective attention is to be explained by internal mediating processes. For the reasons set out above, this seems plausible though at the moment not conclusively proven; it would, in fact, make little difference to the formal operation of our own theory of selective attention if some of the phenomena associated with it were dependent on external orienting responses rather than on internal processes.

In man, the situation is different, since different stimulus dimensions might give rise to different subvocal verbal responses. In the first of the two experiments set out above, a human subject might solve the problem in the following way: at the first stage, he might learn to make a verbal response such as: "Look at whether the rectangles are horizontal or vertical: if they are horizontal go left, if they are vertical go right." At the second stage he might continue to make the early part of the verbal response, since it is already conditioned and merely alter the second part to, "If horizontal approach, if vertical avoid." This was in fact the explanation adopted by Kendler and Kendler (1962) and they would predict in consequence that the phenomenon would not occur in animals or in preverbal children.

The behaviorists in general have denied the existence of the process of selective attention in animals except where the process could be mediated by overt orientation responses. For example, Spence (1936) believed that learning would take place about any stimulus "which impinged on the animal's sensorium at or near the critical moment of response." The realization that there may be mediating processes entirely within the central nervous system, however, immediately opens up the possibility of central responses equivalent to, "look whether it is horizontal or vertical," being conditioned and hence makes it possible to explain, and desirable to study, the type of phenomena instanced. The failure to find any confirmation for muscular responses to mediate thinking has helped to open the way to this type of approach. As we shall see, other influences including the development of information theory and the sheer weight of experimental evidence have led to the study of central mediating processes, and to the reopening of the question of selective attention.

## III. The Gestaltists

The other main experimental school of psychologists in the twenties and thirties were the Gestaltists. Their approach forms almost a complete contrast to that of the behaviorists. Whereas the one school was interested primarily in response processes, the other was interested primarily in perception and thinking. The behaviorists attributed as much behavior as they could to learning, the Gestaltists emphasized innate processes. One school studied behavior, the other had a strong phenomenalist orientation. Even the experimental subjects studied by the two schools were different, the Gestaltists confining their attention mainly to man and other primates, the behaviorists studying mainly the laboratory rat and other lower mammals. On one thing they were agreed: selective attention was not of much importance. The Gestaltists' motives for ignoring this problem are not so clear as those of the behaviorists. They were in general anxious to deny the influence of learning on perception, and since variations in selective attention must involve learning, this may have been a sufficient reason for them to avoid the issue.

It is true that they did not completely ignore the problem. Thus Koffka (1935) distinguished between voluntary attention, "where the origin of the force was in the Ego," and involuntary attention, "where the origin of the force lay in the object," and Gottschaldt (1929) conducted his famous experiments on hidden figures. However, Koffka was anxious to minimize the importance of Gottschaldt's results:

> They "prove that forces originating in the Ego can exert an influence on the behavioural environment of the Ego by influencing its organization. At the same time Gottschaldt proved that there are very definite limits to such an influence . . . the attitudes will not be able to overcome strong internal forces of organization."

In this way, the field was left clear for the majestic, if obscure, operation of Koffka's restraining and cohesive forces. The Gestaltists may have acknowledged the problem, but they did not make much use of the concept of attention in their explanations and did little direct work upon it.

Although the behaviorists and the Gestaltists took little notice of selective attention, there were other workers who were aware of the problem, and more recently, developments in information theory, work on human skills, and some neurophysiological findings have thrust the problem to the fore. We shall now review some of these other trends.

## IV. Pavlov

Pavlov was aware of the role of stimulus selectivity and some of his findings point directly to the importance of attention in learning. For example he tried to condition salivation in a dog to a tactile and heat stimulus presented simultaneously (Pavlov, 1927). When the conditioned response to the two stimuli presented together was appearing regularly, he gave tests with each stimulus presented on its own, and discovered that salivation was now elicited by the tactile stimulus, but not by the heat stimulus. The animals could, however, be conditioned to the heat stimulus given without the tactile stimulus being present. This clearly implies that whether or not learning will occur about a given stimulus depends upon what other stimuli are present at the time. Pavlov also demonstrated that when the paired stimuli were light and tone, conditioning occurred only to the tone, unless the tone were made very faint. Pavlov was also the first to demonstrate the phenomenon of "transfer along a continuum in animals." He showed that to obtain differentiation between a circle and an almost circular ellipse it was best to train the dog first to differentiate between a circle and ellipses differing more widely from the circle. We shall try to show below that this phenomenon is also best explained in terms of selective attention. It was Pavlov who first used the word "analyzers" to refer to the different mechanisms classifying the stimulus input, and having the concept of different analyzing mechanisms is a big step on the way to a theory of selective attention, since it immediately raises the question of what are the relations between different analyzers.

Although some of his experiments strongly suggest that the use of different analyzers can be modified by learning, Pavlov does not seem to have considered this possibility very explicitly. He does, however, conclude, "A definite interaction takes place between different cells of the cortex resulting in a fusion or synthesis of their physiological activities on simultaneous excitation." The context makes it clear that he is here talking about "analyzer" cells. When Hull came to transcribe Pavlov's ideas (adding Thorndike's "law of effect" and some mathematics of his own), this notion of Pavlov's appears as the notorious postulate of afferent stimulus interaction. "All afferent neural impulses ... interact with each other in such a way as to change each into something partially different in a manner which varies with every concurrent associated afferent impulse." In this postulate, Hull acknowledges the problem of how the stimulus is processed without doing anything to solve it. Moreover, Hull does not consider the possibility that the ways in which the incoming stimuli are analyzed might be determined by some central mechanisms influenced by learning. The ways in which the afferent impulses interact are said to be determined only by what else is impinging on the receptors at the same time; the interaction is not altered by central factors. It follows that even this postulate could only be used to cope with variations in attention brought about by stimulus variables operating at the time; it could not be used to account for centrally determined selective attention.

Some of Pavlov's Russian successors have discussed the possibility of efferent tracts in afferent systems influencing the incoming stimuli. For example, Sokolov (1960) argues that such activity can be conditioned in the same way as a conditioned reflex involving motor activity. This would of course allow for the problem of selective attention. Curiously enough, the Russians have been faster than Western psychologists to see that activity at many different levels of the central nervous system might be conditioned in much the same way as the motor response itself. However, as far as we are aware, the Russians have not themselves attempted much experimental work on selective attention.

## V. Structuralists and Functionalists

While the problem of attention could be and was largely ignored by animal psychologists, it was less easy for those interested in human psychology not to pay some regard to it. There is a tradition of work on attentional processes in man that goes back as far as G. E. Müller's doctoral dissertation on the topic and extends until recent times. Titchener (1908), carefully probing the recesses of his own consciousness, felt bound to report that something occurred that corresponded to attention; different aspects of the stimulus appeared in consciousness with different degrees of vividness, and Titchener named this dimension of consciousness "attensity." Some of Titchener's students became interested in the problem of attention (e.g., Pillsbury, 1913), and Dallenbach (1926, 1928, 1930) wrote three vituperative and doctrinaire reviews of the experimental work undertaken in the twenties. Much of this work was somewhat sterile since it was devoted to answering the question, "What is attention?" rather than to attempting to work out explanations for phenomena. Moreover, most of the work was not on selective attention, but on other aspects of the problem of attention.

Among other topics, experiments were performed on the fluctuation of attention both on the sensory side as measured by variations in the limen for a faint stimulus (Taylor, 1901) and on the motor side as measured by fluctuations in the efficiency of continuous work (Bills, 1931, 1937), a problem that was later rechristened "vigilance." Considerable work was also performed on the "span of attention," e.g., on how many dots or other items could be simultaneously perceived. Glanville and Dallenbach (1920) showed that the number of items that could be apprehended simultaneously decreases as the amount to be perceived about each item increases. This result directly anticipates the findings of more recent experiments that were prompted by the concepts of information theory and in particular by the idea of channel capacity. Other problems that were investigated in the twenties and thirties and have recently been posed in an altered form are the ability of subjects to undertake two tasks at once, and the effects of distraction on performance. Some experimenters investigated the stimulus determinants of attention, e.g., how the color of an object or its position on the retina controlled whether or not it would be perceived.

Early investigations of selective attention in man may be briefly summarized under three headings. It was found that attention could influence the time at which a stimulus was perceived, it could influence what elements of the stimulus were perceived, and it could influence how the stimulus was perceived.

Titchener himself (1908) found that when two stimuli were presented simultaneously the stimulus to which attention was directed tended to be perceived as occurring first in time (cf. Needham, 1934, 1936). Mowrer (1941) found that if a subject had to make exactly the same reaction to a light and a sound, the reaction time to the sound was increased if the subject was expecting to receive the light. Since the motor set was the same whether a light or a sound was perceived, this is good evidence for the operation of some sort of perceptual set, and hence, for a process of selective attention.

Of more direct interest for our present purposes, is an experiment performed by Kulpe (1904), which appears to be the first demonstration of selective attention in man. He presented different arrangements and numbers of differently colored letters in a tachistoscope and found that subjects instructed to remember one aspect of the stimulus (number, arrangement, or color) were able to recall that aspect better than the other aspects. If subjects had been set to count the number of letters, they were often totally unable to report on the color of the letters. This experiment suggests that the direction of attention can influence how accurately different dimensions of the stimulus are perceived. Chapman (1932) points out that Kulpe's procedure does not allow us to determine whether selective attention is altering what is *seen* or merely what is *remembered*. He undertook a further experiment of the same sort in which instructions as to which aspects of the stimulus to report were given either before the stimulus was presented or immediately afterwards, and found that subjects did better on the dimension they were instructed to report if the instructions were given *before* the stimulus was presented; however, subjects instructed on which dimension to report immediately *after* the stimulus did no better on that dimension than subjects attempting to report on all the dimensions present. We shall return to the distinction between seeing and remembering in Chapter 3, but both Chapman's and Kulpe's experiments make it certain that attention can influence which aspects of the stimulus situation are stored.

Other experiments undertaken in the thirties showed that what is seen and what is stored about a given stimulus array may be affected by attentional processes. Both Bartlett (1932) and Carmichael, Hogan, and Walter (1932) found that the way an ambiguous object was reproduced depended on what it was the subject thought he was looking at. In the latter experiment the reproduction of an object that could be taken to be either a pair of dumbells or a pair of spectacles was influenced by whether the subjects expected to see one or other object. Zangwill (1937) showed subjects two series of inkblots; the subjects were instructed to look for an *animal* in each of the first series and *mountain scenery* in each of the second and to draw the animals and scenes. When one blot from the first series was introduced into the second, 64% of the subjects failed to recognize it as a blot they had just seen because of the difference in attitude induced by the instructions. Zangwill interpreted his findings in terms of Bartlett's concept of schema which assumes that both recognition and memory are active processes. We shall try to give a more precise interpretation of the concept of schema in Chapter 3, at least insofar as it applies to perception.

## VI. Noncontinuity Theorists

The experiments described in Section V were performed on man. We now turn to a series of animal studies which stem ultimately from some rather casual asides in Lashley's *Brain Mechanisms and Intelligence* written in 1929. He noticed that when rats were learning a brightness discrimination,

> There is in each case a rather sudden drop from a chance number of errors to one or no errors in 10 trials .... In the discrimination box, responses to position, to alternation, or to cues from the experimenter's movements usually precede the reactions to the light and represent attempted solutions. The form of the learning curve is the more significant when considered in relation to such behaviour. In many cases it strongly suggests that the actual association is formed very quickly and that both the practice preceding and the errors following are irrelevant to the actual formation of the association [pp. 134–135].

Lashley's remarks led Krechevsky (1932a) to undertake experiments designed to demonstrate that in learning any problem the rat's behavior is systematic and that rats adopt a series of hypotheses (e.g., go left, alternate between left and right) before finally hitting on the correct hypothesis. Both Lashley and Krechevsky tended to believe that nothing was learned about the correct solution of a problem during trials on which animals were being guided by an incorrect hypothesis. Their work in the thirties led to further experiments and to a protracted controversy that became known as "the continuity–noncontinuity" dispute.

The most vociferous continuity theorist at the time was Spence (1936) and there were several issues at stake between him and his noncontinuity opponents, Lashley and Krechevsky. (1) Spence did not believe that animals attempted solution by forming hypotheses; where hypothesis-like behavior appeared, it could be explained in other ways. (2) Spence believed that learning tended to occur gradually; as the above quotation from Lashley shows, the noncontinuity theorists thought learning could occur very rapidly, perhaps within a single trial. (3) Spence believed that in a simultaneous discrimination involving two stimuli, animals responded to the absolute value of each stimulus; the noncontinuity theorists thought that animals could learn to respond to the relationship between stimuli. (4) According to Lashley (1942), however, the basic issue was whether "all stimuli acting at the time of response become associated with that response." The continuity theorists

> assume that each stimulation combined with a reaction increases the effectiveness of that stimulus for eliciting the reaction; it follows that there will be a continuous strengthening of the association, as a function of the number of times that such combined stimulation has occurred [p. 242].

Noncontinuity theorists on the other hand assume that,

> when any complex of stimuli arouses nervous activity, that activity is immediately organised and certain elements or components become dominant for reaction while others become ineffective . . . . Such an organisation is in part described by Gestalt principles of perception, in part by principles of attention. In any trial of the training series, only those components of the stimulating situation which are dominant in the organisation are associated. Other stimuli which excite the receptors are not associated because the animal is not set to react to them [p. 242].

It is this aspect of the continuity–noncontinuity dispute which is of most concern to us here. It will be noticed that Lashley's statement is in all-or-none terms: dominant elements of the stimulus configuration are associated with the response, nondominant ones are not. This was a fundamental mistake on Lashley's part. Had Lashley admitted that different elements of the stimulus situation could be more or less strongly associated with the response and that which elements become strongly associated and which become weakly associated is determined in part by attention, the way would have been open there and then for experiments to resolve the issue. As it was, Lashley and Krechevsky were committed to showing that *nothing* was learned about the relevant cue at a time when animals were governing their responses in terms of other cues. Thus, if rats are trained on a simple visual discrimination they often start by responding only to position; according to Lashley, nothing should be learned about the visual discrimination until the animal ceases to respond to position. Lashley also undertook to show that if an animal was first trained to respond to one dimension (e.g., size) and then given a problem that could be solved in terms of the same dimension but with a second dimension also relevant (e.g., shape), then nothing would be learned at the second stage about the second dimension. As we shall see in the next chapter, Lashley was proved wrong on both counts. Stating his doctrine in an all-or-none way had two unfortunate consequences.

In the first place, it led Lashley to design experiments to prove his theories on the basis of negative results. In the above-mentioned experiment involving training, first on a size discrimination, and then on size plus shape, Lashley was committed to showing that at the second stage animals would learn nothing about shape. Other examples of experiments in which Lashley relied on obtaining a negative result will be found in Chapter 4. Lashley's reliance on negative results is unfortunately characteristic of much of his experimental work, and, indeed, the whole doctrine of mass action is based on a failure to find qualitative differences in performance with lesions placed in different parts of the brain.

The second difficulty to which his all-or-nothing statement of selective attention gave rise was that it prevented Lashley from designing the right experiments to substantiate his position. Had he believed in a graded effect of attention, he would have been forced to add a control group to the experiment mentioned above in an attempt to show that animals pretrained on size learned *less* about shape at the second stage than animals given no pretraining. This type of experiment has since been performed several times with positive results.

We shall reserve a full description of the experiments to which the continuity–noncontinuity dispute gave rise until Chapter 4 when we shall attempt to explain them in terms of our own theory. Although at the time the continuity–noncontinuity dispute appeared sterile, it gave rise to the series of experiments of D. H. Lawrence conducted between 1948 and 1955. These experiments were beautifully controlled and highly ingenious; the results point directly to selective attention being an important determinant of discrimination learning in animals. Although, until recently, little attempt was made by others to follow up the results of Lawrence's experiments, they did help to bring the problem of selective attention in animal learning alive and later interpretations of discrimination phenomena including our own owe a great deal to Lawrence's formulation of the problem.

Several other influences have led to a revival of interest in the mechanism of selective attention. Three of these will be briefly discussed.

## VII. Information Theory

In recent years, psychologists have attempted to apply to organisms concepts derived from information theory. One of the most important of these concepts is that of channel capacity. As Broadbent (1958) points out, the realization that any information-processing system has a finite channel capacity has raised questions about the channel capacity of the human brain, and this, in turn, leads to the question of whether there are devices built into the brain to offset any limitations in channel capacity. In most automata, making the maximum use of existing channel capacity is extremely important, since the cost of providing surplus capacity can be enormous. Thus, in a television set, information about the brightness value of different points on the picture is sent at successive moments in time down the same channel. It is necessary to transmit information about the brightness value of approximately 20,000 different points. If information about each point were conveyed over a separate channel, television sets would need roughly 20,000 times as many components and the frequency band required for any one station would be impossibly wide. It is, therefore, natural to ask whether in the human brain the same channel is used at different times to transmit different types of information; this would involve simultaneously transmitting information tagging what type of information was being passed over a given channel at a given time, but it is clear that a considerable economy in number of elements could be achieved in this way at the expense of limiting the different kinds of information that could be processed simultaneously.

Sutherland (1959) has elsewhere argued in detail the case for supposing that the brain must achieve economy of pathways by using the same path to transmit on different occasions different kinds of information. It is clear that basically the same types of function will need to be computed in many different sensory discriminations both within one modality and from one modality to another. Now, in a computer, we do not provide a separate arithmetical unit for each block of memory registers; usually, one arithmetic unit is provided to perform calculations on numbers stored in all locations. It is possible that, in the human brain, the same central computing devices are also used to compute similar functions on information of very different kinds, thus effecting an economy in the number of computing devices required. This line of reasoning immediately raises the question of what determines which type of information shall be processed at a given time, and this is part of the problem of selective attention.

A consideration of information-processing machines suggests that as well as there being finite limitations on the information that can be processed at one time, there may also be finite limitations on the information that can be stored. In a given stimulus situation, only some of the stimulus aspects will, from a teleological standpoint, provide useful information to an animal, and these aspects are likely to vary from situation to situation. In order to economize on storage capacity, it would therefore be sensible for organisms to store primarily information that was likely to be of use in a given situation and to discard other information. This again raises the problem of selective attention in an acute form. Since the type of information that is relevant to achieving the animal's goals will vary from one situation to another, it would be desirable to combine economy in *how much* information is stored with flexibility in selecting *what* information to store. The problem of how an animal learns to select what information to store and the consequences of the selective processes is the theme of this book.

## VIII. Human Skilled Performance

A second development that has raised the same type of question is recent research on human skills. In World War II, there arose the problem of training large numbers of men to undertake new and complicated information-processing tasks and of designing the machines they were to use in such a way as best to take advantage of the capacities of the human operator. Many such tasks involved the problem of vigilance—for example, watching a radar screen for protracted periods to detect a signal indicating the approach of enemy aircraft. Broadbent (1958) and others have worked on this problem and have concluded that, among other things, men can attend selectively to particular input channels; that it is difficult to maintain attention to a restricted channel for prolonged periods of time, particularly in the absence of frequent signals on that channel; and that a channel containing novel information will tend to be selected. A related problem studied by Broadbent is that of how much information can be processed simultaneously, and what enables the human operator to process information of one sort rather than another. Although the problem in theoretical form goes back at least as far as William James, today, it has become an acutely practical one, since in tasks like air traffic control the human operator may be flooded with several messages at once and must decide on which to act. The evidence collected by Broadbent and others on this problem shows conclusively that the nervous system is not able to process information in several different ways at once, but selects for processing only some aspects of the information reaching the receptors at any one time.

Thus, Broadbent (1954) found that digits simultaneously presented to the two ears are not accepted by central analyzing mechanisms in their order of presentation, but all digits presented to one ear are first accepted, and then all presented to the other ear. He has found that introducing various differences between two simultaneously arriving verbal messages makes it easier to select one rather than the other; for example, it is easier to select the wanted message if the relevant and irrelevant messages have different sources in space, are spoken in different voices, or are of different loudnesses. Mowbray (1953) showed that when the eye and ear are presented with complex stimuli at the same time, the information presented to one or the other is made effective in the response but not the information presented to both.

Craik (1947), Welford (1952), Davis (1956), and others have obtained evidence suggesting that the human operator acts as a single-channel control system when required to make different responses to signals successively presented. For example, Davis (1956, 1957) found that where two stimuli requiring different but peripherally compatible responses are presented with a time interval of less than 200 msec, the response to the second stimulus is delayed; the length of the delay was approximately the same as the amount of overlap between the time that the first stimulus occupied central pathways and the time that the second stimulus would have occupied central pathways if it had been accepted immediately. This suggests that the second stimulus cannot gain access to central pathways until the first one is cleared. The idea that the same central analyzing mechanisms are used to process many different kinds of incoming information gives a rationale to these findings.

## IX. Some Neurophysiological Evidence

Recent developments in neurophysiology have uncovered some physiological phenomena that seem to be connected with the mechanism of selective attention. Three lines of research are relevant: (1) demonstrations of changes in physiological responses that correlate with selective habituation to a repeated sensory input; (2) attempts to show that the activity in cells analyzing input for a given sensory modality is influenced by whether attention is directed to that modality or to another; (3) demonstrations that central stimulation may both alter activity in nerve cells concerned in sensory analysis and may change the accuracy or speed of a behavioral response dependent upon fine sensory analysis. Under the first two heads, three different types of neurophysiological responses have been studied: the electroencephalographic (EEG) arousal pattern, the sensory evoked potential, and changes in the firing of single units. The latter two responses have been studied at different levels of the sensory system, up to and including the cerebral cortex. Although the whole subject is very lively at the moment, the literature is both confused and confusing. This is not the place for a detailed review, and we shall limit ourselves to describing a few experiments that appear to point towards physiological mechanisms mediating attention (for a good review of the literature see Horn, 1965).

Before detailing the evidence under the above-mentioned headings, it is worth noting that there is good evidence for the existence of efferent fibers associated with afferent systems. Rasmussen (1946) has demonstrated by anatomical means the existence of a descending pathway in the auditory system of cat and pigeon that sends efferents at least as far peripherally as the cochlear nucleus. Cowan and Powell (1963) have demonstrated a similar pathway in the visual system of pigeons ending at the level of the retina and Lettvin (unpublished) has found that efferents are present in the optic tract of octopus; the presence of efferent pathways in the mammalian retina has long been suspected, though there is still no very direct proof of their existence. One possible function for such efferents is to exert control over incoming sensory messages either by reducing firing in a modality not attended to or by selectively controlling the type of information allowed to proceed centrally within a given modality. We now turn to a brief review of some of the neurophysiological evidence.

### A. Habituation

One of the clearest demonstrations of a physiological correlate of sensory habituation at a level above the receptors themselves is that of Sharpless and Jasper (1956). They took EEG recordings in the sleeping cat, and found that the EEG arousal pattern could be selectively habituated to a tone of 5000 Hz followed by a tone of 200 Hz. After habituation to this sequence of tones, the arousal reaction occurred in full strength if the tones were sounded in reverse order. Although there are no other demonstrations showing this degree of specificity in the habituation to a signal of the EEG arousal response, there are many other demonstrations (e.g., Sokolov, 1960) that the response does habituate. Moreover, several investigators have found that habituation may be prevented if the repeated stimulus is followed by a noxious stimulus or if the stimulus is itself made a conditioned stimulus for the animal. For example, Beck, Doty, and Kooi (1958) found that after the arousal pattern had been habituated to a tone, it was reinstated if the tone was paired with an electric shock. Sharpless and Jasper found that habituation still occurred after removal of the auditory cortex but it was no longer specific to a given pattern. This suggests that the mechanisms for analyzing patterns of sound are either in the auditory cortex itself or in the pathways leading to it (cf. Neff, 1961).

In 1955, Hernández-Peón and Scherrer published a study on the effects of habituation on the size of the evoked potential (for a fuller account see Hernández-Peón, Jouvet, and Scherrer, 1957). This study is of considerable historical importance in that it aroused a great deal of interest and much controversy; unfortunately, their experiments were poorly controlled and many of their results cannot be replicated. The authors found that if a click were repeatedly presented to a cat, the size of the evoked potential at the level of the cochlear nucleus decreased with repeated presentations. They interpreted their results to mean that efferent fibers were exerting an influence on incoming activity at a peripheral level of the auditory system. Unfortunately, Hernández-Peón *et al.* did not present any statistics and subsequent investigators have been unable to reproduce their results. Worden and Marsh (1963) were unable to demonstrate any consistent trend in the size of the potential evoked by repeated clicks, though the potentials appeared to be smaller when the animal was in a drowsy state than when it was fully alert. Moreover, Hernández-Peón did not control for the effects of possible changes in the middle ear muscles and the position of the pinna (cf. Hugelin, Dumont, & Paillas, 1960). Garcia-Austt (1963) measured the averaged cortical evoked response to a flash in man and in rats and found a reduction in size with repeated flashes; since the pupils had been dilated by atropine, the effect is not likely to have been due to changes in pupil size. In a very careful study in which the influence of peripheral muscles was excluded, Wickelgren (1968b) showed that the evoked response to a repeated auditory stimulus was considerably reduced after habituation at the cortical level, but its amplitude at the level of the colliculus and below remained unchanged. However, Fernandez-Guardiola *et al.* (1961) were unable to demonstrate any reduction in amplitude in the visual evoked response to a flash in cats whose pupils were dilated by atropine. Although these studies are suggestive it is not possible to draw any very firm conclusions from them because of the apparent contradictions in the literature. Moreover, it is by no means certain that a reduction in the size of the evoked potential is a correlate of decreased attention to a given modality. Thus, Sharpless and Jasper (1956) found that habituation to a stimulus as measured by reduced amplitude of the EEG arousal pattern was actually accompanied by *increases* in the amplitude of the evoked potential, and a similar finding has been obtained by Mahut (unpublished); Wickelgren (1968a) found that the cortical evoked response to a click was of larger amplitude in sleeping than in awake cats.

There is little evidence to show that the response of individual units in the primary sensory pathways decreases with repetitive stimulation; thus Horn (1963) failed to find any decrease in the activity of single units in the cat visual cortex with repeated presentations of a flash. However, Horn and Hill (1964, 1966) have discovered cells in the tectotegmental region of the brain stem that do show a marked reduction in firing when the same stimulus is given repeatedly. These cells have many interesting and unusual properties. The same cell would often respond to stimuli in two or more modalities. Moreover, a cell that had been habituated to a specific stimulus would often fire at full strength if a different stimulus was substituted either in the same or in a different modality. Thus units that had ceased to respond to a tone of a particular frequency would often respond to a tone differing from the original by as little as 100 Hz. Lettvin, Maturana, Pitts, and McCulloch (1961) discovered cells that they have named "newness" and "sameness" units in the visual tectum of the frog. The "newness" neurons have large receptive fields (about 30° in diameter), and adapt readily to an object moved repeatedly across the receptive field in the same direction, but respond again as soon as the object is moved in a different direction through the field. The "sameness" units have even larger receptive fields, and sometimes do not respond when an object is first moved into the field. Once they begin to respond to an object they continue to respond provided the object is kept in motion.

We conclude that there is good evidence for the existence of physiological mechanisms that keep track of the newness of a stimulus. Since nonselective habituation to tones can occur in the absence of the auditory cortex and since the single units exhibiting the greatest changes to repeated stimuli have been found in the reticular formation, it seems likely that such tracking may be carried out by mechanisms outside the classical sensory pathways. Another possibility suggested by Sokolov (1960) is that a copy of a repetitive stimulus is kept at the cortical level and when a matching input occurs, the input to the cortex from the reticular pathways is damped by the action of corticofugal fibers.

### B. Changes in Attention

Adrian and Matthews (1934), first demonstrated a clear-cut connection between changes in EEG and attention. They found that the alpha rhythm was blocked when a subject attended to a visually presented patterned stimulus; the rhythm may reappear if the subject attends to an auditory stimulus, even though the eyes remain open (Adrian, 1944); it may even reappear while the subject is performing a visual task if he becomes bored with it (Oswald, 1959). Normally, the alpha rhythm is present if the subject is in darkness but Adrian and Matthews (1934) found it sometimes blocked if a subject tried to make out some visual detail even when in total darkness.

Studies of the evoked response during changes of attention are beset by the same difficulties as studies of this reponse during repetitive stimulation. Hernández-Peón *et al.* (1956) found that the response to a click at the level of the cochlear nucleus was reduced if the cat was presented with an interesting visual stimulus such as a mouse. Unfortunately, it cannot be concluded that this effect is due to a central discharge blocking activity in the auditory nerve while the cat concentrates on information received through some other modality. The introduction of a mouse results in changes in the orientation of the pinnae that may decrease the response to the click (Marsh *et al.*, 1962); moreover, Simmons and Beatty (1962) have shown that when a cat is presented with a novel stimulus the middle ear muscles contract and this contraction reduces the evoked response to an auditory stimulus.

Hernández-Peón *et al.* (1957) also demonstrated a reduction in the amplitude of the visual evoked response to a flash at different levels of the visual system when the cat's attention was distracted by a whiff of fish. The interpretation of this finding is also in doubt and is further complicated by the fact that Horn (1960) presents evidence to show that the visual evoked potential is actually reduced when the cat is engaged on visual searching behavior. Since Horn observed this reduction in cats whose pupils had been dilated with atropine, it cannot have been caused by pupillary contraction. Jouvet, Schott, Courson, and Allegre (1959) have found increases in the size of the visual evoked potential in human subjects engaged on a visual task and Garcia-Austt (1963) obtained reductions in visual evoked responses to a flash when subjects were instructed to attend to a different stimulus either in the same or another modality; this reduction occurred even when the pupils were fully dilated by atropine. Further difficulties in the interpretation of these and other studies on changes in the evoked potential with changes in the direction of attention are discussed by Horn (1965). No very consistent story emerges. The size of the evoked potential to a stimulus in a given modality seems to be altered by changes in the direction of attention and these alterations cannot fully be accounted for by changes in peripheral muscles controlling the sensory input. The exact circumstances under which the evoked response will be increased or decreased have however still to be determined.

There is some evidence from microelectrode studies suggesting that a number of individual cells in sensory systems are influenced by the direction of attention. Hubel, Henson, Rupert, and Galambos (1959) found some cells in the auditory cortex of cats that fired in response to an auditory stimulus only when the animal appeared to be attending to the stimulus. For example, one cell responded to paper being rustled only if the animal had its head and eyes turned in the direction of the paper. Another cell failed to respond to a repetitive click given over a loud speaker, but began to respond to the click when the experimenter waved his hand before the animal in time to the click.

Wall, Freeman, and Major (1967) have found units in the cat spinal cord that can be fired by a variety of complex inputs; the input needed to fire a given unit changes with alterations in the state of the animal such as changes in posture and in carbon dioxide pressure in the blood. There is thus some evidence for the suggestion made above that the same cells may be used to analyze different types of information.

### C. Central Stimulation

The most direct evidence for changes in physiological activity in sensory pathways occurring under the control of central mechanisms comes from a series of experiments by Desmedt. He showed (1960) that the efferent pathway in the ear passing through the olivo-cochlear bundle has its point of origin in auditory area IV. He stimulated this pathway electrically at various points and found that he could obtain a reduction in the amplitude of the evoked response at all points in the auditory pathway between the cochlear nucleus and auditory area I in the cortex. The reduction in the amplitude of the response during stimulation was equivalent at the most to about a 25 dB reduction in intensity of the sound itself. Desmedt (1962) has demonstrated that the effect is due to inhibition occurring at synapses, since it is still obtained when the peripheral ear muscles are cut. Moreover, the effects are abolished if a small dose of strychnine is given; strychnine is known to prevent segmental inhibition in spinal motoneurons, and it is therefore likely that some inhibitory chemical transmitter is involved in the reduction of the evoked response brought about by stimulation of efferent pathways.

One function of efferents in the auditory pathway has been demonstrated by Dewson (1966, 1967, 1968). He showed that stimulation of the olivo-cochlear bundle tended to inhibit firing of units of low threshold, and inferred that one function of descending efferents might be to reduce the effects of masking noise. In a behavioral experiment he demonstrated that monkeys trained to discriminate between human speech sounds were severely impaired in the discrimination after section of the olivo-cochlear bundle when the speech sounds were presented in noise that masks the sounds (2400-Hz low-pass noise). The monkeys' performance was unaffected by the lesion when they heard the sounds without any superimposed noise and when they heard the sound in noise (2400-Hz high-pass noise) that does not have a masking effect on speech sounds. This is perhaps the most direct evidence we have to date on the function of efferents in sensory systems.

It has for some time been thought that the reticular formation is implicated in the general state of alertness of the animal and in sleeping and wakefulness. It is known that electrical stimulation of the reticular formation can reduce reaction times in the performance of a learned task by animals (Fuster, 1958; Isaac, 1960). Moreover, Lindsley (1961) recorded evoked potentials from the cortex of cats and found that whereas two lights separated by a 50-msec interval normally produce only one evoked potential, they produce two potentials when the reticular formation is stimulated concurrently. Fuster (1958) found that on a behavioral measure the discrimination of different visual stimuli exposed for periods of less than 1 sec was improved when the reticular formation was stimulated during the stimulus presentation. These results could be interpreted in terms of some nonspecific facilitative influence exercised through the reticular formation on the cortex as a whole, and are therefore not direct evidence for selective attention.

We have presented the foregoing studies mainly in order to show that over recent years there has been very considerable interest on the part of neurophysiologists in the physiological mechanisms mediating attention. As we have seen, much of the evidence obtained to date is difficult to interpret and it is even more difficult to make a very direct connection between the physiological findings and the behavioral phenomena of attention. There is no question that important further physiological discoveries in this field will be made over the next few years, and this makes it all the more desirable to pursue behavioral investigations into the process of selective attention in the hope that the results will help in the interpretation of the accumulating body of neurophysiological data.

## X. Scope of Book

We have tried in this chapter to trace in outline the history of work on selective attention and to show why it is that interest in the topic has recently revived after a period of 40 or 50 years during which only sporadic work on the problem was undertaken. Excellent reviews of some aspects of recent work on selective attention in man are contained in books by Broadbent (1958) and by Garner (1962) (see also Treisman, 1964b). We shall mainly be concerned with selective attention in animals. We shall try to show that evidence collected by ourselves and others proves that selective attention plays an important role in animal learning. We shall argue that the evidence compels acceptance of the idea that discrimination learning in animals involves two processes: learning to which aspects of the stimulus to attend and learning what responses to attach to the relevant aspects of the stimulus situation. No single-process learning theory can account for the known facts of animal learning, whereas we believe that a two-process theory can account for many of the more paradoxical findings of learning experiments including both findings to which it has already been applied, e.g., the effect of overtraining on reversal learning, and findings which have not previously been considered in the light of such a theory. The ideas we shall employ, as we have already pointed out, are not new and we are greatly indebted for many of them to our forerunners in this field: Lashley, Krechevsky, and Lawrence. What we have tried to do is to manipulate the ideas in new ways so as to generate a series of experiments the results of which confirm rather strikingly the existence of two processes in animal discrimination learning; we have also tried to extend the ideas to cover a much wider range of phenomena than they have formerly been used to explain.

We shall start by setting out our version of a theory of selective attention in animals, stating the theory initially only at a verbal level (Chapter 2), and attempting to clarify and make plausible the concepts employed (Chapter 3). Chapter 4 shows how the theory reconciles the results of the classical experiments on the continuity–noncontinuity dispute. In Chapter 5 we discuss the applicability of the theory to experiments in which animals are trained with more than one cue present and relevant, and we attempt to show that the phenomena of blocking and additivity of cues are consistent with the model. Chapter 6 reviews experiments in which animals are successively trained to discriminate between different pairs of stimuli; it is argued that results on the transfer of training from one stage to another can only be understood in terms of a model that allows for the modification of attention as the result of learning. Similarly the apparently conflicting results of experiments measuring generalization gradients (Chapter 7) can best be understood in terms of the concept of selective attention. We believe that attention is implicated in the overtraining reversal effect (Chapter 8) and also in non-reversal shifts, and the formation of learning sets (Chapter 9) though a full understanding of the findings involves concepts that are outside the scope of the theory. Chapters 10 and 11 contain a review of the evidence and theories that bear on the partial reinforcement effect and on probability learning; once again our own theory must be supplemented by concepts from other theories to take account of the results. In Chapter 12, we attempt to show how some of the concomitant variation in the behavior of different classes of animals in discrimination learning situations can be readily explained in our own theoretical terms. The final chapter reviews some formalized theories of selective attention and presents results derived from the computer simulation of such theories. In order to avoid raising false hopes, it must be confessed that no formal theory of selective attention (including one put forward by ourselves) is able rigorously to predict all the results that we believe such a theory should predict.

Saving a discussion of formal theories until the last Chapter has the disadvantage that the reader will have to take on trust throughout most of the book our own verbal derivations from the rather loosely stated verbal theory. However, in developing our own experimental tests, we ourselves derived predictions in this way and the fact that the experimental tests for the most part confirmed these predictions (rather to our own surprise) suggests that the method of deriving the predictions was not completely *ad hoc*. Moreover, we believe that the general verbal form of the theory is likely to be approximately correct; the specific form in which we have chosen to cast the theory for the purposes of computer simulation is most unlikely to be correct in many of its details.

It is perhaps worth stressing that selective attention is only one out of many theoretical concepts that are involved in the explanation of animal learning. We believe that many of the theoretical explanations put forward by proponents of frustration theory and stimulus aftereffects theory (to name but two other theoretical formulations) are correct; as we shall demonstrate, such formulations are not in any sense in rivalry with our own.

Because selective attention bears on so many different aspects of animal learning, this book reviews a very substantial part of the literature on the subject and to that extent could be used as a text. The main topics that are not covered in detail center around drive and reinforcement; thus, there is little discussion of the distinction between classical and instrumental conditioning, of latent learning, of the mechanism of avoidance learning, of secondary reinforcement, or of the more recondite schedules of reinforcement used in free-operant situations. With these exceptions, the book provides a reasonable guide to the literature up to the end of 1968, though some more recent references have been included.

CHAPTER 2

# Statement of the Model

## I. Some Phenomena

Before stating our version of a two-process theory of discrimination learning, we shall describe three of the phenomena the model is designed to explain. It is also necessary to define some behavioral terms. It is simplest to understand both the experiments and the terms in the context of a specific experimental situation and we shall therefore start by describing the version of the Lashley stand used in most of our own experiments on rats.

### A. The Jumping Stand

The apparatus is illustrated in Figure 2.1. It is painted a matt gray. Before discrimination training starts, hungry rats are always pretrained to jump from the platform J to one or other of the ledges L and to push down one of the stimulus windows W in order to obtain food in the well F. We use dry food, but water is also made available since it is known that eating dry food increases thirst and thirst, in turn, inhibits hunger. At the start of pretraining, the jumping platform is placed up against the landing ledges and the stimulus windows are removed from the apparatus. Rats are left in groups of four or eight on the apparatus for two 1-hr periods with food available. During this time, they can run freely through the open windows between the jumping platform and the food well, and fear and curiosity responses have a chance to adapt out. Animals are next given 10 trials individually on which they are placed on the jumping platform and allowed to run through the open windows in order to obtain 20-sec food reinforcement in the food well. They are then trained with the jumping platform moved gradually farther and farther from the landing

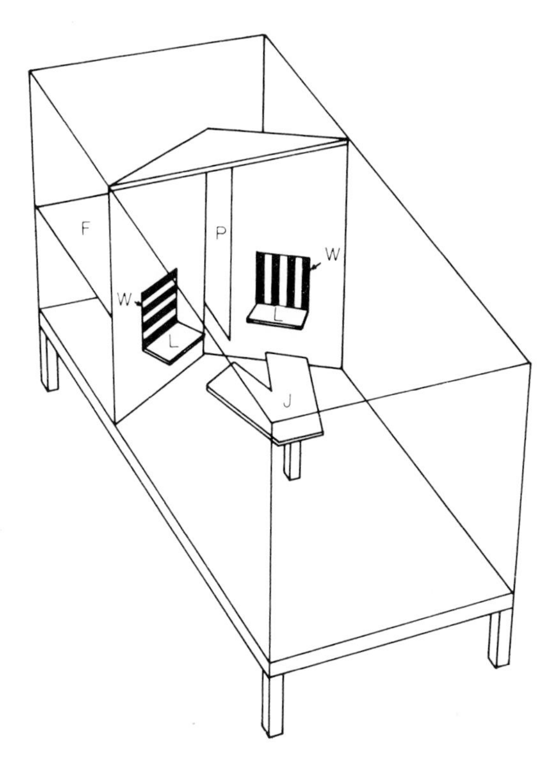


**Fig. 2.1.** The jumping stand. J designates jumping platform; L, ledge; W, window; P, partition; F, food well. [*After* Sutherland (1964).]

ledge until they are jumping freely across an 8-in. gap. This normally takes about 30 additional trials. Next, plain gray or brown stimulus windows are inserted, and animals are trained to continue to jump onto the landing ledge and to push the window down to obtain reward. When the windows are first inserted, it is sometimes necessary to pull one of the windows back slightly in order to encourage animals to continue to jump or it may be necessary to move the platform slightly nearer to the ledges for a few trials. Within a further 30 to 40 trials, almost all animals are jumping to the closed windows with latencies of less than 2 sec. We sometimes pretrain two or three rats more than will be needed in the main experiment and discard those with the longest latency period at the end of pretraining. Throughout the whole of pretraining, a record is kept of which side the animal jumps to, and the number of jumps to each side is equated in the course of a day by forcing animals to jump to their less preferred side by blocking jumps to the other side with a piece of cardboard held in the hand.

On the day following the end of pretraining, the stimuli to be discriminated are placed in the stimulus windows. At this stage animals are rewarded only if they jump to the positive stimulus; if they jump to the negative, they find the window locked and are left on the landing ledge for 10 sec. This is to give them an opportunity to observe the negative stimulus, since Hudson (1950) has shown that animals learn more rapidly to avoid a stimulus followed by a shock if they are given a chance to observe it after receiving the shock than if the lights are turned out immediately after the shock occurs. In addition, it is likely that being left on an exposed ledge for 10 sec serves as a small punishment. Animals are prevented from crossing over to the correct stimulus window by the partition P.

Each rat is normally run for 10 or 20 trials a day, and trials are spaced at intervals of not less than 4 min. The position (left or right) of the positive or negative stimulus is determined from trial to trial by Gellermann-type orders: these are semirandom orders with the constraints that the positive stimulus appears equally often on the left and on the right over each daily block of 10 trials, and that it never appears on the same side for more than 3 trials running. The restriction on the length of a sequence of trials with the positive on the same side is introduced in order to prevent the animal learning always to go to the same side (forming a position habit). In theory, with this type of sequence, animals could do better than chance by learning to switch to the other side if they have been rewarded three times running for a jump to one side. Analysis of the experimental records shows that animals do not learn to do this.

In most of our experiments, animals are taught by the *noncorrection* method; if they make a mistake, they proceed to the next trial with the position of the stimuli determined by the Gellermann sequence. In some experiments we have, however, used a *correction* method in which, when a mistake is made, animals are placed again on the jumping stand and rerun without altering the position of the stimuli. They are rerun in this way until they select the positive stimulus, and only then do they move on to the next trial. When using the correction method, we usually do not allow the animals to make more than three repetitive errors within a trial; animals that make three successive wrong jumps are forced to the correct side on the fourth jump. This forcing procedure rarely needs to be used after the first 20 trials or so, since, by that time, most animals have learned to jump to the correct side after a single error. In the analysis of results obtained by the correction method, it is usual to count only the first error within a trial. The advantage of the method is that it helps to prevent animals from forming strong position habits; under a noncorrection procedure, animals can secure reward on 50% of jumps by always selecting the same side. With the correction procedure the problem cannot be solved in this way, since on trials in which the positive stimulus is on the rat's nonpreferred side, no reward can be obtained until the animal jumps to that side. The disadvantage of the correction procedure is that, over a given number of trials, some animals may make more jumps than others, depending upon how readily they correct their mistakes. We have found that almost all animals form position habits when the noncorrection method is used, but usually break them of their own accord on easy discriminations. Unfortunately, there are no systematic comparisons of learning on a jumping stand using correction and noncorrection methods (but see Sutherland, 1961a). Using a noncorrection technique, we obtained learning of an easy discrimination to a criterion of 18 correct responses out of 20 successive trials, in about 60 trials.

The Lashley jumping stand was a brilliant invention and marked a great step forward in animal training techniques. Its success is due to the following factors:

(1) Since animals look toward the point to which they will jump, they cannot help seeing the stimuli before responding.

(2) Since they normally hesitate before leaping the gap, they have time to register the training stimuli before responding.

(3) The response itself is made very directly to the training stimuli; the subjects can learn an approach–avoidance habit which seems to be one of the easiest kinds of response learning for most species (see page 70).

(4) Immediate reinforcement is given for correct responses, and the subject confronted with a locked door obtains immediate knowledge of results after an incorrect jump.

(5) From the experimenter's point of view, there is no difficulty in classifying which response has been made; it is clear upon which ledge the animal has landed.

The only important modification in our apparatus from that used by Lashley is that we use a landing ledge instead of allowing the animal to jump directly against one or other window as did Lashley. The disadvantage of Lashley's technique is that when animals make incorrect choices, they strike against a locked window and fall into a net slung underneath the apparatus. We have found that under these conditions, animals may become very excited and jump wildly into the air. Alternatively, they may become fearful, crouch on the jumping platform and refuse to jump altogether. Lashley himself sometimes had trouble in keeping his animals jumping and is even reputed to have resorted to a whip on occasion. For these reasons we have always used landing ledges and have obtained fast and efficient learning. Since no punishment is involved for wrong jumps, our animals continue to jump for some time during extinction and in some of our experiments it was important to ensure that animals would do this. Moreover, it is known that rats are prone to develop persistent position habits if they are frightened (Maier, 1949), and it is probably because animals experience so little fear in our apparatus that they break position habits when trained on an easy discrimination by the noncorrection method. Lashley (1938) and most other investigators using a jumping stand always used a correction procedure.

We now proceed to describe three experimental results; all the phenomena are well known and none was discovered by us, but for simplicity we shall describe experiments of our own conducted in the Lashley jumping stand in order to illustrate the phenomena. Each of the results described can be explained in terms of selective attention, and we shall indicate in each case how they can be so explained using anthropomorphic language. More rigorous explanations will be given later in the book.

### B. Intradimensional Transfer

Sutherland and Holgate (1966) performed the following experiment with rats on a Lashley jumping stand: the design used is similar to an experiment performed by Lawrence (1949) and by Mackintosh (1965d). The experiment was conducted in three stages. At Stage 1, rats were trained to go left when two black shapes were presented, and to go right when two white shapes appeared; on different trials the shapes were either vertical or horizontal rectangles, but since there was no correlation between the shape used and the side to which animals had to jump to obtain reinforcement, the shape was an irrelevant cue at this stage of the experiment. At Stage 2, animals were presented with a white horizontal rectangle and a black vertical rectangle; all subjects were trained to jump to the white horizontal rectangle for a fixed number of trials. At Stage 3, transfer tests were run to discover whether, at Stage 2, animals had learned to respond to the black–white cue or to the horizontal–vertical cue. For example, animals were tested with a white vertical and a black vertical rectangle and the percentage of jumps to white were recorded to find whether they had learned the response of approaching white. One group of rats received the sequence of training outlined above; for a second group, Stage 1 training was omitted. Compared with the group that had no Stage 1 training, the rats that experienced this stage performed relatively well during transfer tests on the black–white test and relatively poorly on the horizontal–vertical tests. Put in other words, training on a black–white successive discrimination increased the amount learned about brightness and decreased the amount learned about orientation when animals were subsequently trained on a simultaneous discrimination that could be solved using either cue.

This result is difficult to explain on stimulus–response (S–R) theory, since, at the first stage, animals had not learned to make approach–avoidance responses to black and white that would facilitate the learning of the black–white cue at the second stage. A possible interpretation of the result, which will be spelled out in more detail later in the chapter, is that, at the first stage, animals had learned two things, not one: they had learned both to pay attention to black–white (and to ignore horizontal–vertical) and they had also learned to attach the response of jumping left to black and the response of jumping right to white. Because they had learned at this stage to pay attention to the brightess cue, they learned more about this cue and less about the orientation cue at the second stage than animals that had not first been trained to pay attention to brightness.

### C. The Effect of Overtraining on Reversal

Mackintosh (1962) trained rats to discriminate between a black square and a white square, presented simultaneously in a jumping stand, until they reached a criterion of 18 correct responses out of 20 successive trials with the last 10 responses all correct. Half the animals were trained with black positive and half with white positive. One group of animals was then given *reversal training*; i.e., they were retrained using the same stimuli, but in order to obtain reward they now had to select their original negative stimulus and avoid their original positive. A second group was given an additional 150 training trials on the initial discrimination before receiving reversal training. The group receiving the additional training trials at the first stage of the experiment will be referred to as the *overtrained group*. The nonovertrained group required 125 trials to meet criterion during reversal learning; the overtrained group met the same criterion in an average of 77 trials. This effect is known as the *overtraining reversal effect* and was first clearly demonstrated by Reid (1953), although a somewhat similar result was obtained by Lawrence (1950). The effect is a paradoxical one: the more we train an animal to select stimulus ⟨A⟩[1] and not ⟨B⟩, the faster it subsequently learns to select stimulus ⟨B⟩ and not ⟨A⟩. It is difficult to account for this result on a rigid S–R theory, since, if all that an animal were learning at the first stage was the habit ⟨Approach A and avoid B⟩, we would expect the additional training trials given at this stage merely to strengthen this habit and, thus, make it more difficult for the animal to learn during reversal the habit ⟨Approach B and avoid A⟩.

If we assume that the main effect of overtraining was to strengthen attention to the relevant cue (brightness), then the result can be readily explained. The overtrained animals continue to attend to brightness during reversal and, hence, learn to make the new responses required relatively fast, whereas animals trained to criterion cease to pay attention to brightness when they fail to get reinforced at the beginning of reversal and, hence, take longer to learn the new responses to that cue.

[1] Here and henceforward, where stimuli or responses are named, the names will appear in angle brackets, ⟨ ⟩.

## D. Transfer along a Continuum

Rats were trained in a jumping stand to discriminate between a square and a parallelogram of equal area (Sutherland, unpublished). It is possible to vary the similarity of these two shapes by altering the angles of the parallelogram. Some rats were initially trained to discriminate between a square and a parallelogram that was very similar to the square, having acute angles of 75° ($P_5$ in Figure 2.2); others were initially trained to discriminate between a

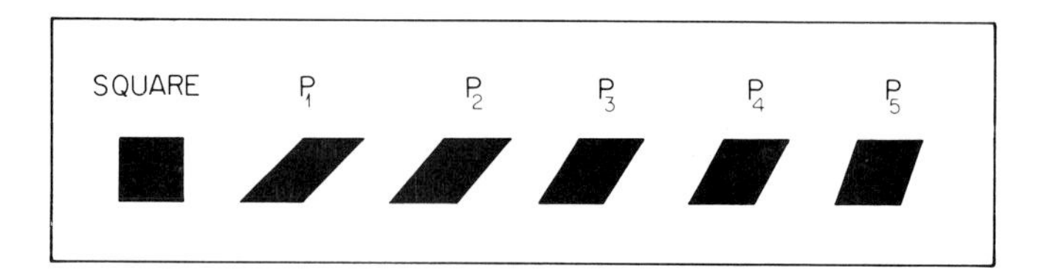


**Fig. 2.2.** Transfer along a continuum.

square and a parallelogram that differed considerably from the square, since its acute angles were 45° ($P_1$ in Figure 2.2). When the second group met criterion on the initial discrimination, they were retrained, substituting for the original parallelogram four parallelograms ($P_2$, $P_3$, $P_4$, $P_5$) that more and more closely approximated to the square. The rats received 20 trials on each in succession; the fifth parallelogram ($P_5$) was the same as the parallelogram used throughout with the first group. By the time the second group started its training with the final parallelogram, both groups had received the same number of training trials, but one group had been trained with parallelograms that were different from its final one and easier to discriminate from a square, the other group had received training throughout with its final parallelogram. It was found that, at this stage of the experiment, the group that had received training on the easier parallelograms performed significantly better over the next 40 trials with the most difficult parallelogram than the group that had been trained on this shape from the outset. This phenomenon was first demonstrated in animals by Pavlov (1927, p.121), and it was systematically investigated by Lawrence (1952) who named it "the transfer of a discrimination along a continuum." Once again, the phenomenon has an air of paradox. The most efficient way of training rats to discriminate between a square and a similar parallelogram is not to reinforce them all the time on the stimuli we want them to discriminate, but to train them first with parallelograms that have a larger difference from the square along the same dimension.

Again it is difficult, although not impossible, for S–R theory to explain this result, but it may be readily explained in terms of selective attention. Presumably it is difficult for a rat to learn to attend to a small difference between two stimuli. One effect of pretraining on parallelograms that exibit large differences from the square might be to focus the animal's attention on the relevant aspect of the stimulus situation and this should assist performance when very similar shapes are substituted since the animal knows what to pay attention to and should therefore more readily detect the small difference between the two shapes.

## II. Some Definitions

It is now necessary to introduce some more terms. In discrimination training, animals are taught to attach one response to one stimulus and another to another. If rats are trained in a jumping stand to discriminate between a black and a white square, they must learn to give the response ⟨Approach⟩ to the black square and the response ⟨Avoid⟩ to the white square (or vice versa). It is true that they could also learn to make a response to the total configuration of stimuli, e.g., ⟨Go left if black is left of white⟩ ⟨Go right if white is left of black⟩, but we do not need to be concerned with this complication at the moment. Where two stimuli are different, they will be said to differ along a particular *dimension* or dimensions; a black and white square differ along the dimension of brightness. Some stimuli differ along several dimensions, for example a horizontal and vertical rectangle centrally mounted on the windows of a jumping stand differ both in the amount of horizontal and vertical contour present in each and in the height of their lowest portion in the visual field; it is known (Sutherland, 1961b) that rats detect differences along both these dimensions when trained to discriminate between a horizontal and vertical rectangle. If rats are trained to select a vertical rectangle and not a horizontal (the pair in Figure 2.3a) and are subsequently given transfer tests with the pairs in Figures 2.3b and c, they select the left-hand member of each pair much more frequently than chance; similarly, if they are trained to select the horizontal member of the first pair, they tend to select the right-hand members of Pairs (b) and (c). Since the members of Pair (b) do not differ in the height of the bottom of the shape in the visual field, but do differ in amount of horizontal and vertical contour present, the performance on this pair indicates that animals had learned this feature of the shapes or something like it during training. On the other hand, the members of Pair (c) do not differ in the orientation of their contours, but do differ in the height of their base lines in the visual field. Thus, even when stimuli seem, at first sight, to differ only along one dimension, they may in fact differ along more than one, and animals may learn to discriminate between them in terms of more than one dimension.

Where there is a dimension of difference that could control an animal's choice responses in a situation, we shall refer to that dimension as a *cue*. For

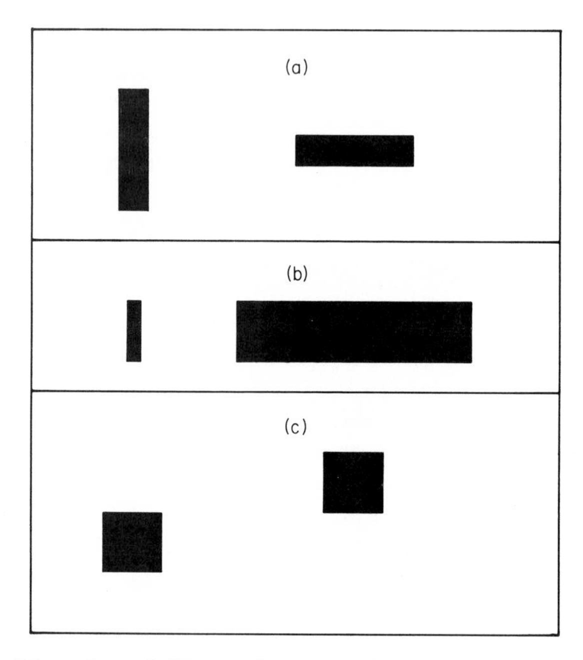


**Fig. 2.3.** Dimensions of difference between horizontal and vertical rectangles.

example, if we present an animal with a white horizontal and a black vertical rectangle, there are two visual cues that could control its behavior—the difference in brightness and the difference in orientation between the stimuli; it is true that the orientation difference can be broken down into a number of other possible cues and we shall take this up at greater length later. By the way he sets the experiment up, the experimenter determines which cues must be used by the animal if it is to solve the discrimination. Thus in order to learn to jump to a black square and not to a white square an animal must learn to use the brightness cue in controlling its behavior.

The experimenter may include in the experimental situation extra cues in terms of which it is not possible for the animal to solve the problem. For example, on some trials we can present a white horizontal and a black vertical rectangle, and on others a black horizontal and a white vertical rectangle. If, using these stimulus pairs, we always reward jumps to the horizontal and never reward jumps to the vertical, then, to solve the problem, the animal must use the orientation cue, and ignore the brightness cue. Any cue, the use of which can lead to a successful solution of a problem, will be termed a *relevant cue*. Cues which are present in a situation but which cannot be used to give a solution, we shall term *irrelevant cues*. In the problem just outlined, orientation is a relevant cue, brightness is an irrelevant cue. It should be noted that in this usage, any feature of the stimulus situation that does not vary is not a cue; if we train an animal to discriminate between a black and a white square, the shape of the stimuli is not a cue, since it could not differentially control which stimulus object the animal selects. When animals are trained on a visual discrimination in a jumping stand, there is always one irrelevant cue present, namely, position. We do not know how an animal discriminates between left and right; we do not even know which modality is used in making this discrimination, but position is, nonetheless, an irrelevant cue that is present. Animals may, of course, fail to solve a problem and respond to the irrelevant cues rather than to the relevant ones; as we have already seen, rats in a jumping stand are particularly prone to respond for long periods of time to the irrelevant cue of position.

It is perhaps worth emphasizing that unless careful transfer tests are performed, it is difficult or impossible to be sure exactly what cue or cues the animal has used to solve the problem. Moreover, some differences between stimuli that to the experimenter appear unidimensional may involve more than one cue for the animal; compare the results of the experiment on horizontal and vertical rectangles mentioned above. This problem will be discussed at greater length in the next chapter.

## III. Some Theory

### A. A Simplified Version of the Theory

We are now in a position to introduce the theory and to illustrate how it functions with reference to the three experiments described above.

The crux of the theory is that it assumes that two processes are involved in learning: animals must learn both to attend to the relevant stimulus dimension and to attach the correct reponses to stimuli having different values on this dimension; the theory can best be understood with reference to Figure 2.4. It is assumed that the stimulus input is fed into a number of different analyzers; each of these analyzers analyzes the stimulus input along a particular dimension. In Figure 2.4, Analyzer 1 ($A_1$) classifies stimuli along the dimension of brightness, whereas Analyzer 2 ($A_2$) classifies stimuli along the dimension of orientation. Each analyzer is shown as having two different possible outputs, but the brightness analyzer will also give other outputs corresponding to different shades of gray, and the orientation analyzer might also give outputs registering intermediate degrees of orientation. It is assumed that not all analyzers can be used effectively at the same time, and one of the things an animal has to learn is which analyzer to "switch in." It must learn to switch in the analyzer detecting the relevant cue. The

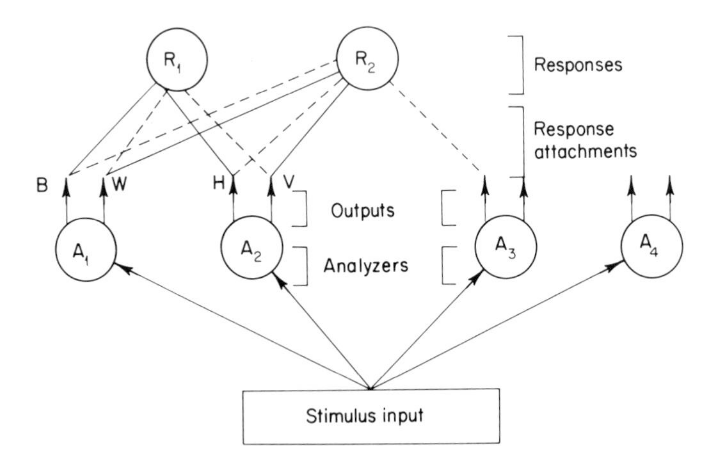


**Fig. 2.4.** Diagram of model. Learned response attachments are indicated by solid lines; other possible response attachments are shown by dashed lines. B designates black; W, white; H, horizontal; V, vertical. [*After* Sutherland (1964).]

second learning process is learning which response to attach to the outputs of the relevant analyzer. Some possible response attachments are shown in Figure 2.4; if we assume that $\langle R_1 \rangle$ is ⟨Approach⟩ and $\langle R_2 \rangle$ is ⟨Avoid⟩, then in the diagram the animal in the state shown will approach if $A_1$ is switched in and the output corresponding to black occurs, it will avoid if the output corresponding to white occurs. If, however, $A_2$ is switched in, responses will be controlled by the orientation analyzer, and the animal will approach horizontal and avoid vertical.

The simplest assumption to make about this type of model is that only one analyzer can be switched in on any one trial. On any trial on which a particular analyzer is switched in, the responses will be completely controlled by the outputs of that analyzer and their response attachments, and learning will occur only about the response attachments to that analyzer. Zeaman and House (1963) and Lovejoy (1966) have proposed models of this kind. Here, we shall consider only Lovejoy's model, which is a simpler version of the Zeaman and House theory; the latter model will be discussed in Chapter 13. Lovejoy assumes that each analyzer has a given probability of being switched in on the first trial of training. Depending on the strengths of the response attachments to that analyzer, the animal will have a probability of making the correct response given that that analyzer is switched in. If the animal has no initial preference for one or other of the stimuli, this probability will be .5. A state diagram of Lovejoy's model is shown in Figure 2.5. At the beginning of each trial the animal is in the state marked "Start" in the diagram. With probability $a$, it then switches in the relevant analyzer ($A_1$); hence with probability $1 - a$, it will select an irrelevant analyzer. If $A_1$ is

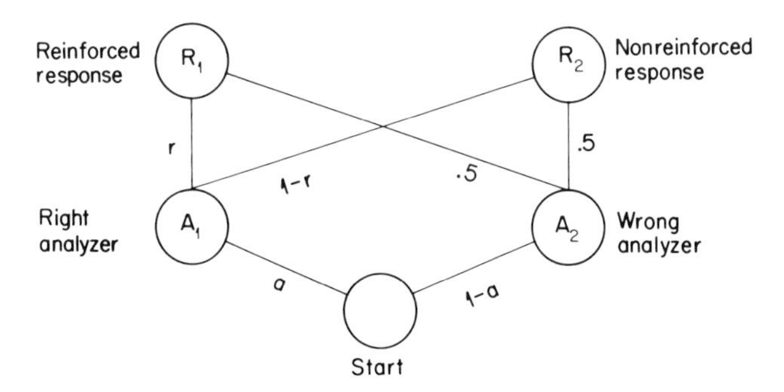


**Fig. 2.5.** State diagram of Lovejoy's model. [*After* Sutherland (1964).]

switched in, the animal selects the right response with probability $r$. If $A_2$ is switched in, the correct response will always be selected with a probability of .5, since, by definition, irrelevant analyzers analyze irrelevant cues, and the values of these cues are not correlated with the occurrence of reward and nonreward. It follows that the probability of an animal making a correct choice is

$$a \cdot r + .5(1 - a).$$

For an animal to perform well, the values of both $a$ and $r$ must tend, during the course of learning, towards 1 and we need a set of rules governing how these values change from their initial values on the first trial.

Lovejoy assumes that the probability $a$ of using the correct analyzer increases on any trial during which the animal either obtains reward by switching in the correct analyzer or fails to obtain reward after switching in an irrelevant analyzer. If the animal either is not rewarded for using the correct analyzer or is rewarded for using the incorrect analyzer, the probability of using the correct analyzer falls. On trials in which the correct analyzer is used, the probability of making the correct response given that this analyzer is used increases, whether or not the animal is rewarded. This is a reasonable assumption since, if the correct response is made, the probability of making it on the next trial increases as the result of reward. Equally, if the incorrect response is made, the animal will not be rewarded and this will weaken the probability of the incorrect response, which will correspondingly increase the probability of making the correct response, since there are only two responses the animal can make on any one trial. If the animal uses an irrelevant analyzer, there will be no change in $r$; since the animal has not attended to the correct cue, it cannot learn to alter the response attachments to that cue. These rules are summarized in Table 2.I.

Having specified the *direction* of the changes in $a$ and $r$ with different trial outcomes, it is necessary to write equations that will govern the *size* of these changes. Lovejoy uses linear equations of the form suggested by Bush and

TABLE 2.I

**Direction of Changes of Variables in Lovejoy's Model**

| States on trial $n$ | | Direction of change of $a_1$ and $r_1$ on trial $n+1$ | |
|---|---|---|---|
| | | $\Delta a_1$ | $\Delta r_1$ |
| $A_1$ | $R_1$ | + | + |
| $A_1$ | $R_2$ | − | + |
| $A_2$ | $R_1$ | − | 0 |
| $A_2$ | $R_2$ | + | 0 |

Mosteller (1951). In these equations, when a probability is incremented it is increased on each trial by a constant fraction[2] ($\theta$) of the total amount of possible increment left. Thus on any trial on which $a$ is increased, it is increased by the amount $\theta(1-a)$. Similarly, when a probability is decreased it is decreased by a constant fraction of the total possible decrease. Thus if the outcome of trial $n$ results in a decrease in $a$, the new value of $a$ ($a_{n+1}$) is $a_n - \theta \cdot a_n$, where $a_n$ is the value of $a$ after the $n$th trial. There is no reason to suppose that the values of $\theta$ are the same either after reward and after no reward or where an analyzer probability and a response probability are operated upon. For simplicity, Lovejoy uses only two different values of $\theta$: $\theta_1$ is used to govern changes in both $a$ and $r$ after reward, and $\theta_2$ after nonreward.

It should be clear that, on such a model, performance will approach an asymptote of 100% correct if enough trials are given. From the outset of training, the value of $r$ will increase after any trial on which the correct analyzer is switched in. As this value increases above .5, the animal will receive reward on a higher ratio of trials in which the correct analyzer is used than of trials in which an irrelevant analyzer is used. Thus, if the value of $r$ is .6, the animal will be rewarded with a probability of .6 on trials when it uses the correct analyzer, and with a probability of .5 on trials on which an irrelevant analyzer is used, since, in the latter case, the probability of receiving reward always remains at .5. Since the animal is rewarded relatively more frequently for using the correct analyzer than for using the irrelevant analyzers, the probability $a$ will increase. In the case where an animal is consistently rewarded for one response and not rewarded for another, the value of both $a$ and $r$ will tend to 1.

[2] In his equations, Lovejoy in fact uses a constant $\alpha$, which equals $1-\theta$; the equations have exactly the same effect whether written in terms of $\theta$ or $\alpha$, but it is easier to gain an intuitive understanding of them when written in terms of $\theta$.

Lovejoy has simulated a model of this kind on a computer, using a procedure known as running "stat rats." He assigns values to the initial probability *a* of using the relevant analyzer and to the initial probability *r* of making the correct response given that the correct analyzer is used. A computer is then used to calculate the trial-by-trial changes in *a* and *r*, which determine the animal's performance. Since changes in these values are themselves determined by the outcome of each trial, and since the trial outcome is itself determined only probabilistically, different stat rats take different numbers of trials to learn up to a given criterion. If, on Trial 1, the probability of using the correct analyzer is .2, then the majority of stat rats will select an irrelevant analyzer on this trial and, hence, the value of *r* will not be changed; a minority will, however, select the correct analyzer, and for them the value of *r* will be increased after the first trial.

Lovejoy has been able to show by this method that with certain parameter values (i.e., with certain constant values of $\theta_1$ and $\theta_2$ and with certain initial values for *a* and *r*), the model does predict an overtraining reversal effect. The reason that this occurs is that the correct analyzer can be strengthened only when the correct response attachments start to be formed. The probability of the correct analyzer being used, therefore, tends to lag behind the probability of the correct response being made given that the correct analyzer is used. Thus animals may reach criterion when the response probability is about .85 on trials when the correct analyzer is selected, and the probability of the correct analyzer being used is only about .80. During overtraining, both probabilities tend to 1. Because the overtrained animals are using the correct analyzer more frequently than the nonovertrained, they learn to reverse their response attachments more quickly after overtraining than do animals trained only to criterion. During reversal, the gain through having the correct analyzer very firmly switched in more than offsets the disadvantage of having the original correct response (the *wrong* response in reversal) more firmly established.

This type of explanation of the effect of overtraining on reversal learning was proposed by Sutherland (1959). Lovejoy by computer simulation has rigorously proved that, given the right parameter values, the overtraining reversal effect can be obtained from this type of model. The Lovejoy model discussed here is a very simple and elegant form of two-process model. Unfortunately, it cannot generate more than a small number of the many phenomena of animal learning that we believe can best be explained in terms of a two-process model embodying the notion of selective attention. Lovejoy (1968) has recently proposed a more complicated model, and our own theory is also considerably more complicated. We shall consider here only one instance of a phenomenon that cannot be explained on the basis of Lovejoy's first model. His assumption that only one analyzer can be used on one trial is implausible and cannot be reconciled with the phenomenon of errorless incidental learning. Terrace (1963b) has shown that an animal can learn about one cue at a time when its responses are completely under the control of another cue. He trained pigeons to peck at a key when it was red, and not to peck when it was green. Next, he superimposed a vertical stripe on the key that was red and a horizontal stripe on the key that was green. The animals continued to respond to red and not to green. After protracted training in this situation, he gradually faded out the colors on the key over successive trials until finally the keys were illuminated by achromatic light. The pigeons continued to respond correctly, pecking when the horizontal bar was present and not pecking when the vertical bar was present. From our point of view, the crucial finding is that some pigeons learned the horizontal–vertical discrimination without making a single error after the horizontal and vertical bars had been introduced. This means that their behavior was being fully controlled at first by the color cue, and that they must have learned about the orientation cue on trials in which the color analyzer was controlling behavior. Hence, it is established that an organism can learn about one cue at a time when behavior is being completely controlled by another. It must therefore, be assumed that more than one analyzer can be switched in on any one trial.

It will be recognized that Lovejoy's first model in some ways resembles Lashley's all-or-none form of noncontinuity theory discussed in the previous chapter. Just as Lashley predicts that when an animal is first trained with one cue relevant and then receives further training with a second additional cue made relevant, it will learn nothing about the second cue, so Lovejoy's model predicts that if behavior is completely controlled by one cue, nothing can be learned about a second. Some of the earlier experiments that disprove Lashley on this point are also relevant to the model under discussion, and they will be described in Chapter 4.

To overcome these difficulties we are driven to assume that more than one analyzer can be used on a given trial. Moreover, it is now necessary to introduce a further distinction. Analyzers function in two ways on any one trial. An analyzer that is switched in on a given trial both determines performance on that trial and determines which response attachments (i.e., its own) are to be changed as a result of the outcome of that trial. In other words, the direction of an animal's attention determines both how it *performs* on a given trial and also what is *learned* from that trial; we cannot assume that the rules governing learning and performance are the same, and what Terrace has shown is that while one analyzer is controlling performance on a trial, learning can occur about the response attachments to a second analyzer. Experiments on conditional discrimination suggest, however, that two or more analyzers can control performance on a given trial, and the same conclusion is suggested by the experiments set out in Chapter 5.

## B. Statement of the Theory

Taking the above conditions into account, we assume that performance and learning are influenced (in different ways) by the strength of an analyzer. Learning occurs about the response attachments to all analyzers on all trials but the amount that is learned about each is proportional to the strength of the analyzer. Performance is normally determined by the highest analyzer, but if there are other analyzers with strengths very near to that of the highest, performance is a joint function of the response attachments to these analyzers. For easy reference, a rough set of rules governing changes in analyzer strength, response strengths, and performance is given in Table 2.II. The rules are stated without any attempt to quantify them and in a form that is (deliberately) rather vague. A quantitative model is described in Chapter 13. The verbal model stated in the table can be manipulated to yield predictions, and some of the vagueness attached to it will be removed in the light of the experimental evidence discussed throughout the remainder of the book.

TABLE 2.II[a]

**Rules for the Operation of the Model**

Rule 1. *Response Strengths*

The strength of a given response attachment is increased by reward, decreased by nonreward. The size of the change in the response strengths attached to a given analyzer on any trial is proportional to the strength of that analyzer.

Rule 2. *Analyzer Strengths*

Analyzer strengths sum to a constant amount. At the start of training, analyzers have different base values: the stronger a given cue, the higher is the base value of the corresponding analyzer. An analyzer is strengthened when its outputs consistently make correct predictions about further events (e.g., trial outcomes) of importance to the animal; when an analyzer is strengthened, the total strength of other analyzers is weakened by the same amount. When no analyzer makes consistently correct predictions, all analyzers revert towards their base level.

Rule 3. *Rate of Change of Analyzer and Response Strengths*

When the base value of an analyzer is low, its strength reaches asymptote more slowly than the strengths of its response attachments reach their asymptote.

Rule 4. *Performance*

Performance is determined by the response attachment strengths of the responses attached to the analyzer with highest strength, and of any other analyzers whose strengths are within a constant amount of the strength of the highest. Where no other analyzer falls within this range, performance is determined entirely by the strongest analyzer.

[a] These rules are referred to throughout the text. For easy reference, they are reprinted on the inside front cover.

It is worth giving here some general justification for the rules, though they were, in fact, worked out on the basis of the experiments that are to be reported in subsequent chapters. It follows from the first rule that the more that is learned about the response attachments to any one analyzer on a given trial the less can be learned about the response attachments to other analyzers. Thus the principle of selective attention during learning is retained, while making allowance for learning about more than one cue on any one trial.

It will be noticed that the rules governing the way in which responses are strengthened and those governing the way in which analyzers are strengthened are different. There is in fact no reason at all to assume that they are the same, though it would be a convenience if they were. The rule governing the changes in analyzer strength (Rule 2) has some plausibility on commonsense grounds. The stimuli to which, from a teleological point of view, an animal must learn to pay attention are precisely those that make predictions of importance to it. If an analyzer were strengthened only when its use were followed by reward, some unfortunate consequences would seem to follow. Supposing an animal receives an electric shock immediately after a bell has sounded: it is important for the animal to recognize the bell the next time it sounds, and from a commonsense point of view, we would expect the animal to learn to switch in the analyzer that would detect the bell. However, it has received no reward for the use of this analyzer, since its use to detect the bell was followed only by pain. It seems likely that some muscular responses known as orienting responses are also learned by the same rule that we have here given to govern the changes in strength of central analyzers. Deer are known to sniff the wind. They are not being rewarded for this activity by the scent of flowers. They presumably sniff in order to detect the presence or absence of predators. At best they will smell nothing of any interest, at worst they will catch a whiff of lion. The action is followed either by a neutral consequence or by fear and the latter is clearly not positively reinforcing for most acts. It is true that this particular reaction may well be innate, but in the absence of reinforcing consequences the question arises on a Hullian type of theory of why the response does not habituate.

There is, moreover, direct experimental proof that some overt reactions can be learned because they provide the animal with information. Prokasy (1956) has found that if rats are run in a T maze with 50% reinforcement on each arm, and with a distinctive signal in one arm that predicts whether or not reinforcement will occur at the end of that arm, they will come to select the arm containing the signal predicting reinforcement despite the fact that the actual frequency of primary reinforcement is the same in each arm. Lockard (1963) (cf. also Knapp, Kause, & Perkins, 1959) exposed rats to unavoidable shock in a box having two compartments. When the rat was in one compartment, shock onset was always preceded by a warning light. When it was in the other, shocks were given with the same frequency but there was no warning light. Rats learned to select the compartment containing the warning light. The results of this experiment are particularly relevant to our present argument: if rats can learn to make an overt orienting response to pick up a warning signal for a painful stimulus, then *a fortiori* we would expect them to strengthen the analyzer that detects such a stimulus. Unfortunately, other interpretations of Lockard's results are possible. While they are probably not due to the rats having learned to make postural adjustments to reduce the pain from a signaled shock, since Perkins, Seymann, Levis, and Spencer (1966) repeated the experiment delivering shock via permanently attached ear clips, nevertheless, the reinforcing factor may not be the signaling of shock, but rather the fact that the rat learns that in the absence of the signal, the compartment is safe (Seligman, 1968).

The results of the experiments quoted provide some support for the idea that analyzers will be strengthened when their outputs provide information about further events of importance to the organism. Moreover, from a teleological viewpoint, it would be highly inefficient to have the direction of attention controlled by any other mechanism. Unlike other biologists, psychologists have tended to ignore arguments from what would be adaptive, although such arguments have considerable force, particularly in the absence of other evidence. It is partly the reluctance to be moved by arguments of this kind that has lead some S–R theorists to adopt the tidy but implausible position that all learning is controlled solely by positive reinforcements.

The assumption that analyzers are brought into play directly by rewards and punishments leads to another unfortunate consequence if we assume, as we must, that learning can proceed about one analyzer at a time when the response is being fully controlled by the outputs from another. Supposing we train an animal to jump to a black circle and not to a white circle; when the problem is solved, the animal's behavior is being controlled by the brightness analyzer. We now alter the shape of the stimuli by making one a triangle and the other a square, and we sometimes make the positive stimulus a square sometimes a triangle, that is, we introduce square–triangle as an irrelevant cue. Provided the animal continues to react in terms of the black–white difference it will continue to be rewarded on every trial. If analyzers were strengthened merely by reward, the square–triangle analyzer should be strengthened and also any other analyzer that can be used in the situation since the animal is being rewarded on every trial. The assumption that an analyzer is strengthened only when it gives different outputs that predict different consequences of importance to an animal avoids this paradox.

There is, in fact, an accumulating body of evidence that learning to respond to a stimulus is not determined solely by the number of reinforcements and nonreinforcements that follow that stimulus. For example, Rescorla (1968) using a classically conditioned response gave all animals the same number of reinforcements [presentations of the unconditioned stimulus (US)] in the presence of a signal, but one group also received the US in the apparatus when the signal was not present, and another never received the US when the signal was not present. Although in both groups the number of reinforcements (and nonreinforcements) given in the presence of the signal was the same, the signal was clearly a better predictor of occurrence of the US for the group that never experienced the US without the signal and Rescorla found that much better conditioning occurred in this group. Our interpretation of this result is that when the signal was a good predictor, the analyzer for that signal was strengthened more than when it was a less good predictor and the conditioned response (CR) accordingly became more strongly attached to it. Wagner (1969a) has performed a highly ingenious series of experiments demonstrating and developing this point and these will be described in Chapter 5.

We have now tried to summarize our main reasons for supposing that an analyzer is strengthened when its outputs predict consequences of importance to the animal and weakened when the outputs do not enable such predictions to be made. What we have not yet tried to do is cast Rule 2 in a precise form. There are various ways in which this could be done, but unfortunately none that we have thought of seems wholly satisfactory. To illustrate some of the problems, consider the case of the rat being trained to approach white and avoid black when white and black cards are simultaneously presented in a jumping stand. In these circumstances, provided the analyzer is in use, the last output given before reward is received will always be that corresponding to white; a black output can only be followed by nonreward since when the animal looks at black it will either jump and fail to be rewarded or it will move over to the white card and then jump in which case the last output before reward is white. To simplify the situation we shall assume that the animal never looks at the same stimulus more than once on any one trial.

It would be possible for the brain to record the number of times black and white are followed by reward and nonreward. This is illustrated in Table 2.IIIA; the figures shown there might be arrived at as follows: on 20 trials, the animal has jumped to white and has been rewarded; on 25 trials, the animal has looked at black and not been rewarded either because it has immediately made an incorrect jump or because it has moved over to the other side of the apparatus before jumping; the latter trials are, of course, the same as some of those included in the 20 on which the animal jumped to white. On 10 trials, the animal has looked at white but then moved over to black; these trials are included in the 25 on which the animal has

TABLE 2.III

**Stimulus-Reinforcement Contingencies**

| | | Reward | Nonreward |
|---|---|---|---|
| **A.** | Sees white | 20 | 10 |
| | Sees black | 0 | 25 |
| **B.** | Sees white | 170 | 20 |
| | Sees black | 0 | 120 |

seen black. We could assume that the brightness analyzer was strengthened only after a difference began to appear between the sums of the entries in the two sets of opposite diagonal cells, and further that it was only strengthened when the trial outcome was such as to increase any existing difference. It should be noted that as response learning proceeds, the difference in the diagonals' totals will increase faster and faster, since there will be fewer and fewer additional entries in the top right-hand column as the animal learns to jump whenever it sees white. One disadvantage of such a mechanism is that no allowance is made for recency. Table 2.IIIB shows how the columns might be filled after 200 training trials. If we now start rewarding the animal for going to black, the analyzer will begin to drop in strength since, on each trial, the difference in the diagonals' totals will decrease. It will clearly continue to drop for many trials, since the animal must see the stimuli on at least 270 occasions before the diagonals' totals become equal. It is known, however, that overtraining on a discrimination often facilitates reversal, and we attribute this effect to the analyzer staying switched in more firmly after overtraining. The present version of Rule 2 (Table 2.II) produces exactly the opposite effect. This difficulty could be overcome if there were some weighting for recency, i.e., if the effects of recent trials on the matrix were greater than the effects of old trials. However, the rule would have to be a rather complex one to encompass such a weighting by recency.

A second method of tightening up Rule 2 is as follows: whenever different outputs from the same analyzer are followed consistently by different consequences of importance to the animal, different responses will be attached to those outputs. In the example we have taken, let us assume that the response ⟨Approach⟩ comes to be attached to the output white and ⟨Avoid⟩ to the output black.[3] Now, since there must be a mechanism in the brain for

[3] This assumption is in fact justified by the evidence (for a review, see Sutherland, 1961a) since, if we first train animals to discriminate between two stimuli simultaneously presented in a jumping stand, and then give tests on which only one of the original stimuli is shown on each trial, the animals continue to jump towards their original positive stimulus and away from their original negative one.

differentially attaching different responses to the outputs of analyzers, we could use this same mechanism to keep account of whether or not different outputs are followed by different consequences, thus effecting some economy of brain hardware. We could represent the strength with which the approach response is attached to each output by the probability that an animal will jump to black or white given that its behavior is being controlled by the brightness analyzer. We could now assume on our model that analyzers will be strengthened when the probabilities of making each response change in the direction in which they are already biased. Thus if $\langle R_W \rangle$ is the response of jumping to white and if it has a probability $r_W$, after some response learning has occurred $r_W$ will be greater than $r_B$ (the probability of selecting black). Where $r_W > r_B$ the brightness analyzer will increase when an animal jumps to white and is rewarded and when it jumps to black and is not rewarded; the analyzer strength will decrease when it jumps to white and is not rewarded and when it jumps to black and is rewarded.

Table 2.IV sets out the directions in which an analyzer will change on this model with different trial outcomes; it can be contrasted with Table 2.I in which the changes on Lovejoy's model are shown. Putting all this in anthropomorphic terms, it means that an analyzer will be strengthened when a response attached to the outputs from that analyzer is carried out and an expectation is confirmed. If an expectation, whether of reward or nonreward, is disconfirmed the analyzer will be weakened. The model could be further refined by assuming that the change in analyzer strength was in some way proportional to the size of the expectation that was confirmed or disconfirmed. Thus we could assume that when the animal makes the response $R_W$ and is rewarded, the change in the strength of an analyzer is multiplied by the factor $(2r-1)$, where $r$ is the probability of the response made on a given trial. This would result in a large increase in analyzer strength when $r$ was near 1, no change when $r = .5$, and a large decrement when $r$ was near 0. Since response strength measured in probability terms has an asymptote of 1, this form of Rule 2 avoids the disadvantage of the first form suggested. If

TABLE 2.IV

**Direction of Change of Variables**[a]

| Response made | Trial outcome | $\Delta a_1$ | $\Delta r_W$ |
|---|---|---|---|
| $R_W$ | Reward | + | + |
| $R_B$ | Nonreward | + | + |
| $R_W$ | Nonreward | − | − |
| $R_B$ | Reward | − | − |

[a] For case where $r_W > r_B$

we overtrain an animal and then give reversal training, the analyzer will start increasing in value as soon as the probability of the correct response in reversal exceeds .5.

In Chapter 13, we shall show that this second version of Rule 2 can, in fact, produce reasonably satisfactory results when it is incorporated in a formal model and the model is simulated by running stat rats on a computer. The main difficulty in casting Rule 2 in this form is that it leaves the relationship between the outputs from the analyzer and the response attachments rather vague; in the formal model the vagueness is circumvented by working with the probability of making an approach response and assuming that whatever responses are learned to the relevant outputs (representing black and white in our example) the difference in the response attached to the two outputs can be represented by the difference in the probability of jumping to black and to white. This ignores the problem of how the response attachments are actually made and in particular of how the difference between two different responses (e.g., approach and avoid) is actually measured. We shall return to the question of how responses are actually attached to outputs in the next chapter.

## IV. Application of Theory to Phenomena

The application of the theory to the first of the reference phenomena described at the beginning of the chapter is very obvious. By Rule 2, when animals are trained on a successive discrimination with brightness relevant and orientation irrelevant, the brightness analyzer will be strengthened and the orientation analyzer weakened. When animals are retrained at a Stage II to jump to a white horizontal rectangle and avoid a black vertical rectangle, they will learn predominantly about the outputs from the strong analyzer, i.e., about black and white.

The justification for Rules 3 and 4 of Table 2.II can best be understood by applying the model to the second of the two experiments given at the beginning of the chapter—the overtraining reversal effect. Its application to the phenomenon is illustrated in Figure 2.6. The lower two graphs show theoretical curves for the strengthening of analyzers and for changes in the probability of the correct response given that the relevant analyzer is controlling the behavior. We assume that the relevant analyzer starts out with a low value; at this stage responses will be controlled by the dominant remaining analyzer (Rule 4) which in the case of the rat in the jumping stand is the position analyzer. Provided the relevant analyzer is not controlling behavior, the animal will perform at 50% correct responses, and, hence, for some time, the performance curve (top graph) does not depart significantly from chance. During this time, however, the animal will be learning to attach the correct

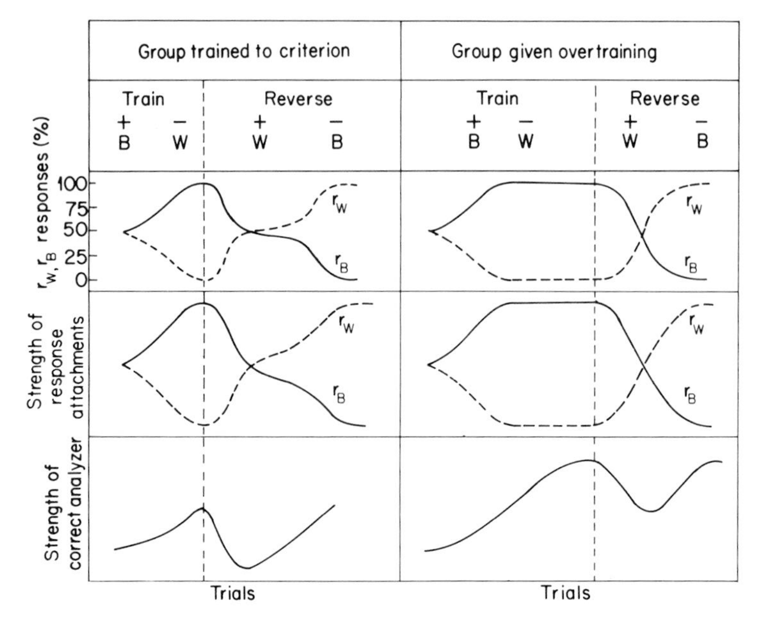


**Fig. 2.6.** Theoretical graphs of learning during training and reversal. B designates black; W, white. [*After* Sutherland (1964).]

responses to the relevant analyzer—slowly at first, but more rapidly as the correct analyzer is strengthened. The correct analyzer is strengthened as the animal learns the correlation between its outputs and reward and nonreward. When the correct analyzer becomes almost as strong as the position analyzer, responses will for a time be controlled by both (Rule 4) and the animal will start to score above chance. When the strength of the correct analyzer exceeds that of the position analyzer by a sufficient amount, responding will be controlled only by the relevant cue. We assume that criterion is reached at a time when the strengths of the correct response attachments are almost at asymptote; at this stage the relevant analyzer is strong enough to control behavior but can be strengthened by further training. At the end of training the group given overtraining (the right-hand curves in Figure 2.6) has the relevant analyzer much more strongly switched in than the group trained only to criterion.

At the start of reversal training the strength of the analyzer will begin to fall since its outputs will no longer correctly predict the occurrence of reinforcement. It will continue to fall until the probability of making the correct response (i.e., approaching the new positive stimulus and avoiding the new negative stimulus) exceeds .5 and at this time it will again start to be strengthened since success and failure will again be correctly predicted by its outputs. Since, however, at the outset of reversal the analyzer for the nonovertrained group is lower than for the overtrained group it should fall to a lower level. This will have two consequences. First the new response attachments will be learned more slowly by the nonovertrained group than by the overtrained since the analyzer is less strongly switched in; hence, the nonovertrained group will relearn the discrimination more slowly than the overtrained group, thus bringing about the overtraining reversal effect. Secondly, the relevant analyzer for the nonovertrained group is more likely to fall below the value of the position analyzer and will stay below that value for longer than in the case of the overtrained group; this should mean that the nonovertrained group is more likely to develop persistent position habits in the course of reversal training than the overtrained group and this is exactly what happens (Mackintosh, 1962).

Rules 3 and 4, therefore, allow us to explain in some detail the course of reversal learning for groups initially trained to criterion or given overtraining. Rule 3 allows the analyzer to be strengthened during overtraining. Rule 4 will result in the development of position habits when all analyzers are low if we assume that the base value of the position analyzer is higher than any of the others, and it also allows for responses being controlled by more than one analyzer when several analyzers are high; we have already seen that this is necessary in order to account for conditional discrimination, but this rule will be expanded in Chapter 3.

The model also accounts for the phenomenon of transfer along a continuum. If, from the outset of training, we train an animal on two stimuli that are very similar, this would presumably mean that the outputs from the relevant analyzer overlapped. Referring back to Figure 2.4, the brightness analyzer is shown to have two possible outputs B and W corresponding to the presence of a black or white stimulus. Two neighboring gray stimuli would produce outputs very close together that could be represented diagramatically as lying between the B and W outputs in Figure 2.4. Now if there is some noise in the sensory system including the analyzer itself, the actual output for a given stimulus will vary slightly from one presentation of the stimulus to another; it is plausible to assume that what makes two stimuli difficult to discriminate is simply that, because the outputs for each stimulus have some scatter and because they lie close together, there is some overlap in the distribution of the outputs for the two stimuli. If this is so, then it will be very difficult for an animal to learn to switch in the right analyzer where the discriminanda are so close together that they produce overlapping outputs. Even when the correct responses are fully attached, animals will not be able to perform with 100% success, since one stimulus will be mistaken for the other whenever its output corresponds to that normally given to the other stimulus. The situation will be even worse when the response attachments are weak.

Since the outputs from the analyzer cannot predict reinforcement with complete consistency, the analyzer will take much longer to be strengthened. If strengthening the relevant analyzer plays an important role in discrimination learning, we might expect to secure better performance by ensuring that the analyzer is already strong before introducing the stimuli which are difficult to discriminate. One way of doing this is of course to train first with stimuli having a large difference on the relevant dimension and only later introduce the stimuli lying close together on that dimension, and this is precisely the procedure that is used in experiments on transfer along a continuum.

## V. Conclusion

We have now set up a rather loose theory that builds selective attention into the discrimination learning process. This theory will be manipulated in the succeeding chapters and we shall try to show that it is possible to make it yield predictions about a wide range of phenomena. One of the interesting things about the model is that many of these predictions are by no means obvious to common sense and many of them appear at first sight to be slightly paradoxical, like the three experimental results with which we started. Before going on to discuss further results that can be explained in terms of the theory, however, it may be as well to say a little more about how we conceptualize the analyzers and the response attachments. The next chapter, therefore, takes up some further questions that arise in connection with the nature of the analyzers and the nature of the response attachments that can be formed to their outputs. To some extent, Chapter 3 is a digression since no actual tests of the model will be presented there, but it is important at this stage to be clear about what is meant by analyzers and response attachments.

CHAPTER 3

# Analyzers and Responses

The theoretical framework within which we are working is that in animal discrimination learning there are two separate processes: changes in analyzer strengths and changes in the strengths of response attachments. It is important to be clear at the outset that there are many problems in perception and learning that cannot be solved in terms of such simple notions. We try, therefore, to spell out in this chapter what is meant by an analyzer and by a response attachment and then discuss these terms in the context of a wide range of problems; we shall stress those problems that cannot be solved in terms of the present theory and make some attempt to show how the theory could be enlarged to accommodate them. We shall also try to show that although we are working with oversimplified concepts, this oversimplification does not invalidate the application of the theory to the phenomena it is designed to explain.

## I. Are Analyzers Innate?

One question that can be asked about analyzers is to what extent they are innate, and to what extent they can be modified or even created as the result of learning. When the present two-process model was first proposed, it was suggested that many of the analyzers were innate. Sutherland (1959) distinguished between two types of theory of shape recognition; those postulating very general analyzing mechanisms and those postulating specific analyzing mechanisms. The former type of theory is exemplified in the work of Hebb (1949), Uttley (1956), and Rosenblatt (1959); examples of more specific theories are those of Deutsch (1955) and Sutherland (1957b). It is worth repeating some of the arguments previously used; although the arguments refer to theories of visual pattern recognition, they apply to the classification of information received through any of the senses.

To simplify the problem of pattern recognition, let us assume that receptors in the eye can only be in one of two states—firing or not firing. One way of achieving pattern recognition would be to connect sets of receptors to further elements using a different element for every possible set of receptors and setting the threshold of each element so that it only fires when all receptors connected to it fire. Each pattern that can be placed upon the retina will now give a unique output since there will be one element fired by that pattern and by no others.[1]

Such a mechanism would require an impossibly large nervous system: it is clearly unrealistic to connect single receptors in this way and we might assume that only discriminably different points on the retina are to be connected. If we take the minimum separable to be 3′ of arc or better out to a peripheral angle of 60° (Polyak, 1941), there would be more than $10^6$ discriminably different points within a radius of 60° from the fovea and there would be $2^{1,000,000}$ different patterns that could be placed on this area. This would already involve far more neurons than exist in the human central nervous system. However, the problem does not end here since animals and men react not simply to patterns but to classes of pattern; they generalize from a triangle of one size to triangles of different sizes, and from a triangle falling on one part of the retina to a triangle falling on another part of the retina. Since each class of pattern is made up by including or excluding each pattern, there are $2^{(2^{1,000,000})}$ possible classes of pattern. If we wish to represent each class of pattern by a separate unit, it would be necessary to use this number of elements at a third stage, connecting each possible combination of the second stage elements up to a separate unit. This method of classification is obviously hopeless since it requires more elements than there are atoms in the universe, yet this is basically the system proposed by Hebb, Uttley, and Rosenblatt, with the modification that instead of connecting all possible receptor units up to one further element, the connections are made randomly. If acuity is to be maintained, it is not clear how much economy could be achieved by a system of random connections.

What is clear is that it is necessary to take advantage of the redundancy of the visual environment; this could be achieved by building specific forms of connectivity into the visual system innately so that it would be able to classify

[1] With such a system, the same element will fire both when a particular pattern and only that pattern is presented to the retina and when the pattern is presented as part of a larger pattern. To overcome this problem, it would be possible to connect all receptors up to each element using both excitatory and inhibitory combinations and to assume that the element only fires when all receptors connected in an excitatory way fire and no receptor with inhibitory connections fires.

only those patterns and classes of patterns that it would, in fact, be useful for an animal to discriminate. Moreover, if connectivity is initially random, it would not be possible to specify in advance what element or elements would correspond to a particular pattern or class of patterns, and this is one of the main disadvantages of Hebb's theory, since it is known that at least in some species there are innate reactions to particular patterns. Walk and Gibson (1961) have shown that rats and some other animals possess an innate perceptual mechanism for detecting differences in the depths of two surfaces; if rats without previous visual experience are made to step down from a ledge onto one or the other of two surfaces at different depths, they choose the less deep surface, thus avoiding a fall. Fantz (1957) has shown that chicks have an innate preference for pecking at round objects rather than at square or triangular objects and Wells (1958) discovered that newly hatched Sepia attack only Mysis, a small crustacean with a specific and complex shape. All these reactions are visually guided. Further arguments against supposing that analyzing systems are of a very general kind can be found in Sutherland (1959). Since that time, new evidence for highly specific and inborn visual analyzing mechanisms has come to light.

Hubel and Wiesel (1962) have found that there are single cells in the striate cortex that respond to specific stimuli on the retina. Most of the cells respond maximally to straight edges, or to bars, and each cell responds maximally to a bar or an edge in a given orientation on the retina and at a given place on the retina. There are also more complex units that respond only to an edge or bar in a given orientation, but respond equally well no matter where, within a limited area of the retina, the stimulus is placed. Moreover, the arrangement of these cells is very specific. If a microelectrode is inserted at right angles to the surface of the striate cortex and driven perpendicularly into the cortex, all cells recorded tend to respond to edges or bars in the same orientation, although some cells within such a column will respond to bars, some to edges, some to a dark bar on a light ground, some to a light bar on a dark ground, and so on. Hubel and Wiesel (1963) have also been able to show that kittens with no previous visual experience have this type of arrangement of cells in the striate cortex. Hence, there is evidence for highly specific systems of visual classification already built into the nervous system at birth.

It so happens that the type of analyzing arrangements found by Hubel and Wiesel are exactly the kind that might be expected if the system were designed to take advantage of the redundancy of the visual environment. If a room is divided up into small points and one systematically goes round the room and tries to predict the color of each point from the colors of those next to it, the prediction can be accurately made for the vast majority of points (Attneave, 1954). The prediction will, in general, fail where there is an edge; it is at edges that the maximum amount of information about the environment is conveyed.

Thus if we know the position of edges, the only additional information we need to know is the color and brightness to either side of the edge and in this way we can clearly specify the visual environment using much less information than if we had to specify the brightness and hue at every point in it. It is presumably because most of the information in the visual stimulus is carried by edges that line drawings can achieve such a faithful representation of everyday scenes.

Clowes (1967) has pointed out that considerable economy in shape recognition can be effected by starting with an analysis of local features (such as straight edges) and then at a later stage analyzing the relation of these local features to one another, discarding information about the exact position of each local feature. The exact localization of such features is not usually important for recognition: a tree can be recognized as a tree provided that there is a more or less upright trunk and provided that some lateral branches join it in the upper portion. Beyond this, the exact relations between the branches and the trunk do not matter for the purposes of recognition. Hubel and Wiesel's "complex" fields that respond to a bar or an edge placed anywhere within a given region of the retina may be part of the mechanism for discarding information about the exact position of local features.

We conclude that, at least at the lower levels of the visual system, there are innate mechanisms for analyzing set features of incoming patterned stimuli. This conclusion is supported by both behavioral and physiological evidence; it also fits with what might be expected if we consider what sort of mechanism is needed to enable animals to make good use of visual information in normal environments.

## II. Perceptual Differentiation

The work we have been discussing has implications for our concept of analyzer. It strongly suggests that some of the analysis of visually presented patterns is done by innate mechanisms: it does not imply that all of it is. It is clear that with practice we become better at identifying members of particular classes of objects. For example, most Westerners have difficulty in remembering Chinese faces, but with practice, they are able to identify and remember Chinamen as well as members of their own community. It is hard to believe that there are innate analyzers with just the properties required to recognize differences in human faces. There are two ways in which such analyzers could come into existence. There might be a control mechanism capable of modification which would come to contain a set of instructions for the sequence of innate analyzers to be used. For instance, if distance between the eyes is an important feature in differentiation between faces, existing analyzers could be programmed to give an output corresponding to this distance, another output corresponding to length of nose, and so on. For a given class of objects, successful recognition of members could only be achieved when such a program had been built up. This follows naturally from our present conceptualization. If, as we have postulated, there is a limit to the amount that can be stored at any one time, then, when we are confronted with a new class of object, we may at first store information about features of the object that are irrelevant for the purposes of recognition either because they remain constant from one member of the class to another or because they vary randomly within each member of the class over time. Thus, in many communities the color of a woman's hair is no longer a good criterion on which to base recognition. In this way, we might achieve perceptual differentiation of complex objects without building up new analyzers. However, the effects of building up a program to control recognition would be very like setting up a new analyzer. If the program for recognizing human faces were called into play, no response would be made to the output from analyzers not switched in by the program. It would be as though a new dimension of perception had been constructed based on using several other dimensions in a particular way. We shall assume that the effects of a control program operating on existing analyzers is to call into play a new analyzer. This complicates matters very considerably since many of these "learned" analyzers will be multidimensional, at least viewed in terms of the innate analyzers.

Moreover, Sutherland (1968) has recently argued that in learning to recognize complex objects, an abstract structural description of the object is formed. Such structural descriptions decompose the input pattern into parts. At the lowest level, these parts would be line (or edge) segments. For example the letter "L" might be described as a vertical line whose bottom point was joined to the left point of a horizontal line, where the vertical line is approximately twice the length of the horizontal line. It should be noted that such a description involves specifying two kinds of relationships: first, spatial relationships between parts exemplified here by "join," and second, comparative relations between the properties of parts exemplified here by "twice as long as." Other examples of the former type of relationship would be relations such as "next to," "inside," "beneath," and of the latter type "darker than," "larger than," and so on. The perceptual units into which objects are decomposed for the purpose of storing a description form a hierarchy. For example, a picture of a man can be decomposed into head, trunk, arms, etc.; head can be decomposed into cheeks, eyes, ears; and at the lowest level, objects are decomposed into line segments (or edges) and textures. At each level, information must be stored not only about the objects concatenated at that level but about the relationships between those objects. The arguments for supposing that object recognition involves the storage of hierarchically organized structural descriptions of the type outlined will not be repeated here, but are to be found in Sutherland (1968). It is clear that the notion of an analyzer does not begin to capture the complexity of what is going on when a representation of an object in the external world is formed in the brain.

In terms of the above account, perceptual differentiation would involve learning into what units the visual input from a member of a class of objects is most appropriately decomposed and also learning what information to store about the relations obtaining between the parts of a unit. This way of thinking would make recognition an active process as has been emphasized by Bartlett (1932). Our learned "program" for decomposing an input picture of a particular kind and analyzing the relationship between its parts corresponds to his concept of the "schema." Some of the reversible figure effects could be interpreted in terms of a model of the sort we propose. In Boring's (1930) wife and mother-in-law figure, it is impossible to see the wife and the mother-in-law simultaneously. Now, if the recognition of faces is based in part on measuring the distance between the eyes, it would be necessary to identify first what constituted eyes. The program would not accept two different sets of eyes and in particular it would not be able to interpret the same feature simultaneously as an eye (in the mother-in-law) and as an ear (in the wife) since once the position of the eyes is fixed the approximate position of the ear is also fixed. Determining what constitutes eyes will in turn determine which part of the figure is to be taken as nose and the appropriate analysis will be made on that part. Hence, it should be impossible to see both the wife and mother-in-law simultaneously.

If the above account of recognition is correct in its outlines, then it would be more appropriate to talk of selective attention as the process whereby an animal learns to select a particular description of an object rather than as "learning to switch in an analyzer." In simple cases, the two ways of thinking are to some extent interchangeable. For example, if an animal is taught to select a white horizontal rectangle and avoid a black vertical rectangle and transfer tests show that it has learned only about the horizontal–vertical difference and not about the difference in brightness, we could say that the animal has learned to attach the response ⟨Approach⟩ to a shape to which the description ⟨Horizontal⟩ can be applied and to attach the response ⟨Avoid⟩ to shapes to which the description ⟨Vertical⟩ applies. In this simple instance (with one exception to be noted below), this way of conceptualizing the processes taking place in the brain is equivalent to saying that the animal has learned to switch in an orientation analyzer and has attached the response ⟨Approach⟩ to the horizontal output and the response ⟨Avoid⟩ to the output corresponding to vertical. In more complex instances, it is no longer appropriate to talk of an animal simply learning to switch in an analyzer; it seems better to talk of an animal learning to form a structural description of an object. Once again, selectivity will be involved since, for any complex object, there will be many possible descriptions.

The problem with which this book deals is the way in which selective attention interacts with response learning and performance. The problem of how descriptions of complex objects are formed is unsolved, and almost all the work we shall discuss has been carried out with simple stimuli such as tones, surfaces differing in brightness or color, and shapes differing in orientation. In such cases, despite the oversimplification involved, the use of the word "analyzer" seems appropriate. Moreover, there is one aspect of perceptual selectivity that is more accurately conveyed by the word "analyzer" than by the word "description." If an animal has learned to make one response to red and another to green, and we describe what is involved in selective attention as learning to switch in a color analyzer, then it follows that the animal should find it easier to solve a subsequent problem involving *different* colors, such as blue and yellow, since the detection of these differences involves the use of the same analyzer. This notion is not caught by the use of the word "description," since, until we can specify more closely how the process of forming descriptions works, there is no reason to suppose that if an animal has learned to store the descriptions ⟨Red⟩ and ⟨Green⟩ in one problem this will facilitate the use of the descriptions ⟨Blue⟩ and ⟨Yellow⟩ in another. This point will be taken up again at the conclusion of the chapter. For the time being, it is enough to emphasize that describing the process of selective attention as learning to switch in an analyzer is a gross oversimplification, but that we believe that the oversimplification is legitimate and useful for our present purposes.

Before we leave the problem of what constitutes an analyzer, there are two more caveats that should be entered. Even in dealing with simple dimensions, such as brightness or orientation, it must not be supposed that analyzers are located at the level of the primary visual cortex nor that they involve, except at an early stage, the units at this level identified by Hubel and Wiesel. Although the orientation analyzer might receive inputs from some of the units discovered by Hubel and Wiesel, it is conceived as a superordinate center in which the decision is taken as to whether a horizontal or vertical rectangle is present.

Secondly, when an animal learns to switch in a given analyzer, it learns to switch it in in a given situation. The rat that has learned to respond in a jumping stand to a black–white difference will not show an increased tendency to control its responses by responding to brightness cues in totally different situations such as its home cage. In other words, analyzer strength is itself conditioned to input stimuli. In order to switch in the black–white analyzer, the rat must first recognize that it is in a jumping stand, and this will involve the use of other analyzers or the formation of a structural description of the jumping stand that can be matched to a stored description.

## III. Absolute Judgments

It is known that, with practice, the ability to discriminate between very similar stimuli improves. We have already suggested that stimuli that are very hard to discriminate may on our model produce overlapping outputs from the common analyzer. It is possible that with practice the degree of overlap between these outputs is reduced; i.e., in some way the signal-to-noise level in the analyzing system becomes greater. There is not enough evidence to decide whether or not such a process occurs. It may simply be that, when an analyzer is switched in weakly, less information can be stored about the exact position of a given output and that the effect of extra training is to switch in the analyzer more strongly so that more information can be stored about the position of a given output. The results of work on absolute judgments make it likely that some modifications in the outputs from a given analyzer can occur.

Pollack (1952) found that when subjects were asked to judge the frequency of tones by assigning different tones to categories, they were capable of transmitting only about 2.0 to 2.5 bits per judgment. The range of tones used had very little effect on how much information was transmitted. Subjects tested with tones ranging between 100 and 8000 Hz transmitted only .2 more bits of information per judgment than subjects tested with a range of 100–500 Hz. Similar findings on absolute judgments of size have been obtained by Eriksen and Hake (1955a) and by Alluisi and Sidorsky (1958). Schipper (1953) has shown that absolute judgments of loudness are little affected by the range of loudness used. These findings can be interpreted on our present model in the following way.

It may be that within one analyzer it is possible to stabilize the outputs over a limited range, at the expense of losing discriminability elsewhere. We could compare a unidimensional analyzer to a voltmeter, on which 5–10 different positions can be read. By throwing a switch, we can make these positions read in volts, tens of volts, or hundreds of volts, but with any one position of the switch, there will only be a limited number of categories that can be read off and the number of categories will not vary from one position of the switch to another. Such a device is built into voltmeters in order to economize on the number of parts, and a similar device may be incorporated in the nervous system for the same reason. The range of stimulus intensities with which the nervous system has to deal is enormous. Thus, the ratio of the least discriminable intensity of light to the maximum bearable intensity is of the order of 1,000,000,000 : 1. Normally, however, we need only make discriminations within a part of this range, depending on the illumination conditions. A device of the sort instanced would take advantage of this by using the same central mechanism to detect differences at many different points on the scale, depending on how it is preset. It will be seen that a switch to determine how the outputs from a given analyzer are to be interpreted is a very similar type of device to that which we have suggested to account for perceptual differentiation. Within one analyzer, the same computing mechanisms are used but on different ranges of incoming information, depending on a switch controlled by central mechanisms.

It should be noted that if this explanation of the limits of absolute judgments is correct, it implies that there is no means of switching rapidly between one range and another. If such a device were present and if the range setting could be read off centrally, then subjects would be able to operate with two different ranges at once and thus increase the number of categories that could be successfully used with a large range. We could, for instance, teach subjects to discriminate accurately between loud tones of varying intensity and between a second range of soft tones. If now the subject could rapidly adjust the switch determining which range would be read and if he could read off the position of the switch and use this information together with the output produced to determine the category of the stimulus, then he would be able to discriminate 14 different loudnesses. So far as we are aware, no experiments have been undertaken to discover whether, with extensive practice on problems of this sort, subjects might not become more proficient in making absolute judgments. For example, it would be interesting to see whether a subject could work accurately with two ranges of loudness if a visual stimulus were given before each tone informing him from which range the tone would be selected; such an experiment would answer the question whether the position of the switch determining which range was to be read could itself be conditioned to an external stimulus. It may be that no record of the state of the range switch is stored, simply because it is normally not important for survival to know what is the range within which the system is operating. In most visual environments, the intensity of light from different points does not differ by more than a factor of about 100 : 1 and, so far as recognizing objects is concerned, it does not much matter where in the possible range this restricted range of intensities is located. To a large extent, the operating range might be set by the input at a given time, and no record kept of what it is. If this is correct, it would be predicted that there would be considerable falling off in the ability of subjects to make accurate absolute judgments when a time gap was inserted between practice and testing.

If information about the range setting cannot be used effectively to determine the response, we would also have an explanation of why relative judgments are so much better than absolute judgments. If two stimuli are presented simultaneously or in very close succession, the range setting will not change from one stimulus to the other and, hence, it would be possible to determine which gave the higher output. Another way of looking at this problem is that whereas an output from an analyzer can be accurately read, there are very severe limitations on the accuracy with which it can be stored. Our account of absolute judgments gives a rationale for why this should be so by suggesting that the range setting is determined by the incoming stimuli, and there is no mechanism for storing accurately what the range setting was on a given occasion.

It is worth mentioning here a second finding that has emerged in the course of work on absolute judgments. If more than one dimension is used at a time, the amount of information transmitted is increased, but the increase is less than the sum of the information that can be transmitted using each analyzer separately. Thus, Ericksen and Hake (1955b) found that subjects making absolute judgments separately of brightness, size, and hue could transmit, respectively, about 2.3, 2.8, and 3.1 bits of information. When judgments were made on pairs of these dimensions about 3.4 bits could be transmitted; when all three dimensions were used, they transmitted about 4.1 bits rather than the 8.2 bits that they would have been able to transmit had they been able to sum the information that can be transmitted with each dimension alone. Pollack (1953) obtained similar findings on the amount of information that could be transmitted using pitch and loudness alone or in combination, and Pollack and Ficks (1954) found that when eight different auditory dimensions were combined, subjects were still only able to transmit 7.0 bits of information. These findings fit very well with our present theory: they suggest that more than one analyzer can be used to determine the response on a single trial, but that the more analyzers used, the less efficiently any one of them can control behavior.

There is, of course, no way of knowing at present whether information from more than one analyzer can be processed simultaneously; information from several analyzers could be made effective in determining the response even if the actual processing of information from each were done at successive moments of time. It would be interesting to have reaction-time data available for human subjects judging in terms of several categories at once and also for animals solving problems involving a combination of cues. Unfortunately, such data might not be decisive since the processing times involved might be a very small part of the reaction time. Moreover, if the use of one analyzer produced worse performance than the use of several, organisms might offset this disadvantage by processing information several times from one analyzer before responding. It is, of course, well established (e.g., Hick, 1952) that under many circumstances, reaction times increase linearly with increases in information transmitted. This is a strong argument against supposing that all processing is conducted in parallel.

Although we have argued that much information processing must be conducted in successive operations rather than in parallel in the human nervous system, there is certainly a degree of parallel processing in peripheral systems that is greater than in any automaton yet built. Hubel and Wiesel (1962) have shown that the same operations on the input are conducted in parallel at many different points of the retina. If we were building a machine to do the same type of processing we would almost certainly have a smaller number of processing units and scan these successively over different parts of the receptor surface using the same channels to transmit the information, but coding place of origin in terms of time. The reason that parallel processing appears to be more common in the nervous system than in automata may be that, although manmade electronic transducers have faster operation times than those made by nature, they occupy much more space than elements in the nervous system, and are more expensive to produce. (A generous estimate of the cost of a neuron in a young adult would be .0001 American cents and this compares very favorably even to a Japanese-built transistor.) Despite this difference between the human brain and existing automata, it is certain, for reasons we have already given, that not all processing performed by the human brain is conducted in parallel.

## IV. Stimulus Control of Analyzers

In Chapter 2, we discussed some of the central determinants of analyzer strength and the way in which analyzer strength can be altered by learning. It is clear, however, that the incoming stimulus itself can also determine analyzer strength to some extent, and we have already shown that the strengths of particular analyzers must be determined by the outputs from other analyzers. There is a second way in which the stimulus input itself determines whether or not the outputs from a given analyzer will gain control of central mechanisms. It is well known that, in animals (Berlyne, 1960) and men, both novel stimuli and very intense stimuli readily capture attention. It is, in fact, obvious that no matter how much someone concentrates on a task, they are likely to notice sudden loud noises or painful stimuli, although there is evidence that in states of intense excitement or fear (e.g., in battle) even painful stimuli may not be noticed.

The fact that attention can be controlled by the novelty of a stimulus suggests that some analysis of the stimulus is made before the decision is taken as to whether or not the stimulus will gain control of response or be stored in memory. In Chapter 1, we discussed an experiment of Sharpless and Jasper (1956) showing that the EEG arousal pattern habituates in sleeping cats to two tones, one of 5000 Hz and one of 200 Hz, sounded successively in that order; the habituation is specific to the order in which the tones are given, and the arousal pattern reappears if the tones are given in the reverse sequence with the 200-Hz tone sounded first. Sokolov (1960) has shown that the physiological orienting reaction may be habituated selectively to a strong tone: after such habituation, the reaction is reinstated if a weak tone is sounded. Such findings imply that considerable analysis of the stimulus is made before it is accepted or rejected by central mechanisms. In the absence of such analysis, it would be impossible to decide whether or not a stimulus was novel. It also appears that at least a temporary record is kept of what outputs have recently occurred, otherwise it would not be possible to allow novel stimuli to gain control of attention.

The same type of conclusion is forced upon us by the extensive work that has been performed recently on the problem of selective attention to auditory messages in man. As already mentioned, subjects can select efficiently one of two auditory messages distinguished by different physical characteristics, such as point of origin in space, intensity, and frequency range. Cherry (1953) has shown that if two verbal messages are sent simultaneously, subjects can also use the probability of occurrence of words in a given context to select one message, but selection is now much less efficient than if the two messages are differentiated by different physical features. Moreover, Treisman (1964a) has shown that if messages with different physical characteristics are presented to bilingual subjects, they do very little better in selecting one of the two messages if they are presented in two different languages than if both messages are in the same language. It seems clear that central analyzers can be set to receive one of two imputs, where the inputs differ in terms of some simple physical characteristic. Where the messages do not vary along a simple physical dimension but vary in a multidimensional way, for example, when the two messages differ only in content, it becomes very difficult to set the central analyzers to receive one and not the other. Moreover, when subjects are attending to one message they can often report simple physical features of a second (e.g., whether it is louder or softer than the one attended to, whether it is higher or lower in pitch, and whether it differs in point of origin in space). When subjects attend to one of two simultaneously presented messages differentiated by physical features, they normally cannot report any of the words in the second message nor can they report changes in the actual language used in this message (e.g., from English to French).

All this suggests that central analyzers can be preset to accept only messages having certain physical characteristics; if they are preset in this way, no information is available about messages having characteristics that are rejected by the central analyzing mechanisms, except what these characteristics were. This is in line with the physiological evidence quoted to show that the initial stages of analysis are conducted by the nervous system in parallel. Some of these peripheral channels can be excluded from the central analyzing mechanisms. When they are excluded, information is available about whether a message was present on these channels, but no information is available about the contents of the message.

Since, however, it is possible to select (albeit with difficulty) one of two messages not differentiated by physical features by following sequential probabilities,[2] it seems that central analyzing mechanisms can be biased to accept complex features of the message. As Treisman (1964b) points out, if a message is picked out in terms of sequential probabilities, there is a very wide range of words that are probable, and to know whether or not one of these words has occurred, it might be necessary to analyze all the distinguishing features of the words before the decision was made as to whether to accept or reject a given word. If this were so, it would mean that analysis was being made in parallel up to this stage, and we could no longer argue, as we did earlier, that one reason that the more one analyzer is used, the less others are used is that some of the computational mechanisms may be common to more than one method of analysis. Treisman (1964b) makes a suggestion to overcome this problem:

> A possible system for identifying words is a hierarchy of tests carried out in sequence and giving a unique outcome for each word or other linguistic unit. The decision at each test point could be thought of as a signal detection problem: a certain adjustable cut-off or criterion point is adopted on the dimension being discriminated, above which signals are accepted and below which they are rejected as "noise". The criteria determining the results of the tests would be made more liberal for certain outcomes if favoured by contextual probabilities, by recent use or by importance [p. 14].

Treisman includes "recent use" and "importance" as well as contextual probabilities here because Cherry has shown that if two identical messages are presented simultaneously but out of phase, the identity is detected, and Howarth and Ellis (1961) have shown that where two competing messages are presented on the same channel, a subject's own name appearing as part of the rejected message may be detected.

In terms of our model, it is likely that the same low-level analyzers have to be used in the analysis of different spoken messages coming through the same channel. If this is right, words from the rejected message could only be rejected after their features had been analyzed. Hence, more information would have to be processed than in the case of two verbal messages sent in over different channels, since, in the latter case, the rejected channel would not be given access to the analyzing mechanisms. This would explain why it is more difficult to follow one of two messages when they are differentiated only by contextual probabilities than when they are differentiated by some physical feature.

[2] Miller and Isard (1963) have provided evidence suggesting that more may be involved in such selection than following sequential probabilities. In particular, the subject is following grammatical rules in determining what to accept and what to reject. This additional complication does not affect our present argument.

The issue of where the rejection of a signal occurs and, in particular, how fully a message must be analyzed before being rejected is important in its own right, but it does not affect our basic theory. It may be that an analysis is made in parallel of most incoming stimuli, and that it is only possible to reject before analysis messages coming in on discrete physical channels in the nervous system. Although this would not effect much economy in the analysis of the incoming stimuli, there would still be considerable economy in storage space. Which outputs gained control of the response and gained access to storage would be determined partly by central factors influencing which analyzers were switched in for these purposes and partly by the actual outputs yielded by different analyzers. Outputs that corresponded to novel or important events would be given priority, and could determine responses as well as the outputs from those analyzers that were switched in by central mechanisms.

## V. Analyzer Strength and Response Attachments

We have now seen that the way in which analyzers were conceptualized in the previous chapter was much oversimplified. The same sort of considerations apply to the way in which we have so far conceptualized response attachments. In the present section we consider further the relation between response attachments and changes in analyzer strength. In the following section we shall consider in more detail the problem of the relations between outputs from analyzers and response attachments and we shall conclude the chapter by taking up the more general issue of what is a response.

Rule 2 of the model as stated in Chapter 2, Table II contains the statement: "An analyzer is strengthened when its outputs consistently make correct predictions about further events (e.g., trial outcomes) of importance to the animal." We examined two versions of this rule and showed that both could account for the strengthening of the relevant analyzer, in the case where an animal is learning a simultaneous discrimination. The first version ran as follows: as the animal learns the correct response attachments, the sight of the positive stimulus will more and more frequently be followed by reward, the sight of the negative stimulus will never be followed immediately by reward. The sight of the positive stimulus will sometimes be followed by nonreward, but, as the animal learns to make the correct response, the number of occasions on which this happens will decrease. The animal will, therefore, be able to learn that reward follows the output yielded by the relevant analyzer when the positive stimulus is viewed, whereas the output produced by the negative stimulus is not followed by reward. Hence, the relevant analyzer will be strengthened, since its different outputs are consistently followed by different events of importance to the animal.

Unfortunately, this version of Rule 2 will not satisfactorily account for why the relevant analyzer is strengthened in certain other discrimination learning situations. Consider, for example, what happens with the following training procedure. The animal is shown a single stimulus on each trial. When the positive stimulus is shown, the animal receives a reward if it approaches within a certain time. When the negative stimulus is shown, the animal receives reward only if it refrains from approaching it for the same length of time. Rosvold and Mishkin (1961) and others have used this situation to train monkeys and have christened it a "Go–no-go" problem. Pigeons have been taught to solve essentially the same problem in Skinnerian situations; in Skinnerian language the problem involves training on a Mult. CRF–DRO schedule. With this procedure, the outputs from the relevant analyzer do not predict different events of importance to the animal even after it has learned the correct response attachments (i.e., ⟨Go⟩ in the presence of the positive stimulus, ⟨Don't go⟩ in the presence of the negative). Early in learning, both the positive and negative stimulus will sometimes be followed by reward and sometimes not; when the animal has mastered the problem, both stimuli will always be followed by reward. The application of the first version of Rule 2 in this situation would therefore not result in the relevant analyzer being strengthened. In Chapter 2 we gave other reasons for rejecting the first version of Rule 2; the present argument provides an even more cogent reason.

The second (revised) version of Rule 2 given in Chapter 2, page 43 is also best suited to dealing with a simultaneous discrimination situation and it is difficult or impossible to apply it to certain other discrimination learning problems. According to this version of the rule, the strength of an analyzer would be increased both when the more probable response occurs and is rewarded and also when the less probable response occurs and is not rewarded. After the other two trial outcomes (the more probable response followed by nonreward and the less probable response followed by reward), the analyzer strength is decreased. This formulation depends upon there being only one response connected with different strengths to two different stimuli. The argument no longer holds in the go–no-go situation, since the probabilities of the response ⟨Go⟩ to the positive stimulus and of the response ⟨No-go⟩ to the negative are independent. There are in fact four stimulus–response (S–R) connections that must be considered in a go–no-go situation, namely, ⟨Go⟩ to positive stimulus, ⟨Don't go⟩ to the positive, and ⟨Go⟩ and ⟨Don't go⟩ to the negative. If, at the outset of training, the probability of ⟨Go⟩ to the positive stimulus is .9 and the probability of ⟨No-go⟩ to the negative is .2 the reward given after the animal has correctly failed to ⟨Go⟩ to the negative would decrease analyzer strength rather than increase it, since an improbable response occurs and is rewarded. This means that the strength of the relevant analyzer will be decreased when the animal makes the correct response to the negative until such time as the probability of that response occurring exceeds .5, and this is a most improbable outcome.

The difficulty could be circumvented by giving a strength to the response ⟨Go⟩ and adopting the rule that changes in analyzer strength depend upon there being a response attached with a different strength to two different outputs. When the effect of reward outcome is to increase the existing difference in the strength of a given response to two different outputs, the analyzer would increase in strength, when the effect is to decrease the existing difference the analyzer would be decreased in strength. This version of Rule 2 is really a more generalized version of the second version given in Chapter 2. It could be applied to the simultaneous training situation in the following way. If the response in question is that of jumping and it is attached to the output corresponding to the positive stimulus with strength $R_1$ and to the negative stimulus with strength $R_0$, we could apply Luce's rule governing the relation between response strengths and probability of making a given response in a two-choice situation; thus if $p_1$ is the probability of jumping to the correct stimulus then $p_1 = R_1/(R_1 + R_0)$.

There are, however, learning situations which it is difficult to handle even with this version of Rule 2. We can regard a "go–no-go" situation as one in which only one response is involved, the animals learn to make the response ⟨Go⟩ to the positive stimulus and not to make it to the negative. Suppose, however, we train an animal in a successive situation of the following kind. If $S_1$ is presented the animal is rewarded if it makes the response of turning its head to the left; if $S_2$ is presented the animal is rewarded for turning its head to the right (Grindley, 1932). We now have two different responses being learned and in this case it is not possible to measure directly the change in the difference in the strength of a single response attached to two different outputs. Rule 2 could be generalized still further to take account of this case by assuming that, on a given trial, the change in analyzer strength is in proportion to the sum of all changes in the difference between existing response strengths to different outputs. To put this formally, let $R_{i1}$ be the strength of response $i$ to output 1 and let $R_{i2}$ be the strength of response $i$ to output 2. Then if $\Delta A_n$ is the change in analyzer strength after trial $n$,

$$\Delta A_n \propto \sum [(R_{i1} - R_{i2})_{n+1} - (R_{i1} - R_{i2})_n][(R_{i1} - R_{i2})_n].$$

In this rule $(R_{i1} - R_{i2})_n$ is the difference in the response strength of the $i$th response at the start of trial $n$. If this difference is positive and if it increases from trial $n$ to trial $n + 1$, then both expressions in square brackets will be positive, and the analyzer strength will increase. The larger the change in the difference between response strengths from trial $n$ to trial $n + 1$, the more the analyzer will be strengthened: this is intuitively plausible. Moreover, the bigger the difference in response strengths at the start of trial $n$, the bigger the increase in analyzer strength. This means that the stronger the animal's expectations, the bigger will be the effect of a single trial on analyzer strength. It will be seen that if the expression $(R_{i1} - R_{i2})_n$ is negative, analyzer strength will only increase if $(R_{i1} - R_{i2})_{n+1}$ is a larger negative number. Moreover, whether $(R_{i1} - R_{i2})_n$ is positive or negative, analyzer strength will always decrease if the size of the difference disregarding sign decreases.

This generalized version of Rule 2 covers all the experimental situations we have so far considered. Moreover, if we make a further assumption, it leads directly to a further important prediction. It is plausible to assume that when a response strength is altered to one output, there is a generalized change in the strength of that response to other outputs and that the amount of generalization increases the closer together are the outputs. It now follows that the change in the difference between the strengths of a single response attached to different outputs will be greater, the bigger the difference in outputs. When animals are given discrimination training with two stimuli lying close together along the relevant dimension, the relevant analyzer would, therefore, be strengthened relatively slowly. This would clearly help to explain the results on transfer along a continuum discussed in the previous chapter.

Although the present version of Rule 2 is general enough to cover a wide variety of situations, it still does not necessarily cover all situations in which analyzer strength must be increased. Solomon and Turner (1962) performed the following ingenious experiment. First, they taught dogs to press a panel in order to avoid shock using light offset as the conditioned stimulus. The dogs were then given curare, replaced in the apparatus, and exposed to two tones of different frequency sounded in a random order. The onset of one tone was always followed by unavoidable shock, the other was never followed by shock. Since the animals were paralyzed by curare, they were unable to make any skeletal responses at this stage. Nevertheless, when subsequently tested without being curarized, these animals made the avoidance response learned at the first stage to the tone associated with shock and did not make it to the other tone. The animals had, therefore, learned to use the analyzer detecting the difference between the two tones although no skeletal responses were made during the period of learning. However, since the discriminated response appeared as soon as animals were tested without the drug, the response attachments must have been formed during training. These results could be interpreted as showing that for the relevant analyzer to be strengthened it is sufficient that differential response attachments be formed at a central neural level without the response ever having been made. Therefore, this experiment is not in contradiction to the present version of Rule 2, provided that we think of response strengths as reflecting some central neural process and not necessarily depending upon an overt motor response for their formation. Moreover, differential emotional responses may well have been conditioned to the two tones, and such responses, insofar as they are mediated by smooth muscles, would not have been blocked by the curare.

In classical conditioning, reinforcement does not depend upon the animal's response, and it could, therefore, be questioned whether the present rule can explain the strengthening of the relevant analyzer in such situations. However, where the difference between the reinforcing conditions following two outputs from an analyzer is great enough, we would normally expect some emotional response to be differentially conditioned to the two outputs. For example, if one stimulus is always followed by pain and the other is not, we would expect fear to be conditioned to the former stimulus and not to the latter, and this would be enough to strengthen the analyzer detecting the difference between the two stimuli. Similarly, if one stimulus is always followed by food and the other is not, it is likely that some emotional response may be conditioned to the former stimulus but not to the latter. At the very least, salivation will be conditioned to one stimulus but not to the other. It is worth noting that it is impossible to account for the strengthening of the relevant analyzer in a classical conditioning situation, if we assume with Zeaman and House and Lovejoy that analyzers are strengthened by reward and weakened by nonreward. Since in classical conditioning, reinforcement is not contingent on a response being made, their theories provide no mechanisms for the strengthening of an analyzer in classical conditioning, and this is a very powerful argument against them.

It has sometimes been argued that the occurrence of a salivary conditioned response (CR) is, in fact, instrumentally rewarded because when an animal salivates to the conditioned stimulus (CS) it can eat and digest the food faster. Sheffield (1965) has shown that this explanation is probably false since animals cannot be instrumentally trained not to salivate to a signal by witholding food when they do salivate. This suggests that the reason an animal learns to pay attention to a signal for food is not that it is reinforced for so doing as a result of making a preparatory response. It would appear then that the present formulation of Rule 2 covers the case of classical conditioning as well as the different types of instrumental conditioning that we have considered. It should be noted, however, that Rule 1 of our model cannot account for response learning in classical conditioning. Since almost all the evidence on selective attention that we shall cite comes from experiments on instrumental learning, we shall not attempt to rewrite Rule 1 in order to include the case of classical conditioning. It seems clear (Sheffield, 1965) that a different rule is needed to cover instrumental and classical conditioning, but no one has yet succeeded in specifying what class of responses can be classically conditioned.

We now go on to consider whether Rule 2 will cover the phenomena of latent learning. If rats are allowed to explore thoroughly a five-unit T maze with no primary reinforcement present, and are then placed several times in the goal box thirsty with water present and with no access to the remainder of the maze allowed, as many as 50% of the animals will run the maze without error on the first trial on which they are placed thirsty in the start box (Kimball, Kimball, & Weaver, 1953). It seems likely that the analyzers detecting the cues present in the maze were strengthened during the preexposure to the maze, although it is not clear that any differential responses were attached to the different outputs from these analyzers at this stage. As we shall see, there is evidence that the position cue is very dominant for rats and it is possible that no strengthening of this analyzer would be required during the preexposure period. It is not, in fact, certain how much animals can learn in the course of latent learning about cues requiring analyzers of low initial strength for their detection.

Another type of situation in which it seems possible that an analyzer is strengthened without differential response attachments being formed is the sensory preconditioning paradigm. As an example of such an experiment Brogden (1939) presented a bell and a light simultaneously to dogs 200 times. He then made the light the CS for a conditioned forelimb flexion, with shock serving as the unconditioned stimulus (US), and found that the CR occurred when the bell alone was presented. This suggests that the analyzer detecting the bell was strengthened at a time when no differential response was learned. It is enough that the stimulus be accompanied (or followed) by some other stimulus. Hoffeld, Thompson, and Brogden (1958) have shown that sensory preconditioning is greater when the preconditioning stimulus precedes the other one than when they are presented simultaneously (see also Silver & Meyer, 1954).

For our purposes, one of the most interesting results is that obtained by Hoffeld, Kendall, Thompson, and Brogden (1960) (cf., Prewitt, 1967): they varied the number of presentations of the two stimuli during the preconditioning phase, and found that the amount of sensory preconditioning was maximal after only four preconditioning presentations; animals given additional presentations showed less preconditioning. This suggests that when one stimulus predicts the occurrence of another, the analyzer detecting the first may initially be strengthened, but if the event being predicted is of little importance to the animal, further presentations result in a decline in strength of the analyzer for the first stimulus (cf. Lubow's results quoted below).

The question to which we must now address ourselves is whether one implication of these experiments is that the analyzer detecting the preconditioning stimulus is strengthened without any differential responses being formed to its outputs. Here, we run into the same problem that we met in interpreting the results of latent learning experiments. In order to demonstrate the sensory preconditioning effect, investigators have used very conspicuous stimuli. This means that the analyzer for these stimuli would already have high strength at the outset of the paired presentations, and learning could, therefore, occur about its outputs without any further strengthening being necessary.

It is, in fact, impossible to resolve the issue of whether an analyzer can be strengthened under conditions where its outputs are followed by different events and no different response strengths are associated with its outputs. If different outputs from one analyzer are consistently followed by events of importance to the animal, it is hard to exclude the possibility that some emotional response or an alerting response is being differentially conditioned to the outputs. The rule that any increase in the existing difference between response strengths attached to different outputs from one analyzer increases the strength of that analyzer is therefore sufficient for all practical purposes, and we shall use either that rule or the more specialized rule that applies to simultaneous discrimination training throughout the remainder of this book. If a more general rule were required, then it might be stated in the form that an analyzer increases in strength whenever any existing difference between the neural events following two outputs is increased. This would cover both the case when there were changes in response tendencies and the case when there were no changes in response tendencies but only changes in the animal's expectation of a further stimulus input to different outputs from the same analyzer.

We should perhaps add a brief statement about the second half of Rule 2. When different outputs are not consistently followed by events of importance to the animal, analyzer strengths revert to their base level. The results obtained by Hoffeld *et al.* give some support for this claim. The theory as it stands does not, however, cope readily with all instances of habituation. Consider, for example, some experiments conducted by Lubow and Moore (1959) and Lubow (1965). Goats and sheep were preexposed to a flashing light or a turning rotor. After preexposure trials of this sort, the stimulus to which the animals had been preexposed was then made the CS for a shock to the leg. The more preexposure trials given, the more slowly leg flexion developed as a CR to the stimulus. Now, according to our model, this result could only be explained if the strength of the analyzer detecting the light or rotor was weakened relatively to the other analyzers operative in the situation, since by Rule 2 analyzer strengths sum to a constant amount. Since neither the light nor the background stimuli were followed by events of any consequence to the animal there is no reason to expect any *differential* weakening of the analyzers for either set of stimuli. It may be that when a stimulus is presented for a brief time against a constant background, its novelty at first gives it control of behavior but that with repetition it ceases to be novel and hence the output yielded from its analyzer becomes weaker. There is in fact no postulate in our theory that accounts for such habituation.

The rule states that analyzers return to a "base level" since it seems certain that when a cue is very dominant in a given situation it is impossible to reduce the value of the analyzer for that cue below a certain point. It is for this reason that position habits are so difficult to eradicate in a jumping stand. No matter how much training on other cues is given in this apparatus, if all else fails, the animal is always likely to revert to a position habit.

## VI. Analyzer Outputs and Response Attachments

In the last section, we considered in some detail how analyzers are strengthened; in order to do this, we examined some of the possible interactions between response strengths and analyzers. In the present section we discuss the relationship between the response attachments and the outputs from analyzers.

So far, we have assumed that in two-choice discrimination learning problems, animals attach one response to one output from the relevant analyzer and a different response to a second output. This is indeed the only analysis that can be given of a go–no-go problem. In the case of a two-choice simultaneous discrimination problem, however, there are two distinct rules that the animal can learn. Consider a rat being trained to discriminate between black and white in a jumping stand with black positive. The animal can either learn the rule ⟨Jump to black, avoid white⟩ or it can learn the rule ⟨When black is left of white, jump left; when white is left of black, jump right⟩. Our own theory assumes that the difference between black and white and the difference between left and right positions are detected by different analyzers. On such a theory the simplest assumption to make is the one we have in fact made up to this point, namely, that the animal learns the rule ⟨Approach black and avoid white⟩. This is the simplest assumption on our model because the problem is solved once the two appropriate responses are attached to the two appropriate outputs from a single analyzer. To learn the configurational rule, however, according to our own theory, the animal would have to make use of two analyzers and the outputs from these would have to be combined before any response attachments could be made. Figure 3.1 depicts the difference in the two mechanisms that would be needed to master each rule. An extra mechanism is needed in order to attach the response ⟨Jump left⟩ to the combination of outputs signaling either ⟨Black on left⟩, or ⟨White on right⟩ and the response ⟨Jump right⟩ to the combination ⟨White on left, black on right⟩.

Many psychologists who have proposed formal mathematical models of discrimination learning have assumed that animals learn the second rule (e.g., Gulliksen & Wolfe, 1938; Bush & Mosteller, 1951; Sternberg, 1963). They describe the situation as though there were only two stimuli that need to be considered: $S_1$ ⟨Black left, white right⟩ and $S_2$ ⟨Black right, white left⟩. The animal learns to make one response to $S_1$ ⟨Go left⟩ and another to $S_2$ ⟨Go right⟩. Lovejoy (1968) has pointed out that this description of the situation simplifies the form of a mathematical model since only two stimuli

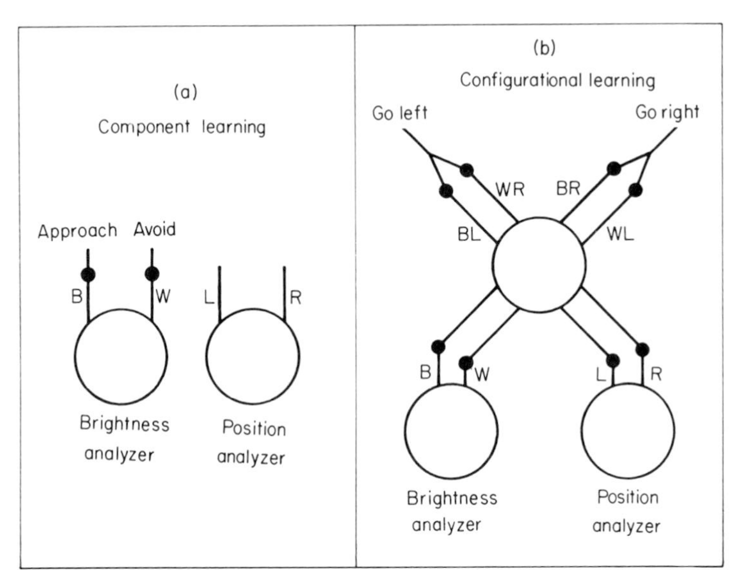


**Fig. 3.1.** Component versus configuration learning; B designates black; W, white; L, left; R, right.

and only two response tendencies need be considered. This simplification is, however, only achieved by begging the whole question of perception. Models of this type throw no light on the problem of how the animal analyzes the stimulus situation. The model assumes that somehow the animal knows there are two and only two different stimulus configurations, that the differences between them are detected from the outset of training, and that all the animal need do is build up different responses to the two situations.

Moreover, there is compelling evidence that under many circumstances animals solve simultaneous discrimination problems by learning the component rule. It has been found that chimpanzees (Nissen, 1950) and monkeys (Miller & Murphy, 1956) having learned to select one member of a pair of stimulus objects presented side by side, will transfer to the same stimulus objects arranged one above the other. Such transfer could only occur if the animals had learned to approach one object and avoid the other, rather than learning to go to one place with one stimulus configuration and to go to another with a different stimulus configuration. Similarly, Fields (1935) found that rats having learned to jump to one of two cards simultaneously exposed in a jumping stand, continued to jump correctly when the positive card was exposed with four negative cards. Sutherland (1969) has shown that octopuses show excellent transfer from a simultaneous discrimination problem to a go–no-go problem; such transfer could occur only if the animal had solved the simultaneous problem by learning to approach one stimulus object and avoid the other.

Parenthetically, it should be noted that we have assumed that animals usually learn in a simultaneous discrimination situation both to approach the positive and to avoid the negative. The problem could of course be solved if the animal merely learned to approach the positive stimulus and learned nothing about the negative. We have insisted that animals usually learn both responses, since the evidence that this is normally true is compelling. Sutherland, Carr, and Mackintosh (1962) have shown that rats trained in a simultaneous problem will subsequently transfer to a situation in which a single positive or negative stimulus is presented on one or other side of the apparatus. When retrained with single stimuli in this way, animals performed significantly above chance from the outset of retraining both on trials when they had to select the positive stimulus and on trials when they had to run to the opposite side to the negative stimulus. Similar results have been obtained by other investigators (for a review, see Sutherland, 1961a).

While it is certain, then, that animals solve some simultaneous discrimination problems by learning to approach one stimulus and avoid another, not all problems that animals can learn are soluble in this way. Rats learn to solve the following problem: on some trials two black cards are presented, on others two white cards; animals are rewarded for jumping to the left-hand card when both are black, to the right-hand card when both are white. Here, the animal cannot learn any of the simple rules ⟨Approach left⟩, ⟨Approach right⟩, ⟨Approach white⟩, or ⟨Approach black⟩. The problem can only be solved by learning the rule ⟨If black, go left; if white, go right⟩. We shall refer to such problems as "successive conditional positional problems" (or, for brevity, simply as "successive problems"). It will be seen that the rule involved is of the same basic type as the configurational rule. If animals solve a simultaneous problem by learning the configurational rule, then a successive conditional position problem should be easier for the animal than a simultaneous problem. In the simultaneous problem, the animal must detect both the presence of black and white, and the side of each, whereas in the successive situation it only need detect whether the stimuli are both black or both white. The responses learned are the same in both cases, namely, jump left and jump right. If, on the other hand, animals learn the component rule when confronted with a simultaneous problem, the successive problem should be more difficult than the simultaneous. In the simultaneous problem, the animal must detect only whether black or white is present and then attach the response ⟨Jump⟩ to one stimulus and the response ⟨Don't jump⟩ to the other. In the successive problem, the animal again has to detect the presence of black or white, but it has to attach the responses of ⟨Jump left⟩ to one stimulus and ⟨Jump right⟩ to the other. Since executing these responses will involve the use of a second (positional) analyzer, we might expect the successive problem to be harder to learn than the simultaneous. Hence an answer to the question which type of problem is solved more readily will provide further evidence about which rule an animal normally learns in a simultaneous discrimination problem.

Numerous experimenters have compared the ease with which simultaneous and successive problems are learned. Bitterman, Calvin, and Elam (1953), Bitterman and McConnell (1954), Bitterman, Tyler, and Elam (1955), MacCaslin (1954), North and Jeeves (1956), Spence (1952), and Wodinsky, Varley, and Bitterman (1954) all found that simultaneous discriminations were easier for rats than successive, provided the simultaneous problem could be solved by the animal learning to approach the positive stimulus object and to avoid the negative. Warren and Baron (1956) obtained the same result in cats.

There are two circumstances under which it has been found that successive discriminations are easier to learn than simultaneous. They are in fact exceptions that prove the rule since the exceptions would be predicted by our own account of discrimination learning.

(1) Wodinsky *et al.* (1954) trained rats in a four-window jumping stand. For two of their groups, the stimuli were placed in the center windows; animals were trained to discriminate either between vertical and horizontal stripes or between a large and a small circle. The simultaneous group had to learn to jump to the extreme left window when vertical was left of horizontal, and to the extreme right when vertical was right of horizontal; the successive group had to learn to jump to the extreme left window when two verticals were shown and to the extreme right when both center cards bore horizontal striations. The relations between the stimuli and the correct response are shown in Figure 3.2a. In this situation the successive group learned faster than the simultaneous, a result which was confirmed by Bitterman *et al.* (1955). In terms of our analysis, the simultaneous group can no longer learn the rule ⟨Approach vertical, avoid horizontal⟩. It must learn ⟨When V is left of H jump to the extreme left-hand window, when H is left of V jump to the extreme right-hand window⟩. This involves more complex processing than that required by the successive group, since both groups must detect horizontal and vertical, but the simultaneous group must use a position analyzer twice—once to determine the relative positions of horizontal and vertical and once to determine which response to make, whereas the successive group only has to use the position analyzer once—to determine which response to make. One condition, therefore, that makes successive problems easier than simultaneous is if the response to be learned is not made directly to one of the stimulus objects.

(2) The other condition is when the within-pairs difference between the two members of the simultaneously presented stimuli is much less readily discriminable than the difference between the two pairs of successively

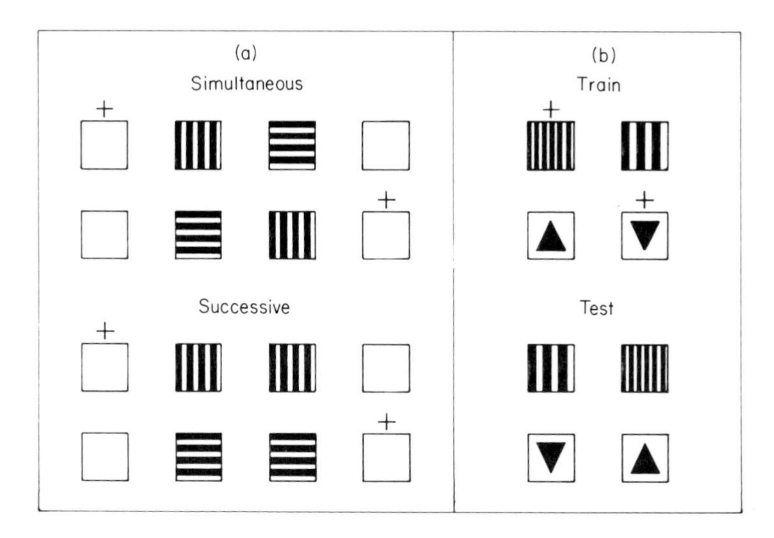


**Fig. 3.2.** Simultaneous and successive discriminations.

presented stimuli. The first experiment to demonstrate this point clearly was conducted by Teas and Bitterman (1952). They trained rats on the problem illustrated in Figure 3.2b. Two pairs of stimuli were presented; animals had to learn to jump to narrow stripes shown to the left of broad stripes, and to jump to an inverted triangle shown to the right of an upright triangle. In the training phase of the experiment the position to left or right of each member of each pair remained constant. After training, animals were tested with the positions of the two sets of stripes and the two triangles switched. If the original problem had been solved by animals learning the rule ⟨Approach narrow stripes, and approach inverted triangle⟩, then the animals should have transferred to the test shapes by continuing to follow the narrow stripes and the inverted triangle. Instead of performing in this way, they continued to jump to the left when stripes were presented and to jump to the right when triangles were presented. This demonstrates that they had solved the original problem by learning the configurational rule ⟨Go left if stripes, go right if triangles⟩. The reason for this is, presumably, that it is easier for the rat to discriminate presence of stripes from presence of triangles than to discriminate the difference between the two sets of stripes and the difference between the two triangles. In our terminology, it is likely that the analyzer for performing the former discrimination has higher initial strength than the analyzers required for the latter two discriminations.

White and Spiker (1960), using children, have obtained results similar to those obtained by Teas and Bitterman. By varying the stimuli used, Zeiler and Paul (1965) have shown directly that whether a problem of this type is solved by learning a component or a configurational rule depends upon the within-pairs and between-pairs similarity of the stimuli.

The finding that the relative ease of successive and simultaneous discrimination learning depends upon the relative discriminability of between-pair and within-pair differences resolves certain other apparently contradictory results which appear in the literature. Several experimenters (e.g., Weise & Bitterman, 1951; Calvin & Seibel, 1954) have trained rats on T or Y mazes in which the animal must either learn to follow the darker or brighter of two alleys (simultaneous problem) or select one side if both alleys are lit, the other if both are dark. The most common finding in this situation is that the successive problem is easier than the simultaneous. However, when the alleys are differentially illuminated within a single trial, it is likely that some light will spill across into the "dark" alley. Hence, the simultaneous discrimination required in this situation might well be very much more difficult than the discrimination between both alleys lit and both dark. It might therefore be more difficult to strengthen the brightness analyzer in the simultaneous situation than in the successive and it would follow that successive learning would be faster than simultaneous. Unfortunately, these experiments are confounded by a further factor, which makes it difficult to interpret the results with certainty. When offered a choice between a bright and a dark alley, rats have a strong preference for the darker side (Muenzinger & Evans, 1957) and it has sometimes been found, in fact, that the simultaneous problem is easier than the successive if rats are learning to approach the darker alley, but more difficult if they are learning to approach the brighter alley (Calvin & Williams, 1956).

One further finding on the relative ease of simultaneous and successive problems deserves mentioning here. Wodinsky *et al.* (1954) and Bitterman and McConnell (1954) taught rats to solve two problems one after the other; for some animals both problems were simultaneous, for others both were successive. The first problem was mastered faster when it was a simultaneous problem, but the difference tended to disappear in the second problem. This suggests that once the rat has learned the configurational rule necessary for solution of a successive problem, it is able to use the rule in the solution of further successive problems even though a different visual analyzer is required.

The upshot of this discussion on simultaneous and successive discrimination problems is, then, that although animals tend to solve simultaneous problems by learning the rule ⟨Approach the positive, avoid the negative⟩, where the problem allows this solution, animals can and do solve problems involving very much more complex relationships between response attachments and outputs from analyzers than the simple attachment of one response to one output and a different response to another output. Moreover, even simultaneous problems may be solved by learning a more complex rule if the within-pair differences are less discriminable than the between-pair differences.

We conclude that responses may be attached to the outputs from analyzers in complex and varying ways. This conclusion is reinforced by the results of experiments on transposition. Numerous experiments suggest that animals may sometimes learn the relative values of two stimuli simultaneously presented rather than the absolute value of each stimulus. For example, if an animal is trained to select a stimulus of brightness value ⟨2⟩ presented with one of brightness ⟨3⟩, it may in subsequent tests consistently select ⟨1⟩ of the pair ⟨1, 2⟩ and ⟨3⟩ of the pair ⟨3, 4⟩.

Spence (1936, 1937) has suggested a different interpretation of this result, namely, that because of generalization of habit strength and inhibition, stimulus ⟨1⟩ may actually have more effective habit strength than stimulus ⟨2⟩, and ⟨3⟩ may have more than ⟨4⟩. One prediction made by Spence's theory is that if the test stimuli are sufficiently different from the training stimuli, relational transposition will no longer occur. For example, if we train an animal on stimuli with values ⟨4⟩ and ⟨5⟩, with ⟨4⟩ positive, the animal may show relational transposition to the pair ⟨3, 4⟩ by selecting ⟨3⟩ but not the pair ⟨1, 4⟩ since ⟨1⟩ is so far removed from ⟨4⟩ that it has gained little or no habit strength by generalization. This prediction is, in fact, supported by the evidence: relational transposition does usually break down when the test stimuli are far removed from the training stimuli (Spence, 1937; Kendler, 1950; Ehrenfreund, 1952).

Notwithstanding these results, other findings have unequivocally demonstrated that, at least in certain circumstances, animals learn to respond to the relationship between two stimuli rather than to their absolute values on the relevant dimension. Gonzales, Gentry, and Bitterman (1954) trained chimpanzees on a size discrimination. Stimulus objects of nine different sizes were used each one being 1.15 times the area of the next smaller one. If we call these stimuli ⟨1–9⟩ in order of ascending size, than the procedure was as follows. The stimuli were always presented in triplets: the subjects were first trained to select ⟨5⟩ when presented with ⟨1⟩ and ⟨9⟩, and were next trained to select ⟨5⟩ when presented with ⟨3⟩ and ⟨7⟩. The animals were then given tests with the triplets ⟨2, 4, 6⟩ and ⟨4, 6, 8⟩. Spence must predict that the ratio of choices of ⟨4⟩ to ⟨6⟩ will be the same whether ⟨2⟩ or ⟨8⟩ is included in the test set since the inclusion of one or other of these stimuli cannot affect the existing habit strengths to ⟨4⟩ and ⟨6⟩. In fact, ⟨4⟩ was chosen on 77% of the trials when ⟨2⟩ was presented, ⟨6⟩ was chosen on 62% of the trials when ⟨8⟩ was presented. This finding suggests that animals had solved the original problem not by learning always to select ⟨5⟩ but by learning always to select the stimulus of intermediate size. Gentry, Overall, and Brown (1959) obtained basically the same result in an experiment on monkeys.

It would, of course, be dangerous to generalize from primates to rats, but an experiment by Lawrence and DeRivera (1954) makes it certain that rats too sometimes learn to solve problems in terms of stimulus relationships rather than absolute values. They trained rats on a successive conditional position discrimination. Each window bore a stimulus that had one brightness value at the top and another at the bottom, During training the bottom halves of the cards were always a mid gray ⟨brightness 4⟩. If the top half was brighter than this (values ⟨5–7⟩) the animals had to select the left-hand window, if it was less bright (values ⟨1–3⟩) they had to select the right-hand window. After the animals had been trained, they were given tests with cards bearing all other combinations of brightness values, e.g., ⟨3 at top, 2 at bottom⟩, ⟨1 at top, 6 at bottom⟩. On any one test both windows still contained identical cards. Differential predictions about the results of the tests follow from the assumption that animals had solved the original problem in terms of absolute values from those that follow if they had solved it in terms of relative brightnesses. For example, on the absolute assumption we would predict that if ⟨3 at top, 2 at bottom⟩ is shown, animals would jump to the right-hand window since when either of these values was present in training, choices of the right-hand window were rewarded. On the relational theory, however, the animals should jump left when stimuli with these values are presented, since, in training, left was correct whenever the top half of the cards was brighter than the bottom half. In fact, on 65% of trials for which a differential prediction can be made, rats followed the prediction derived from the relational theory. If we consider only trials on which the top differed from the bottom by a brightness interval greater than ⟨1⟩, then 80% of the rats followed the relational predictions. This experiment clearly demonstrates that rats sometimes solve problems in terms of relational differences between the stimuli, rather than by learning to respond to their absolute values.

Further results demonstrating that rats sometimes respond to relational values of the stimuli were obtained in two experiments undertaken by Riley (Riley, Ring, & Thomas, 1960; Riley, Goggin, & Wright, 1963). Rats were trained on a brightness discrimination under two conditions: the stimuli were either arranged in such a way that animals could see them both at once before jumping, or they were so arranged that only one stimulus at a time could be seen, although both were present on each trial and could be viewed successively by VTE. When both groups were subsequently *tested* with both stimuli simultaneously visible, rats *trained* with both stimuli simultaneously visible showed much better transposition than animals that had only seen one stimulus at a time in training. The criticism can be made that the rats trained with only one stimulus visible at a time were tested in a different situation and therefore the poor transposition was due to generalization decrement. This argument is rendered less plausible by the fact that the same rats showed little decrement when they were tested with the original training stimuli simultaneously viewed. Moreover, Riley *et al.* found that the animals learned faster with both stimuli visible particularly when the discrimination was a difficult one. A similar result has also been obtained by MacCaslin (1954). Again these findings imply that rats can learn to respond in terms of a relational difference between two stimuli rather than to the absolute values of the stimuli. It would indeed be very surprising if this were not the case, in view of the difficulty human subjects have in learning to respond to the absolute value of a stimulus and the comparative ease with which they can learn to select one or other of two very similar stimuli presented together (compare the results reviewed in Section III page 56).

The fact that rats can respond to a relational difference between stimuli implies once again that the relationship between the outputs from analyzers and response attachments may be a very complex one. Responses cannot always simply be attached to a single output; there must be some mechanism to determine the relative strengths of two different outputs and the response attachment must be made to the output from this mechanism. Very little attention has been given by psychologists to the sort of mechanism that would be necessary; the problem is one of the most obscure and important in experimental psychology. Unfortunately, we have no solution to offer. Although it is often possible to state, as we have tried to do in the preceding discussion, the rules that are learned by an animal in a given situation, it is more difficult to formulate a plausible mechanism that would mediate the learning of such rules. The situation is analogous to the present state of psycholinguistics. Chomsky and Halle and their co-workers have succeeded in working out in some detail many of the formal rules that govern the capacity to generate grammatical sentences. The success of the linguists in this enterprise, however, has hardly advanced the deeper problem of how such rules are, in fact, learned by the human child nor has it as yet cast much light on the mechanism in the listener's brain that recovers the grammatical structure of a sentence input to it.

The first section of this chapter showed that it is a gross oversimplification to suppose that there are separate analyzers already in existence before the subject enters a given situation; it is likely that central control programs are established, which govern the sequence in which analyzers are used in any complex situation. Similarly, we have now shown that to think of responses as attached to separate outputs from an analyzer is also a gross oversimplification. Nevertheless, we believe that pending solution of these problems it is possible to investigate selective attention, particularly if such investigations use simple situations that minimize interaction between analyzers and allow responses to be directly attached to the outputs from analyzers. It is for these reasons that in our own experiments we have used rather simple stimuli varying along a limited number of dimensions and have employed predominantly the simultaneous discrimination situation.

## VII. What Is Learned?

The problem of what is learned, although important in its own right, is again not of crucial importance to our own model. As so often, one of the first experiments demonstrating the importance of the problem was performed by Lashley: he showed that rats that had learned to run a maze for food reward were subsequently able to roll through the maze without error after extensive lesions to the cerebellum (Lashley & Ball, 1929). In the following year, MacFarlane (1930) published a paper showing that, having learned to run through a maze correctly, rats could swim through when it was flooded with very few errors. It is clear that whatever is learned, it is not a sequence of specific muscular contractions. It is a depressing fact that after a generation of further experimentation and theorizing, we are no nearer to having an explanation of response equivalence than was Lashley. Hull acknowledges the problem by postulating the "habit family hierarchy," though this postulate is in fact no more helpful than his complementary postulate on the stimulus side, that of "afferent stimulus interaction." Today the problem is largely ignored; by giving an arbitrary definition of a response and by not considering experimental situations in which response transfer is called for, mathematical learning theorists can afford to ignore it. We confess that we are not able to provide any solution, but we would like to make a plea for taking some of the old skeletons of experimental psychology out of the cupboard to which they have been relegated and having a fresh look at them.

In order to account for the phenomena of response equivalence and latent learning, Tolman (1932) has argued that animals learn stimulus–stimulus sequences. Much of Tolman's account was anthropomorphic and vague, but more recently, Deutsch (1960) put forward a theory that translates many of Tolman's principles into a suggested mechanism of the learning process.

Deutsch assumes that when an animal moves around an environment, it learns that stimuli follow one another in a certain sequence. Their internal representations are linked together in a chain in an order corresponding to the sequence in which they were encountered by the animal. In the case of the rat learning a T maze for food reward, the final stimulus represented in the chain will correspond to the stimulus the animal obtains when it receives the reward. Deutsch postulates that when the animal is in the drive state appropriate to the reward, this final stimulus representation will be activated and activation will be passed back down the chain decreasing in strength as it passes from link to link. When a chain has been formed in this way and the drive is on, the animal will indulge in searching behavior until it receives an input from one of the stimuli on this chain. As soon as such an input occurs, firing is cut off in lower links and the animal's behavior is controlled by that link until a stimulus represented higher up the chain occurs. Deutsch is vague about the way in which responses are learned, and talks almost entirely in terms of approach responses. If a given stimulus is controlling behavior, the animal will continue to approach this stimulus until such a time as a stimulus higher up the chain is received. Strictly speaking, animals cannot approach a stimulus, but only a stimulus object, and Deutsch does not specify any mechanism to control the behavior of approaching.

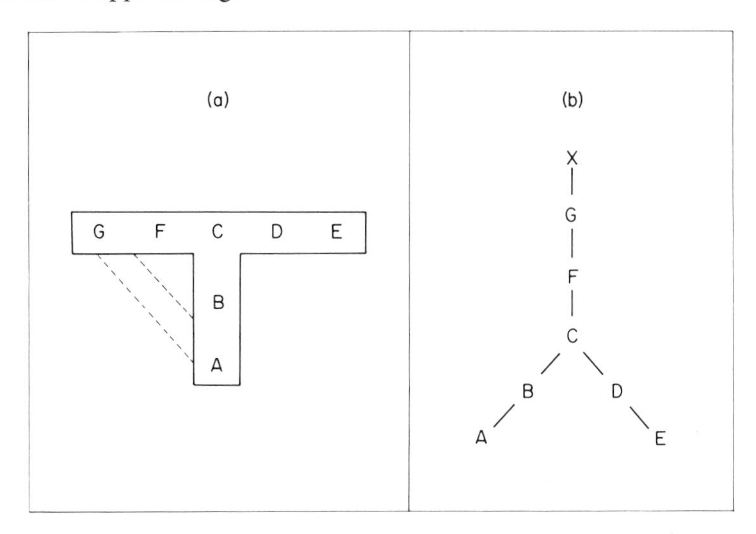


**Fig. 3.3.** Illustration of Deutsch's theory.

The working of the theory with respect to simple T maze learning is illustrated in Figure 3.3. Suppose the animal is being trained to run down the arm terminating with stimulus G (for goal box). On arrival at G, it will receive stimulus X, the stimuli arising from eating. If it has run correctly, this will give the ordering X–G–F–C–B–A since stimuli are ordered on the chain in the opposite order to that in which the animal encounters them. If the animal makes a mistake by entering the wrong arm E and then retraces until it reaches the goal box, then the order in which the stimuli are encountered (from last to first) is X–G–F–C–D–E–D–C–B–A. When a stimulus is encountered twice, it is dropped from the lower position in the chain, and the stimuli coming after it are moved up to follow its first occurrence on the chain. Thus the second D and C will be dropped and B–A will follow the first C giving rise to a chain branching at C as shown in Figure 3.3b. When the rat runs up the stem of the maze, C will fire, and he will approach it. Since excitation to links lower than C is now cut off, behavior after C has fired cannot be controlled by D, but only by F or G; hence, the rat will learn to take the correct turn. Moreover, if a new pathway (shown by dotted lines in Figure 3.3) is opened up leading directly to the goal box G, the rat will take that pathway if it sees the stimulus G at the end of it, since the firing of G will prevent responses to stimuli (such as C) lower down the chain. Since chains are formed whenever the animal moves around, Deutsch's model would account for some of the data on latent learning. If the animal has already formed the chain G–F–C–B–A, and is then placed in the goal box G and fed, the connection X–G will be made. When the rat is next placed hungry in the start box, the whole chain will be activated and it will run the maze correctly (for a critique of the application of Deutsch's model to latent-learning experiments, see Watson, 1959).

Despite its ingenuity, Deutsch's theory is inadequate in some respects.

(1) No attempt is made to postulate a mechanism for learning or controlling approach responses. By taking these for granted the theory fails to account for many aspects of response equivalence just in the same way as more conventional S–R theories.

(2) As Deutsch himself observes, the theory can give no account of avoidance learning.

(3) The theory appears to make some incorrect predictions. For example, if a rat is placed at F and then returned to the start, it should not run since excitation to links lower than F will have been cut off by the animal seeing F.

(4) The theory has difficulty accounting for learning in a Skinner box where the animal has to press a bar and then approach a food magazine. It is not clear why the animal does not always approach the food magazine without first pressing the bar since the magazine approach is controlled by a higher link than the bar-press response and provided the animal can see both at once, the sight of the magazine should cut off the excitation to lower links on the chain.

(5) The theory is weakest just where S–R theories are strongest—it fails to account for response learning.

Animals can learn to make one response to one stimulus, another to another (e.g., Grindley, 1932, taught guinea pigs to turn the head to the left when one stimulus occurred, and to turn the head to the right when another occurred; see also Sheldon, 1967). Nor are such responses necessarily approach responses towards a particular condition of proprioceptive input; Taub and Berman (1968) have shown that monkeys can learn to make remarkably dexterous hand and arm movements when they have been deprived of proprioceptive input from the limb and when they cannot see the position of the limb.

Despite these criticisms, there is almost certainly a considerable element of truth in Deutsch's theory, and he has succeeded in translating Tolman's anthropomorphic "cognitive maps" into a workable mechanism. Whereas traditional learning theorists have postulated only S–R connections, Deutsch postulated only S–S connections. It seems certain that both can be learned and that animals must, in addition, learn S–R–S connections.

It is likely that any adequate theory of learning will have to make use of "expectancies" in some form. This concept need not be anthropomorphic. An expectancy can be the internal representation of an event that has some likelihood of occurrence if a given response is made. Thus the animal in stimulus situation S might learn that response $R_1$ is followed by event $E_2$, and so on (see Figure 3.4). Events could be given a value, and their value at any one time would depend upon the drive condition of the animal. Events corresponding to primary reinforcement would have high values, and these values would be transmitted back to other events on the same chain, with a decrease in value at every stage. Hence, an animal at $E_2$ will come to make $R_5$ rather than $R_6$ if $E_5$ is a primary goal although it can be reached by either response. In addition, the lower the expectation (subjective probability) of an event given a response, the lower would be the transmission of value backwards down the chain. The animal's task would now be to search out a path leading to a terminal event with high value taking into account the probability of actually arriving at this event by a particular path. This account is not dissimilar to that of Newell and Simon (1963) of human problem solving in which subgoals are set up and the subject (or program) searches for a route to these subgoals.

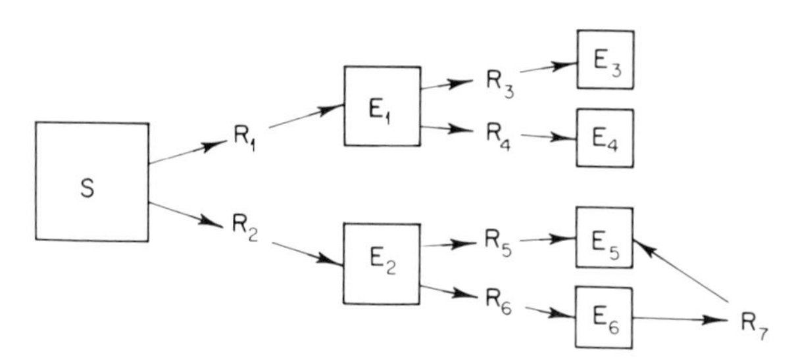


**Fig. 3.4.** The learning of expectancies.

It is likely that, within such a system, many of the response items would be very general, e.g., ⟨Approach $S_1$⟩, where $S_1$ is a visually perceived stimulus object. Where a response was coded in this way, a subroutine would be called for which would have general application to visually guided approach responses. Such subroutines might involve steps such as: measure the distance away from the object; orient the body towards the object; determine whether there is a flat surface leading to the object, if so maintain orientation and run, if not measure gap between self and nearest flat surface in line to object and, depending on distance of gap, jump with such and such a force, and so on. The details of such routines are enormously complex, and working them out for particular cases constitutes one of the major tasks of experimental psychology. Proposals similar to our own have been made by Miller, Galanter, and Pribram (1960) and, alas, they have been made at a similar level of vagueness.

Some of these general subroutines may be innate or partially innate. For example, Lashley and Russell (1934) showed that the force with which a visually inexperienced rat jumps is correlated with the width of the gap to be crossed. However, in mammals at least, it is likely that many of the general routines are learned. It is characteristic of young children and animals brought up in a severely restricted and pain-free environment to show very slow avoidance learning. William James gives several examples, admittedly anecdotal, of infants receiving a painful stimulus and immediately putting themselves back into the situation in which they received it. Melzack (1954) found that dogs reared without pain would give reflex withdrawal response to a burning match placed under their nose, but would immediately place their nose back in the vicinity of the match. Later in life, normally reared animals can learn some avoidance responses in a single trial (Madsen & McGaugh, 1961). Presumably, the animals have learned a very general routine to be applied when pain is received; the routine must result in their identifying the situation and withdrawing from it. The withdrawal response itself would involve subroutines for maximizing the distance of the organism from the identified situation.

Although we believe that a model of the type outlined is the only one that will eventually cover the facts of learning and performance, we shall not be concerned with expectancies in what follows. In terms of our own model, the diagram in Figure 3.4 represents the situation when a particular output occurs from a given analyzer or a particular combination of outputs from one or more analyzers. We have assumed that a particular analyzer is strengthened when there is an increase in the difference between the response attachments to different outputs. For our purposes, it is largely irrelevant how this change in response strength occurs. It does not directly affect our model whether the underlying change is in the strength of a connection between the output and the response attachment or whether it is a change in the expectation of the consequences of a given response which in turn will render that response more or less probable in a given situation.

There is one more topic on which we should touch. In the absence of primary drives most organisms make exploratory responses. The frequency of such behavior as compared with behavior aimed at satisfying primary drives increases as we move up the phylogenic scale. In terms of the learning mechanism schematized in Figure 3.4, this means that the building up of chains of the type there shown is itself rewarding for the animal. An animal in a novel stimulus situation tries out various responses and hence establishes a series of expectations. Many of these responses are overt orienting responses. If we assume that such responses are reinforced when they decrease the animal's uncertainty about the situation, we will have very much the same rule for the strengthening of overt orienting responses that we have in our own model for the strengthening of analyzers. An overt orienting response will be made when the animal cannot predict with certainty which of two or more events it will lead to, and presumably the bigger the difference between the values of these events as determined by the chains attached to their internal representations, the more likely is the observing response to be made.

## VIII. Conclusions

This chapter has complicated the simple picture of what is meant by "analyzer" and "response" as sketched in Chapter 2. Before proceeding further, it may be worth recapitulating these complications and spelling out why they are ignored in the remainder of the book.

It was suggested above that man and animals often learn to attach a response not simply to the outputs from a single unidimensional analyzer but to a more complicated "description" of objects in the external world; such descriptions involve not merely the specification of values along a number of dimensions, they may also involve the storage of structural relationships (e.g., "join," "inside," "above") obtaining between the parts of an input picture, and also the storage of comparative relationships holding between the parts of a picture (e.g., "brighter than," "smaller than"). It was argued that, in perceptual differentiation, animals may have to learn what type of description to store of the members of a given class of objects. All this clearly carries us a long way beyond the simple notion of an analyzer as set out in Chapter 2.

Neither we nor anyone else has clearly formulated ideas of how such complex descriptions are arrived at nor of the learning processes involved in the perceptual differentiation of complex objects. This problem, the problem of how the external world is represented in the animal's brain, is not the topic with which this book is concerned. We are asking the question: given that animals learn to select which aspects of the stimulus situation to which to attach responses, how does this selective process influence performance under a variety of experimental manipulations? Some examples of the variables with which we are concerned are the number of training trials given, the initial dominance of the stimulus aspect to which the response has to be attached, the number of stimulus dimensions that are relevant or irrelevant for success in a given task, and the schedules of reinforcement used. We believe that the influence of perceptual selectivity on response learning and performance can be studied before a full solution is found to the problem of how the external world is represented. The problem of how the animal forms complex structural descriptions can be bypassed, provided the stimuli used to control the animal's behavior are kept simple. It must be admitted that choosing "simple" stimuli such as shapes that exhibit a difference in their ratio of horizontal and vertical extent, or in brightness or color involves to some extent an act of faith. However, the proof of the pudding is in the eating: if, when we use stimuli such as these and manipulate the type of variable listed above, consistent results predicted by theory are obtained, we are safe in ignoring the more general problem of the representation of patterned stimuli.

In view of the above discussion, it may well be asked whether it is appropriate to use the word "analyzer." Should we not rather talk of the organism attaching responses to a particular description of the external situation or to a particular feature of it? Although the notion of an analyzer does not cover the formation of complex descriptions of objects, it is appropriate for the types of stimuli with which the rest of the book is mainly concerned. We assume that an animal that has learned to control a response according to whether a stimulus is red or green has not merely learned to look for the feature red or the feature green, it has learned, in some sense, to pay attention to wavelength differences so that training on a red–green discrimination should facilitate training on other color discriminations. The word "analyzer" is appropriate where it can be demonstrated that there is some physical dimension on which stimuli may have a value and animals tend to order stimuli in generalization tests along this dimension. This notion is not captured by the terms "feature-detector" or "description." If an animal has learned a red–green discrimination, it must, in some sense, have stored the descriptions "red" and "green" and have attached the appropriate responses to these descriptions. Some further mechanism is needed, however, to explain why, after such training, it will tend to attach the description "blue" to a blue triangle rather than the description "triangle," and this further mechanism is encapsulated in the notion of an analyzer whose outputs correspond to variations in the stimulus along a particular dimension. Although there are regrettably few experiments on this problem, evidence will be presented in Chapter 6 suggesting that animals do learn to attend to a dimension and not merely to particular values on a given dimension.

There is a further problem that our present account bypasses. It could be asked whether, when an analyzer is strengthened, the animal sees the world in a different way or whether it sees it in the same way, but merely stores information about a particular aspect of it. All that is necessary for our purposes is to assume the latter (weaker) alternative, and we therefore do not need to discuss the difficult question of what is meant by "seeing." There are, of course, cases alluded to earlier (page 54) where two different and conflicting descriptions may be formed of the same stimulus input (e.g., of reversible or ambiguous figures). In these instances, the description that is in fact formed affects the way the object is seen. In the cases we are dealing with, however, there is no such conflict. The rat may in some sense see that the shape in front of it on a jumping stand is both white and horizontal but depending upon whether it stores information about the whiteness or the horizontality its responses may come to be controlled by one or other aspect of the situation.

Some further caveats should be entered before closing this chapter and proceeding to the meatier sections of the book. Where there is a strictly limited number of dimensions along which stimuli differ, there is no great difficulty in suggesting plausible ways in which an animal or man might come to select the dimension that is important. College students faced with concept-formation tasks involving differences in up to seven attributes appear to consider on any one trial differences along only a single dimension and when they hit on the dimension that correctly predicts the category in which to place the object they stick with that (Bower & Trabasso, 1964). There is evidence, some of which was reviewed in Chapter 2, suggesting that in animals the process is less all or none; response learning may proceed about the outputs from several analyzers simultaneously, but in proportion to the strength of each analyzer, and as the response attachments are strengthened, the analyzer gains in strength at the expense of others. This account is sufficient where there is a strictly limited number of dimensions that can be processed. It is hopelessly inadequate to cover the more complex phenomena involved in perceptual differentiation where the number of possible relevant dimensions and relationships does not form a small, closed set, but forms a large and possibly unbounded set. Our account does not deal with the problem of how we learn to form appropriate structural descriptions for complicated objects: such descriptions must almost certainly take advantage of redundancies in the input and must be appropriate for individuating members of a class. Since our main interest is in the relationship between response learning and learning to attend selectively to simple aspects of the stimulus, we can again avoid this problem by using only simple stimuli.

We have shown that the concept of a "response attachment" is as oversimplified as that of an analyzer. Indeed, if, by a response, we mean a set of muscular movements, it is not clear that in most instrumental situations animals learn to make responses at all. Rather, they may learn what state of affairs they must bring about in order to obtain a particular goal, and translate learning into performance by computing from the stimulus input they receive how to respond in order to obtain this state. Again, this complication can be ignored for our present purposes, and for simplicity we shall continue to talk of the learning of response attachments.

Finally, it should be pointed out that we have a predilection for reifying concepts. We use the term analyzer because we believe that there are mechanisms in the brain that correspond to this notion. However, the formal working of our model would be little affected by substituting any of the terms "mediating response" (Kendler & Kendler, 1962), "observing response" (Zeaman & House, 1963), "hypothesis" (Lashley, 1929) (provided that this is understood as a hypothesis about what stimulus dimensions are relevant and not a hypothesis about the correct S–R bond), or "coding-response" (Lawrence, 1959). The basic ideas behind our thinking we owe entirely to others—particularly to Lashley, Krechevsky, and Lawrence. What we have tried to do is to specify more carefully than our predecessors the way in which changes in selective attention interact with response learning and thus to produce a more coherent theory, to manipulate this theory to provide and test predictions, and to apply the theory to a wider range of phenomena. To these phenomena we now turn.

CHAPTER 4

# Continuity and Noncontinuity

## I. Origins of Continuity–Noncontinuity Controversy

The idea that it might be profitable to apply the concept of selective attention to certain learning phenomena is not a novel one. In particular, as we noted in Chapter 1, the suggestion that animals learning to solve discrimination problems do not attend to all stimuli at once can be traced back to the writings of such noncontinuity theorists as Lashley and Krechevsky. Nor was there only talk about the role of attention in learning; a series of experiments was initiated to examine at least one way in which attention might be supposed to affect animal discrimination learning. Although the amount of information provided by these early experiments now seems disappointing, they did show that it was possible to derive testable predictions from an attentional analysis.

The impetus for the so-called continuity–noncontinuity dispute, and for the experiments performed in an attempt to resolve the dispute, was provided by Lashley (1929), when he noticed that if the daily scores of individual rats learning a brightness discrimination were examined, there was usually a sudden change from a chance score to virtually perfect performance:

> In many cases the form of the learning curve . . . strongly suggests that the actual association is formed very quickly, and that both the practice preceding and the errors following are irrelevant to the actual formation of the association [p. 135].

Irrespective of the validity of Lashley's interpretation of his discovery there is no doubt that the finding is a reliable one: a simultaneous visual discrimination is usually solved relatively abruptly by the rat, after a longer or shorter series of chance responses. We present our own contribution to this evidence in Figure 4.1, which shows the backward learning curve of 60 rats trained by noncorrection on a black–white discrimination in a jumping stand. These rats do not in fact show an instantaneous transition from 50 to 100% correct, and we discuss below some reasons for this; nevertheless, the learning curve is certainly stationary for much of the range. Over a relatively long run of early trials, rats typically show no increase in the probability of a correct response.

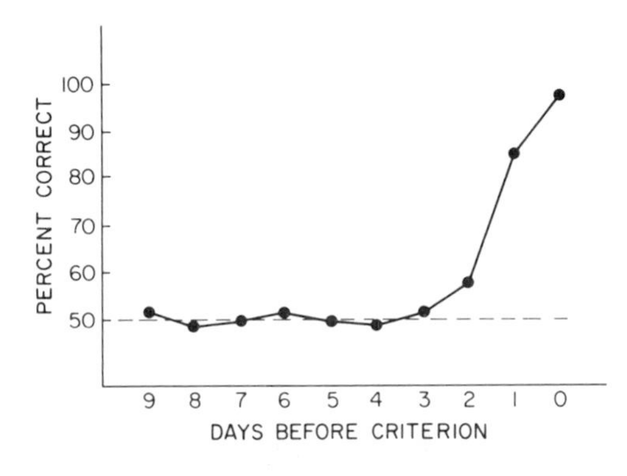


**Fig. 4.1.** Backward learning curve for 60 rats training by noncorrection on a brightness discrimination.

How are such data to be interpreted? Lashley argued that stationarity of a learning curve implies absence of learning This is certainly a possible interpretation. Recent years have seen a revival of all-or-none views of learning; Bower and Trabasso (1964) and Trabasso (1963), for example, have provided several clear-cut instances of an abrupt transition from chance to perfect performance in experiments on concept identification by college students, and have interpreted their data in terms of an all-or-none theory (see Trabasso and Bower, 1968, for these authors' latest position). An incremental theory, however, can assume that correct response strength increases over early trials, but is not evidenced in choice behavior until a threshold is passed; a subject may have *learned* something about the relevant cue, but still *perform* at chance. While therefore stationarity of the learning curve is consistent with an all-or-none view of the learning process, and while such a view finds a ready interpretation within an attentional framework, stationarity *per se* is not sufficient to prove that attentional effects are operating.

## II. Hypotheses and Position Habits

### A. Development of Noncontinuity Position

Lashley in fact provided one further line of evidence to support an attentional analysis. If rats fail to score above chance during the "presolution period" (as it came to be called) because they fail to attend to the relevant cue, it is reasonable to assume that they are attending to other, irrelevant cues. Lashley noticed that,

> . . . responses to position, to alternation, or to cues from the experimenter's movements usually precede the reactions to the light and represent attempted solutions which are within the rat's customary range of activity [*loc. cit.*].

The analysis of these systematic "attempted solutions" or "hypotheses" was undertaken more thoroughly by Krechevsky (1932a). He used a four-unit serial maze with a light and a dark door at each choice point, and in effect extended the presolution period indefinitely by providing the animal with a problem that had no solution at all. At each choice point, one door was locked, the other open, but the brightness and position of the open door was randomized from trial to trial. Krechevsky (1932a) found that in both soluble and insoluble problems rats tended to respond consistently now to one cue, now to another. "Learning," he therefore claimed, "consists of changing from one systematic, generalized, purposive way of behaving to another and another until the problem is solved [p. 532]."

At a descriptive level, the performance of a rat in a discrimination learning problem thus falls into two stages: an initial period of 50% correct responding, followed by a more or less abrupt shift to 100% accuracy. During the initial period, when the curve of correct responses remains at the 50% level, the animal is typically selecting, say, a given position on 100% of trials. (We present below some additional evidence on the prevalence of such position habits.) It was left to Krechevsky (1938) to put an extreme interpretation on these data in terms of an attention theory, an interpretation that has come to be regarded as embodying the noncontinuity view of discrimination learning. Krechevsky (1938) suggested that while the rat was responding to one stimulus, it

> . . . learns (wrongly perhaps) something about the significance of this particular stimulus . . . but does not learn anything about the "correctness" or "wrongness" of the to-be-finally-learned set of discriminanda [p. 111].

This represents an all-or-none view of attention: attention to one cue completely excludes the possibility of attention to any other cue. It is in fact misleading, as well as unfair to Krechevsky, to quote this extract in isolation, since it would be equally easy to find other statements in his writings implying a far less extreme position (see page 104). The point is that noncontinuity theory was usually characterized, at least by its opponents, as maintaining such a point of view, and most of the early experiments were directed towards testing this position (which is why they produced a misleading impression of the plausibility of continuity theory).

## B. Spence's Theory of Discrimination Learning

We may grant that Krechevsky's was a fairly natural interpretation to place upon the available data, and certainly was far more satisfactory than most alternative formulations. The theoretical work of Spence (1936), however, showed that the data could, in principle, be explained without making any use of the concept of attention. Although we disagree with Spence's views on discrimination learning, we equally believe that this early paper represents an outstanding contribution to psychological theory. It is not too much to say that it is the only formal model developed before 1960 that begins to come to grips with the problems of discrimination learning.

In what follows, we shall take as an illustrative example the case of a rat being trained (by noncorrection) on a simultaneous black–white discrimination, with black positive. Spence argues that the stimulus situation can be analyzed into four components—left, right, black, and white—and that the response of approaching each of these four component stimuli has some measurable strength. The component stimuli can be thought of as combining to form compounds. More exactly, response strengths to components combine additively to determine choice between the compounds. If the response strengths to left, right, black, and white are, respectively, 10, 5, 4, and 4 units, then on a trial when black is on the left and white is on the right, the response of approaching black on the left has a total strength of $4 + 10 = 14$ units, while the response of approaching white on the right has a total strength of $5 + 4 = 9$ units. Confronted with the choice between these two, the rat would select black on the left. In the original version of the theory, performance was not determined probabilistically: the animal was always assumed to select the compound having the greater total response strength. It would, of course, be possible to assume some trial-to-trial oscillation that would allow for some indeterminacy in response selection when the difference between two competing response tendencies was small.

In this example, it is obvious that the rat would go left irrespective of whether black or white was on the left. This would remain true even if the strength of the black response was increased to 8 units; the combined response strengths of white and left $(4 + 10)$ would still be greater than the combined strengths of black and right $(8 + 5)$. An animal will continue to select the same position, i.e., have a position habit, so long as the difference between left and right response strengths is greater than the difference between black and white response strengths.

These are the rules determining performance within Spence's theory. The rules for learning will be discussed shortly, as soon as we have considered some empirical evidence.

## C. Evidence Relevant to Spence's Theory

### 1. *The Development of Position Habits*

So far, we have discussed the conditions for the occurrence of position habits, not the reasons for their development. Within the framework of Spence's theory, we may ask how the difference between left and right response strengths can become bigger than the difference between black and white response strengths. One simple hypothesis is that the position difference is larger to begin with; rats might enter the experimental situation with a marked preference for left over right and no preference between black and white. Position habits would simply reflect this initial bias.

However simple, this hypothesis does not fit most of the facts. In the first place, although many rats show a position preference during pretraining, the correlation between the side of this preference and the side of any subsequent position habit is often extremely small. Turner (1968) has analyzed the data of 78 rats trained on visual problems in a jumping stand, all of whom had strong position habits during acquisition (they selected the same position for at least 20 consecutive trials). Thirty-five of these rats developed a position habit on the *opposite* side to the preference they showed during pretraining. In Table 4.I, we present an analysis of certain features of the performance of

TABLE 4.I

**Summary of the Performance of Rats Learning a Simultaneous Brightness Discrimination**

| Total number of subjects | Number with 20/20 position habits | Number breaking position habit to S− [a] | Number making errors after breaking position habit |
|---|---|---|---|
| 204 | 154 | 2 | 74 |

| Total number of errors after break | Number making errors to nonpreferred side after break | Total number of errors to nonpreferred side |
|---|---|---|
| 192 | 37 | 53 |

[a] A rat making 20 consecutive responses to the left is said to break this position habit on the trial terminating this run of left-side responses.

204 rats trained in a jumping stand. (The data come from a variety of experiments, but all the experiments shared the following features: the problem was a black–white discrimination, 10 noncorrection trials were given each day, the criterion of learning was 18 correct responses in 20 consecutive trials.) The majority of these rats developed position habits; even those not meeting an arbitrary criterion of two days of responding entirely to the same side usually showed marked position biases. By Day 3 of training, an average of nearly 89% of all responses was made to the preferred side; but over the first five trials of Day 1, only 68% of responses were made to each subject's subsequently preferred side. There is no doubt that rats *develop* position habits during the course of discrimination training, and that we cannot account for such habits simply in terms of preexisting preferences.

Rats, therefore, frequently enter a black–white discrimination problem without any marked preference for either side. Despite the fact that choice of one brightness is always reinforced, and choice of either position is equally often reinforced and not reinforced, preference for one or other position appears to develop faster than preference for the positive brightness. It is not too difficult for an attentional theory to explain these results, by assuming that the strength of the position analyzer is initially much greater than the strength of the brightness analyzer. Since rate of response change depends upon analyzer strength, it is possible for position responses to change faster than brightness responses, despite the difference in reinforcement schedules. In order to explain why one position "runs away," we must assume, as does Spence (see page 93), that reinforcement is a more effective event than nonreinforcement, so that the net effect of an equal proportion of rewards and nonrewards is an increase in response strength.

It is less easy to explain these results without recourse to the concept of attention. For example, Spence must explain how the rate of development of the left–right response strength difference is greater than the rate of development of the black–white difference, even though the reinforcement schedules for left and right are the same and those for black and white are different, and even though his theory assumes that rewarding an animal for a response to black on the left increases both black and left response strengths. One possible solution would have been simply to assign permanently different rate parameters to spatial and visual stimuli, and it would be interesting to see how such a theory would fare. Spence, however, chose to explain the development of position habits by making a number of assumptions about the effect of absolute response strength on the magnitude of changes in response strength. Specifically, he assumed:

(1) that the effect of a single reinforcement is a bell-shaped function of the absolute value of response strength such that the effect of reinforcement is at first small, later increases, and yet later decreases again;

(2) that the effect of a single nonreinforcement is an increasing linear function of this absolute value;

(3) that below a certain (moderately high) absolute value of response strength, the effect of reinforcement is greater than the effect of nonreinforcement, but that above this value, nonreinforcement is more effective than reinforcement.

These assumptions are illustrated in Figure 4.2.

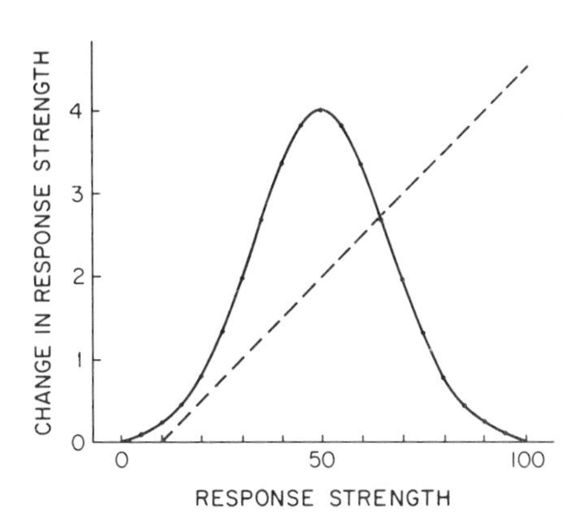


**Fig. 4.2.** Magnitude of the incremental effect of reinforcement and decremental effect of nonreinforcement as a function of current value of response strength (—): effect of reinforcement; (- - -): effect of nonreinforcement. [*After* Spence, 1936. Copyright by the American Psychological Association. Reproduced by permission.]

Given these assumptions about reinforcement and nonreinforcement, it follows that if initial response strengths to black and white are low, while initial response strengths to left and right are higher, then one or other position may increase faster than the positive brightness despite the disparity in reinforcement schedules. For, if position response strengths are such that the effect of an unrewarded trial is small by comparison with the effect of a rewarded trial, then response strength to one or other position will run away. Although any one position is equally often rewarded and not rewarded, selection of that position will, on average, increase its strength, and any temporary imbalance between the two sides will rapidly increase.

Meanwhile, response strength to black (positive) will be increasing steadily, at first more slowly than response strength to the stronger position, so that a position habit will develop, but eventually more rapidly, so that the animal will finally break its position habit and solve the discrimination.

In order to derive these predictions, Spence has made a relatively large number of assumptions, many of them arbitrary. Furthermore, the exact values of initial response strengths that are chosen, as well as the precise form of the functions shown in Figure 4.2, are all crucial to the successful outcome of the predictions. (With initial values of about 10 for brightness and about 30 for position, the development of a position habit is predicted; if these values are increased to about 20 and 40, no position habit would occur.) The fact remains that Spence can now account for the two most salient features of brightness discrimination learning in rats—the development of position habits and the rapidity of learning once the position habit is broken. Neither of the facts, therefore, can be taken as evidence critically supporting an attentional analysis.

### 2. *The Learning of the Discrimination*

In order to decide between Spence's account and a strict noncontinuity account, it is necessary to derive some further predictions. The following three predictions would seem to follow from Spence's theory (although it should be pointed out that Spence himself only derives the second):

(a) On the assumption that latency of response depends, *inter alia*, on the difference between the response strengths of the two compounds, in such a way that the less this difference the longer the latency, it follows that while a rat is still responding systematically to position, the latency of its responses to the negative stimulus should gradually increase.

(b) Subjects will first break their position habit by selecting the positive stimulus on their nonpreferred side.

(c) Such further errors as are made after the breaking of the position habit and before criterion is reached, will still be to the preferred side.

The rationale for these predictions is fairly simple. Latencies to the negative stimulus (white) will increase because, as the difference between response strengths to black and white approaches the magnitude of the difference between the response strengths to the two positions, so the combined strength of black on the nonpreferred side will approach the combined strength of white on the preferred side. Hence, prediction (a) follows. Now, selection of the nonpreferred side can occur only when the combined strength of the compound of which it is a part is larger than the combined strength of the alternative compound; this can happen only when black is on the nonpreferred side, hence, prediction (b) follows. Prediction (c) is a further consequence of this point; any subsequent selection of the nonpreferred side when white is on that side would appear to be out of the question, since each of the two components that go to make up this compound has less strength than the two components of the alternative compound.

An extreme version of noncontinuity theory makes the opposite prediction in all three cases. If, while a rat is responding systematically to position, it attends exclusively to position and therefore learns nothing about brightness, then clearly its speed of response to black and white must remain the same. Similarly, if *throughout* the time that a rat is selecting a single position it learns nothing about brightness, then it should be a matter of chance whether it breaks its position habit by selecting black or white. Finally, if the breaking of the position habit represents cessation of attention to position, any further errors before criterion should be made equally often to either position. We now turn to the data bearing on these three predictions.

*a. Latency Data.* Eninger (1953) and Mahut (1954) recorded speed of responding during simultaneous discrimination learning. They both reported that, while their rats continued to respond to position, responses to the negative stimulus slowed down, while speed of positive choices was maintained. We have not usually recorded latency data in our experiments, but did obtain latencies from one group of 16 rats learning black–white in our jumping stand. Thirteen of these animals developed strong position habits, selecting the same position for at least four consecutive days. All animals responded more rapidly to their positive stimulus than to their negative stimulus on the day before breaking their position habit. More extensive data have been obtained by Turner (1968), also using a jumping stand. Figure 4.3 shows latencies to positive and negative stimuli of 59 rats trained on a brightness discrimination. All these rats responded consistently to one position for at least 20 trials preceding the break, but the differences in speed to the positive and negative are highly significant over these trials.

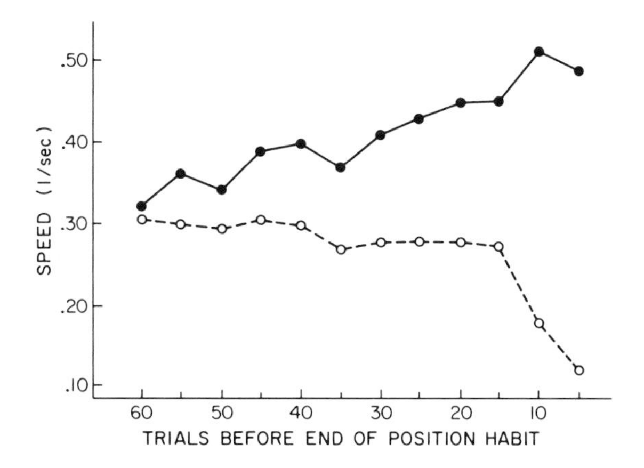


**Fig. 4.3.** Speed of responding to positive and negative stimuli during consistent position responding. Data are plotted as backward learning curves, with each subject contributing scores for the duration of his position habit. (●): speed to $S+$; (○): speed to S−. [*After* Turner, 1968.]

These results are consistent with our interpretation of Spence's theory. An attentional theory is confronted with unmistakable evidence that animals, whose choice behavior is consistently controlled by one cue, are learning which value of another cue is rewarded. Rats can attend to brightness (in order to learn about it) while simultaneously attending to position (in order to respond to it), and any version of noncontinuity theory which assumes that animals attend to only one cue at a time must be rejected.

These results show that rats may learn about a cue other than the one which controls their choice behavior. There is a clear sense in which they show more than this, namely, that while choice behavior is controlled by one cue, other aspects of behavior are simultaneously controlled by another cue. In this respect, they are reminiscent of data obtained by Maier (1949) on "abortive" jumping. Maier reported that if rats were trained on an insoluble problem in a jumping stand (i.e., one where each position and each visual stimulus was equally often rewarded and not rewarded), they tended to adopt rigid position habits, which were often maintained even when the problem was made soluble. Many rats, however, showed that they had learned which of the two visual stimuli was rewarded, by jumping normally to that stimulus but "abortively" to the negative.

*b. Breaking Position Habits.* That rats responding consistently to position learn about brightness is also shown by an analysis of the trial on which the position habit is broken, i.e., of the first response to the nonpreferred side. Our own data are given in Table 4.I. Of 204 rats, 154 developed position habits, and all but two of these animals first responded to their nonpreferred side when the positive stimulus was on that side. This agrees well with Spence's predictions.

Like the latency data, these results show that brightness learning can occur while choice behavior is being controlled by position, and are thus incompatible with a strict noncontinuity theory. They also imply one or other of two further conclusions. The probability of an error on the trial on which the position habit is broken is virtually nil. If behavior on this trial is controlled only by brightness, then brightness learning must already be virtually perfect, and no further errors would occur if behavior continued to be controlled by brightness. In other words, any further errors after the break must occur under the control of position. The only alternative to this conclusion is to assume that behavior at the time of the break is under *joint* control of brightness and position. This, of course, is the line taken by Spence, and it is clear that Spence's theory will predict that the first response to the nonpreferred side will occur when the positive stimulus is on that side, even if only a moderate amount of brightness learning has occurred.

*c. Errors after the Break.* We have already noted (see Figure 4.1) that the learning curves of rats solving simultaneous brightness problems do not show an instantaneous transition from 50 to 100% correct performance. That is to say, some errors occur after animals have broken their position habit. An analysis of these errors casts some doubts on the adequacy both of Spence's theory and of a strict noncontinuity analysis. Again, our data are given in Table 4.I. Of the 154 rats with a position habit, 80 made no further errors between breaking their position habit and reaching criterion. The remaining 74 rats made an average of 2.59 further errors. Of these 74 rats, 37 (or exactly half) made errors to their nonpreferred side.

The occurrence of these errors, totaling 27.6% of all errors after the break, poses something of a problem for any theory of discrimination learning, and is indeed intuitively somewhat surprising. All the rats involved had responded entirely to one side for at least two days prior to the break (the average duration of position habits in these subjects was, in fact, 4.7 days); virtually all had first responded to their nonpreferred side only when the positive stimulus was on that side; and yet, within a few trials (in some cases on the very next trial), they selected the negative stimulus on that side.

On the face of it, it seems improbable that position habits that have been maintained so rigidly could disappear so rapidly, and it is tempting to assume that control of behavior switches to brightness, leaving the former position preference intact but exerting no control over choice behavior. At least three lines of evidence, however, suggest that such a conclusion is too simple. In the first place, most errors occurring after the break (over 70%) consist of choices of the old preferred side: if behavior after the break is always controlled by brightness alone, errors should occur equally often on either side. Secondly, the first preferred-side error after the break occurs sooner than the first nonpreferred-side error. From our data, 55 animals made their first preferred-side error in an average of 2.95 trials after breaking their position habit, while 37 animals made their first nonpreferred-side error in 7.92 trials from the break. This difference is consistent with the idea that it takes some time to break down the old position preference, and, therefore, implies that the errors may be occurring partly under control of position. Finally, Turner (1968) has analyzed the latency data of animals making only preferred-side errors and of animals making errors on their nonpreferred side. Figure 4.4 shows latency of response to the two sides over the final 10 trials of training; animals making nonpreferred-side errors respond at the same speed on either

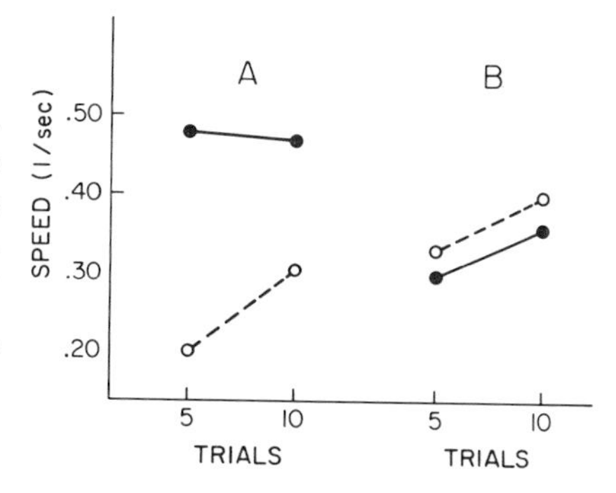


**Fig. 4.4.** Speed of responding to preferred and nonpreferred sides during 10 criterion trials. A, data from subjects making no errors to nonpreferred side; B, data from subjects making errors to nonpreferred side. (●): speed to preferred side; (○): speed to nonpreferred side. [*After* Turner, 1968.]

side, while animals making no such errors respond faster to their preferred than to their nonpreferred side, thus showing that they still retain a preference for that side. These data suggest that nonpreferred-side errors occur only if the old position habit has disappeared. In other words, a position preference sufficiently strong to ensure consistent selection of one side for over 40 trials can disappear surprisingly fast (in an average of less than 8 trials). Although surprising, this may not be unreasonable, for rats also learn position responses very rapidly. The question remains, however, whether the data can be handled by a theory, such as Spence's, which assumes joint control of behavior on all trials. We stated above that Spence's theory predicts that any errors after the break must occur on the preferred side. Since this is the first prediction from the theory to be disconfirmed, it will be as well to examine the data and the implications of the theory rather closely.

For over 40 trials before breaking their position habit, the rats in question responded consistently to one side. In terms of Spence's theory, this implies that the difference between left and right response strengths was sufficient to control behavior against the growing difference between black and white response strengths. On the trial on which the position habit is broken, the black–white difference must be (at least momentarily) greater than the left–right difference. Since the left–right difference was large enough to ensure consistent selection of one side for over 40 trials, so now the black–white difference should be great enough (other things being equal) to ensure consistent choice of one brightness. It should also be rapidly increasing. Further errors may occur because other things are not equal; on some trials the pre-existing left–right difference acts in opposition to the black–white difference, and, assuming some trial-to-trial oscillation, errors to the old preferred side may occur. However, even with a threshold or oscillation assumption, the prediction of errors to the nonpreferred side is a different matter. Let us assume a threshold $T$, such that when the difference between competing response strengths is less than $T$, choice behavior is determined probabilistically. During most of an animal's position habit, the difference between $L$ and $R$ must now have been greater than $T$ by an amount greater than the difference between $B$ and $W$: $(L - R) - (B - W) > T$. As soon as this inequality is reversed, the position habit may be broken. An error to the preferred side can now occur if $(B - W) - (L - R) < T$, and this is probable enough; but an error to the nonpreferred side can only occur if $(B - W) + (L - R) < T$. In other words, a nonpreferred-side error demands that $(L - R)$ become less than $T - (B - W)$, although, until recently, $(L - R)$ was greater than $T + (B - W)$. To compound the difficulties, $(B - W)$ must be increasing on every correct trial after the break, and although $(L - R)$ will decrease on 50% of such trials (when the subject jumps to the right and is rewarded), on the remaining 50% of trials it also will increase. Given the rate parameters shown in Figure 4.2, the prediction would appear to be impossible, for it is almost certain that the break will occur at a point where $B$ is changing very rapidly. Whether a different set of assumptions (such as those suggested above, which assigned permanently different rate parameters to position and brightness) would enable one to predict nonpreferred-side errors in sufficient quantity, it is difficult to say. We repeat that it would be interesting to see such a theory in action.[1]

For these reasons, the occurrence of nonpreferred-side errors is at worst incompatible with, at best difficult to reconcile with, any class of theory that assumes that choice behavior is always controlled by a combination of all cues present. It is possible that, despite its many successes, Spence's theory may not provide the best conceivable description of the acquisition of a simple visual discrimination.

### D. Modified Noncontinuity Theory

Although the behavior after position habits have been broken is difficult to account for in terms of Spence's theory, most of the evidence we have been discussing is even less consistent with a strict version of noncontinuity theory. Both latency data during position habits, and the fact that virtually all animals choose correctly on the trial on which they break their position habit imply that rats learn about the relevant visual stimuli while responding systematically to position. This is entirely consistent with our theory; although from Rule 1 (on the inside front cover) it follows that the amount learned about the relevant brightness cue will be decreased by a strong position habit, with sufficient trials appropriate responses to brightness will be formed. Our theory, in common with Spence's, does not actually have any direct way of representing latencies, but we can clearly explain the main fact indicated by differential latencies, namely, that some brightness learning occurs under position control.

Rule 4 of our model states that choices are determined by the responses attached to the strongest analyzer and to any other analyzers with strengths within a constant amount of the strongest. Early in training, the relevant brightness analyzer will be low relative to the position analyzer, and only the position analyzer will control behavior. Position habits occur because one or another position response will run away. Nevertheless, the initially weak brightness analyzer will increase in strength on all trials (because its different outputs are consistently correlated with differences in reinforcement), while the initially strong position analyzer will gradually weaken.

[1] This analysis of nonpreferred-side errors has been greatly influenced by discussions with C. Turner (see Turner, 1968).

When the brightness analyzer gets within a constant amount of the position analyzer, choice behavior will depend (in much the same way as in Spence's theory) on both brightness and position response strengths. If the position habit is broken under joint control, then we can predict (again as can Spence) that the choice of S− on the nonpreferred side is virtually impossible, i.e., that subjects will break their position habit to S+.

After the break, both analyzers will continue to control behavior until the brightness analyzer becomes sufficiently stronger than the position analyzer and gains exclusive control. This means that there are two ways in which our version of attentional theory could explain nonpreferred-side errors.

If the brightness analyzer gains sole control before its response attachments reach asymptote (i.e., before the probability of making the correct response reaches 1.0, given that behavior is entirely controlled by that analyzer), then we would expect some nonpreferred-side errors to be made at this stage. The first nonpreferred-side error should occur later than the first preferred-side error, because it would take several trials for the brightness analyzer to exceed the strength of the position analyzer sufficiently enough to gain sole control.

Secondly, as soon as the animal starts jumping to the positive stimulus on the nonpreferred side, the response attachment made to the position analyzer for jumps to that side will be increased: the increases in the strength of this response might occur sufficiently quickly to reduce the preference for the preferred side to a point at which some errors might be made to the nonpreferred side before the brightness analyzer gained sole control. Once again, this account would predict that the first error to the nonpreferred side, made after the break, would tend to occur later than the first error to the preferred side.

Some subjects may satisfy neither of the conditions necessary for the appearance of errors to the nonpreferred side. Before the brightness analyzer gains sole control, they may both fail to reduce their side preference and have effectively completed learning of the response attachments to brightness. Such subjects will make no errors to the nonpreferred side; the experimental evidence suggests that the majority of subjects are of this type.

Although it is possible with attentional theory to explain nonpreferred-side errors in the above two ways, the evidence suggests that most of these errors occur while behavior is controlled by both analyzers, and that they are due to equalization of position preferences. Data on position habits show that latency differences to brightness may occur at a time when choice behavior is fully under control of position. This means that if animals were making nonpreferred-side errors at a time when behavior was fully under the control of brightness, there would be no correlation between the tendency of an animal to make nonpreferred-side errors and the tendency to respond faster to the preferred side than to the nonpreferred side. As we have seen, however, such a correlation exists. We therefore conclude that most, if not all, nonpreferred-side errors occur at a time when behavior is controlled by both analyzers, and they are due to a reduction in the difference between the response strengths attached to the left and right outputs of the position analyzer.

### E. Conclusions

To summarize the discussion so far, the early experiments on hypothesis behavior, originally claimed as evidence for an extreme noncontinuity theory, can (albeit with a number of arbitrary assumptions) be explained without recourse to the concept of attention. Their results can also, of course, be explained in terms of the modified type of noncontinuity theory exemplified by our own model. The original data do not provide decisive tests of the relative merits of different theories. Certain further observations, however, show that the extreme noncontinuity theory cannot be correct. It is quite certain that learning occurs about the relevant cue while animals are still systematically responding to position, i.e., while their choice behavior is controlled by the position analyzer. Observations on the errors made after the abandonment of a position habit, however, cast some doubt on Spence's theory; it is difficult to account for these data while, at the same time, preserving the assumption that behavior is always jointly controlled by all available cues. A modified noncontinuity theory is capable of explaining all the phenomena so far observed.

We have not yet considered any of the specific experiments which, historically, were designed to test the continuity and noncontinuity theories. By now, it should be obvious why we have postponed discussion of this work: no special experiments were needed, over and above the simple observations just described, to prove that the extreme noncontinuity theory is quite certainly incorrect, and that the original continuity theory is not entirely satisfactory. Nor indeed did the early experiments add very much to the general sum of knowledge. That is to say, they were good tests of noncontinuity theory, and in virtue of their results we can have added confidence in our belief that this theory was exaggerated. But as tests for continuity theory they left nearly everything to be desired: not only could their results never have satisfactorily disproved Spence's position, they could equally never have provided any evidence in its favor. These experiments fall into two main classes: those on presolution-period reversal, and those on blocking and incidental learning.

# III. Presolution-Period Reversal

## A. Early Experiments

The logic of the presolution-reversal experiments is simple and obvious. The argument runs as follows: if, while animals respond systematically to position, they learn nothing about brightness, then, if two groups are trained either with black or with white positive for as long as position responses predominate, both groups will subsequently learn to select black and avoid white at the same rate. During the presolution period, it should not matter whether the subsequent positive stimulus or subsequent negative stimulus is the one rewarded; since the animal is not attending to either, it will learn nothing about their correlation with reward. A series of experiments of increasing sophistication has been performed using this basic design. The results have been reviewed on numerous occasions (most recently by Goodrich, Ross, & Wagner 1961) and there is no need for us to cover in detail such well-trodden ground. Listed in Table 4.II are all published experiments performed with rats, divided into those that found that presolution reversal retarded subsequent discrimination learning, and those that found no effect.

TABLE 4.II

**Summary of Experiments on Presolution Reversal with Rats**

| Presolution reversal retards subsequent discrimination learning | Presolution reversal has no effect on subsequent discrimination learning |
|---|---|
| McCulloch and Pratt (1934) | Krechevsky (1938), Group II |
| Krechevsky (1938), Group III | Ehrenfreund (1948), Experiment I |
| Bollinger (1940) (quoted by Spence, 1945) | |
| Spence (1945) | |
| Ehrenfreund (1948), Experiment II | |
| Ritchie, Ebeling, and Roth (1950) | |
| Gatling (1951) | |

It is clear on which side the weight of the evidence lies, and this is, of course, not surprising. The position is even less favorable to noncontinuity theory than it appears, since Ehrenfreund's results do not provide any evidence for a central selective mechanism that is responsible for failure to learn about the relevant stimuli. Ehrenfreund tested Spence's suggestion that if animals *see* the relevant stimuli, they will learn about them during the presolution period, but that they may not necessarily see the stimuli at the outset of discrimination training. On the assumption that, in a jumping stand, rats initially look at the bottom half of the doors, Ehrenfreund placed the stimuli on the top half of the doors and found that no presolution-period learning occurred. This is taken to provide evidence for the importance of learning appropriate orienting responses, and obviously provides little support for noncontinuity theory. On the other hand, it has been claimed that Ehrenfreund's results support Spence's interpretation of Krechevsky's results, and this seems distinctly less reasonable. Krechevsky used a difficult pattern discrimination, and found that 20 trials of reversed pretraining did not retard subsequent learning. Spence (1940) claimed that this was because the animals failed to see the relevant stimuli during these 20 trials. However, the stimuli were placed normally on the doors, and since the subjects must have looked at the doors in order to jump to them, it is hard to see how they could have failed to receive differential retinal stimulation.

This is not to imply, however, that Krechevsky's results provide very good evidence against continuity theory. As pointed out, he gave only 20 reversed trials (when 40 trials were given, subsequent learning was significantly retarded), and used a difficult discrimination. On any theory, therefore, the amount of retardation to be expected would be small. Failure to find a small predicted difference cannot be regarded as providing crucial evidence against a theory, for the good reason that it is impossible to prove the null hypothesis. The general conclusion to be drawn from these experiments can only be that they confirm the suggestion that animals do learn about the relevant cue even when responding to position, and the most that can be said in favor of an extreme noncontinuity theory is that Krechevsky's own results cast a slight, but only a slight doubt on Spence's position.

### B. Later Developments

There were six experiments in which reversed pretraining did retard subsequent learning; it is usually assumed that these results provide positive support for Spence's continuity theory. If this were so, then it would seem reasonable to dismiss a single negative result as inadequate grounds for doubting Spence's theory. By now though, it should be clear that this is a totally unfounded assumption: the fact that animals learn *something* about the relevant cue while responding to position is no more than what we predict and, in fact, very much less than what Spence predicts. For Spence should not only predict that animals will learn something about the relevant cue when responding systematically to position, but that, since systematic responses to one cue are not to be taken as evidence of attention to that cue, such systematic behavior should have no effect whatsoever on the amount learned about other cues. Our own position was stated over 30 years ago by Krechevsky (1938) in the same article as that from which the extreme noncontinuity quotation was taken.

> An animal might be reacting to an "irrelevant" set of discriminanda, as well as to the relevant set, and would thus be learning something about the final solution even while seeming to be carrying out an "irrelevant" hypothesis. . . . However . . . since, during the so-called presolution period, the animal does spend some time responding to "irrelevant" discriminanda, the residual neurological effect of a "rewarded" or a "punished" response is not the same during this period as during the period the animal is paying attention to the relevant discriminanda [pp. 108–109].

It is a depressing reflection on the nature of theoretical controversies that, until recently, there has been no single published attempt to test predictions derived from this statement of noncontinuity theory, while there have been six separate reports testing the extreme noncontinuity position.

The type of prediction that follows from such a modified noncontinuity position would be that the greater the strength of the relevant analyzer, the greater will be the interfering effect of a given number of reversed pretraining trials. In a small-scale attempt to test this prediction, Mackintosh manipulated the strength of the relevant brightness analyzer by varying first the number of irrelevant cues, and secondly whether black and white stimuli were presented simultaneously or successively. The design of the experiment is shown in Table 4.III. All animals were initially allowed to develop a position preference, and all continued to select this position consistently during 40 presolution period trials. For both groups, a black and a white vertical rectangle were presented in this position; for nonreversal subjects, black was rewarded and white not rewarded, while for reversal subjects, white was rewarded and black not. For the control group, the stimuli presented on the opposite side during the presolution period were rectangles of the same orientation but of opposite brightness; for the experimental group, the stimuli were of opposite orientation but of the same brightness. Thus, throughout the presolution period, the control group was, in effect, confronted with a conventional simultaneous brightness discrimination without extra irrelevant cues, while the experimental group was presented with a problem in which the relevant brightness difference was only shown successively, while an extra irrelevant cue was shown simultaneously. The prediction was that there would be a greater difference between reversed and nonreversed subjects in the control group than in the experimental group, since the former were more likely to attend to the relevant cue. The results obtained from eight rats (two per subgroup) are shown in Table 4.III. We do not wish to lay much stress on results, which, with such a small number of subjects, were not significant, but they were at least in the predicted direction, and to

TABLE 4.III

**Design and Results of a Presolution Reversal Experiment**[a]

| Groups | | Stage 1 | Stage 2, 40 trials. Stimuli presented on | | Stage 3 | Trials to criterion in Stage 3 |
|---|---|---|---|---|---|---|
| | | | Preferred side | Non-preferred side | | |
| Control | Non-reversal | | BV+ WV− | WV BV | | 62.5 |
| | Reversal | Position training | WV+ BV− | BV WV | Simultaneous brightness discrimination training (B+) | 147.0 |
| Experimental | Non-reversal | | BV+ WV− | BH WH | | 115.0 |
| | Reversal | | WV+ BV− | BH WH | | 132.0 |

[a] B designates black; W, white; H, horizontal; V, vertical

this extent, they suggest that, contrary to Spence's claim, the amount learned about a relevant cue during the presolution period depends upon attentional factors.

Sperling (1967) has attempted to test a more general assumption of noncontinuity theory, namely, that the amount learned about the relevant visual cue will be inversely related to the strength of attention to position. More specifically, Sperling compared the rate at which two groups of rats learned a simultaneous visual discrimination, one group trained in such a manner as to eliminate position habits, the other trained with a standard noncorrection procedure. In order to eliminate position habits in the first group, the position of the positive stimulus was maintained after an error, and changed after a correct choice. An animal consistently selecting the same side would therefore never be reinforced. The procedure did eliminate position habits; both groups, however, reached criterion in about the same number of trials, and both showed appropriate latency differences in subsequent tests to the positive and negative stimuli presented singly.

The results show that the occurrence of a position habit (defined as consistent selection of one side) does not interfere with learning about a visual cue. They are, therefore, inconsistent with any noncontinuity theory that postulates that attention to position is directly related to the strength of position habits. They are also, however, incompatible with Spence's theory. Although Spence assumes that the amount *learned* about visual stimuli would be unaffected by a position habit, he also assumes that the realization of this learning in performance will be directly affected by the strength of a position habit. Above-chance choice of the positive visual stimulus will occur only when the difference between the two visual response strengths is greater than the difference between the two position response strengths; it follows that the stronger the position habit, the greater the visual difference must be before criterion can be reached. In order to explain Sperling's results, therefore, Spence would have to assume that both groups had competing spatial habits: in one group, the competing habit was a simple preference for one side; in the other group, it was a tendency to alternate sides. A similar assumption can be easily be made by noncontinuity theory: both groups could have been attending equally to position—one group because they had an ordinary position habit, the other group because they were learning to alternate.

# IV. Incidental Learning and Blocking

## A. Incidental Learning

The design of the presolution-period experiments was suggested by the observation that rats solve simultaneous visual discriminations by initially responding systematically to spatial cues, and abruptly shifting from chance to near-perfect performance. Noncontinuity theorists interpreted the systematic behavior as evidence that the animal was attending to certain cues, and from the notion of attention certain further predictions followed. In 1942, Lashley pointed out that since it was the concept of attention that seemed important to noncontinuity theories, it was not necessary for them to rely on this spontaneous expression of attention provided by the presolution behavior of the rat; it should be equally possible to preset an animal's attention by specific training procedures. To illustrate the argument, consider an animal that has been trained on a size discrimination between a large and a small circle If, after the animal has learned this problem, a large triangle is substituted for the large circle, then a second incidental cue (shape), has been added to the original, primary cue (size). If the original size discrimination training established attention to size, and if attention to one cue precludes learning about another, then animals should learn nothing about the incidental shape cue.

Until very recently, fewer experiments had been performed using this paradigm than had been undertaken on the presolution period, but once again, most early studies shared two main inadequacies with the presolution studies. First, in the design used, noncontinuity theory could only be proved correct by proving the null hypothesis. Secondly, any results obtained could not distinguish between a modified noncontinuity theory, such as our own, and Spence's continuity theory. For, once again, Lashley was committed to proving that no learning whatsoever occurred about an incidental cue, and the finding that some learning occurs is far from providing convincing evidence in Spence's favor.

### 1. *Does Any Incidental Learning Occur?*

Not surprisingly, the results of early experiments were not particularly favorable to Lashley's point of view. It is true that Lashley himself, when he performed the experiment outlined above, claimed to find no evidence that incidental learning had occurred, but he used one insensitive and one inappropriate measure of learning. Having trained his rats on the size discrimination and then on the size and shape discrimination, he gave two sets of transfer tests. In the first set, he presented a triangle and a circle of equal size. All his subjects performed at chance on these transfer tests, but this is quite insufficient to prove that they had learned nothing about shape, for it is well known that transfer tests in a jumping stand are not particularly sensitive measures of learning, since rats often relapse into a position habit upon any change in the stimulus situation. Lashley's second test involved opposing the primary and incidental cues, but the finding that the primary cue now controlled behavior hardly proves that animals had learned nothing about the incidental cue. It merely suggests that they had learned less about the incidental cue, and this is not surprising since they had received less training on it.

Blum and Blum (1949) performed an experiment similar to Lashley's, and claimed that some incidental learning occurred. Although they eliminated some undesirable features present in Lashley's design, they introduced several of their own. They trained rats on a large–small circle discrimination, and then substitited a large erect triangle for the large circle, and a small inverted triangle for the small circle. After further training, the rats were tested on triangles of equal size; they continued to select the erect triangle, and appeared, therefore, to have attended to the orientation difference between the stimuli during the second stage of the experiment. However, ever since Lashley's systematic series of experiments in 1938, it has been generally accepted that the behavior of rats in jumping stands is controlled more by the bottom than by the top half of the stimuli. An erect triangle with a long base would presumably look larger to a rat than an inverted triangle; thus, selection of the erect triangle need not indicate that the subjects had ever learned anything other than to select the larger of the two stimuli.

In two more recent experiments, the situation has been considerably simplified (the cues were brightness and position) and the design used has been more satisfactory (the test of whether subjects learned about the incidental cue was to train on this cue in isolation, with half the subjects reversed and the other half not reversed). In both of these experiments (Bitterman & Coate, 1950; Hughes & North, 1959), a significant amount of incidental learning was detected.

### 2. *Incidental Learning under Control by Another Cue*

The studies just discussed show unequivocally that some incidental learning occurs. They do not, however, prove that learning about one cue occurs on a trial when behavior is controlled by another cue, for it is possible that animals did not attend to the original cue on every single trial after the introduction of the incidental cue, and only learned about the latter on trials when they were not controlled by the former. In order to prove that animals can learn about an incidental cue while behavior is controlled by the original cue, it is necessary to show that the original cue must have been controlling the subject's choices. Three recent studies provide good evidence on this point.

Riopelle and Chinn (1961) trained monkeys on a series of six-trial object discrimination problems. On Trials 2–6 of each problem, the positive stimulus was randomly in one or other position but on Trial 1 it was always in the same position. During the course of the experiment, the monkeys' Trial-2 performance attained an 85% level of accuracy (as, of course, is common with monkeys); more importantly, all subjects also reached an 85% level of accuracy on Trial 1 of each problem. To do this, they must have learned that the correct stimulus was always in a particular position on Trial 1, and on subsequent problems their Trial-1 choice must have been usually controlled by position. Since they also scored well above chance on Trial 2, they must have learned about the *visual* stimuli on a trial when their behavior was being controlled by *position.*

Terrace (1963b) in an experiment discussed in Chapter 2 (see page 38), trained pigeons to peck at a red key and not to peck at a green key. He next superimposed a horizontal and a vertical stripe on the two keys, vertical on red, horizontal on green, and then, over a series of trials, faded out the color on the keys. He was able to show that some birds performed with perfect accuracy on the incidental orientation cue in isolation. What is more important, some birds did not make any errors during the period when both cues were present. In order to perform perfectly at this stage, their behavior must always have been controlled by color; but in order to perform even above chance on the horizontal–vertical discrimination, they must, on some trials, have learned about this cue.

An even more striking result utilizing this general procedure has been reported by Schusterman (1966). A single sea lion, extensively pretrained to select the larger of two stimuli, was used. When given a free choice between a circle and a triangle of equal size, the animal selected the circle; but this preference could be outweighed by increasing the size of the triangle and reducing the size of the circle; the animal now consistently selected the large triangle. Training continued with the size difference being gradually reduced, until the subject, without having made any errors, was consistently selecting a triangle in preference to an equal size circle. Two further reversals were learned without any errors by initially making the former negative shape very much larger than the former positive, and then gradually eliminating the size difference.

All this evidence establishes two points: first that animals may learn something about one cue while controlled by another cue (a conclusion already suggested by our analysis of position habits), and second that explicit discrimination training with one cue relevant does not necessarily prevent some learning occurring about an additional, incidental cue. While both conclusions argue against early versions of noncontinuity theory, and are consistent with Spence's theory, they are equally consistent with our own position, and do not, therefore, provide crucial evidence for a strict continuity theory.

## B. Blocking

According to our understanding, one of the main issues separating attentional and nonattentional analyses of discrimination learning is this: any continuity theory such as Spence's may assume that the amount learned about a given cue on a given trial is a function of a number of variables, such as drive level, reinforcement schedule, magnitude and delay, the nature of the cue, and the current strength of responses to that cue; but learning about one cue should not normally depend upon the number of other cues in the situation, their correlation with reinforcement, or the subject's prior experience with those cues.[2] The crucial assumption implicit in any theory of selective attention is that there is some limit to the amount of information processed on a trial, that cues may compete for the subject's attention, and therefore that the amount learned about one cue is critically dependent upon the number and nature of other cues in the situation. Within this general position, a number of possibilities may be stressed. According to early versions of noncontinuity theory, the relationship between the amount learned about different cues is an all-or-nothing one: attention to one cue precludes any

[2] See, however, page 119 : Wagner (1969b) has recently proposed a "modified continuity theory," whiich assumes precisely this.

learning about another. It is also possible to assume that the amount learned about a cue is weighted by the strength of attention to that cue, and that the various weights add up to one. This allows learning about more than one cue at a time but assumes strict reciprocality between the amount learned about different cues. This is the position adopted in our own theory (see Rules 1 and 2, inside front cover). A third possibility is that the amount learned about a given cue depends upon the corresponding analyzer strength, but analyzer strengths are independent of one another.

Regardless of which of these possible versions of an attentional theory is correct (and we shall argue that existing evidence supports our own position), they all differ in testable ways from any continuity theory. Furthermore, it should be clear what kinds of experimental test are required. In order to decide between our own position and Spence's, it is not, for example, enough to know that some learning about Cue A occurs in the presence of Cue B, even if the subject has received prior training on Cue B alone. We need to know whether the presence of Cue B reduces learning about Cue A, and whether pretraining on Cue B alone reduces such learning still further. There are, in fact, two possible types of experiment here, and for reasons of convenience, we shall discuss them separately. For the remainder of this chapter, we shall deal with studies concerned with the blocking of incidental learning, i.e., with studies investigating whether pretraining on one cue reduces learning about another. In the following chapter, we discuss whether the mere presence of a second relevant cue influences the amount learned about another (experiments on "overshadowing").

The general design of experiments on blocking is as follows: the experiment consists of three stages; in Stage 1, the experimental groups learn a discrimination on the basis of Cue A; in Stage 2, both experimental and control groups are trained on a discrimination in which both Cue A and a second Cue B are relevant; in Stage 3, animals are given test trials designed to show how much they have learned about Cue B. Continuity theory in general predicts that the two groups should perform equally well on these test trials; noncontinuity theories predict that the control group will perform better on the test. A number of qualifications will be added to this straightforward assertion in a moment.

### 1. *Blocking in Operant Discrimination Learning*

An experiment by D. F. Johnson (1966) illustrates both this general design and some of the problems of interpretation that arise. Johnson initially trained an experimental group of pigeons to peck at a white vertical line on a dark key and not to peck at a white horizontal line. In Stage 2, both experimental and control groups were reinforced for pecking at the white vertical line on a blue-green (510 nm) background, and not reinforced for pecking at the white horizontal line on a yellow-green (551 nm) background. In Stage 3, rate of responding was tested in extinction to all possible combinations of five different orientations (ranging from vertical to horizontal) and five different wavelengths (ranging from 510 to 551 nm). The general finding of Johnson's study was that, by comparison with the control group, the experimental birds (various differences in treatment between different experimental groups will be discussed later) responded at relatively similar rates regardless of the wavelength. In other words, birds initially trained on the orientation discrimination alone showed less discrimination of different wavelengths than control birds.

The results are obviously consistent with an attentional analysis. There are, however, at least three conceivable ways in which they might be reconciled with a continuity theory.

(1) Pretraining on the horizontal–vertical discrimination may have reduced wavelength learning, not because it established attention to one cue and thereby interfered with the development of attention to another, but because it established a specific pattern of receptor orientation that prevented the birds from seeing the background wavelength. The argument is not particularly plausible, since the whole key was only 1 in. in diameter (the line was 1 × 3/16 in.). Nevertheless, the possibility that overt receptor orientation rather than attention was responsible for reducing incidental learning cannot be completely ruled out in the present instance.

(2) The difference between the test performance of experimental and control groups may have been due not to differences in the amount *learned* about wavelength during Stage 2, but to differences in the *control* exercised by orientation during the test. It is possible to construct the following argument: animals had been reinforced for pecking at a vertical line, therefore, a vertical line controlled a high rate of responding; they had not been reinforced for pecking at a horizontal line, therefore, a horizontal line controlled a low rate of responding; by generalization, intermediate orientations controlled intermediate levels of responding. Experimental animals had received more orientation training than had control animals, therefore, this orientation control was better established. If, on the test, an animal always responded at a high rate to vertical lines, a low rate to horizontal lines, and an intermediate rate to intermediate lines, it would inevitably respond at an equal average rate to different wavelengths, and show no sign that it had learned anything about wavelength.

Once again, the force of the argument can be disputed. It is, perhaps, reasonable to suggest that a ceiling effect might have masked differences in response rate to different wavelengths when the vertical line was present, but the argument depends upon much more than this. To exclude the argument altogether, it would be necessary to devise a test situation where the initially relevant cue was absent.

(3) Experimental animals may indeed have learned less about wavelength than control animals, but this may have been because the conditions of reinforcement they experienced during Stage 2 were unfavorable for discrimination learning. Specifically, to the extent to which these subjects had already learned to respond to vertical and not respond to horizontal, they could continue to behave appropriately from the outset of Stage 2, and would, therefore, make few nonreinforced responses to the negative wavelength. If we assume that discrimination learning depends upon the occurrence of nonreinforced responses to the negative stimulus, then the experimental group would be expected to learn less about wavelength than the control group. We have, of course, only recently been discussing two experiments showing excellent discrimination learning in the total absence of errors or nonreinforced responses (Terrace, 1963b; Schusterman, 1966); nevertheless, experiments on generalization gradients (see Chapter 7) have suggested that nonreinforced responses may play some part in the formation of a discrimination.

The first and second of these three possible explanations of Johnson's results have been tested in other experiments with pigeons; it is reasonably certain that pretraining on one cue reduces learning about a second cue, and that this effect is not due to the establishment of overt orienting responses. Miles (1965) trained pigeons on a light intensity go–no-go successive discrimination, and then added a second cue (presence versus absence of a tone). A control group received training only on the combined light-plus-tone problem. Stage 3 consisted of nonreinforced test trials to 10 stimulus compounds made up of five different light intensities each presented with or without the tone. The presence or absence of the tone had a very much greater effect on test responding in the control group than in the experimental group. The incidental cue in this experiment was the presence versus the absence of a diffuse sound; it would be straining the meaning of the term "receptor orientation" to suppose that animals learning to peck at a key of a particular brightness are so orienting their receptors as to fail to hear the stimulus in question. Thus we have a case of reduced incidental learning that cannot plausibly be ascribed to any orienting response.

It is still just conceivable, however, that Miles' experimental group responded equally on tone and no-tone test trials, not because they had failed to learn about the tone, but because test responding was controlled by light intensity. The argument is even less plausible here than it was in the case of Johnson's results, since several intermediate light levels were used, and there was ample room for response rate to change. Nevertheless, it would be nice to know whether similar results can be obtained if test trials are given not only with intermediate values of the Stage 1 cue, but also when it is removed from the situation. Variation in light intensity is not, of course, a cue that can be removed from a situation, since darkness is presumably at one end of the intensity continuum. Other stimulus dimensions, however, could be used that would permit one to provide a test situation containing no particular value of the Stage 1 cue; if, for example, experimental animals were initially trained on a frequency discrimination (between two equal intensity tones), tests for learning about an incidental visual cue could be conducted in silence. Vom Saal (1967) has conducted a study that would seem to meet these requirements.[3] His experimental group of pigeons received Stage 1 training on a wavelength discrimination: responses to a pale red key were reinforced, and responses to a pale green key were not reinforced. In Stage 2, both experimental and control groups were trained on a combined red–green plus tone versus noise discrimination; and in Stage 3, both groups were tested for the amount learned about the auditory cue. Test trials consisted of the nine possible combinations of a red, green, or white key, and tone, noise, or silence. Vom Saal calculated an index of auditory control that consists essentially of the proportion of test responses that occurred on tone trials; the higher this proportion, the greater the extent to which test performance was appropriately controlled by the auditory cue. For the experimental group, this proportion was .578; for the control group, it was .752. The difference was significant.

These results were averaged over all test trials. On some of these trials, however, the key was red and on others it was green. If prior wavelength training did not interfere with auditory learning so much as increase visual control, then the index of auditory control should have been lower on trials when wavelength was present than on trials when the key was white, and the difference between experimental and control groups should have varied as a function of the presence or absence of a color on the key. Neither of these suggestions was supported by the data. Although the index of auditory control was lower on red than on white trials, it was actually highest of all on green trials; furthermore, the experimental group's auditory index was still lower than that of the control group, even on white trials. Vom Saal's results, therefore, not only provide in themselves a convincing demonstration that prior wavelength training can interfere with the learning of an incidental auditory cue; they also show that neither Miles' nor Johnson's results can reasonably be attributed to differences in control. Vom Saal included in his study two further control groups receiving Stage 1 training (either to red and

[3] Johnson's, Miles', and Vom Saal's studies have all been recently published. See Johnson (1970), Miles (1970), and Vom Saal and Jenkins (1970).

green keys on a nondifferential partial reinforcement schedule, or to a red key only), neither of which showed the reduction in auditory learning shown by the experimental group. This rules out a number of improbable explanations of the main results, and we can safely conclude that pigeons, trained to solve a discrimination problem in terms of one cue, learn very little about a redundant cue added at a later stage.

The security of this empirical conclusion is not matched by any certainty of explanation, for in none of these pigeon experiments is it possible to distinguish between two possible causes of the effects of pretraining. Experimental animals may have learned less than controls about the incidental cue, either because their attention was directed mainly to the Stage 1 cue, or because they made fewer nonreinforced responses to the negative incidental stimulus. The design used in these experiments inevitably confounds differences in attention to the Stage 1 cue with differences in the number of errors made during Stage 2.[4]

2. *Blocking in Classical Conditioning*

There are two ways in which these factors can be disentangled. The first, as has been pointed out by Wagner (1969a), is to investigate incidental learning in a classical conditioning situation. In classical conditioning, the occurrence of reinforcement is entirely under the experimenter's control, and the subject's responses have no effect on trial outcome. In such an experiment, therefore, it should be immaterial whether the subject does or does not respond "appropriately" at the outset of Stage 2 training. An extensive and important series of experiments by Kamin (1968, 1969) has established that in at least one classical conditioning situation, subjects learn remarkably little about a redundant, incidental cue. Kamin used a conditioned emotional response (CER) procedure. In this situation, rats are initially trained to bar-press for food, until a stable rate of responding has been attained. Conditioning trials consist of the pairing of a relatively long CS (white noise or light) with a brief electric shock. Within a relatively small number of trials, the onset of the CS is sufficient to stop the animal bar-pressing. The measure of conditioning is the extent to which bar-pressing is suppressed during CS periods, and the usual index of learning is to express the number of bar-presses that occur during CS periods as a proportion of total bar-presses occurring both in CS periods and in equal periods of time preceding each CS. This ratio is .50 when the CS has no effect on bar-pressing, and declines to zero with complete suppression of bar-pressing during CS periods.

[4] A recent experiment by Mackintosh and Honig (1970) provides a further instance of blocking in pigeons, which cannot be attributed to differences in control. This study also suggested that blocking is not due to differences in the number of errors in Stage 2.

TABLE 4.IV

**Design and Results of Kamin's Incidental Learning Experiment**[a]

| Groups | Stage 1, 16 training trials[b] | Stage 2, 8 training trials | Stage 3, 4 training trials | Suppression ratios on first test trial |
|---|---|---|---|---|
| Control | — | N + L | L | .05 |
| Experimental | N | N + L | L | .45 |
| Control | — | N + L | N | .25 |
| Experimental | L | N + L | N | .50 |

[a] Data obtained from Kamin (1968).
[b] L designates light; N, noise.

Kamin's basic experimental design and results are shown in Table 4.IV. After the establishment of a base line rate of bar-pressing, the experiment consists of three stages. In Stage 1 the two experimental groups received 16 training trials in which either a light or a white noise CS was paired with shock. In Stage 2, both experimental and both control groups received eight reinforced training trials with a compound (light plus noise) CS. In Stage 3 all animals received nonreinforced test trials, either to the light alone or to the noise alone. The results reported are based on the suppression ratios for the first test trial only. As can be seen, when trained with the compound CS, control animals showed relatively complete suppression to the light, and rather less, but still substantial suppression to the noise. The experimental animals pretrained on one component, however, showed no signs of suppression to the other (the suppression ratio of .45 to the light shown by the experimental group pretrained on noise, compares with a ratio of .44 to light shown by a group given 24 trials to noise alone without ever having received pairings of light and shock).

With this procedure and this test of learning, therefore, Kamin appears to have confirmed Lashley's prediction that an animal given a set to respond in terms of one cue will learn nothing about a second, incidental cue. Furthermore, there is no possibility that his results can be accounted for by any of the three alternative explanations that we have been considering. The stimuli in question are not such as to require elaborate receptor orientation for their perception; the Stage 1 stimulus is not present on test trials, and cannot have been controlling behavior; and the reinforcement schedule experienced by experimental and control subjects in Stage 2 is identical. These results stand as a definitive demonstration that the amount learned about a given cue in a given situation is partly determined by the subject's prior experience with other cues in the situation.

Kamin has extended the analysis of this blocking effect considerably further. The first point to note is that animals do, in fact, appear to learn something about the incidental cue; although this learning cannot be detected in performance on unreinforced test trials, it produces slight but significant savings on a relearning measure. An experimental group given training first on one cue, then on both cues, and finally on the second cue alone, learns to suppress to the second cue somewhat faster than does a control group given trials with the first cue only. A further indication that experimental animals do at least register the addition of the second cue on the first Stage-2 trial, comes from the observation that suppression is significantly decreased on this transitional trial. This disruption of suppression may be attributed to the novelty of the compound CS resulting in either external inhibition or generalization decrement; either way it implies that subjects have noticed the added component.

Suppression ratios return to normal (near zero) on the second Stage-2 trial; it is only on the first trial of its introduction that animals show evidence of having perceived the incidental cue. This suggests the possibility that the weak association between this cue and shock occurs entirely on the first or transitional Stage-2 trial. To test this, Kamin added a third group to the experiment just described. The complete design and results are shown in Table 4.V. In Stage 1, all animals received 16 reinforced trials to the noise.

TABLE 4.V

**Design and Results of Kamin's Experiment on Transitional Trial Learning**[a]

| Groups | Stage 1, 16 training trials[b] | Stage 2, 8 training trials | Stage 3, 4 training trials | Average suppression during Stage 3 |
|---|---|---|---|---|
| Control | N | N | L | .38 |
| $E_1$ | N | N + L | L | .28 |
| $E_2$ | N | Trial 1: N + L<br>Trials 2–8: N | L | .28 |

[a] Data obtained from Kamin (1968)
[b] N designates noise; L, light

In Stage 2, control animals received a further eight trials to noise alone; Group $E_1$ received eight trials with a compound CS of noise plus light; Group $E_2$ received a single trial with noise plus light followed by seven noise trials. Stage 3 consisted of four reinforced trials to the light for all animals. The two experimental groups showed exactly the same suppression to the light over these four trials, in both cases significantly more than the control group. Since subjects in Group $E_1$ showed no more suppression than those in Group $E_2$, their extra light–shock pairings contributed nothing to the strength of the association; all learning about the light occurred on the trial when it was first presented, i.e., on the only trial when an independent indicator showed that they had noticed it.

Kamin has also obtained evidence suggesting that it is the redundancy of the incidental cue that causes the blocking effect. For example, if in Stage 1, the noise is paired with shock of a given intensity, pairing noise plus light with the same shock in Stage 2 results in virtually no association between light and shock, but pairing noise plus light with a stronger shock results in considerable suppression to the light alone. This is not simply the consequence of the use of a strong shock, for if the strong shock was used in Stage 1 as well as in Stage 2, complete blocking still occurred. It implies that the *change* in shock level is critical in producing learning. Learning about the light also occurred when noise alone predicted shock, and noise plus light predicted no shock. Under these conditions, animals showed a much more rapid decrease in suppression than control animals simply extinguished to noise alone; and when the light was removed and noise again presented alone, they showed an increase in suppression. They had learned that noise plus light was not associated with shock, without extinguishing to noise alone. This result, of course, is reminiscent of Pavlov's conditioned inhibition, and to that extent is not surprising. But it must be viewed in the context of Kamin's other results demonstrating that the added light has virtually no effect when shock level remains unchanged.

While Kamin's initial findings provide important evidence for the basic assumption of a theory of selective attention, these subsidiary results have equally important implications for the details of any such theory. They suggest that an animal learns to ignore a stimulus when it discovers that the stimulus does not predict new information; the animal attends only to informative stimuli. The processes responsible for deciding which stimuli to attend to are central ones; this suggests that the blocking is also central. Thus far, the behavorial evidence that has led us to talk of switching-in analyzers, has been concerned with variations in what animals learn about different cues under different conditions. As pointed out in Chapter 3, however, failure to learn about a given cue can be interpreted in terms of a block at a number of different points in the nervous system. An animal might not learn about a given cue because it failed to look in the appropriate direction. Many learning theorists have seriously offered this as a comprehensive account of attention in subhuman animals. Moving further into the nervous system, animals might fail to "notice" a stimulus even though they were "looking at" it. This corresponds to the notion of attention as an internal but still relatively peripheral gating system: if an analyzer is not switched-in, then stimuli may be fed into it, but no outputs are produced. Moving still further into the nervous system, an analyzer might always produce different outputs for different inputs, but when we talk of an animal failing to switch-in to a given analyzer, it may have failed to store information about the consequence of selecting one output rather than another. There is no need to draw further distinctions, since the problem of inference from behavioral data towards structural organization is a familiar one in psychology, and, indeed, this particular inference has been known to be a difficult one for a long time. Investigators of attention in human beings have long been pre-occupied with the question whether "attention" is a perceptual or response phenomenon, and have devised a number of experiments (usually not easily adaptable to animal situations) designed to distinguish between these and other possibilities (e.g., Harris & Haber, 1963; Treisman & Geffen, 1967). Kamin's results fit more easily into a theory that conceives of attention as a decision about what information gets stored, than into one that stresses peripheral gating systems. Other evidence discussed below supports this conclusion. In Chapter 5, we shall discuss in greater detail the implications of these results for specifying the nature of the events that reinforce the switching-in of analyzers.

### 3. *Asymptotic Learning as a Cause of Blocking*

We argued above that Kamin's use of a classical conditioning procedure eliminated one possible explanation of blocking effects. In the instrumental discrimination learning situations discussed earlier, animals that were pre-trained on one cue made very few errors on the compound discrimination problem. Pretraining may, therefore, have reduced the amount learned about the incidental cue, not because it locked in attention to the Stage 1 cue, but because it prevented nonreinforced responses to the negative value of the incidental cue. In a classical conditioning situation, the subject's responses do not have any effect on trial outcomes, and the reinforcement schedule is controlled by the experimenter. This particular explanation, therefore, cannot account for Kamin's results. It is still the case, however, that in all of these experiments, the animals given Stage 1 training on one cue enter into Stage 2 having already learned all that it is necessary to learn. It is possible to argue that, just as the pigeons in the discrimination experiments may have failed to learn about the added cue because they made no further errors, so Kamin's rats may have failed to condition suppression to the added cue because they had already learned to suppress bar-pressing completely. Conditioning of suppression may occur only on trials when suppression is less than perfect. Kamin has suggested this as a possible explanation of his results—although noting that it cannot easily account for his findings on the effects of changing shock level. It is perhaps ironical that Kamin's findings, which may reasonably be regarded as the most definitive demonstration of the operation of a selective attention mechanism, should be described in a paper entitled, "'Attention-like' Processes in Classical Conditioning" (see Kamin, 1968).

The type of explanation suggested by Kamin differs from an attentional analysis in this sense: we should say that a subject might fail to learn about an incidental cue because it is already occupied in analyzing another cue; Kamin is suggesting that no further learning occurs because response strength is already at asymptote. This idea has also been taken up by Rescorla (1969) and Wagner (1969b). Wagner, indeed, has, in a preliminary manner, shown how a formal model could assume that changes in response strength to a particular stimulus depend not only on the current value of response strength to that stimulus (as, for example, in Spence's theory), but on the current value of response strength to the entire compound of which the stimulus forms a part. If, in Kamin's situation, noise has been paired with shock, and light is then presented with noise, little increment in response strength to light will occur because the response strength of the noise-plus-light compound is already nearly at asymptote. The idea that increments in response strength to *components* should depend upon present response strength to a *compound* seems to us to be a more drastic departure from simple continuity theory than Wagner implies.[5] It would appear to imply, furthermore, that overtraining in the CER situation should have no further effect on performance; in fact the evidence suggests that it increases subsequent resistance to extinction (Brimer & Dockrill, 1966). At any rate, one way to test this suggestion, utilizing an instrumental discrimination learning procedure, is so to arrange conditions that a pretrained experimental group, although having learned to attend to one of the cues relevant in Stage 2, still makes a substantial number of errors during this stage because the response requirements have been changed from Stage 1 to Stage 2. This is illustrated in Table 4.VI, which shows the design and results of an experiment using rats in a jumping stand (Mackintosh, 1965d; very similar results have been obtained by Sutherland and Holgate, 1966). The experiment involved two experimental groups, both of which were trained in Stage 1 on a successive (conditional) brightness discrimination, e.g., to go right on trials when two black stimuli were shown, to go left on trials when two white stimuli were shown. For Group BW, the stimuli were squares; for Group Bhv Whv, the stimuli were rectangles whose orientation was irrelevant. In Stage 2, both experimental groups together with a control group received 60 training trials on a simultaneous discrimination with two relevant cues, brightness and orientation; and in

[5] Rescorla and Wagner have very recently elaborated their theoretical analysis in more detail. We discuss the most recent version of their theory in Chapter 13.

TABLE 4.VI

**Design and Results of Incidental Learning Experiment**[a]

| Groups | Stage 1, 240 trials (successive brightness training) | Stage 2, 60 trials (simultaneous brightness training) | Errors in Stage 2 | Stage 3 20 test trials | Percent correct responses in Stage 3 |
|---|---|---|---|---|---|
| Control | — | Half of each group: | 15.6 | | 92.5 |
| BW | B, W squares[b] | BH + versus WV− | 12.4 | Gray H and V rectangles[b] | 83.1 |
| Bhv Whv | B, W rectangles[c] | Remainder of each group: WH + versus BV+ | 11.9 | | 60.3 |

[a] Data obtained from Mackintosh (1965d).
[b] B designates black; W, white; H, horizontal; V, vertical.
[c] Orientation cue irrelevant.

Stage 3, all animals received transfer tests to black and white squares, and gray horizontal and vertical rectangles.

Although the experimental subjects had learned to switch-in a brightness analyzer by the end of Stage 1 training, they had to learn new response attachments in order to solve the Stage 2 problem, and consequently made a substantial number of errors in Stage 2. Furthermore, with the stimuli actually used, the horizontal–vertical discrimination is in fact somewhat easier than the black–white (Sutherland & Holgate, 1966). As a consequence of this, the differences between the experimental and control groups in the total number of errors made in Stage 2 were small. An analysis of variance revealed an overall group effect that just attained significance, but only one of the three comparisons between pairs of groups was significant: that between Group Bhv Whv and the control group ($t = 2.87$, $p < .01$). Group BW did not differ significantly from the control group ($t = 1.96$, $p > .05$), and the difference between the two experimental groups was negligible ($t = .37$).

Despite these relatively small differences in number of Stage 2 errors, the differences in the amount learned about orientation, as revealed in the Stage 3 scores, were substantial. Both experimental groups made significantly fewer correct responses than the control group, and Group Bhv Whv made fewer correct responses than Group BW. Pretraining on brightness and pretraining with orientation irrelevant both affected test performance. Group Bhv Whv did not, in fact, score reliably above chance during the series of test trials. It is unlikely that differences in Stage 2 performance can account for differences in test performance, and it is quite certain that the difference *between* the two experimental groups cannot be accounted for in this way. Since the test for orientation learning utilized stimuli of an intermediate value of brightness, it is also unlikely that orientation learning can have been confounded with brightness control. Finally, we are skeptical of the possibility that the stimuli could have so impinged on the subjects' retinas that brightness differences, but not orientation differences, could be detected. The results of this experiment, therefore, seem to require an attentional interpretation; specifically, they imply both that prior strengthening of a brightness analyzer will interfere with subsequent learning about a second cue, and that prior weakening of an orientation analyzer is even more effective in preventing subsequent learning about orientation.

An experiment with octopuses also obtained a significant blocking effect despite similar Stage-2 error scores (Sutherland, Mackintosh, & Mackintosh, 1965). In this experiment, unlike most others, the control group also received training during Stage 1. Specifically, Stage 1 training consisted of a size problem with shape irrelevant for experimental animals, and a shape problem with size irrelevant for control animals. In Stage 2, both groups were trained for 40 trials on a problem with two relevant cues, size and another pair of shapes. During these trials, the experimental and control groups averaged, respectively, 76 and 78% correct responses. Final tests with stimuli differing only in shape revealed significantly superior performance by the control group. While this result cannot conceivably be accounted for by appealing to differences in Stage 2 errors, it is possible that the control group's superiority was due less to the experimental group's having been trained to attend to size than to the control group's having been trained to attend to shape. In other words, the result may be a demonstration not of blocking, but of acquired distinctiveness (see Chapter 6). This is possible, just as it is also possible that both effects contributed to the results; it may be, however, that the shape discrimination problems of Stages 1 and 2 were entirely independent and that no positive transfer occurred for the control group. This is suggested by the fact that the performance of the control animals in Stage 2 was not affected by which stimulus had been positive in Stage 1.

### C. Conclusions

We have reviewed a substantial body of data in this section. Two conclusions are suggested. First, animals may learn something about a second cue even though the discrimination problem can still be solved on the basis of an already learned cue, and even though it is certain that the previously learned cue is consistently controlling their behavior. But secondly, it is equally certain that animals trained in a situation with two relevant cues perform less well in a subsequent test with one of the cues if they have received prior training with the other. Several experiments have observed such blocking of incidental learning under circumstances where it is not possible for differences in test performance to be attributed to differences in *control* of behavior during the test, and one study (Vom Saal, 1967) has explicitly shown that the magnitude of the difference is not affected by the conditions of the test. So, it is reasonable to conclude that these studies have shown that the amount *learned* about one component of a compound cue is influenced by prior experience with the other component. Most of the experiments utilized stimuli which do not require elaborate receptor orientation. So it is not reasonable to attribute differences in learning about the incidental cue to its failing to stimulate the appropriate receptors. Finally, two experiments have obtained differences in test performance despite similarities in performance during compound cue learning. It is, therefore, unreasonable to ascribe the blocking effect to differences in the reinforcement schedules experienced in the presence of the incidental cue, or to differences in the amount of learning still required to master the problem. All the results follow very simply from Rule 1 of our model: the magnitude of changes in response strength to a cue depends upon the strength of the analyzer detecting that cue relative to the strength of all other analyzers in the situation. As we shall see, this rule also provides a satisfactory explanation of another body of data—that on "overshadowing" (see Chapter 5).

## V. Variables Affecting Incidental Learning

We conclude this chapter by discussing four factors that appear to influence the amount learned about an incidental cue. The experimental findings are relevant to two issues. First, they add weight to the case against an extreme continuity theory; e.g., according to Spence the variables affecting incidental learning should be exactly the same as those affecting ordinary learning. Second, they help to specify the requirements of a satisfactory attentional theory.

The following factors have been shown to affect the amount learned about an incidental cue when it is introduced conjointly with a primary cue on which subjects have already received some training: the amount of previous training on the primary cue, drive level, the discriminability of the primary cue, and the abruptness of the introduction of the incidental cue.

### A. Number of Prior Training Trials

Two experiments suggest that in the typical incidental learning design, overtraining in Stage 1 reduces the amount learned about the cue added at

Stage 2. D. F. Johnson (1966) varied the duration of Stage-1 orientation training received by his pigeons from .5 to 10 hr. Although there was an initial *improvement* in performance on the wavelength tests as Stage-1 training increased from .5 to 1 hr, further overtraining markedly impaired wavelength control.

These results may possibly be explained by supposing that overtraining on orientation increased orientation control rather than decreased wavelength learning. No such explanation can be offered for the results of the second relevant experiment. Bruner, Matter, and Papanek (1955) initially trained rats on a brightness discrimination in a four-unit maze. In Stage 2, half the subjects (experimental groups) were given 20 trials during which the positive brightness stimulus appeared in a regularly alternating position, and, in Stage 3, all subjects learned this alternating pattern with the brightness cue removed. The measure of incidental learning, therefore, was the difference between experimental and control subjects in the number of trials taken to learn the alternating pattern in Stage 3. Either 30 or 100 Stage-1 trials were given, and, in Stage 3, the difference in the performance of experimental and control groups was significant for subjects trained for 30 trials but not for subjects trained for 100 trials. Although the requisite statistical results are not provided, the implication is that overtraining reduced the amount learned about the incidental cue.

The result follows naturally from the rules of our model. From Rule 1, the amount of learning about the incidental cue will be inversely related to the strength of other analyzers used by the animal in the situation. Rule 3 states that analyzers normally reach asymptote more slowly than responses. Hence, after a relatively small number of Stage 1 training trials the strength of the analyzer relevant to the Stage 1 problem will be substantially below asymptote, even though the subject is performing quite accurately. Overtraining will increase the strength of this analyzer, and thereby decrease the amount learned about response attachments to other analyzers.

## B. Drive Level

### 1. *Incidental Learning*

Bruner *et al.* also showed that incidental learning was a function of drive level. Half their subjects were trained throughout Stage 1 and Stage 2 under a 12-hr drive; the remainder were under a 36-hr drive. In Stage 3, the difference between experimental and control subjects was significant for those run under the low drive, but not significant for those run under the high drive. Increasing drive level reduced incidental learning. In terms of our theory this implies that the difference between strong and weak analyzers is exaggerated by increases in drive level: the higher the drive, the less learned about the weaker (incidental) cue in the presence of a stronger (primary) cue.[6]

### 2. *Latent Learning*

While there are no other studies directly concerning the effects of drive level on incidental learning, there is further evidence to support the addition of some such corollary as this to the rules of our model. First, there is the effect of drive level on latent, or irrelevant incentive learning. Johnson (1952) ran rats satiated for food in a T maze with food at the end of one arm, and then tested them hungry to discover whether they had learned the position of the food. During the first phase of the experiment, different rats were run under different levels of thirst; the lower the level of thirst at this stage, the more correct choices animals made when tested under hunger drive. Both the results of this experiment and those of experiments by Spence and Lippitt (1940, 1946) suggest that the higher the level of one drive, the less is learned about the location of an object relevant to the satisfaction of a different drive.

### 3. *The Yerkes–Dodson Law*

A further line of evidence is enshrined in one of the few laws of animal learning, the Yerkes–Dodson law. Yerkes and Dodson (1908) trained mice on a brightness discrimination using escape from shock as the incentive, and varying drive level by varying shock intensity. They found that the more difficult the discrimination problem, the weaker was the shock that produced fastest learning. The general statement of the Yerkes–Dodson law is that, for the learning of any task, there is an optimal drive level, and the harder the task, the lower this optimal level is. The most recent and satisfactory confirmation of this generalization was provided by Broadhurst (1957), training

[6] It might seem that the simplest way to represent this notion would be to assume that analyzer habit strength was multiplied by some factor $D$ (corresponding to drive level) to produce analyzer reaction potential. This is, of course, exactly the same as Hull's and Spence's multiplicative rule for drive and habit strength. Unfortunately, in our case it would not have the same consequence. Rule 1 of our model states that the amount learned about one cue depends upon the strength of that analyzer relative to the strengths of all available analyzers. The formal statement of this rule involves assigning each analyzer a strength $A$, and weighting the amount learned about cue $k$ by the expression $A_k/\sum A$. Unhappily for present purposes, this type of expression is insensitive to multiplicative transformations; if all analyzer strengths are multiplied by a constant, the weight for each analyzer will be unaffected. In order to provide a formal representation of the effect of drive level, we should have to assume something like a power function relating drive level and analyzer strength.

rats on a brightness discrimination. At the start of each trial, the rat was immersed in water, and the incentive for correct choices was escape from water. Drive level was manipulated by varying the length of time that animals were kept under water before they were allowed to make a choice. Problem difficulty was manipulated by varying the brightness differences between the alternatives. The results were entirely consistent with the Yerkes–Dodson law; drive level significantly affected the speed with which all problems were learned, but the harder the problem, the lower the drive level required to produce fastest learning.

In terms of our model, one source of variation in problem difficulty is variation in the base-level strength of the relevant analyzer: the easier the problem, the greater this initial strength. If increases in drive level exaggerate the difference between strong and weak analyzers, and if any simultaneous brightness problem contains a relatively strong irrelevant cue (position), then high drive will only facilitate the learning of very easy brightness problems, and for any harder problems, drive level will be inversely related to speed of learning.

Very much the same sort of prediction, in fact, follows from Spence's theory (Spence, 1956, Chapter 8). If a high drive level increases the discrepancy between strong and weak response tendencies, then the learning of any problem requiring the extinction of strong competing habits will be interfered with by a high drive. As we outlined above, Spence's analysis of the learning of a simultaneous visual discrimination assumes that the subject will continue to respond at chance until the difference between positive and negative response strengths is greater than the differences between left and right response strengths. The higher the drive level, the greater will be the initial discrepancy between response strengths to position and those to a weak visual cue, and the longer it will be before the latter overcome the former. The explanation of the Yerkes–Dodson law attributes it not to any effects on learning, but to effects on performance. A high drive level does not so much retard the rate at which appropriate differences in response strength to the positive and negative stimuli are established; it prevents this learning from being translated into performance by exaggerating the effect of irrelevant position habits.

None of the relatively few studies of the Yerkes–Dodson law has distinguished between these possible interpretations. In order to decide between our own account, which attributes the result to the effect of drive on learning, and Spence's account which attributes it to the effect of drive on performance, it would be necessary to train animals on problems of varying difficulty under one drive level, and test under another. If Spence is right, animals trained on a difficult problem at too high a drive level should immediately perform as well as animals trained under a lower drive, as soon as they are shifted down to the lower drive. In the absence of any direct evidence of this sort, however, it is reasonable to appeal to the results of less ambiguous experiments. The other studies mentioned in this section (Bruner *et al.*, 1955; Johnson, 1952) are unambiguous enough; neither of their results can be attributed to the effects of drive on performance. It is more probable than not, therefore, that the effect of drive on the learning of problems of different levels of difficulty, may also represent an effect on learning not on performance.

## C. Abruptness of Incidental Cue Introduction

One experiment has shown that the more abruptly an incidental cue is introduced, the more is learned about it. Mackintosh (1965d) trained rats to discriminate between gray horizontal and vertical rectangles, then added an incidental cue of brightness by making one rectangle black and the other white, and finally gave transfer tests to determine the amount learned about brightness. All animals were trained to criterion on the horizontal–vertical discrimination; over the next 40 trials, the rectangles were changed in eight steps from mid gray to black or white for one group, while for the second group they remained gray. Stage 2 training (with black and white rectangles) followed immediately for both groups. Animals for whom the brightness of the rectangles was not changed until the start of Stage 2 learned more about brightness than animals experiencing the gradual introduction of brightness differences over the preceding 40 trials.

Animals confronted with a new cue introduced abruptly learned more about that cue than did animals for whom the cue was introduced gradually. The result is not particularly surprising; the idea that a sudden change attracts attention is sufficiently familiar to everyday observation to have been sanctified as a law of structuralist psychology (Pillsbury, 1908). It has one interesting implication, however. In order to explain this result in terms of our theory, we must assume that an abrupt change in the brightness of the rectangles resulted in a greater strengthening of the brightness analyzer than did a gradual change. For the subsequent strength of an analyzer to be determined by the rate of prior change along the dimension detected by that analyzer, however, it is necessary to assume that *detection* of a given stimulus dimension does not depend upon a particular analyzer being "switched-in." For if it did, we would be implying that a change, only detectable when a particular analyser is switched in, will cause that analyzer to be switched-in—and that is clearly absurd. The necessary conclusion is one that was also suggested by some of Kamin's data (see page 118), namely, that the term "switching-in an analyzer" refers to events that occur after stimuli have been detected. To say that an animal has not switched-in a brightness analyzer is not to say that brightness differences are not detected, only that they do not control choice behavior, and that information about their correlation with reinforcing events is not stored.

## D. Difficulty of Original Discrimination

### 1. *Evidence*

A final variable that appears to influence the amount learned about an added incidental cue is the difficulty of the initial discrimination problem. Both D. F. Johnson (1966) in his study of operant discrimination learning in pigeons, and Kamin (1968) utilizing the CER situation with rats, have presented evidence bearing on this.

Johnson initially trained his experimental groups of birds to discriminate between a horizontal and a vertical white line on a black blackground. The difficulty of this discrimination was manipulated by varying the brightness of the white line; for different groups, the intensity of the line was either .75, .15, or $\bar{1}.55$ log fL, and this caused substantial differences in the speed with which the orientation problem was solved. All groups were then trained on a combined orientation and wavelength problem, with the intensity of the line maintained at its Stage 1 value, and were finally given generalization tests to various values of orientation and wavelength. Johnson found that animals trained with the less intense white lines showed steeper gradients of wavelength generalization. The implication is that the more difficult the initial problem, the more is learned about a subsequently added cue.

Similar results have been reported by Kamin (1968). Three groups of rats received 16 conditioning trials with noise alone acting as the CS, then eight trials with a compound noise plus light CS, and were finally tested for suppression to the light alone. Each group was run with a different noise intensity, 50, 60, or 80 dB; on test trials to the light, the suppression ratios for the three groups were, respectively, .21, .34, or .42. The less intense the noise, the more completely animals suppressed to the light when tested with it alone. Again, a difficult Stage 1 problem resulted in more learning about an added cue than did an easy Stage 1 problem.

There are several difficulties, however, that prevent an uncritical acceptance of this conclusion. The first is that the difficulty of the Stage 1 problem is confounded with level of performance at the outset of Stage 2. Kamin specifically noted that animals pretrained on the 50-dB noise showed significantly less suppression over early Stage 2 trials than animals pretrained on the 80-dB noise. Although Johnson trained different subgroups for different lengths of time in Stage 1, so that it is doubtless the case that the

Stage 2 performance of one subgroup trained for a long time on the difficult problem was about the same as the performance of another subgroup trained for a shorter time on the easier problem, the data are not presented in a way that allows the necessary comparisons to be made easily. The problem is that differences in Stage 2 performance (either in the number of errors made, or the amount of room for further suppression) may have been responsible for the differences in the amount learned about the added cue. This explanation is, of course, the same as one we considered above as a general explanation of blocking effects.

The second difficulty lies in the nature of the test situation. In Johnson's experiment, the test for the amount learned about wavelength took place in the presence of specific values of the Stage 1 cue. In Kamin's experiment, the test for the amount learned about the light took place in silence. In either case, there are considerable problems. It is possible to argue that Johnson's results are due not to differences in the amount learned about wavelength, but to differences in the extent to which orientation controlled reponse rate. The orientation of a bright line would be more likely to control behavior than the orientation of a dim line. On the other hand, it is possible to argue that Kamin's results are due to generalization decrement; the louder the noise used in Stages 1 and 2, the greater the change between training trials with the noise, and test trials without the noise. We shall come across similar problems in experiments on overshadowing in the next chapter.

In this context, the generalization decrement explanation seems more plausible than the suggestion of a confounding of learning with control. We have already pointed out that this latter account has considerable difficulty explaining apparent blocking effects obtained when intermediate values of the ostensibly controlling cue are used. Unless, therefore, the Stage 1 cue is such that its removal will not represent a greater change for animals trained on an easy discrimination than for animals trained on a difficult discrimination (wavelength would be a suitable example), it seems safer to conduct an experiment in this area without removing the Stage 1 cue on test trials.

An experiment by Mackintosh (1965d) provides reasonably satisfactory controls for some of these problems. The study has already been partially described: rats were initially trained to discriminate gray horizontal and vertical rectangles, a second cue (brightness) was added after they reached criterion, and they were finally tested for the amount learned about brightness. The stimuli for test trials were black and white rectangles, on alternate trials both were horizontal or both were vertical. The difficulty of the Stage 1 problem was manipulated by varying the ratio of long to short sides of the rectangles; for the easy problem, the rectangles measured $10 \times 2$ cm; for the difficult problem, the rectangles measured $6.66 \times 3$ cm. This difference was sufficient to cause no overlap between the two groups in the number of trials to learn the horizontal–vertical problem. The difference between the test scores of the two groups was also substantial: animals trained on the easy problem made 54% correct responses over 20 test trials; those trained on the difficult problem made 88% correct responses. Again, there was no overlap between the two groups.

In this experiment, all animals were trained to the same high criterion in Stage 1 (90% correct over 40 trials), and differences in Stage 2 performance were extremely small. It is true that animals in the difficult discrimination group continued to make more errors than animals in the easy discrimination group, but the difference (an average of 2.4 errors in 100 trials as against an average of 1.1 errors) was hardly large enough to have been responsible for the very large difference in test scores. It is also true that there were small differences between the two groups in latency of response over early Stage 2 trials, but these fell far short of significance. That either of these differences in Stage 2 performance caused differences in test performance is made even less likely by the results of correlations calculated between Stage 2 errors and Stage 3 test performance, and between Stage 2 latencies and Stage 3 performance. Over all subjects in the experiment, neither correlation approached significance. By contrast, the correlation between initial learning scores (i.e., the direct measure of problem difficulty) and final test scores was highly significant.

The remaining possibility is that differences in test performance were determined more by differences in the extent to which performance was controlled by orientation than by any differences in the amount learned about brightness. It is not entirely clear how this could happen in a choice situation, but one possibility is that animals controlled by orientation would look at one alternative first and if they saw the positive value of orientation would respond before looking to see that the positive orientation was on the other side also. Such behavior would presumably not occur on those test trials when an animal's negative orientation was on both sides. The obvious prediction from this account is that the differences between animals trained on the easy and the hard problem should be greater on "positive" test trials than on "negative" test trials. In fact, although the difference was not large, it was in the opposite direction.

These results, therefore, tend to rule out alternative explanations of the effect of problem difficulty on incidental learning, and suggest that the finding must be interpreted within an attentional framework. Two further studies have also obtained results which are certainly not attributable to any differences in performance during Stage 2. We have already discussed the basic experimental procedure used by Terrace (1963a) in which pigeons are trained on a go–no-go discrimination without errors, and his finding of errorless transfer of the discrimination habit from one dimension to another (Terrace,

1963b). One of the conditions necessary for completely errorless transfer was that the original color cue be faded out gradually; when the color cue was switched off abruptly, the animals made a number of errors on the orientation discrimination. The difference between these two conditions would seem to involve the difficulty of the originally trained discrimination. When that discrimination remained easy, animals failed to learn completely about the added orientation cue; when it was made difficult by fading the colors out over a long series of trials, animals learned much more about orientation. Since, under neither condition were any errors made before test trials with orientation, it is not possible to account for differences in test performance by differences in the number of previous unreinforced responses to the negative orientation. Similar results have been obtained by Schusterman (1966) in his procedure for errorless reversal learning by sea lions. When the size difference between the stimuli was reduced abruptly, errors occurred; when the difference was reduced gradually, performance was perfect. Unfortunately both Terrace's and Schusterman's results could be attributed to generalization decrement effects. In both experiments, one would expect performance to be more disrupted by the abrupt removal of the controlling cue than by gradually fading it out.

2. *Interpretation*

Although there are considerable problems of interpretation, all the available evidence goes to show that animals pretrained on a difficult problem perform more accurately on subsequent tests of the amount learned about an incidental cue than animals pretrained on an easy discrimination. It is unlikely that all of these results can be entirely attributed to differences in opportunity for learning about the incidental cue, or to differences in control by the initial cue during testing. It seems that the difficulty of the initial discrimination problem directly affects the probability that animals will attend to a subsequently introduced incidental cue. How is this effect to be interpreted?

One conceivable interpretation is that it could be cited as an example of one of the more hallowed generalizations in psychology, namely, the law of least effort. Animals that have learned an easy discrimination do not need to attend to further cues, since the one they are using already is simple and effortless; but animals that have learned a difficult discrimination might (by analogy) be supposed to find this exhausting, and would attend to any new cue that might reduce the burden imposed on them. The sense in which the burden of analyzing the relevant cue of a difficult discrimination could be said to be exhausting is presumably different from the sense in which pulling in a heavy weight is exhausting, and it is up to the proponent of such an account to define what sense he does imply. One candidate would perhaps be information processing; one might suggest that an easy discrimination imposes a slighter demand on the information processing capacity of an organism than does a difficult discrimination.

This suggestion seems at first sight to fit in with the rationale for selective attention provided by information theory. In Chapter 1, we briefly discussed the role played by the development of information theory in focusing interest on the selectiveness of perception. The concept of a limited channel capacity has suggested to Broadbent among others that the nervous system must contain some sort of filter to protect central decision mechanisms from being overloaded by incoming information. Applying these ideas to animal discrimination learning and the continuity–noncontinuity dispute, Broadbent (1961) suggested that animals can take in only so much information at a time and that, since discriminability may be measurable in terms of information, while animals may be able to attend to more than one simple cue at a time, if the discrimination is made difficult enough then attention will be limited to a single cue only.

This analysis immediately reveals the ambiguity of this application of informational concepts. Animals trained on a difficult discrimination learn more about a subsequently introduced incidental cue than animals trained on an easy discrimination. The result could be interpreted as showing that animals are continually searching for ways of reducing the amount of information to be processed, but it cannot be accounted for by a causal model of the type proposed by Broadbent. Broadbent supposes that an animal solving a difficult discrimination has little or no spare capacity left for attending to additional cues, while an animal solving an easy problem has ample capacity to spare. Thus the animal trained on the *easy* discrimination should be more likely to attend to a newly introduced incidental cue, and this is the exact opposite of the result obtained.

Rule 1 (inside front cover) of our model states that the amount learned about one cue is inversely related to the strength of analyzers detecting other cues. This rule accounts for the basic blocking effect because animals pretrained to attend to one cue will switch-in the analyzer detecting that cue more strongly than animals not so pretrained. It also predicts that if animals are trained from the outset on a two-cue discrimination problem, the amount learned about one of the cues will depend upon the discriminability of the other. The more discriminable a cue, the greater the base-line strength of the analyzer detecting it, and the less will be learned about any other cue present. As we shall see in the next chapter, although further problems of interpretation arise, there are several studies that suggest that the discriminability of a cue determines the extent to which it will overshadow another. In order to explain how the discriminability of a cue also determines the extent to which prior training on that cue will block learning about an added cue, all that is necessary is to assume that discriminability not only affects base-line analyzer strength, but also continues to influence the strength of an analyzer even after a common criterion of learning has been reached. This assumption seems reasonably plausible and given that we are talking about strengths of analyzers, not about the probability of switching an analyzer in, there does not seem to be any reason why the relevant analyzer must always have attained the same strength at the time a particular performance criterion is reached.

## VI. Summary

In this chapter, we have traced the development of the so-called continuity–noncontinuity dispute, and suggested that the main questions asked by the original antagonists were wrongly formulated. Lashley and Krechevsky seemed to claim that animals only attend to a single cue at a time, while Spence and other continuity theorists claimed that, in some sense, animals learn equally about all cues impinging on their receptors. Evidence intended to decide between these views has been derived from three sources:

(1) analysis of systematic "hypothesis" behavior, more specifically of the position habits displayed by rats solving visual discrimination problems;
(2) experiments on presolution period reversal;
(3) experiments on incidental learning.

While the majority of results from all three types of experiment has argued against the noncontinuity theory of Lashley and Krechevsky, there is no justification for concluding that Spence's position has thereby been supported. For all three types of experiment were directed towards answering the question: do rats learn about only a single cue at a time? The discovery that they learn something about other cues does not imply necessarily that attention has no effect on learning.

Certain features of the data on position habits are extremely difficult, if not impossible, to account for, without postulating some selectivity of control. Experiments on presolution reversal are certainly consistent with a continuity theory, but it can be argued that this is largely because they have not been the right sort of experiment. Numerous studies of incidental learning have shown that pretraining on a single cue blocks (partially or, more rarely, completely) learning about a second cue. While alternative explanations for some of these results cannot always be ruled out, there remains a large body of evidence that is incompatible with a strict continuity theory. A large proportion of these data is accounted for economically by Rules 1 and 2 of our model, which state that the amount learned about one cue is inversely related to the strength of other analyzers in the situation.

CHAPTER 5

# Learning and Performance with More Than One Relevant Cue

In Chapter 4, we discussed experiments in which animals first learned a discrimination involving only one relevant cue and then received further training with a second cue added. The present chapter is devoted to experiments in which two or more cues are made relevant from the outset of training. The experiments will be discussed under two main heads. First we consider how the presence of additional cues affects performance during acquisition. Second, we take up the question of how much is learned about each cue separately when an animal is trained with more than one cue; the amount learned about each cue is represented not in performance during acquisition but in performance during transfer tests with single cues usually given after training is completed.

## I. Acquisition with Multiple Cues

### A. The Phenomenon

For some time, it has been known that animals trained with two or more relevant cues usually acquire a discriminatory habit faster than animals trained with only one relevant cue. Eninger (1952) trained three groups of rats on different discriminations in a T maze. One group had to learn to turn left or right, according to whether the maze stem was black or white, a second group was trained to base direction of turn on whether a tone was on or off, and the third group was trained with both the visual and auditory cue present and relevant. The first two groups, trained with single cues, took an average of 234 and 148 trials to reach criterion, whereas the third, trained with both cues relevant, reached the same criterion in 55 trials. Warren (1953) found that monkeys learned a discrimination with both form and size cues relevant faster than when either one of these cues was relevant on its own. The tendency for animals to learn faster with several relevant cues than with only one has come to be known as "additivity of cues."

Additivity of cues has also been frequently found in experiments on place and response learning in a T maze. This is illustrated in Figure 5.1. If an

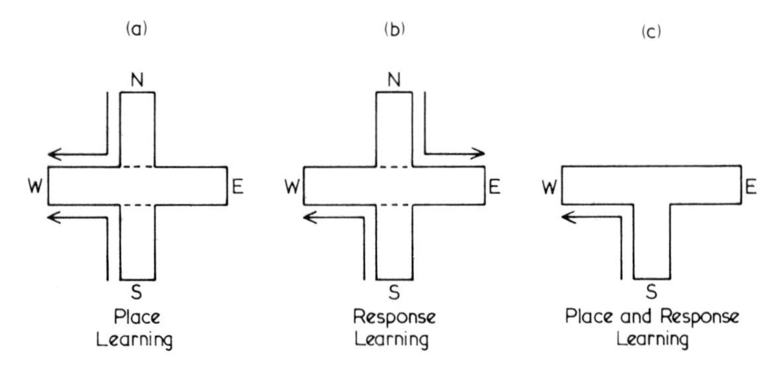


**Fig. 5.1.** Place and response learning.

animal is started on different trials during training from the North and South positions on a cross maze and trained always to select the West arm, it must learn to go to a place but the direction of turn made (left or right) is irrelevant (Figure 5.1a). As Figure 5.1b shows, animals can also be trained always to make the same turn with place irrelevant, or they can be trained on a problem that can be solved in terms of both place and turn (Figure 5.1c). Learning to go to a place is presumably achieved by using external cues and making a response in a particular direction relative to those cues. A series of studies (excellently reviewed by Restle, 1957), has shown that animals learn a place habit most readily when the environment is well differentiated. We can regard a turning habit as also under the control of cues, but the cues are presumably internal ones arising from kinesthesis. Several studies (e.g., Galanter & Shaw, 1954; Scharlock, 1955; Blodgett, McCutchan, & Mathews, 1949) have shown that learning is faster when both place and response cues are relevant as in Figure 5.1c than when only one of the two cues is relevant as in Figures 5.1a and b.

There is one exception to the rule that acquisition is faster with two cues than with one (see Chapter 4, page 131). If one cue is considerably weaker than the other, then performance during acquisition improves at the same rate as when animals are trained on the stronger cue alone. Warren, in the experiment cited above, found that monkeys learned discriminations involving both color and size or both color and form at the same rate as discriminations with color alone relevant. Similar findings were obtained by Harlow (1945). With the values of each dimension used in these experiments, color is a much stronger cue for monkeys than either form or size.

Having given a rather general account of the phenomenon to be explained, we now turn to consider different explanations of it.

## B. Restle's Models

The most thoroughgoing attempt to explain the data on additivity of cues was made by Restle (1955, 1957). His theory makes no use of the concept of selective attention and it is, therefore, important for us to see how well it copes with the data. He assumes that there are two types of stimulus elements present in the learning situation: those that are relevant to the solution of a problem and those that are irrelevant. The notion of a stimulus element is highly abstract, and there might be many stimulus elements going to make up one cue in our sense of the word "cue." In what follows, we use the word "cue" in Restle's sense to mean stimulus element. On each trial, a constant proportion of the unconditioned stimulus elements is conditioned: i.e., the correct response is attached to them. If $c_n$ is the proportion of all relevant cues conditioned by the $n$th trial then the proportion conditioned by trial $n + 1$ is assumed to be

$$c_{n+1} = c_n + \theta(1 - c_n), \tag{5.1}$$

where the constant $\theta$ is the proportion of stimulus elements sampled on any one trial. Irrelevant cues are said to be adapted; once a cue is adapted it will no longer contribute towards the response. The proportion of irrelevant cues adapted on any one trial is again $\theta$, so that if, on trial $n$, the proportion of all irrelevant cues adapted is $a_n$, then the proportion adapted before the next trial will be

$$a_{n+1} = a_n + \theta(1 - a_n). \tag{5.2}$$

It is assumed that if either an unconditioned relevant cue or an unadapted irrelevant cue is sampled, the probability of a correct response is .5. This leads to the following equation governing performance

$$p_n = c_n + .5(1 - c_n) + .5(1 - a_n), \tag{5.3}$$

where $p_n$ is the probability of correct response on trial $n$. Restle does not provide a complete rationale for this equation, since strictly speaking it follows from his assumption only if not more than one cue is sampled on each trial.

Restle next makes the simplifying assumption that the constant determining the proportion of cues sampled on one trial is

$$\theta = r/(r + i), \tag{5.4}$$

where $r$ represents the number of relevant stimulus elements and $i$ the number of irrelevant ones. It follows from Eqs. (5.1)–(5.3) that if the probability of a correct response on trial $n$ is $p_n$, then the probability on the next trial is

$$p_{n+1} = p_n + \theta(1 - p_n). \tag{5.5}$$

If we know the number of errors made in reaching a high criterion of learning in a given discrimination, it is possible to estimate the quantity $\theta$. Restle's additivity-of-cues assumption postulates that when two different cues are both made relevant, the total number of relevant stimulus elements is simply the sum of the number in each cue and the number of irrelevant stimulus elements does not change. Thus, if we have two relevant cues with numbers of stimulus elements $r_1$ and $r_2$, then the two learning rate constants for each on its own will be

$$\theta_1 = r_1/(r_1 + i), \tag{5.6}$$

and

$$\theta_2 = r_2/(r_2 + i). \tag{5.7}$$

Moreover, the learning rate constant when both cues are relevant will be

$$\theta_3 = (r_1 + r_2)/(r_1 + r_2 + i). \tag{5.8}$$

These equations can be solved for $\theta_3$ in terms of $\theta_1$ and $\theta_2$, yielding,

$$\theta_3 = (\theta_1 + \theta_2 - \theta_1 \cdot \theta_2)/(1 - \theta_1 \cdot \theta_2). \tag{5.9}$$

We can estimate $\theta_1$ and $\theta_2$ by running experiments with each cue relevant on its own. We can then use Eq. (5.9) to determine $\theta_3$, the learning rate constant when both cues are relevant; hence, acquisition performance with two cues can be predicted from performance with each cue relevant on its own. Restle (1955, 1957) and Bourne and Restle (1959) have applied Eq. (5.9) to a number of different results and shown that a reasonably good fit is obtained.

As Restle says elsewhere (1962), there are a number of difficulties in his model, despite its success in predicting the result of experiments on additivity of cues. First, there is no rationale for the assumption that $\theta = r/(r + i)$. It is, indeed, very difficult to see how it possibly could be equal to this quantity, since only *after* learning has occurred can the animal "know" what is the proportion of relevant and irrelevant cues. Second, he assumes that the same amount is learned both on rewarded and nonrewarded trials, and this makes it difficult to explain how animals can learn a discrimination under conditions of partial reinforcement (Sutherland, 1966b). A third criticism (related to the first) is that no mechanism is postulated to explain how the animal can sort out cues to be adapted from cues to be conditioned and it is difficult to see how the animal can know, particularly on early trials, which cues are to be conditioned, which are to be adapted. Fourth, according to the model, once a stimulus element is adapted, it cannot be conditioned; hence, no mechanism is provided to account for how it is that after an animal has been trained on a particular discrimination with a given cue irrelevant it can subsequently learn to discriminate in terms of that cue when it is made relevant. Although pretraining with a cue irrelevant may retard subsequent learning of that cue (see Chapter 6, page 199), it does not make such learning impossible. Finally, Restle's assumption that the rate of learning about a given cue is not affected by the presence or absence of a second relevant cue is inconsistent with the results of experiments on overshadowing (see Section 5.II).

Restle (1962) subsequently developed a second model of discrimination learning that was also intended to cope with the data on additivity of cues. In this model he assumes that in any situation animals have at their disposal a number of "strategies"; the different strategies correspond roughly to what Krechevsky meant by "hypothesis." Thus a rat in a jumping stand might adopt the strategy of jumping left, or jumping right, or following white, etc. Restle considers three models based on the notion of sampling strategies; the models differ in the assumptions made about how many strategies are sampled at once and how the correct strategy is discovered. Since he shows that the different models make the same predictions, we shall only consider one of the three here. In this model, Restle assumes that the organism picks a strategy at random from those available to it and keeps the same strategy until a mistake occurs; it then resamples with replacement from all the strategies available until one is picked that no longer yields errors, i.e., until the correct strategy is sampled. On such a model it is possible to obtain estimates of the proportions of correct ($c$), wrong ($w$), and irrelevant ($i$) strategies from the number of errors made in learning and their distribution. Additivity of cue data can be predicted on the assumption that if $c_1$ and $c_2$ are the number of correct strategies available when either of two cues is relevant, then the number of correct strategies when both cues are relevant is $c_1 + c_2$. Restle again shows that the model fits some aspects of additivity-of-cue data in animals; given the number of trials taken to reach criterion by two groups of animals, each trained with one of two single cues, the model predicts reasonably accurately the number of trials to criterion taken by a third group of animals trained with both cues relevant.

This model cannot be a correct account of discrimination learning in the rat for three reasons. First, as we saw in Chapter 4, it can be shown that some learning about the correct response occurs at a time when the animals' responses are entirely determined by position habits. On Restle's model, an animal should be just as likely to break a position habit by jumping to the negative visual stimulus as by jumping to the positive, but in practice, over 90% of rats break position habits when the correct stimulus is on the non-preferred side.

Second, Restle's model implies that learning takes place on one trial and that, until the correct strategy is selected, performance will remain at or below chance since, at this stage, the animal is sampling only irrelevant or wrong strategies. When the probability of making an error does not change over a given number of trials, the learning curve is said to be stationary. Restle's model implies that the learning curve for individual subjects is stationary until the trial at which the last error occurs. As we saw in the last chapter, rat learning curves are not completely stationary. In general, rats make fewer errors in the blocks of trials immediately preceding criterion than they do in earlier trial blocks. This effect can be seen from an examination of behavioral learning curves. Such a curve is shown in Figure 4.1 (page 88), and further data on this point from an unpublished experiment of Sutherland and Holgate are shown in Figure 5.2. Group Or learned an orientation discrimin-

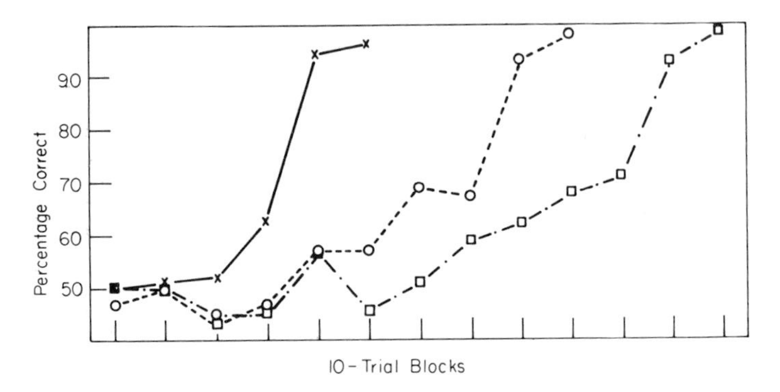


**Fig. 5.2.** Backward learning curves for rats trained with two relevant cues [Group C (×)] and with only one relevant cue [Groups Or (○) and Br (□)]. Or designates orientation; Br, brightness.

ation, Group Br a brightness discrimination, and Group C a discrimination with both cues relevant. It will be seen that none of the backward learning curves is stationary immediately before criterion is reached, though they are all stationary during the earlier stage of training. Restle's second model fails to account for this lack of stationarity; nor can it account for the fact that the more difficult the discrimination the less steep is the slope of the learning curve once performance departs from chance.

A third difficulty with the model is that it assumes that when two cues are relevant, learning will take place about one or other cue, but not both, and as we shall see in the next section this is incorrect. Bower and Trabasso (1964) made some simplifying assumptions to Restle's second model and found that it fitted results obtained from college students in a concept-formation situation involving additivity of cues. Unlike rats, the human subjects did learn about only one cue when two cues were relevant and did exhibit complete stationarity of the learning curve up to the trial of the last error. Bower and Trabasso's version of Restle's model, therefore, does not fit animal data. Subsequently, Trabasso and Bower (1968) have found that, under some conditions, some human subjects do learn more than one relevant cue. They have modified their model to take this into account. Since their model assumes that learning occurs in a single trial and leads to stationarity of the learning curve up to the trial of the last error, it does not fit the animal data, and will not be discussed further. Their recent book gives some examples of simple mathematical models which, depending on the parameters used, can be made to fit apparently very different results.

Despite the elegance of his formulation and his success in predicting the number of trials taken to acquire a habit with single and combined cues, we conclude, therefore, that Restle's theories do not correctly describe discrimination learning in rats. We have discussed his two basic models here because, despite their inadequacy, they represent one of the most thorough and elegant attempts yet made to put forward a formal model of the discrimination process. We must now turn to our own model to see how well that explains the findings.

### C. Two-Process Models

Rule 1 of our theory states, "The size of the change in the response strengths attached to a given analyzer on any trial is proportional to the strength of that analyzer." Since analyzer strengths sum to 1.0, this implies that when two cues are present, the amount learned on any one trial about each will tend to be less than would be learned about a single cue presented on its own. When only one relevant cue is present, it is only the analyzer detecting that cue that can successfully predict reward and nonreward and therefore only that analyzer will be strengthened. If two cues are present and relevant, the analyzers for both will start to be strengthened, and as a result, less will be learned about the response attachments to each than if only one of the two cues had been relevant. When both cues are relevant the increasing strength of one analyzer will reduce the amount learned about the other. Our theory therefore conflicts with the additivity-of-cues postulate, if by this is meant the type of assumption made by Restle that the amount learned on one trial about any one cue is independent of the amount learned about any other. In the next section, experiments will be described that directly support the conclusion that the amount learned about one cue is inversely related to the amount learned about other cues. In this section, we are, however, mainly concerned with the effects on trials to criterion during acquisition of making different numbers of cues relevant.

We begin by considering an extreme form of noncontinuity model in which only one analyzer is used on any one trial. Suppose that there are two cues always present, but one or other or both may be made relevant; let the probabilities of the analyzers for each cue being used on the first trial both equal .3. Performance should improve more rapidly with two cues than with one, since from the outset of training, learning will occur on .6 of trials when two cues are relevant, whereas it will only occur on .3 of trials when only one cue is relevant. To obtain perfect performance (given that response learning has occurred) each analyzer will have to be strengthened to a probability of being switched in of about .5, whereas with only one cue relevant that analyzer must be strengthened from .3 to about 1.0. This argument is not dissimilar to those used in Restle's first model. Contrary to what has sometimes been implied, the phenomena of additivity, therefore, is not inconsistent even with an extreme form of noncontinuity theory.

We have, however, hitherto argued that an extreme form of noncontinuity model postulating attention to only one cue on each trial is wrong and that, while it is true that the more an animal attends to one cue the less it attends to any other, learning can occur about more than one cue on one trial. Following Rules 1 and 2 of our own model, let us assume that when only one cue is present the base analyzer strength for that cue is .25 and the total analyzer strength for all irrelevant cues is .75. If we now add a second relevent cue of equal weight to the first, this cue can be chosen either from among the irrelevant cues present with the single cue or it can be an additional cue not present among the irrelevant cues. In the former case, each relevant cue will have a base value of .25, and the irrelevant analyzers will sum to .5; in the latter case on the assumption that the added cue detracts from the strengths of the other cues in proportion to their previous strengths, each relevant cue will have an initial value $.25/(.25 + .25 + .75) = .2$ and the irrelevant analyzers will sum to .6. In both cases, it is clear that one or other of the relevant analyzers will overtake the strength of the dominant irrelevant analyzer more quickly when two relevant cues are present than when there is only one. There are two reasons for this: First, when two relevant cues are present, there is a smaller difference between the starting values of each relevant cue and the dominant irrelevant cue (usually position); second, the gains in strength made by the added relevant cue will at first be made mainly at the expense of the dominant irrelevant analyzer. Moreover, by Rule 4 both relevant cues should contribute to performance at the time their strengths approach that of the dominant position analyzer provided that at that time there is not too big a difference between their values.

While this argument is watertight for the case where the added relevant cue in two-cue learning is present but irrelevant during single-cue learning, there is an added complication in the case where it is not present in single-cue learning. In this instance, it will be seen from the numerical example given above that although the combined strengths of the relevant analyzers in two-cue learning (.4) are higher than the strength of the relevant analyzer in single-cue learning (.25), the strength of each relevant analyzer on its own (.2) is slightly lower than the strength of the relevant analyzer when only one cue is present. This implies that the learning of response attachments to each analyzer on its own will proceed more slowly when two cues are relevant than when only one is relevant. It seems certain that with some parameter settings the effects of the slower learning rate of response attachments to the individual analyzer would be more than offset by the three factors mentioned above that contribute to relevant response learning affecting performance after fewer trials with two relevant analyzers than with one. Moreover, we shall show later that there is a further reason why learning in terms of trials to criterion should be faster with two cues than with one.

Our model, therefore, predicts that the break in a position habit will occur earlier with two relevant cues than with one and that (provided the two relevant cues are of roughly equal strength) there will be fewer errors made between breaking a position habit and meeting criterion, since performance will be a joint function of the response attachments made to both analyzers. In an unpublished experiment, Sutherland and Holgate compared learning rates when rats were trained with either one or two visual cues and found that not only did animals trained with two cues reach criterion faster, they also broke their position habit earlier and made fewer errors between this point and reaching criterion; backward learning curves from this experiment are presented in Figure 5.2.

Lovejoy (1968) has performed an ingenious experiment that bears on the question of whether when there are two relevant cues, both relevant analyzers contribute to the breaking of a position habit. He trained rats on two cues, size and orientation, giving all rats three kinds of trials, some with both cues present, and some with each cue present on its own; equal numbers of the three types of trial were given throughout training. Lovejoy argued that if behavior were controlled by only one cue on any one trial, then position habits should be broken by a correct response to the non-preferred side equally often on each type of trial. He obtained the surprising result that position habits were more frequently broken when either cue on its own was presented than on trials on which both cues were simultaneously present. This result is puzzling. It may be that when both cues were simultaneously present, they masked one another; the cues used were large and small circles and oblique lines in opposite orientations, and when both cues were presented simultaneously the oblique lines were superimposed on the circles.

Turner (1968) repeated Lovejoy's experiment using rats in a jumping stand and orientation and brightness as cues. He obtained a very different result. To overcome the problem of the single cues masking one another when presented in the compound, he used the stimuli shown in Figure 5.3.

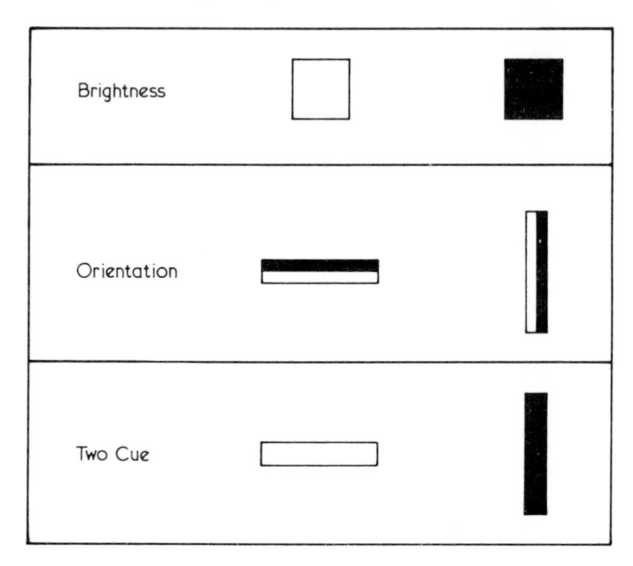


**Fig. 5.3.** Shapes used by Turner (1968).

The orientation stimuli were split black-and-white rectangles rather than gray rectangles, because gray rectangles against a brown background tend to be very inconspicuous. Of seven rats freely breaking a position habit to the positive stimulus, six broke on a compound trial. In another experiment, Turner ran equal numbers of compound trials and brightness-only trials. Since brightness was present on all trials and the orientation cue on only half the trials, Turner argued that the probability of an animal attending to brightness and therefore having its behavior controlled by the brightness cue must be at least as high as the probability of its attending to orientation. If we make the unlikely assumption that the probabilities are in fact equal, it can be shown that the hypothesis of single-cue control predicts that at most two-thirds of the subjects will break their position habits on compound trials. Let $p$ be the probability of attending to brightness, then the probability of attending to either brightness or orientation on a compound trial cannot be more than $p + p(1-p)$ and the maximum ratio of breaks when both cues are present to all breaks will be $[p + p(1-p)]/[p + p + p(1-p)]$ or $(2-p)/(3-p)$, which is always less than $\frac{2}{3}$. Fourteen subjects, in fact, broke their position habit to the positive stimulus on compound trials, and only one subject broke when the single brightness cue was presented on its own. Turner's results strongly suggest, therefore, that behavior can be controlled by more than one analyzer within a trial.

Our arguments up to this point have been based on the assumption that the two relevant cues are of roughly equal strengths and that they have the same base values for all subjects. The latter assumption is, in fact, highly implausible, and it is much more likely that the base values of the two cues will vary from one member of a population of subjects to another. The occurrence of such variation would, in fact, provide yet a further reason for expecting to find an additivity-of-cues effect on our own model.

Response learning will occur most rapidly to whichever analyzer is higher at the start of training; the greater the degree of response learning, the faster the rate of gain in strength of the analyzer and, therefore, the slower will be the learning of response attachments to the weaker analyzer and hence the slower its gain in strength. This should mean that the stronger analyzer will "run away" and will suffer little retardation in increments of strength due to the growth of the weaker analyzer. The point at which the stronger analyzer of the two overtakes the strength of the position analyzer will now clearly be earlier on average in a group of animals trained on two cues than in a group trained on one cue, since half the one-cue animals will, by chance, be receiving training on their weaker cue, whereas the two-cue animals will always solve the problem by strengthening whichever analyzer is stronger at the outset of training. One prediction follows from this account, namely, that the standard deviation of errors to criterion should be larger for animals trained on one cue than on two; this prediction was tested in an unpublished experiment run by Sutherland and Holgate. Rats were trained on simultaneous discriminations and different groups learned with brightness alone as the relevant cue, with orientation alone, or with both brightness and orientation relevant. The trials to reach criterion taken by the three groups were, respectively, 117, 85, and 59 and the corresponding standard deviations were 34, 16, and 10. The result cannot be unambiguously interpreted since although, as predicted, the standard deviation for the two-cue group was smaller than for either of the single-cue groups in absolute terms, it was about the same if plotted as a ratio of the mean number of trials to learn.

Our account of additivity of cues yields two further predictions about rates of learning. If animals are trained with two relevant cues of which one is very strong and the other very weak for all members of the group, then we would not expect the weak cue to make any contribution to performance before criterion is reached: the dominant analyzer will come to control behavior before the weaker one is strong enough to contribute. It follows that in this situation animals should not reach criterion faster with two relevant cues than with one. As we saw above, this is precisely what was found by Warren (1953) and Harlow (1945). Moreover Kamin (1969), using a classically conditioned CER in rats with light-plus-noise as the CS, found that when intense noise was used, learning was faster than with light alone, but it was no faster when the noise was of low intensity. However, Medin and Davis (1967) have recently demonstrated some additivity of cues in monkeys with a weak cue added to a dominant cue.

Second, one counterintuititive prediction can be drawn from the model: if there are many weak relevant cues and one dominant relevant cue, animals may actually take longer to reach criterion than if the dominant cue is present on its own. This arises because the weak relevant analyzers will each gain small increments of strength and, hence, reduce the amount learned about the dominant one; but if they are not strengthened to the point at which they influence performance, the net result will be to reduce the speed of acquisition as measured by trials or errors to learn.

Our model also makes predictions about how much will be learned about each cue when there are two or more relevant cues; these predictions cannot be tested during acquisition but can be tested by measuring performance on the separate cues after learning with both cues relevant and it is to this question that we now turn.

## II. What Is Learned with Multiple Cues

In this section, we discuss what animals learn when there is more than one relevant cue. Several different problems arise and they will be considered seriatim. First, there is the problem of the extent to which a strong relevant cue will reduce learning about a weak cue simultaneously presented. Second, we consider how the presence of two cues of roughly equal strengths affects the amount learned about each. Third, we consider how far performance with two cues present can be predicted from performance with each cue presented alone. Finally, we take up the question of whether animals learn "compounds" in addition to components.

### A. Overshadowing

When two cues are relevant, the reduction in the amount learned about one cue should depend critically upon the relative strengths of the two analyzers, as we showed in the last section. If one analyzer is very strong and the other very weak, the tendency for the strong analyzer to increase in strength rapidly should reduce learning about the weak analyzer. Since the weaker analyzer will only increase in strength slowly, the presence of a weak relevant cue will have little effect on the amount learned about a stronger cue. This effect is, of course, rather similar to that discussed in Chapter 4. In the experiments on blocking therein described, the first analyzer is deliberately made strong by training on one cue before a second cue is introduced and, therefore, learning about the second "incidental" cue is retarded. We now turn to experiments in which both cues are introduced from the outset of training, but one is much stronger than the other; the effect of the strong cue in suppressing learning about the weaker one is known as "overshadowing."

In Chapter 1, we described an experiment by Pavlov in which he found that when a strong stimulus and a weak stimulus were given together, there was less learning about the weak stimulus than when the weak was presented on its own throughout training. This result has since been confirmed using both classical and instrumental conditioning procedures. Using the CER technique outlined in Chapter 4, Kamin (1968, 1969) showed that if an 80-dB noise is presented with a light during training, animals learn to make the CER to both components. When a weaker noise (50 dB) was used, it was impossible to demonstrate any learning about the noise as measured by trials to relearn the conditioned response to noise alone. As already mentioned, Kamin also showed that when learning about the noise was completely overshadowed in this way, acquisition to the compound cue was no faster than to light alone, whereas it was considerably faster when the stronger noise was used.

Clearly, Kamin's results cannot be explained by any of the three alternative explanations suggested in Chapter 4 for blocking experiments; learning to orient the eyes in such a way as to pick up the light can hardly have prevented the animals from hearing the noise; since testing with noise was carried out in the absence of light, failure to find conditioning to the low noise cannot have been caused by the light exercising control over behavior; finally, since classical conditioning was used, the reinforcement schedule with light plus 50-dB noise was the same as with 50-dB noise alone.

There is one further finding of Kamin's that is somewhat disquieting for our own theory. In the experiment quoted above, he used a 1-mA shock. With this shock intensity, acquisition of the CER to 50-dB noise alone proceeds more slowly than to 80-dB noise alone. With a 4-mA shock, however, the CER is learned as quickly to 50-dB noise as to 80-dB noise. Kamin therefore tested the overshadowing effect of light on 50-dB noise using the stronger shock level and found that now no overshadowing occurred. It is difficult for us to explain this finding and, indeed, if increased shock level increases drive, then it should actually narrow the range of attention rather than increasing it (see Chapter 4). It may be that the heightened shock reduced overshadowing because it increased the difference in the consequences of the presence versus absence of noise (see Chapter 2). In the absence of more evidence, it is scarcely worth speculating on this further, but it would be interesting to manipulate independently level of drive and level of reinforcement in appetitive learning experiments on overshadowing; we would expect that increases in drive level should decrease the amount learned about nonsalient cues, whereas increases in amount of reinforcement should increase learning about weak cues.

Experiments on overshadowing using instrumental situations fall into two classes: those involving successive discrimination training and those using a simultaneous situation. Miles (1965) provides a good example of the first type of experiment. He rewarded pigeons for pecking when a very bright light ($L_1$) was on and gave no reward for pecking in darkness ($L_6$) or when different lights ($L_2$–$L_5$) all less intense than $L_1$ were presented. All animals were trained with $L_1$ as S+ (stimulus signaling possibility of reinforcement) but different groups received training with $L_2$–$L_6$ as S– (stimulus signaling nonreinforcement). $L_1$ was always coupled with a tone, and the other lights were always presented in a background-masking noise. After a fixed number of training sessions, all birds were tested with lights $L_2$–$L_6$ both presented with the tone and without. The tone exerted very little control over the behavior of animals trained with $L_1$ and $L_6$, but exerted increasing control on animals trained with smaller differences in the intensity of the two training lights. This experiment is particularly interesting since it was not the strength of the overshadowing stimulus presented with the tone that varied; what varied was the strength of the light presented in the absence of the tone. Nevertheless the easier was the light discrimination, the greater was the blocking of the tone. We can explain this on the grounds that the more readily an animal detects differences in the consequences of different outputs from an analyzer the more strongly that analyzer will be switched in. The smaller the difference in light intensity, the more slowly will the light analyzer be switched in and therefore the less the overshadowing of the tone cue. The Miles experiment shows, therefore, that overshadowing decreases as the difficulty of discriminating the overshadowing cue increases; it is not merely dependent on the absolute intensity of the overshadowing stimulus.

Eckerman (1967) has also presented evidence on pigeons suggesting that overshadowing occurs in a successive instrumental discrimination, and Thompson and Van Hoeson (1967) using rats showed that the effect occurs in avoidance learning. It is implausible to suppose that any of these results can be explained by animals learning orienting responses to the overshadowing cue that blocked reception of the weaker cue or that the control exercised by the stronger cue prevented control by the weaker cue in tests, since in all these experiments animals were tested in the absence of the stronger cue. It could, however, be argued that the reduced learning about the weaker cue may have been due to animals making fewer errors during acquisition because the strong cue was rapidly learned. The plausibility of this argument is, however, reduced by the finding discussed above that overshadowing occurs when classical conditioning is used.

Finally, there are four experiments showing overshadowing in instrumental simultaneous discriminations. Babb (1957) trained rats on a simultaneous discrimination with two relevant cues, black–white, and the presence versus absence of chains. He used three groups of animals: one with both cues present and relevant, one with the brightness cue alone present and relevant, and one with the chains cue alone present and relevant. All animals received 30 training trials at this stage. Each group was split in two and half the animals were trained with one cue alone present and relevant, half with the other cue alone present and relevant. Where the same cue was present at both stages of the experiment, the same stimulus remained positive at both stages. The results are given in Table 5.I. From the errors made in

TABLE 5.I

**Babb's Results** [a, b]

| Stage I, initial training | Br and Ch | | Ch | | Br | |
|---|---|---|---|---|---|---|
| Errors in 30 trials | 6.7 | | 10.0 | | 7.7 | |
| Stage II, relearning | Br | Ch | Br | Ch | Br | Ch |
| Errors to criterion | .92 | 2.6 | 4.5 | 1.2 | .7 | 3.9 |

[a] Data obtained from Babb (1957).
[b] Br designates brightness; Ch, chains.

initial training it will be seen that brightness was the more dominant cue. As we would expect, it also overshadowed the learning of the chains discrimination; the two-cue group made significantly more errors in relearning with chains alone than did the group trained from the outset with chains alone. Almost as much, however, was learned at the first stage about brightness by the two-cue group as by the group trained initially on brightness alone; the difference in errors to relearn was not significant. It should be noted that both this result and the results quoted above are in conflict with Restle's first model, since that assumes that when two cues are present as much is learned about each as when either is present on its own.

Unfortunately, all the results quoted so far could be explained without recourse to selective attention. We illustrate this with reference to Babb's experiment. His group trained with both cues and shifted to chains alone was trained with black and white alleys at the first stage and with gray alleys at the second stage, whereas the group trained at both stages on chains alone experienced gray alleys throughout; the poorer performance of the two-cue groups at the second stage could therefore be explained in terms of generalization decrement. To control for this possibility, Sutherland and Andelman (1967) repeated Babb's experiment using a jumping stand and a modified experimental design. All animals were initially trained on a horizontal–vertical discrimination for 70 trials: one group (E) was trained with a black–white cue simultaneously present and also relevant, and two single-cue groups were trained with the black–white cue present but not relevant; one of these groups (C1) was trained with stimuli that differed in brightness within a trial, the other (C2) with stimuli differing in brightness between trials. The experimental design and the results are shown in Table 5.II. After training was completed, all animals were tested on the orientation discrimination with an intermediate value of the brightness cue (gray). Since both the two-cue groups and the single-cue groups were shifted from black and white to gray rectangles after training, there should have been the same amount of generalization decrement for each. As will be seen from Table 5.II, the two-cue group actually did significantly better in tests on orientation than the single-cue groups, which is exactly the opposite result to that predicted by our model.

The cues used in this experiment were of approximately equal difficulty; as will be seen from the results of transfer tests on black–white given to the two-cue group, this cue was only slightly better learned than the orientation cue. Nevertheless, our theory predicts that even with cues of approximately equal difficulty the presence of a second cue will result in less learning about the other. There are three ways in which this anomalous result can be explained.

(1) If, as we argued earlier, the presence of two relevant cues causes a dominant nonrelevant analyzer to decrease in strength faster, then the presence of black–white may have caused the position analyzer to drop more quickly in strength and hence allowed more learning about orientation to show in the tests.

(2) Lovejoy has pointed out that, in the jumping stand, rats may have to learn an external orienting response to the visual stimulus. It may be that the presence of a relevant brightness cue facilitated such learning and hence allowed the two-cue group to learn the orientation cue faster than the single-cue groups.

(3) Sutherland has suggested a third possibility, namely, that analyzers may be hierarchically arranged. This is illustrated in Figure 5.4. If strengthening attention to brightness involved strengthening attention to vision, then

TABLE 5.II

**Experiment of Sutherland and Andelman**[a]
**A. Experimental Design**[b]

| Group | Training shapes: Experiment I + | Experiment I − | Experiment II + | Experiment II − |
|---|---|---|---|---|
| E | VW | HB | VR | HM |
| C1 | VW | HB | VR | HM |
| | VB | HW | VM | HR |
| C2 | VW | HW | VR | HR |
| | VB | HB | VM | HM |
| | **Transfer shapes** | | | |
| | VG | HG | VR | HR |
| | | | VM | HM |

**B. Percentage Correct Responses during Transfer**

| Experiment | Group | Retraining | Transfer |
|---|---|---|---|
| I | E | 99 | 84(91)[c] |
| | C1 | 66 | 57 |
| | C2 | 66 | 68 |
| II | E | 91 | 68(75)[d] |
| | C1 | 83 | 85 |
| | C2 | 82 | 80 |

[a] Data obtained from Sutherland and Andelman (1967).

[b] V designates vertical rectangle; H, horizontal rectangle; W, white; B, black; G, gray; R, rubber; M, metal.

[c] Score on B–W tests.

[d] Score on R–M tests.

the orientation analyzer might actually gain in strength (relative to analyzers for hearing, touch, etc.) by strengthening the brightness analyzer.

To test the third possibility, Sutherland and Andelman ran a second experiment in which once again the orientation cue was always present at both stages but the second cue was a tactile one (rubber versus metal on the jumping platform). The design and results of this experiment are shown in the

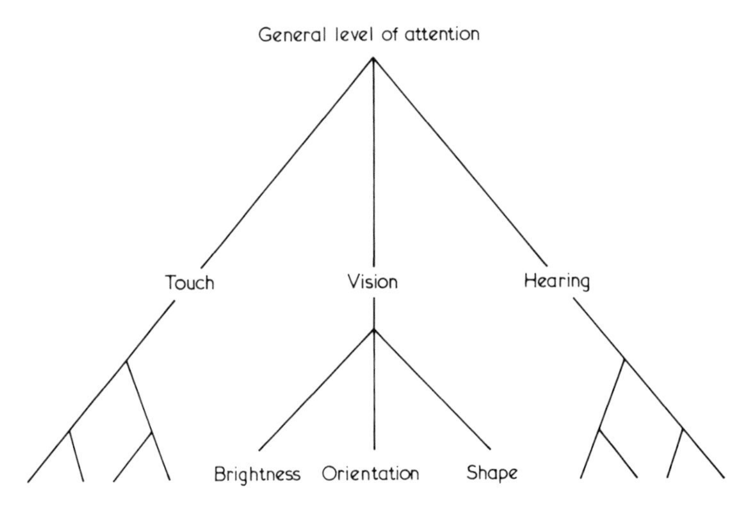


**Fig. 5.4.** A hierarchical arrangement of analyzers.

right-hand columns of Table 5.II. It will be seen that when widely separated cues were used, the predicted result was obtained. The group trained with two cues learned significantly less about the orientation cue than either of the groups trained with only this cue relevant. This result was not due to rubber–metal being a very dominant cue since, when animals in the two-cue group were tested on rubber–metal, they performed very little better than in the tests on orientation and much worse than in the tests given in the first experiment on black–white. It must be pointed out, however, that this result does not discriminate between explanations (2) and (3) above since the presence of rubber–metal as a relevant cue would not facilitate orienting responses to the horizontal–vertical cue and might indeed inhibit them.

Lovejoy and Russell (1967) used the same design as Sutherland and Andelman, but made the relevant cue in the single-cue problem a very weak orientation cue. Whereas Sutherland and Andelman used horizontal and vertical rectangles measuring 10.0 × 2.5 cm, Lovejoy and Russell used rectangles measuring 6.0 × 3.3 cm. All groups received 360 training trials, and in the subsequent transfer tests on orientation the single-cue group only performed at 78% correct. However, when animals were trained with both brightness and orientation relevant, the strong brightness cue completely overshadowed the orientation cue since animals performed at chance on orientation tests.

Finally, D'Amato and Fazzaro (1966) rang an interesting variation on the design of overshadowing experiments. Two monkeys were trained with both color and orientation relevant. On some trials, a center key was illuminated; when this happened the animal could press the center key, which resulted in the full compound stimuli (color plus orientation) being displayed. Equal numbers of four different trial types were given. (1) The full compound stimuli were presented with the center key unlit. (2) The center key alone was lit; the animal had to press it in order to produce the compound stimuli. (3) The orientation stimuli were displayed and the center key was lit; the monkey could either respond immediately to the orientation stimuli or first press the center key to obtain the full compound stimuli. (4) The color stimuli were displayed and the center key was lit; again the animal could either respond immediately to the discriminative stimuli or first press the center key to obtain the compound stimuli. Both animals relied almost entirely on the color cue; they pressed the observing key when the orientation cue was present on its own, but rarely when the color cue was presented alone. One monkey did learn the orientation discrimination, but very slowly, and even after learning it, continued to press the center key when only the orientation cue was present. The other monkey had a strong initial preference for the negative orientation stimulus; after training, this subject was tested with the orientation cue alone and the center key unlit; it responded at a level of 20% correct. This result cannot have been due to the animal's performance being disrupted by seeing an unfamiliar stimulus and is unlikely to be due to generalization decrement. Unfortunately, no control was run to discover how long it would take to learn the orientation discrimination presented completely on its own.

From the experiments reviewed in this section, we can conclude with confidence that the presence of a strong relevant cue retards learning about a weak relevant cue and indeed may on occasions suppress such learning completely. When two cues are of approximately equal strength, the presence of one cue may either facilitate or depress learning about the other. The exact conditions for obtaining one or other effect cannot be specified with any certainty at the moment, and this is clearly a problem that deserves further experimental investigation. The same type of problem arises in connection with the effects of overtraining with one cue on the subsequent learning of a different cue. As we shall see in Chapter 9, overtraining sometimes retards the learning of an extradimensional shift, as we would expect if the main effect of the overtraining is to strengthen the analyzer for the first stage problem. However, there are now several results in the literature showing that overtraining sometimes actually benefits the learning of an extradimensional shift problem. The three possibilities outlined above could explain why this sometimes occurs; the overtraining might weaken a dominant analyzer that is not relevant to the solution of either problem, it might strengthen external observing responses common to both problems, or, if there is a hierarchy of analyzers, it might result in some gain in strength of the analyzer relevant to the solution of the second-stage problem. It must be admitted, however, that both the finding that adding a second cue may sometimes assist learning of another relevant cue and findings showing a beneficial effect of overtraining on extradimensional shift learning represent something of a crux for our own model.

Recently, a further variable that affects the amount of overshadowing has been identified. Wagner, Logan, Haberlandt, and Price (1968) varied the reliability with which an overshadowing cue predicts reinforcement. All their animals received a light and a tone on all acquisition trials. On 50% of presentations, the light was always paired with a reinforcing event: in one condition, tone 1 was paired with reinforcement whenever it occured, and tone 2 was never paired with reinforcement; in a second condition tones 1 and 2 occurred randomly, and each was paired with reinforcement 50% of the time. When the different tones predicted reinforcement with complete reliability (i.e., under the first condition), almost no learning occurred to thelight. When, however, reinforcement was no better predicted by the tones than by the light alone, the light came to exert strong control over behavior. This finding was replicated in three very different situations: with rats given discrete trial bar-press training for food, with rats given CER training, and with rabbits given classical eyelid conditioning. As the authors point out, the fact that the effect is obtained under classical conditioning eliminates the possibility that it is caused by a change in reinforcement schedule brought about by the animals' own pattern of responses, when reward can be accurately predicted by the tone. We have already suggested that analyzers are maximally strengthened when their outputs *consistently* predict different events of importance to the animal and the above results are a nice demonstration of this assumption, since the stronger the tone analyzer becomes the more it should suppress learning about the light analyzer.

The same point is demonstrated in a further experiment by Wagner and Munoz, reported in Wagner (1969b). They investigated the extent to which a light would gain control over bar-pressing for food under three different conditions. In all conditions, the onset of light always signaled food availability, and light onset was always accompanied by the onset of a tone. In Condition 1, the tone sometimes occurred without the light but tone on its own was never followed by food becoming available; hence the tone was a poor predictor of food availability. In Condition 2 the tone never occurred without the light; tone and light now predicted food equally well. In Condition 3, the tone sometimes occurred on its own without the light and was always followed by food becoming available; in this condition the tone predicted food availability better than the light since food was only available when the tone was sounded but it was sometimes available when the light was not on. The amount of control acquired by the light over bar-pressing was greatest in Condition 1, and least in Condition 3. Hence, the more reliably food is predicted by the tone, the less control is acquired by the light. We shall return, in Chapter 7, to a further experiment of a similar sort undertaken by Wagner on generalization; it is this experiment that provides the most direct evidence for our claim that in some circumstances, learning about one cue may be facilitated by training given on a second cue because the learning of the second cue depresses the strength of a third analyzer that is relevant to both problems.

We conclude, therefore, that the presence of a strong relevant cue during acquisition depresses learning about a weaker relevant cue. One cue may be stronger than another either because the values of the cue used are more readily discriminable from one another (or from the background) or because it predicts reinforcement conditions more consistently.

### B. Multiple Cues of Equal Strength

We have been unable to demonstrate unequivocally that when two cues of roughly equal strength are present, the presence of a second cue reduces the amount learned about the first. There is, however, a further prediction that can be drawn from our model about this situation. If a group of animals is trained with two relevant cues for a fixed number of trials, it can be predicted that the more an individual animal learns about one cue, the less it will learn about the other. If the initial strengths of the analyzers are not equal for a given animal, response learning will proceed faster about the outputs from the stronger analyzer than about those from the weaker. Moreover, this will be a runaway process, since the faster the correct responses are attached, the faster the analyzer itself will be strengthened. By Rule 2, as soon as one analyzer becomes stronger than the other, the amount of learning about the response attachments to the weaker analyzer will be reduced; hence the analyzer will only gain strength slowly, thus leading to even faster learning of the correct response attachments to the stronger analyzer.

There are three experiments whose results support this prediction. Reynolds (1961) trained two pigeons to peck when a key was illuminated by a white triangle on a red background and not to peck when the key was illuminated by a white circle on a green background. After training, he tested the birds by illuminating the key on separate trials with a white triangle, a white circle, and with unpatterned red light and green light. During testing one bird gave differential responses only when the two shapes were presented, the other only when the two colors were presented. It seems that one bird had solved the problem by strengthening the color analyzer, the other by strengthening the shape analyzer. Mahut (1954) found that when rats were presented in a jumping stand with two pairs of stimuli, one above the other, they tended to learn about one or the other but not both; since, however, the stimuli were spatially separated, this result could be explained in terms of a peripheral orienting mechanism rather than in terms of inhibitory relationships between analyzers.

To test the prediction further, Sutherland and Holgate (1966) undertook a series of experiments on rats. The animals were trained in a jumping stand with two relevant cues (brightness and orientation) simultaneously present. They were then given transfer tests with each cue presented on its own. The stimuli used in training and testing are shown in Figure 5.5.

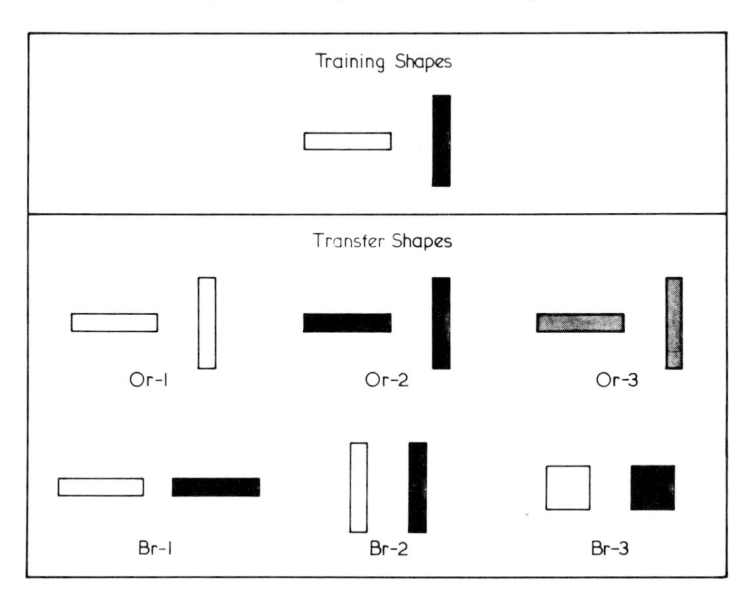


**Fig. 5.5.** Shapes used by Sutherland and Holgate (1966). Or designates orientation; Br, brightness.

It was found that after 60 training trials on the two-cue discrimination, there was a negative correlation between the number of correct choices made by individual animals on the brightness transfer tests and the number made on orientation transfer tests. This bears out the prediction: the more an animal learns during two-cue training about one cue, the less it will learn about the other. Negative correlations were obtained in three different experiments, and although not all the correlations were significant they did not depend upon how the values of the two cues were paired during training; i.e., negative correlations were obtained both when the original stimuli were a white horizontal and a black vertical rectangle and when they were a black horizontal and a white vertical rectangle.

Although the overall results were highly significant, the values of some of the individual correlations obtained were rather low. The reason for this may well be that individual differences in general learning ability, in health, emotionality, and visual acuity tend to make different rats learn at different rates about both cues. To test this possibility, Sutherland and Holgate trained animals on the two single-cue problems, presenting each problem to each subject on alternate trials. In this situation, one analyzer cannot gain in strength at the expense of the other, since the outputs from neither analyzer alone consistently predict success and failure. It was found that there was now a positive correlation between the numbers of trials that individual animals required to reach criterion on each problem, as would be expected if there are individual variations in general learning ability.

In the same series of experiments, it was found that if animals were given 200 training trials instead of 60, their performance on the less preferred cue improved very markedly. Whereas after 60 training trials animals averaged 67% correct on transfer tests with each subject's weaker cue (brightness for some, orientation for others) and 90% correct on their stronger cue, after 200 training trials with two cues they averaged 82% correct on their weaker cue and 89% correct on their stronger cue. This result would again be predicted, since the rules of our model allow for response attachments to the weaker of two analyzers to be strengthened at a time when performance is being predominantly controlled by the stronger analyzer. We have already described in Chapter 4 experiments by Terrace and Schusterman in which the same type of result was obtained in a stronger form. Terrace found that if pigeons are first trained with only one cue relevant, and a second relevant cue added later, the birds may learn the second cue perfectly at a time when performance is being fully controlled by the first cue.

### C. Relationship between Performance on Single and Multiple Cues

When an animal has been trained with several cues present and relevant, we can ask what will be the relationship between its performance on the multiple cues and its performance on the individual cues. Similarly, an animal could be trained independently on the single cues and we could determine the relationship between performance on the cues presented singly and performance on the cues presented together. It has been found that, whether animals are originally trained with multiple cues presented together or whether they are trained with the individual cues, performance with multiple cues bears a lawful relationship to performance on the individual cues.

Data on this problem are available from the experiments by Sutherland and Holgate already described. Subjects were trained with two cues present simultaneously and subsequently tested with each cue on its own; retraining trials with both cues present simultaneously were interposed among these tests. Table 5.III shows the results analyzed in two ways: the first two columns show average percent correct responses on brightness and on orientation tests; Columns (3) and (4) show the data tabulated by the cue on which each animal

TABLE 5.III

**Percentage of Correct Responding during Testing**[a]

| | (1) | (2) | (3) | (4) | (5) | (6) | (7) |
|---|---|---|---|---|---|---|---|
| | | | | | | Predicted | |
| Group | Brightness | Orientation | Weak cue | Strong cue | Both cues | 1 | 2 |
| 1 | 72 | 86 | 67 | 90 | 96 | 94 | 93 |
| 2 | 83 | 88 | 82 | 89 | 97 | 97 | 96 |
| 3 | 86 | 94 | 84 | 95 | 99 | 99 | 99 |
| 4 | 90 | 92 | 87 | 94 | 97 | 99 | 99 |
| 5 | 69 | 83 | 66 | 86 | 93 | 92 | 90 |
| 6 | 72 | 71 | 66 | 77 | 87 | 87 | 84 |
| 7 | 90 | 59 | 57 | 92 | 94 | 94 | 92 |

[a] Data obtained from Sutherland and Holgate (1966).

performed better and worse, irrespective of whether a given subject's better cue was brightness or orientation. The right-hand columns show how well two different formulas predict performance with two cues from the performance on each single cue.

The first formula is

FORMULA 1: $$P = (p_1 \cdot p_2)/[p_1 \cdot p_2 + (1 - p_1)(1 - p_2)],$$

where $P$ is the probability of a correct response with both cues present and $p_1$ and $p_2$ are the probabilities of a correct response with each cue present on its own. One rationale behind formulas of this type is as follows: suppose we have two urns filled with black and white balls, where the proportion of white balls in each urn represents the probability of correct response ($p_1$ and $p_2$) with either cue on its own. Next, assume that when both cues are present, the animal samples from both urns simultaneously. If two white balls are drawn, it responds correctly, if two black are drawn an incorrect response is made. If, however, a black and a white ball are drawn they are replaced and the urns are resampled until two balls of the same color are drawn. The probability of drawing two white balls on the first draw is $p_1 \cdot p_2$ and that of drawing two black balls is $(1 - p_1)(1 - p_2)$. The probability of drawing neither two white nor two black balls on any draw is therefore $1 - p_1 \cdot p_2 - (1 - p_1)(1 - p_2)$: if we call this expression $X$, the probability of a terminal draw consisting of two white balls is $p_1 \cdot p_2(1 + X + X^2 + \cdots)$. Since the series in brackets sums to $1/(1 - X)$, the probability of terminating with a correct choice (drawing two white balls) will be

$$P = (p_1 \cdot p_2)/[p_1 \cdot p_2 + (1 - p_1)(1 - p_2)].$$

In terms of our own model, the formula can be interpreted to mean that when two analyzers are controlling behavior, the animal samples the responses attached to each until the two response attachments agree and then it makes that response. This is a not altogether implausible assumption; however, many samples would have to be made before getting agreement between the responses if the cues were combined in opposition, i.e., if the negative value of one cue were combined with the positive value of the other. This would then be a conflict situation and it is known that in such situations reaction times do increase (Berlyne, 1960). It is clear that this formula involves multiple cue control of behavior: both cues are used simultaneously to determine the response.

A different formula can be derived from the following urn model. Assume again that we have two urns, one for each cue. When only one cue is relevant, we sample from the urn representing that cue; if a white ball is drawn, the correct response is made with probability 1, if a black ball is drawn the correct response is made with probability .5. (The white balls may be regarded as representing conditioned stimulus elements and the black balls as unconditioned elements.) In this model $p_1 = w_1 + .5(1 - w_1)$, where $w_1$ is the proportion of white balls. We now assume that when both cues are present, the animal samples from the first urn. If a white ball is drawn, the correct response is made, if a black ball is drawn the second urn is sampled. If a white ball is now drawn from the second urn, the correct response is made, but if a black ball is drawn, the correct response is made with a probability of .5. On this model, the probability of making a correct response with both cues present is

$$\begin{aligned} P &= w_1 + (1 - w_1)[w_2 + .5(1 - w_2)] \\ &= .5[w_1 + w_2 - (w_1 \cdot w_2) + 1]. \end{aligned}$$

In order to predict $P$ from $p_1$ and $p_2$, we must rewrite the equation in terms of $p_1$ and $p_2$. Since $p_1 = w_1 + .5(1 - w_1)$, $w_1 = 2p_1 - 1$ and similarly, $w_2 = 2p_2 - 1$. Substituting in our equation for $P$, we obtain

FORMULA 2: $$P = 2[p_1 + p_2 - (p_1 \cdot p_2)] - 1.$$

It should be noticed that the final formula is the same whichever urn is sampled first. One disadvantage of this formula is that it cannot be applied in cases where we combine cues in opposition; if an animal is trained to go to a white shape and not to a black and to go to a horizontal rectangle and not to a vertical, the formula cannot be used to yield a prediction about the probability of a subject selecting a white vertical rectangle when presented together with a black horizontal. The reason for this is that it now makes a big difference which urn is sampled first. This can be most clearly seen in the case where $w_1 = 1$ and $w_2 = 0$; if the first urn is sampled first the probability of choosing the stimulus with the positive value on the first dimension is 1, whereas if the second urn is sampled first the probability is 0.

Formula 2 is based on a similar assumption to that underlying Restle's first model, in that it incorporates the notion of two sets of elements, one of which, if sampled, leads to the correct response with probability 1, the other with probability .5. The formula is, however, not directly derivable from Restle's model. A mathematically equivalent formula, stated in different terms, was used by Hara and Warren (1961) to describe data on additivity from an experiment they performed with cats.

Since, according to the above rationale for Formula 2, on many trials both cues are sampled before the response is determined, this formula also involves multiple cue control of behavior; it is a weaker version than Formula 1 since the sampling could be successive rather than simultaneous so that at the moment of choice only one analyzer is determining behavior.

Dawkins (1969a,b) has produced a different model from which Formula 2 may be derived. Consider three stimuli *a*, *b*, and *c*, where *a* is preferred to *b* and *b* to *c*. The problem with which Dawkins was concerned was to predict percent preferences for *a* over *c* from the known percent preference for *a* over *b* and *b* over *c*. He suggests that for a response to any stimulus to occur a threshold must be exceeded by some process varying over time: the threshold for response to *a* will be the lowest and that for *c* the highest. He further assumes that when the variable process exceeds the threshold values for two stimuli both simultaneously present a response to each stimulus is equally likely. From this threshold model Formula 2 can be derived. Dawkins shows that this formula fits results obtained from experiments on the pecking preference of chicks, though the results were almost as well fitted by Formula 1 which he also tested. Many of the results quoted below are comparable to those discussed by Dawkins. Consider animals trained in the Sutherland and Holgate experiments with a horizontal white rectangle (HW) positive and a black vertical rectangle (BV) negative. These stimuli are equivalent to *a* and *c* in Dawkins' formulation, while the stimuli BH or WV are equivalent to *b*. In the transfer tests Or-1 and Br-1 shown in Figure 5.5 each of these intermediate stimuli *b* is paired with the original positive stimulus *a* while in transfer tests Or-2 and Br-2 each is paired with the original negative stimulus *c*. However, it is not possible to predict performance with two cues on Dawkins' formulation from the results with pairs Br-3 and Or-3; since the constant second cue takes a neutral value, there is now nothing to correspond to the intermediate stimulus *b*. As we shall see below, both Formula 1 and Formula

2 give a reasonable fit to our own and others' data on the relationship between performance with two cues simultaneously present and performance with each cue on its own. Dawkins' results are important since they suggest that the same formulas also predict preference scores in experiments where only one cue with three separate values is involved.

The right-hand columns of Table 5.III show how the formulas predict two cue performance from the results with single cues in the Sutherland and Holgate experiment. The formulas were applied to results analyzed by individual subject's weak and strong cues rather than to results analyzed by whether the cue was the orientation or the brightness cue. The formulas also produce good predictions if applied to scores on orientation and brightness cues, but the results are less meaningful if analyzed in this way. We are dealing with averaged scores and individual variability is much higher within type of cue than within the weak or strong cue for each animal, since some animals learned the two-cue discrimination mainly in terms of brightness, others mainly in terms of orientation. It will be seen that Formula 1 and Formula 2 yield very similar predictions, though Formula 1 predicts consistently higher two-cue performance than Formula 2. Formula 1, in fact, yields remarkably accurate predictions of two-cue performance.

Hara and Warren (1961) performed an experiment, the results of which can be analyzed in a similar way. Whereas Sutherland and Holgate trained on two cues and then tested with single cues, Hara and Warren trained on single cues and tested with multiple cues. Moreover, Sutherland and Holgate prevented performance reaching 100% correct by giving a limited number of training trials while Hara and Warren picked values on each dimension such that with single cues, performance would be around either 70 or 80% correct at asymptote. They trained their cats to discriminate between three different single cues: brightness, size, and orientation (horizontal–vertical). All cats were trained on all cues, but during training the stimuli differed along only one dimension on each trial. For each dimension the experimenters determined values of the two stimuli that would yield for individual animals 70 and 80% correct performance at asymptote. For example, the difficulty of the orientation discrimination was manipulated by varying the ratio of horizontal to vertical extent for each shape within a range of 1 : 1 to 1 : 4. Finally, animals were tested with stimuli that differed along one, two, or three dimensions using both values of each dimension. We shall refer to values yielding 70% correct performance as "weak" values, and to those yielding 80% correct as "strong" values. Hara and Warren presented stimuli differing weakly along two dimensions (brightness and size, brightness and orientation, size and orientation), differing strongly on one dimension and weakly on the other (giving six possible combinations), differing strongly on two dimensions, and similarly for all possible combinations of three dimensions. Hara and Warren found that the results of the tests on all the various combination cues were well predicted from performance levels with the single cues on Formula 2. Formula 1 yields predictions that tend to be slightly too high. McGonigle (1967) recently repeated Hara and Warren's experiments using rats and obtained almost identical results.

Whether or not either formula is correct in the sense that the rationale behind one or other formula corresponds to the mechanism at work in the animal's brain, these experiments do establish that there is a lawful relationship between performance on multiple cues and performance on the component single cues.

It is clear that models assuming that within a single trial behavior is controlled by only one analyzer can also predict that performance will be better with two combined cues than with either cue on its own. Let $pA_1$ be the probability of control by the $i$th analyzer, let $a_1$ and $a_2$ be the strengths of two relevant analyzers for single cues, and let $a_{irr}$ be the sum of the strengths of the irrelevant analyzers, then on single-cue trials, the probability of control by the relevant analyzer will be either

$$pA_1 = a_1/(a_1 + a_{irr}) \qquad \text{or} \qquad pA_2 = a_2/(a_2 + a_{irr}).$$

If we assume that the strength of irrelevant cues does not change when both relevant cues are present then the probability of control by one or other relevant cue on such trials would be

$$pA_{1+2} = (a_1 + a_2)/(a_1 + a_2 + a_{irr}),$$

which is greater than the probability of control by a single relevant analyzer when only one relevant cue is present. Assuming that the probability of a correct response is .5 when performance is controlled by an irrelevant analyzer and 1.0 when performance is controlled by a relevant analyzer, Turner (1968) derives the following formula to predict the probability of a correct response with both cues present from the probabilities with each present on its own,

FORMULA 3: $P = (2p_1 + 2p_2 - 3p_1p_2 - 1)/(p_1 + p_2 - 2p_1p_2)$.

When this formula is used to predict two-cue performance from the results with single cues in the experiments of Holgate and Sutherland, Hara and Warren (1961) and McGonigle (1967), the predictions are consistently too low, though the actual difference between predicted and obtained results is fairly small (about 2–5%). This is further evidence in favor of multiple-cue control.

Lovejoy (1968) produced the following argument in an attempt to design a test for single or multiple cue control. Consider animals trained with two cues and tested on two pairs of stimuli, one pair containing the positive values of the second cue (cf. pairs Or-1 and Br-1 in Figure 5.5), the other the negative values (cf. pairs Or-2 and Br-2). He argued that, if only one cue can control behavior, then where both stimuli have the positive value of the constant cue, animals switching in the analyzer for the constant cue should select whichever stimuli they happen to be looking at and on such trials will score 50% correct; on trials on which they switch in the other analyzer they will perform correctly (assuming perfect response learning). Hence performance in this situation should not average more than 75% correct provided the single cue is not stronger than the constant cue. When, however, the stimuli have the negative values of the constant cue, an animal switching in that analyzer will be unable to respond to that cue; it should therefore switch in the other analyzer and always respond correctly. Lovejoy (1968) did, in fact, find that performance was better when the constant cue took negative values than when it took positive values, but two of his animals performed at very much better than 75% correct when the weaker of two single cues was present and the stimuli took the positive value of the constant cue. However, Sutherland and Holgate (1966) found that it made no difference whether the constant cue took positive or negative values. More recently, Turner (1968) has repeated Lovejoy's experiment and shown that under some conditions, rats may perform at very much higher than 75% when the weaker cue is present and the stimuli both take the positive values of the stronger cue.

The failure of Lovejoy's prediction from a model positing single-cue control does not, in this instance, disprove single-cue control. If animals sample both stimuli before responding, and only switch in an analyzer yielding different outputs for the two stimuli, then a theory based on single-cue control can predict performance levels much higher than 75% in the situation considered above.

One basic problem in interpreting the results given in this chapter on our own theory is the difficulty of knowing what proportion of mistakes made both with multiple cues present and with a single cue present is due to failure to learn fully the correct response attachments to a given analyzer and what proportion is due to failure to strengthen sufficiently the correct analyzer. In the formalized version of our model we, in fact, assume that once the relevant analyzer strength exceeds all others by a critical amount, all mistakes are due to failure to learn fully the correct response attachments to that analyzer. Where more than one analyzer is strong and relevant, we make use of Formula 1 to predict the probability of correct responses in the multiple-cue situation from performance in the single-cue situation. Our main reason for preferring Formula 1 is that, unlike Formula 2, it can be used to predict results when two cues are set in opposition to one another. It is, therefore, worth looking at such experiments to see if the formula still applies.

Nissen and Jenkins (1943) trained chimpanzees on a two-cue discrimination involving brightness and size. The animals were then given tests both on the two single cues presented separately and with the two cues paired in opposition to one another; e.g., if an animal had been trained with black and large positive and white and small negative, it was given an opposition test in which a small black stimulus was paired with a large white. Formula 1 predicts reasonably well the results obtained with cues in opposition from the scores on single cues, provided averaged scores are used; animals performed with an accuracy of 76.5% on size, and 81.1% on brightness and they selected the two-cue shape with the positive brightness values and the negative size value on 58.7% of trials as compared with 56.9% predicted from the formula. However, the subjects showed a great deal of individual variability and many of the individual results cannot be predicted by the formula. For example, one subject scored 96% correct with brightness alone and 60% correct with size alone, but chose the stimulus with the positive brightness value on 98% of trials in two-cue tests with the cues in opposition. Out of eight subjects, three in fact gave this type of result; i.e., when the cues were combined in opposition their scores did not show a compromise between the two cues since the animals responded as though only the stronger of the two cues was present. The results of these three subjects suggest a strong form of noncontinuity at work.

Hara and Warren (1961), and McGonigle (1967), in the experiments mentioned above, also ran opposition tests. In neither experiment could the results be predicted from the two formulas we have proposed. Both cats and rats tended to follow the positive value of the brightness cue when it was paired in opposition to either size or form. When form and size cues of the same strength were opposed to one another, animals chose the positive value of each cue on about 50% of trials as predicted by Formula 1. It is difficult to explain why brightness should predominate over form or size in opposition tests when the level of performance with each cue on its own was identical. Hara and Warren suggest that the brightness cue might to some extent mask the form and size cues, but if this happened in opposition tests, it should also have happened when cues were combined additively and a different result should have been obtained with brightness added to either size or form and with size and form added together. Moreover, McGonigle found that when brightness alone was present latencies were lower than when either size or form alone was present. Although, therefore, the correct explanation of the result is in doubt, the phenomenon would appear to necessitate the postulation of some form of selective attention: all three cues exert exactly the same amount of control over behavior when presented on their own, yet the animal's behavior is selectively controlled by brightness when this cue is put in opposition to either of the other two cues.

We have already seen that one of the difficulties in comparing performance on single cues with performance when two or more cues are present is the problem of deciding how far errors are to be attributed to failure of the relevant analyzer to control behavior and how far to the response attachments not having been fully established. Unfortunately, there are two further problems of interpretation. The first problem has already been referred to in our discussion of overshadowing: if an animal learns a problem with two cues present, it might perform poorly on transfer tests with one cue, owing to generalization decrement. The second problem is that in learning a two-cue problem animals may to a greater or lesser extent learn to respond not merely to the components but to the compound cue. These issues will be discussed in the next two sections. It is, however, worth remarking that none of the formulas mentioned above takes any account of either generalization decrement or of the possibility of compound analyzers being formed and it would therefore be unwise to take any of these formulas too seriously.

### D. Generalization Decrement

We argued in Chapter 2 that animals must learn not merely to alter the strengths of analyzers, but to condition the strength of a given analyzer to outputs from other analyzers that identify the situation in which an animal finds itself. Now, if an animal is trained with black–white and horizontal–vertical simultaneously present and relevant, the presence of horizontal and vertical rectangles could well form part of the stimulus situation to which an increase in strength of the brightness analyzer gets conditioned. Hence, when the animal is subsequently tested on the brightness cue with the orientation cue omitted, some drop in performance may be caused because one cue to which the brightness analyzer has been conditioned is no longer present. The existence of this possibility makes it more difficult than ever to propose with any certainty a formal model accounting for the relationship between performance on two cues together and on each cue on its own.

It has been known for some time that if animals are trained to discriminate between two different black (or white) shapes, they rarely perform perfectly in transfer tests with the same shapes presented in the opposite brightness value (for a review, see Sutherland 1961a). It may be that the shape analyzer has been conditioned to some extent to the constant brightness of the two stimuli. That animals can learn to process shape independently of brightness is nicely demonstrated in an old and little quoted experiment by Fields (1932). He trained rats to discriminate between a white upright triangle and a white circle, each on a black background. Animals failed to show any transfer when figure–ground brightnesses were reversed. Moreover, they failed to transfer when the triangle was presented in the original brightness but a new orientation.

Fields next retrained the animals to discriminate between a black upright triangle and a black circle. He also retrained then to discriminate between a white inverted triangle and a white circle. In the final stage of the experiment, he obtained immediate transfer when his animals were tested with black inverted triangles presented with the black circle. This result suggests that the original failure to transfer to shapes when their brightness was changed may have been caused by the shape analyzer having been conditioned to the brightness cue. Once the animals have been trained to ignore brightness with one orientation of the triangle, they transfer when the brightness value of the triangle in other orientations is changed. Again, it is difficult to account for this result except in terms of selective attention.

Sutherland (1966b) performed an experiment in which rats were trained on a discrimination with seven relevant cues. When tested with each of the seven cues presented on its own, animals performed above chance on only one or two of the single cues. Moreover, when Formula 1 was applied to the results with single cues, performance on all seven presented conjointly was grossly underestimated. It seems likely that although animals had not learned to attach the correct responses to the appropriate values of most of the cues, they were using the analyzers for these cues to switch in the analyzers detecting the one or two cues that they gave evidence of having learned. Hence, when the cue which had been learned was presented during testing, the analyzer for that cue may have been insufficiently switched in since the situational cues provided by the other relevant cues were now absent.

## E. Compound Learning

The results of the two experiments just quoted could equally well be interpreted as showing that animals, under some circumstances, learn to respond to compounds, not components; if this is the explanation, the Fields experiments at least show that animals can be trained to respond only to a component and that this training transfers to a situation in which the value of the component is changed.

By "component learning," we mean learning to make responses to each single cue present; by "compound learning" we mean learning to respond to the combined values of all cues presented together. Figure 5.6a illustrates how compound learning can be conceptualized in terms of our model. Consider an animal being trained to jump to a white horizontal rectangle and to avoid a black vertical rectangle. Component learning would occur if the response of jumping is seperately attached to an output signaling white from the brightness analyzer and one signaling horizontal from the orientation analyzer. Compound learning takes place if the outputs from the two analyzers are mixed in a higher level analyzer before response attachments are formed;

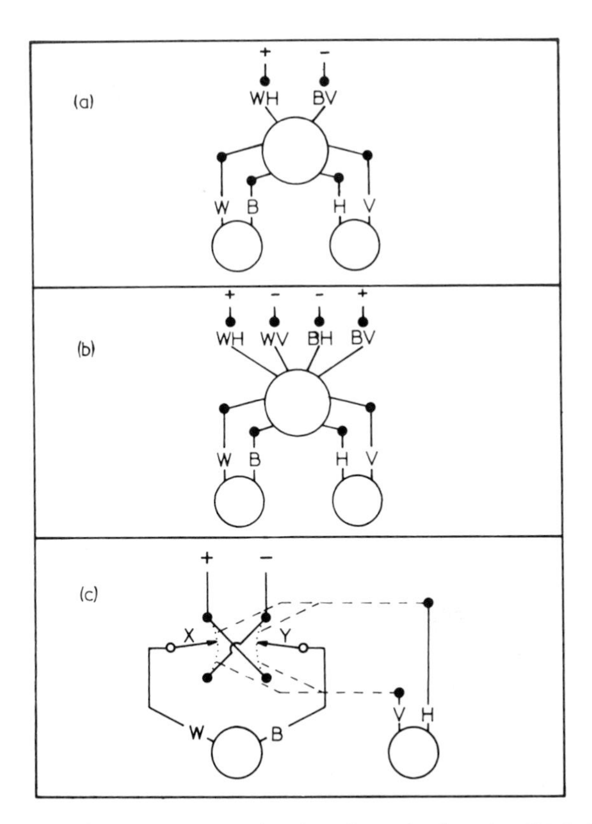


**Fig. 5.6.** Mechanisms for compound and configuration learning. W designates white; B, black; H, horizontal; V, vertical; X and Y, switches controlled by outputs V and W.

the animal would now learn to attach the response of jumping to the output from the higher level analyzer signaling white horizontal. It is possible, and indeed likely, that both types of learning occur. This raises the question of how far the drop in performance when animals are trained on two cues and tested on one or other component is due to the omission of the compound cue rather than to the omission of one component.

Some of the data from Sutherland and Holgate (1966) are relevant to this issue. Referring back to Figure 5.5, all our groups were tested with the transfer pairs Or-1 and 2 and Br-1 and 2, except for Group 4, and this group was tested with pairs Or-3 and Br-3. Now, as we have already seen, for an animal trained with a white horizontal rectangle positive and a black vertical rectangle negative, at least one member of the stimulus pairs Br-1 and 2, and Or-1 and 2 is one of the original compound stimuli. If, however, pairs Br-3 and Or-3 are used, none of the compounds appears as a member of the single-cue test pairs. If rats have learned compounds we might, therefore, expect that single-cue performance will be worse in the group tested with Or-3 and Br-3 than in groups tested with Or-1 and 2 and Br-1 and 2. In fact, inspection of Table 5.III shows that Group 4 actually performed better on single-cue tests than any other group, and it is the only group for which application of Formula 1 to the results of single-cue transfer tests predicts higher performance on two-cue trials than was actually obtained. The evidence suggests, therefore, that in this experiment, rats learned components not compounds. House and Zeaman (1963) showed that retardate children do learn something about compounds: they trained children to select a white triangle paired with a black triangle. With this design, component learning gives no basis for correct choice when a white triangle is paired with a white circle. Nevertheless, in this type of situation, children selected their positive stimulus more often than chance at the second stage; showing that they had learned a compound.

Williams (1967a) carried out an experiment with the same design using pigeons. He found that birds trained to select a red square presented with a red cross continued to select a red square when it was presented with a blue square. He also found that when two examples of the negative shape were presented, one in the color used in original training (red) and the other blue, birds had a smaller but still significant preference for the original shape; i.e., they selected their original compound negative stimulus. The latter finding suggests that birds may not have learned to respond to the compound stimulus since if they had learned to approach the positive compound they should also have learned to avoid the negative. The pigeons appear to have learned to associate both the positive shape and the constant irrelevant color with reward. That some learning occurred about color is not surprising since color tends to be more a dominant cue than shape for pigeons, judging by the number of trials required to train birds on shape and color discriminations.

Perhaps the best proof that animals may learn compounds comes from a recent experiment by Guth (1967). He trained rats to press a bar for water reward when a stimulus was on (S+), and not press when the stimulus was not present (S−). One group of animals was trained with light and tone simultaneously presented as S+, another was also trained to respond to both stimuli but each stimulus was presented on its own in a randomized order. All subjects received the same total number of stimulus presentations in training, and were then given a fixed number of extinction sessions, but some subjects were extinguished only on the light, others only on the tone. At the third and final stage of the experiment, the animals were tested with all three stimulus combinations (light and tone together, tone alone, and light alone). When rats trained on the compound cue were tested, bar-pressing rates were very much higher on that cue than on the component cue not used during extinction. It could be argued that this occurred because extinction to the component cue had not been complete and therefore when the extinction cue was added to the other cue, its presence as a component increased the rate of responding to the combined cues. This is very implausible since bar-pressing rates to the extinction cue presented on its own in tests were very low indeed. Moreover, the results from the subjects trained on the component cues completely rule out this possibility: when they were presented with the compound cue in tests, their response rate was actually lower than on the component cue not used in extinction. This experiment, therefore, provides clear-cut evidence that animals may learn compounds as well as components.

### F. Conditional Discrimination Learning

We conclude this chapter by reviewing briefly a further source of evidence showing that animals can learn to use more than one cue to control their behavior on a single trial. It is well known that animals can solve conditional discrimination problems and, indeed we have already mentioned one such problem in Chapter 4. If rats are trained to go left when shown two black squares and to go right when shown two white squares, they must make use of both a position analyzer and a brightness analyzer, in order to solve the problem. Lashley (1938) was, so far as we are aware, the first investigator to demonstrate that rats could solve a conditional discrimination.

In what follows, we shall take as a paradigm case that of a rat trained to select the white rectangle when presented with two horizontal rectangles differing in brightness, and to select the black rectangle when presented with two vertical rectangles differing in brightness. Such a problem could be solved in two ways which are diagramed in Figures 5.6b and c. The outputs from the first-stage analyzers could be mixed in a higher level analyzer: four different outputs will be yielded, and the animal could learn simply to approach when outputs WH and BV occur and to avoid when outputs WV and BH occur. This type of solution is shown diagrammatically in Figure 5.6b. A second possibility is shown in Figure 5.6c. Both approach and avoidance response attachments could be formed to each output from the brightness analyzer, and the outputs from the orientation analyzer could be used to throw a switch that would determine which response attachment would be used on a given trial. Thus, in our example, when the output H occurred from the orientation analyzer, the output W would be connected to the approach response and the output B to the avoidance response, whereas when the output V occurred, a switch would be thrown causing the response to W and

B to be reversed. It is difficult to distinguish between these two possibilities by behavioral experiments but they clearly represent two different information processing mechanisms each capable of mediating conditional discriminations.

North, Maller, and Hughes (1958) and North and Lang (1961) performed two experiments, the results of which help to decide which method of solution the rat employs. They trained rats on a conditional discrimination; the relevant dimensions were different in the two experiments, and differed from those we have used here for illustrative purposes. For simplicity, however, we shall continue to use the same lettering to indicate values on the two dimensions with one dimension taking the values W and B, the other H and V. Animals are again trained to select W in the presence of H, and B in the presence of V. After training, tests were given on which the stimuli were paired in a different way: the positive member of one of the original pairs was presented with the negative member of the other pair; i.e., WH was paired with WV, and BV with BH.

On the basis of our second hypothesis (Figure 5.6c), animals would be expected to perform practically at chance in these tests. In the first test pair, they may detect either an H or a V. If they pick up the H, a switch will be thrown on the first analyzer connecting W to approach but since both stimuli contain a W there is no way to choose between them; similarly, if a V is detected, W will be connected to avoid, but since again W is present on both sides, this cannot form a basis for choice. Similar reasoning can be applied to the second pair of test stimuli.

On the first model, however, which postulates the learning of compounds, animals should continue to perform well on test trials; they have learned the response ⟨Approach WH and avoid WV⟩ (see Figure 5.6b) so that they should continue to select WH when the first test pair is presented and to select BV when the second test pair is presented.

In both experiments, animals did, in fact, perform well on the test pairs, and this suggests that in this instance rats had learned compounds and had solved the problem in accordance with Figure 5.6a.

It should be noted that North's own theoretical analysis differs from ours, and in particular, he uses the words component and compound in a different sense. North's experiments were, in fact, directed to a different issue, and they do not conclusively settle the problem in which we are here interested. It could be argued that when WH and WV are presented, even if rats learn according to the second model, they will still tend to select WH; they will only receive the output H when they are looking at the WH stimulus and, therefore, they should tend to approach that rather than WV. If they look first at WV, they will be unable to find a B and hence will not jump until H is detected. This argument is rather tortuous and in the absence of other evidence, these experiments at least suggest that conditional problems may be solved by the rat in terms of compounds.

# III. Conclusions

In the early part of this chapter, we reviewed evidence on the speed of learning with multiple cues. It was shown that both our own two-process model and also a two-process model that allows control only by a single cue on any one trial can account for more rapid acquisition with two cues than with one. The evidence on the way position habits are broken when trials with two cues and trials with a single cue are intermingled favors a model of our own type, allowing for multiple-cue control. The findings cited at the end of the chapter on compound learning and on conditional discrimination learning necessitate some degree of multiple-cue control.

Evidence was presented that the presence of a strong cue reduces the amount learned over a fixed number of trials about a weaker cue, and this result is clearly predicted from our model. However, the presence of one cue of equal strength to a second should also retard learning about the second cue and we have been unable to demonstrate unequivocally that this is true. It is possible that when the two cues are in the same modality, the learning of an observing response to one cue may facilitate learning about the other; in addition, analyzers may be hierarchically arranged so that the strengthening of an analyzer within a modality may strengthen (to a lesser extent) other analyzers within that modality. Evidence was presented that, at least in some circumstances, when there are two relevant cues of roughly equal strength, there is an inverse relationship between the amount that individual animals learn about each cue; once again this is predicted by the model.

We discussed at some length ways in which performance with two relevant cues might be related to performance with either cue presented on its own. Although the results of many experiments on this problem can be fitted by the application of either of two rationally derived formulas, other results cannot be so fitted; moreover, since none of the formulas take into account generalization decrement nor the possibility of compound learning, it is not possible to have much confidence in their usefulness.

Although most of the results given in this chapter are consistent with a two-process model, they do not necessitate one. Many of the results could be interpreted without postulating that selective attention was a separate process from response learning: the results on overshadowing and on the negative correlation between the amount learned about two simultaneously presented cues are an exception to this, in that it is difficult for any single-process model to account for these findings. In Chapter 6 we discuss experiments on the way in which learning a discrimination based on one cue affects transfer performance when the task is changed. The results of many experiments of this type are not merely consistent with a two-process theory, they appear to necessitate one, insofar as it is difficult or impossible to account for them if the only learning process is response learning.

# CHAPTER 6

# Learning to Switch-In Analyzers

## I. The Analysis of Attentional Learning

The evidence reviewed in Chapters 4 and 5 implies that any complete account of animal discrimination learning must incorporate, in some form or other, the concept of selective attention. What is learned about a particular cue is determined not only by the salience of, and reinforcement associated with, that cue, but also by the presence of other cues in the situation and the salience and reinforcement history of those cues. Animals do not necessarily utilize all analyzers simultaneously, and the variables determining the relative strengths of different analyzers are open to empirical investigation.

The next step follows very simply from acceptance of this position and forms the basis for much of the further argument in this book. If an animal does not automatically utilize all analyzers in a given situation, it is possible that it will not, at the outset of a discrimination problem, consistently switch-in the appropriate or relevant analyzer for solving that problem. In order to solve the discrimination (i.e., to perform at a near 100% level of accuracy) the animal's choice behavior must be predominantly controlled by the relevant analyzer on all trials; thus, solution of a discrimination implies that the strength of the relevant analyzer (or analyzers) may have to increase from some low level to some high level. It is reasonable to assume that this increase in strength is due to learning and Rules 2 and 3 of our system are rules governing how animals learn to alter the strengths of different analyzers. (It should be noted that this argument does not depend on any extreme assumptions about attention. We have seen that Lashley and Krechevsky were wrong if they supposed that rats never attend to the relevant cue of a visual discrimination problem at the outset of learning; but even if rats switch in the relevant analyzer on some trials, this will not be sufficient to allow accurate performance, and it remains true that learning must ensure that the relevant analyzer has a high strength.)

This then is the rationale for the belief that learning should be regarded as a two-stage process: animals must learn both which features of the experimental situation to attend to and what responses to make. At this point, it becomes clear that there is a further sense in which the type of theory which we are putting forward is a compromise between the original continuity and noncontinuity theories. This compromise has been expressed by Lawrence, who was largely responsible for shifting the emphasis in animal discrimination learning research away from the presolution period and incidental learning type of experiment towards experiments aimed at separating out the two postulated stages of learning. Lawrence (1950) wrote that his own two-stage account of discrimination learning

> . . . is in many ways analogous to Lashley's theory. Both formulations conceive of the relationship between the stimulus situation and the instrumental responses as depending on at least two variables . . . . The Lashley type or noncontinuity theory, however, . . . tends to emphasise the all-or-none characteristic of shifts in attention . . . . The present formulation assumes that the changes . . . are of a gradual and continuous nature [p. 186].

The break between Lashley's and Krechevsky's noncontinuity theory and that of Lawrence and of ourselves is that we are attempting to analyze what is involved in learning to attend to a given cue and to discover the rules governing attentional learning (which may or may not be the same as those governing response learning). Krechevsky (1932a), on the other hand, was inclined to say that the learning of a discrimination problem "consists of changing from one systematic, generalized, purposive way of behaving to another and another until the problem is solved [p. 532]," but his attempts to analyze what was involved in the change from one hypothesis to another were such as to arouse little but opposition in the context of American behaviorist psychology:

> When we say that an individual is working on an "hypothesis" we at once imply that he is behaving in a purposive, "if–then" manner. . . . If the hypothesis does not lead to certain expected results it is soon dropped. "*If* this attempt is correct, *then* I should get such and such results, if I do not get such and such results then I must change my behavior". . . . The animal's behavior does meet this requirement of the term "hypothesis". . . . The animal systematically enters every lighted alley, which attempt, he discovers does not lead to the desired effects. . . . He *changes* his behavior and tries another response and another until he finally gets one which *is* verified by the efficient attainment of the goal-object [*op. cit.* pp. 529–530].

This purposivist interpretation, along with the notion of insightful reorganization which would lead abruptly from one attempted solution to another, came to be regarded as one of the hallmarks of noncontinuity theory, and little further analysis of attentional learning was undertaken by noncontinuity theorists. Lashley (1942) was interested more in perceptual problems—what constituted the effective stimulus for an animal in the experimental situation—than in problems of learning, and although clearly unenthusiastic about the anthropomorphic overtones of the theory, did little to provide an alternative analysis. "I am willing," he wrote of what Spence had called Lashley's theory of discrimination learning, "to accept the responsibilities of paternity, though from a certain purposive cast in the infant's features I suspect that I am cuckold [p. 241]." It was this purposive cast, however, together with the insistence that attention was confined to only one feature at a time, that led their opponents to ignore Lashley's and Krechevsky's immensely important insights.

Lawrence was the first to suggest that the existence of selective attention does not necessarily imply that shifts in attention are abrupt or unanalyzable. He was also the first to follow up this insight by performing the appropriate experiments.

## II. The Acquired Distinctiveness of Cues

If successful performance on a discrimination problem depends upon both learning to switch-in the relevant analyzer and learning to attach appropriate responses to the outputs of the analyzer, then speed of learning a given problem will depend upon the amount of each type of learning that is needed, and therefore, on the amount of each that has already occurred and will transfer to the present situation. *Ex hypothesi*, animals cannot learn to attach responses to the outputs of an analyzer unless that analyzer is switched in, but there is no reason why they should not learn to switch-in a brightness analyzer without learning the response attachments necessary for solution of a normal simultaneous brightness discrimination. Hence, if an animal is trained to switch-in a given analyzer by learning to make one set of responses to stimuli varying along the dimension in question, its learning of a subsequent problem involving the same relevant dimension but different responses should be facilitated.

### A. Transfer between Simultaneous and Successive Discriminations

#### 1. *Lawrence's Experiments*

The first experiments designed to investigate attentional transfer in animals were those of Lawrence (1949, 1950). The design of his experiments rested on the premise that the response requirements of simultaneous and successive conditional discrimination problems are sufficiently different that transfer from one to the other, based on the responses learned in the first, would be impossible. In the first experiment (Lawrence, 1949), rats were initially trained on a simultaneous discrimination with a single cue relevant. Three separate groups were trained: one with brightness relevant, one with the width of the two goal alleys the relevant cue, the third with the texture of the floor as the relevant cue. To simplify the exposition, we need only consider a single testing situation. All three of the above groups were, in Stage 2 of the experiment, trained on a *successive* discrimination: to turn right if both goal arms were black, and left if both goal arms were white, the width of the alleys being irrelevant to correct solution, and there being no difference in the texture of the floor of the positive and negative arms. Thus the first group had, in Stage 1, been trained to switch-in a brightness analyzer, which was also appropriate for solution of the second problem; the second group had been trained to switch-in an analyzer irrelevant to the solution of the second problem, while the third group had been trained to switch-in an analyzer that did not detect any cue in the second stage of the experiment. Lawrence's argument was that this was the only possible basis for transfer, that, for example, animals that had learned to approach black and avoid white could derive no benefit from such *response* learning when required to go right to two black stimuli and left to two white stimuli. Since significant transfer did occur, it must have been due to the transfer of analyzers; although there was no difference in the Stage 2 performance of the second and third groups, both learned the successive problem more slowly than the first group, the one initially trained with brightness relevant. In a third stage of the experiment, animals were shifted back to a simultaneous discrimination; for one group, the cue relevant in this Stage 3 problem had been relevant in the successive problem of Stage 2, for another group it had been irrelevant. Once again, there was significant transfer between successive and simultaneous problems, apparently based on the nature of the relevant cue.

Two more recent studies have established the reliability of these results in slightly different situations, while Lawrence himself, in a second experiment, devised a number of different tests for the presence of transfer between the two types of problem. Mackintosh and Holgate (1967) trained rats either on a successive brightness discrimination (with orientation irrelevant), or on a successive orientation discrimination (with brightness irrelevant). In Stage 2 of the experiment, both groups were trained on a simultaneous brightness probability discrimination, with one alternative rewarded on 75% of trials, and the other alternative rewarded on the remaining 25% of trials. A probabilistic reinforcement schedule of this sort usually results in very slow learning, to an asymptote that is usually less than 100% choice of the more favorable alternative (see Chapter 11). In this particular experiment, Stage 2 continued for 500 trials; throughout this time, animals that were pretrained with brightness relevant performed significantly more accurately than those pretrained with brightness irrelevant.

Mumma and Warren (1968) have replicated Lawrence's results with cats. In their first study, an experimental group pretrained on a successive discrimination, learned a subsequent simultaneous discrimination significantly faster than a naive control group. In their second study, they used two groups of cats with extensive prior experience on visual discriminations (with the exception of size problems). Both groups were tested on a successive size discrimination after the experimental group (but not the control group) had been pretrained on a simultaneous size discrimination. Again, the experimental group learned the test problem significantly faster than the controls.

Finally, in his second experiment on the acquired distinctiveness of cues, Lawrence (1950) again utilized simultaneous and successive discriminations, but this time, he used rather different measures of transfer. The rather complex design of this experiment is illustrated in Table 6.I. In the first stage of

TABLE 6.I

**Design of Experiment by Lawrence**[a, b]

| Groups | Stage 1, Successive discrimination | Stage 2, Two-cue simultaneous discrimination | Stage 3, Opposition test | Stage 4, Single-cue simultaneous training |
|---|---|---|---|---|
| Brightness | B+ B−<br>W− W+<br>(C/NoC irrelevant) | BC+, WNoC− | BNoC, WC | B+, W− |
| Chains | C+ C−<br>NoC− NoC+<br>(B/W irrelevant) | | | C+, NoC− |

[a] Data obtained from Lawrence (1950).
[b] B designates black; W, white; C, chains; NoC, no chains.

the experiment, animals were trained on a successive discrimination with one relevant and one irrelevant cue. The cues were the brightness of the goal boxes and the presence or absence of chains at the entrance to the goal boxes; half of the subjects were trained with brightness relevant and chains irrelevant, while the other half were trained with chains relevant and brightness irrelevant. In the second stage of the experiment, all animals were trained on a simultaneous discrimination with both cues relevant. Such a problem can be solved in terms of either cue alone or both together, and the prediction was that animals would be more likely to solve it in terms of the cue that had been relevant for them in the first stage of the experiment, rather than in terms of the cue that had been irrelevant. In other words, if they had learned to switch-in one analyzer and not another, it would be easier to attach new responses to the first analyzer rather than to the second. Lawrence then gave a number of separate tests to discover which cue had been more used for solution of the second problem. In the third stage of the experiment, animals were given an *opposition* test; e.g., for a subject whose simultaneous discrimination training had involved learning to approach a black goal box with chains and to avoid a white goal box without chains, an opposition test consists of dissociating the two cues, i.e., presenting a black goal box *without* chains versus a white goal box *with* chains. These subjects showed a significant preference for the cue that was relevant in Stage 1. In the fourth stage of the experiment the subjects were divided into four subgroups, each of which learned a simultaneous discrimination with a single relevant cue; for one group this was their Stage 1 relevant cue, for a second it was their Stage 1 irrelevant cue. The first group learned this problem significantly faster than the second. Both this and the opposition test show clearly that the combined problem had been solved predominantly in terms of the cue to which subjects had originally been trained to attend.[1] The third and fourth subgroups were trained on a single-cue simultaneous discrimination, but with the sign of the stimuli reversed. The results of this test will be discussed in Chapter 8.

These experiments establish the generality of both positive and negative transfer effects between successive and simultaneous discrimination problems. Although, as we discuss below, and as Lawrence himself recognized, questions can be raised about the precise nature of the transfer being demonstrated, it does not seem easy to dispute the conclusion that such transfer depends upon subjects having learned to attend to or ignore particular cues. If this is true, the results are of fundamental importance for theories of discrimination learning. Indeed, they establish the same basic point as do experiments on incidental learning or blocking: that is, the amount learned about a given cue in a given number of trials (rate of learning that cue) depends not only

[1] Unfortunately, as he recognized himself, Lawrence did not, in this experiment, use two cues of initially equal difficulty; brightness was clearly a more dominant cue for the rats than was the presence or absence of chains (see also Babb, 1957). The result of this imbalance was that, although animals pretrained to attend to brightness showed an overwhelming preference for brightness over chains on the tests, those pretrained to attend to chains did not, in fact, show a preference for chains over brightness. This last fact has been seized upon by those anxious to dispute the validity of Lawrence's results (e.g., Kelleher, 1956); but the criticism is not valid, since, when the results for chains and brightness groups are combined, an overall preference for the former relevant cue emerges as statistically significant.

on such static variables as stimulus–response-reinforcement contiguity, but also on the subject's prior experience with that and other cues.

### 2. *Criticisms of Lawrence's Experiments*

These conclusions have not, however, gone undisputed. All the experiments just discussed have been predicated on the assumption that the instrumental responses required in a simultaneous brightness discrimination (e.g., ⟨approach black, avoid white⟩) are different from those required in a successive brightness discrimination (e.g., ⟨go right to black, left to white⟩). In a recent, important paper, Siegel (1967) argued that this assumption may not be justified, and supported his argument with the results of an ingenious experiment. Siegel's argument is most easily understood with reference to Figure 6.1.

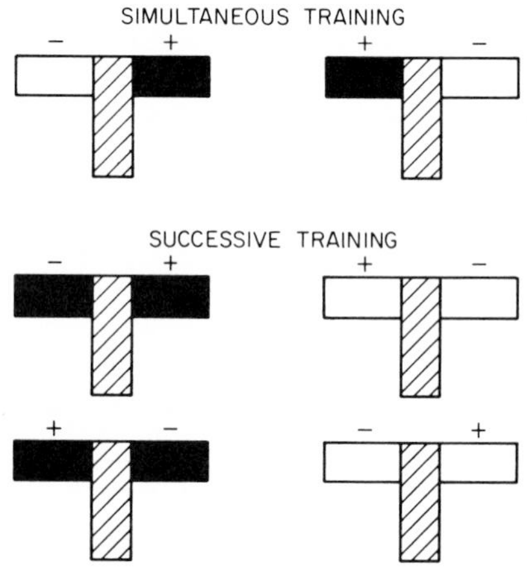


**Fig. 6.1.** Stimulus configurations and reinforcement assignments for a simultaneous brightness discrimination (top panel), and two possible arrangements for a successive brightness discrimination (lower panels).

In the top panel are shown the two stimulus presentations of a simultaneous black–white discrimination in a T maze: on 50% of trials, black is on the left and white on the right; on the remaining trials, black is on the right and white on the left. Let us suppose that instead of simply learning to approach black and avoid white irrespective of their positions, a rat solves this problem in the following manner: on every trial, it looks at one alternative only, this alternative always being on the same side, say, right. On 50% of trials the stimulus on that side is black; to solve the problem the animal must enter the goal arm, i.e., approach black. On the remaining trials, the stimulus on that side is white; to solve the problem, the animal must retrace and enter the other goal arm, i.e., avoid white. If an animal were to adopt this orienting strategy during simultaneous discrimination training, then one of two things might happen when the problem is changed to a successive brightness discrimination. If the reinforcement contingencies of the successive problem are such that the right-hand alternative is correct when both stimuli are black and the left-hand alternative is correct when both stimuli are white (as shown in the second panel of Figure 6.1), then the animal has already learned the responses required for solution of the problem. When it sees black on the right, it enters the right goal arm; when it sees white on the right, it retraces and enters the left goal arm. It is equally possible, however, either that the contingencies of the successive problem will be as shown in the bottom panel of Figure 6.1, requiring the animal to go left to black and right to white, or that the animal happened to orient to the left during simultaneous discrimination training; in either of these cases, the behavior learned in Stage 1 will lead to consistent errors when transferred to the successive problem.

Animals that have not been trained on a brightness problem in Stage 1, will show no such immediately appropriate or inappropriate behavior at the outset of Stage 2. By assuming that it is relatively easy to shift initial orientation from one goal arm to the other when the orientation bias established in Stage 1 leads to consistent nonreinforcement, Siegel is able to account for the overall superiority of a brightness pretrained group to other controls. The critical features of his analysis, however, are the predictions it makes about animals' patterns of orienting responses, and about the distribution of learning scores in Stage 2. In order to provide unequivocal data on orienting responses, Siegel used an unusual apparatus. Although basically a simple T maze with the discriminanda placed in the goal arms, the apparatus also contained opaque doors at the entrance to each goal arm. These doors had to be pushed open by the subject before the discriminanda could be seen. In this way, it was possible to record the side that each subject chose to inspect at the outset of each trial. Siegel reported the development of stable preferences for orienting towards one side rather than the other during original training. Once it is known on which side a particular animal samples the stimuli, and also what are the reinforcement contingencies for the two problems, Siegel's analysis predicts whether that animal will learn Stage 2 rapidly or slowly. Dividing his subjects into those whose sampling strategy matched the reinforcement contingencies of Stage 2, and those whose strategy did not match, Siegel found that subjects with an appropriate strategy learned the second problem in 33.33 trials, while those with an inappropriate strategy required 164.67 trials to reach criterion.

These results provide striking confirmation of Siegel's analysis. Given such a bimodal distribution of Stage 2 scores, it is an inescapable conclusion that some subjects had learned something in Stage 1 that was appropriate to Stage 2 and other subjects had not. Given also that it was possible to predict whether a particular subject would be a fast or slow Stage-2 learner from a knowledge both of the reinforcement contingencies of the two problems and of that subject's sampling strategy, there can be no reasonable doubt that the basis for the transfer was (at least in part) as Siegel argues. The crucial question is whether Siegel's analysis applies equally to experiments other than his own, i.e., whether it constitutes a complete explanation of the results of studies of transfer between simultaneous and successive discriminations, or whether it applies only to the particular situation used in his experiment. Siegel has pointed to the overall similarity between his own results and those of Lawrence, arguing that similar effects are likely to have similar causes. While this argument has some force, an alternative possibility must be recognized, namely, that in order to record initial orienting responses, Siegel devised an apparatus that forced upon his subjects the very strategy postulated by his analysis. The presence of opaque doors at the choice point meant that Siegel's rats could not possibly see more than one alternative goal arm at a time. In view of the rat's propensity to develop position habits during visual discrimination training (see Chapter 4), it is not surprising that Siegel's rats came to make their initial sample always on the same side. Once this had happened, it is not surprising that they should solve the simultaneous discrimination by learning to enter the goal arm sampled if it contained the positive stimulus and to retrace if it contained the negative.

The question, therefore, is whether, in all types of apparatus in which significant transfer has been demonstrated, rats learn to sample a single alternative and make their choice in the manner outlined above, at least to the extent necessary to account for the amount of transfer obtained. Part of the problem is to obtain an unequivocal answer without forcing such a pattern of behavior onto the animals. Nevertheless, there are two arguments which incline us to doubt the general validity of Siegel's analysis. The first is a plausibility argument, and may not, therefore, convince everyone. However, it does seem questionable whether, in view of the nature of the apparatus he used, Lawrence's rats would have been likely to develop this sort of strategy. Lawrence used one apparatus for Stage 1 simultaneous discrimination training, and a second apparatus for Stage 2 successive training. Floor plans of the two are shown in Figure 6.2. In both cases, animals had to cross an air gap to descend to the choice area located $4\frac{1}{2}$ in. in front of and 1 in. below the start box. As Lawrence (1949) points out, the apparatus used in Stage 1 ensured that "all choices were made from the same position with both compartments visible at the same time [p. 772]." It is also worth noting that each apparatus required somewhat different left and right turns. In the absence of any evidence to the contrary, the most reasonable conclusion would surely be that Siegel's analysis does not, in fact, apply to Lawrence's situation.

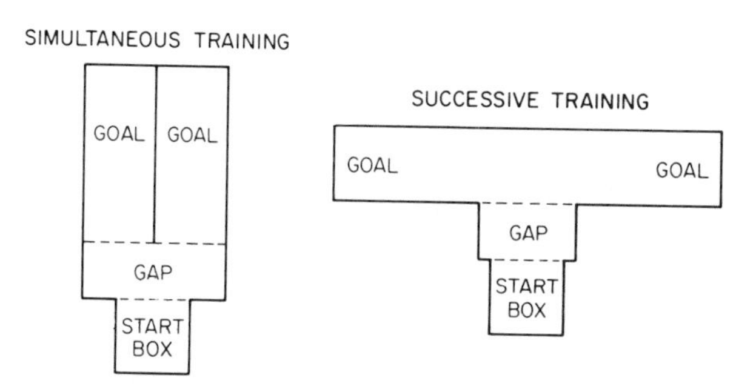


**Fig. 6.2.** Floor plans of the apparatus used by Lawrence (1949).

Fortunately it is not necessary to rely solely on this plausibility argument, for, although there is no direct evidence available, it is possible to bring indirect evidence to bear on Siegel's analysis. While it is true that data on the sampling habits of individual subjects are necessary to predict which particular subjects will learn the transfer task rapidly and which subjects will learn it slowly, nevertheless such data are not necessary in order to predict that *some* subjects will learn very rapidly, and others more slowly. Siegel's analysis predicts a bimodal distribution of Stage 2 scores, and this prediction can be tested in the absence of any data on the specific patterns of orienting responses shown by specific subjects. Although Lawrence did not, of course, specifically look for evidence of a bimodal distribution, he did report that the variances of the Stage 2 scores for different groups were not significantly heterogeneous. If the group given relevant pretraining in Stage 1 had shown a bimodal distribution, and the other groups a normal distribution, such heterogeneity should have been observed. That it was not reinforces the conclusion that Lawrence's results may not be explicable in terms of Siegel's analysis.

In two more recent experiments on transfer between successive and simultaneous discriminations (Mackintosh & Holgate, 1967; Mumma & Warren, 1968), the data for a direct examination of this point are available.[2] Figures 6.3 and 6.4 show the distribution of Stage 2 scores for experimental and control groups in these two experiments. In the Mackintosh and Holgate experiment (Figure 6.3), the Stage 2 problem was a simultaneous, probabilistically reinforced, brightness discrimination. Since there was no set criterion of learning, the data shown consist of the proportion of errors over Trials 1–50 of this problem. This should provide at least as sensitive a test of Siegel's analysis as the trials to criterion scores provided by Siegel for his own experiment, and shown in Figure 6.4 for Mumma and Warren's experiment.

[2] We are grateful to J. M. Warren for providing us with the individual data from one of his studies.

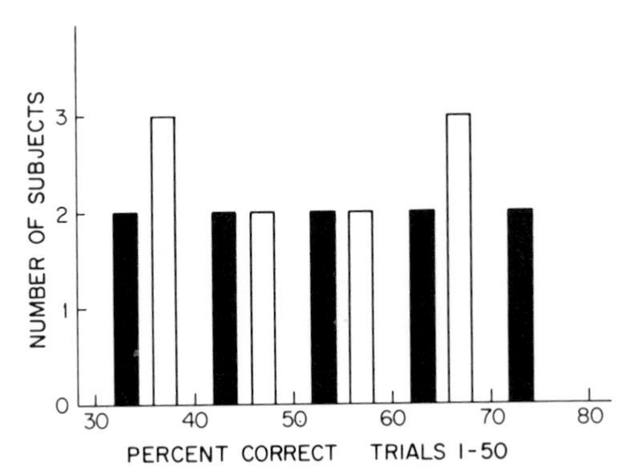


**Fig. 6.3.** Distribution of scores over Trials 1–50 of a simultaneous brightness discrimination for rats pretrained either on a successive brightness discrimination or on a successive orientation discrimination with brightness irrelevant. (■): relevant pretraining; (□): irrelevant pretraining. [*After* Mackintosh and Holgate (1967).]

Although no statistical tests for bimodality (if there are any) have been applied to the data shown in these figures, none of these distributions is self-evidently bimodal, and it is quite clear that the experimental groups show no greater tendency towards bimodality than the control groups.

While there are some problems inherent in this argument (e.g., the extent to which Siegel's analysis predicts a bimodal distribution presumably depends upon the the rate at which subjects with inappropriate orienting responses abandon them, and this may not be easy to specify), the results of our analyses certainly do not support Siegel's position. We conclude, therefore, that while the rats in Siegel's experiment must have been basing their choice behavior on the sort of conditional reorientation strategy he proposed, and while this must have influenced the transfer effects observed in that situation, there is no evidence that this particular strategy occurs in other types of apparatus, and some evidence that the transfer effects obtained in other experiments

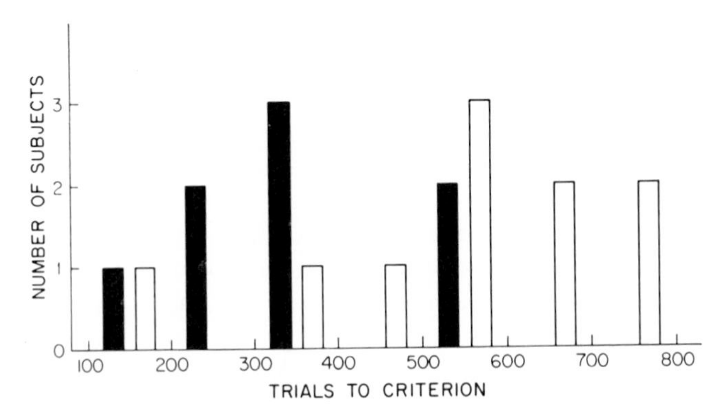


**Fig. 6.4.** Distribution of learning scores on a simultaneous discrimination for cats pretrained on a successive discrimination and for controls. (■): relevant pretraining; (□): control. [*After* Mumma and Warren (1968).]

cannot be accounted for in Siegel's terms. Lawrence's experiments, therefore, stand as convincing demonstrations that the relative distinctiveness of different cues can be altered by prior training.

### B. Intradimensional and Extradimensional Shifts

As we have said, the assumptions underlying Lawrence's acquired distinctiveness experiments were that the response requirements for simultaneous and successive discriminations are different, and, therefore, that even with identical stimuli, there will be no direct transfer between the two problems. Although it would be rash to underestimate the importance and appeal of Siegel's analysis, and although it certainly applies to some situations, the weight of the evidence suggests that it does not provide a complete explanation of the results obtained by Lawrence and others. To this extent, therefore, the data on transfer between successive and simultaneous discrimination support an attentional analysis. There are, however, other ways in which the basic prediction of an attentional theory—that the relative dominance of different analyzers can be altered by prior training—should be open to test. It is possible to devise two discrimination problems which share the same relevant dimension, but between which (because the values of the relevant stimuli have been changed), there can still be no direct (response-based) transfer. In Lawrence's experiments, direct transfer was hopefully eliminated by changing the response requirements; in the experiments to be discussed in this section, the same end is achieved by changing the stimuli.

One method for doing this is illustrated in an experiment by Mackintosh (1964). In Stage 1 of this study, one (experimental) group of rats was trained on an absolute brightness discrimination; specifically, the subjects were presented with two kinds of trials (black versus gray square, or white versus gray square) and were reinforced either for selecting black and white, or for selecting gray. A second (control) group was trained on a square–diamond discrimination in which the stimuli were either black or white; these subjects, therefore, were also reinforced either for selecting or for avoiding black and white squares. For all subjects, the Stage 2 problem was a standard simultaneous brightness discrimination between black and white squares. Although both experimental and control groups had received comparable experience with the Stage 2 stimuli, and although no subject in either group had received *differential* reinforcement on black versus white, the experimental group had been trained with brightness relevant in Stage 1, and the control group had been trained with brightness irrelevant. Confirming the predictions of an attentional theory, the experimental group learned the Stage 2 problem in 42.6 trials, while the control group took 71.0 trials to learn.[3]

[3] A recent study by Singer, Zentall, and Riley (1969) failed to confirm this result.

A more commonly used experimental design also involves comparing the learning of intradimensional and extradimensional shifts (IDS and EDS). The design is similar to that of Mackintosh's experiment just described, with the experimental group of that study corresponding to an IDS group, and the control group corresponding to an EDS group. The distinguishing feature of the IDS–EDS experiment is that the values of the relevant and irrelevant dimensions are changed from Stage 1 to Stage 2. Suppose, for example, that subjects are trained on a red–green discrimination in Stage 1 with horizontal–vertical irrelevant. The stimuli for Stage 2 would be blue and yellow, and left and right oblique rectangles; IDS subjects would learn the Stage 2 problem with color (blue–yellow) relevant, while EDS subjects would learn with the orientation of the rectangles relevant. With proper counterbalancing of the relevant cue in Stage 1, it is possible to compare the learning of both color and orientation discriminations in Stage 2 when they are either IDS or EDS problems.

Most comparisons of the speed of IDS and EDS learning have been undertaken with human subjects. Campione, Hyman, and Zeaman (1965) used retarded children with a mental age of 4 years; Trabasso, Deutsch, and Gelman (1966) used 4-year-old children; Dickerson (1966) used 4- and 5-year-old children; Eimas (1966) used both $5\frac{1}{2}$- and 8-year-old children; and Furth and Youniss used 7- to 8-year-old children. Finally, Harrow (1964), Isaacs and Duncan (1962), P. J. Johnson (1966), and Uhl (1966) used college students. With the exception of one of two experiments by Trabasso *et al.*, all these studies found IDS problems to be significantly easier than EDS problems. In the situation used by Trabasso *et al.*, subjects were presented in Stage 1 with stimuli differing, e.g., only in color; in Stage 2, an IDS problem would involve two new colors, while an EDS problem would involve two stimuli differing in shape only. Throughout the experiment, only a single irrelevant cue (position) was present. In contrast to this, in all other studies stimuli have differed along at least two dimensions (other than position); an IDS has involved maintaining the same dimensions relevant and irrelevant, while an EDS has involved switching the relevant and irrelevant dimensions. In other words, in the EDS problem, the dimension relevant at Stage 1 becomes irrelevant at Stage 2, while the dimension irrelevant at Stage 1 becomes relevant.

It is clear that, in general, human subjects find a discrimination easier to learn if they have been previously trained to attend to the relevant dimension, rather than trained to ignore that dimension. It is also clear that this conclusion applies as well to young children as to college students, and that there is little sign of the age trend predicted by a developmental theory of the role of verbal mediation in discrimination learning (e.g., Kendler & Kendler, 1962). There is, nevertheless, no guarantee that animals will behave in a manner similar even to very young children and, unfortunately, the evidence from animal studies is less substantial.

The earliest study of EDS and IDS learning by animals is by Shepp and Eimas (1964). They trained one group of rats on a shape discrimination with the width or orientation of background black and white stripes irrelevant, and a second group on the stripes discrimination with shape irrelevant.[4] After extended training (including overtraining), animals were trained in Stage 2 either on an IDS problem (from one shape discrimination to another, or from one stripes discrimination to another) or on an EDS problem (e.g., from a shape discrimination to a stripes discrimination or vice-versa). The IDS problem was learned in an average of 23.2 trials; the EDS problem was learned in 56.0 trials. In this experiment, therefore, rats, like humans, showed appropriate effects of prior relevant or irrelevant training on current discrimination learning.

Two further animal experiments, one with pigeons, the other with monkeys, have obtained similar results. Mackintosh and Little (1969) initially trained pigeons either on a red–yellow discrimination with horizontal–vertical irrelevant or on the horizontal–vertical discrimination with red–yellow irrelevant. For Stage 2, the values of the stimuli were changed to blue, green, left oblique, and right oblique, and half of each group was trained on an IDS problem (e.g., from color to color), while half was trained on an EDS problem (e.g., from color to orientation). The IDS problem was learned significantly faster than the EDS problem. Shepp and Schrier (1969) trained monkeys on an initial discrimination (either shape or color), and then on a series of three IDS or three EDS problems. They found the following significant effects: only IDS subjects showed positive transfer from the original discrimination to the first shift problem; only IDS subjects continued to show improvement over the series of shift problems; and finally, by the end of the series, the IDS problem was learned significantly faster than the EDS problem.

The combined evidence of these three studies indicates clearly that animal subjects, like human subjects, will typically learn IDS problems more rapidly than EDS problems. However, just as there are conditions under which human subjects fail to show such an effect (as in the study by Trabasso *et al.*

[4] The two stimulus dimensions, therefore, were shape and stripes. The stimuli on the shape dimension were a rectangle, triangle, cross, circle, and T; the stimuli on the stripes dimension were horizontal, vertical, and obliques (differing in orientation), and three different widths. The assumption that variations in either width or orientation constitute variations along a single "dimension" (and this is borne out by the results of the experiment) implies either that a single analyzer detects both sets of differences (this is not at all implausible, see, e.g., Sutherland 1963b), or that although different analyzers are involved, they appear close together within some hierarchy (see the discussion in Chapter 5).

above), so there are two studies in which rats have learned an EDS problem as rapidly as an IDS problem. In the first (Sutherland & Andelman, 1969), rats were initially trained to discriminate either between black and white squares or between horizontal and vertical rectangles. In Stage 2, all subjects were trained to discriminate between two oblique rectangles. This problem was learned as fast by animals previously trained on the brightness discrimination as by those trained on the orientation discrimination; in other words, there was no evidence of an IDS being easier to learn than an EDS problem. As in the study by Trabasso *et al.*, only one irrelevant cue (position) was present throughout the experiment. It is tempting to conclude that EDS problems are learned more slowly than IDS problems only when the dimension relevant at Stage 1 becomes irrelevant, or the irrelevant dimension becomes relevant [an alternative possibility is that the absolute number of irrelevant cues is a critical variable: in his experiment with college students, for example, P. J. Johnson (1966) found that an increase in the number of irrelevant cues from to four to six significantly retarded EDS learning, but had little effect on IDS learning].

The simplest interpretations of these possibilities are as follows: if EDS problems are learned slowly only when subjects are required to attend to a dimension that has previously been irrelevant, the implication is that the main effect of Stage 1 training is to weaken irrelevant analyzers; if, on the other hand, EDS problems are difficult only when a previously relevant, but now irrelevant, cue is present, the implication is that Stage 1 training does strengthen relevant analyzers, but that these analyzers are only functional when the discriminanda differ along the dimension detected by them.

A first step towards deciding between these two possibilities could be taken by comparing IDS and EDS learning when the irrelevant cue is held constant either in Stage 1 or in Stage 2. If subjects are trained either on color or on orientation in Stage 1, and are shifted to a different color problem in Stage 2, we can ask whether the difference between the two groups is affected by the presence of irrelevant color or orientation cues in Stage 1, or by the presence of an irrelevant orientation cue in Stage 2. The only animal experiment to throw light on this question is a study by Turrisi, Shepp, and Eimas (1969); they repeated the Shepp and Eimas experiment described above, but held the irrelevant cue constant in Stage 1, and found no difference between IDS and EDS scores in Stage 2.[5] This result is consistent with the idea that the main difficulty in EDS learning is that subjects have learned to ignore the cue that is made relevant in Stage 2. Turrisi *et al.*, recognize, however, that there are other possible interpretations. If, for example, orientation were

[5] Bryant (1967) has reported similar results with young (5 years old) children. Older children ($6\frac{1}{2}$ years old), however, learned IDS problems more rapidly regardless of whether the irrelevant cue was held constant or varied in Stage 1.

held constant in Stage 1 and then made relevant for EDS subjects and irrelevant for IDS subjects in Stage 2, its very novelty might attract attention, thus facilitating EDS and retarding IDS learning.

There is, therefore, some ambiguity about the exact interpretation of IDS and EDS comparisons. It is equally possible to ascribe the differences normally obtained either to the strengthening of attention to a cue relevant in Stage 1, or to the weakening of attention to a cue irrelevant in Stage 1; attempts to decide between these two possibilities are complicated by the intrusion of possible confounding factors. Nevertheless, these experiments serve both to reinforce and to extend the conclusions derived from studies of transfer between successive and simultaneous discriminations. Like those studies, the IDS–EDS comparison implies that transfer between two discrimination problems depends not only on the relationship between the response requirements of the two tasks, but also on the relationship between the relevant and irrelevant stimuli. They also, however, show that such transfer does not depend upon using identical physical stimuli, but will occur when other values of the same stimulus dimension are used. Therefore, they support the theoretical conclusion that the relative strengths of different analyzers, each appropriate to a different stimulus dimension, may be changed by suitable training procedures.

## C. Transfer along a Continuum

A third experimental situation has provided evidence consistent with this conclusion. In studies of transfer along a continuum, it has been found (see Chapter 2) that animals pretrained on an easy discrimination may perform more accurately on a difficult problem than animals trained from the outset on the difficult problem. This difference in test performance is predicted by an attentional theory, not because one group has been trained to attend to the relevant cue and the other has not, but because one group has been trained *more effectively* to attend to the relevant cue.

### 1. *Attentional Analysis*

If an animal is trained on a difficult discrimination, where the difference between the positive and negative stimuli is so small as to ensure that completely accurate performance is very difficult or impossible, then, according to our theory, the causes of inaccuracy should be twofold:

(1) Even if the animal does switch-in the relevant analyzer, the stimuli will not always be correctly identified. The analyzer can be viewed as a detection device for discriminating signals from noise. Here, the discriminability of the stimuli would be the signal-to-noise ratio; as this signal-to-noise

ratio decreases, so the probability of a misidentification increases and performance will be correspondingly less accurate.

(2) From (1) it follows that even when the relevant analyzer is used, the animal will not be consistently rewarded. The outputs from the analyzer do not consistently predict different events of importance to the animal; therefore according to Rule 2, the analyzer will never be very strongly switched in.

If there are two separate reasons for the inaccuracy of performance in this situation, there should be at least two ways of improving performance. First, any procedure that temporarily increased the signal-to-noise ratio would lead to less chance of an error due to misidentification; secondly, any procedure that strengthened the relevant analyzer should also lead to higher accuracy. In order to test the first prediction, Sutherland, Mackintosh, and Mackintosh (1963) trained one group of octopuses on a discrimination between a square and parallelogram with 70° acute angles for 170 trials, during which they reached an asymptote of 60% correct responses (see Figure 2.2, page 30 for an illustration of the stimuli). Two additional days of training were given, interspersed with transfer trials to a square and a parallelogram with 45° angles; the subjects were not differentially rewarded in the transfer tests. Performance on these test trials immediately improved to 70% level of accuracy, while performance on the original stimuli remained at the 60% level.

We interpret the phenomenon of transfer along a continuum (as already mentioned in Chapter 2) as the second method of improving accuracy of a difficult discrimination, one which, according to us (and to Lawrence who coined the term), involves effectively pretraining subjects to switch-in the relevant analyzer. Lawrence trained one group of rats on a dark-gray, light-gray discrimination, while a second group was trained initially on a black–white discrimination and only then gradually transferred (via a series of intermediate steps) to the difficult discrimination. The performance of the second group at this last stage was significantly more accurate than that of the group trained from the outset on the difficult discrimination. Figure 6.5 shows similar results from our experiment with octopuses. In addition to the group trained throughout on the difficult square–parallelogram discrimination, a second group was initially trained on an easy square–parallelogram discrimination and then shifted, via a series of intermediate steps, to the difficult discrimination. Training on the easy discrimination led to a high level of accuracy and therefore ensured that the relevant analyzer was strongly switched in. Provided the analyzer is not rapidly extinguished when subjects are transferred to the more difficult discrimination, their performance is bound to be more accurate than that of subjects who have never learned to switch-in the relevant analyzer consistently.

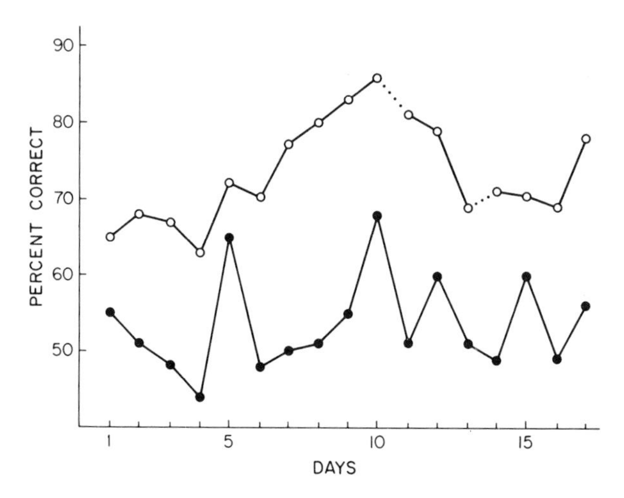


**Fig. 6.5.** Performance of two groups of octopuses either trained throughout on a difficult shape discrimination or trained on an easier problem and shifted gradually to the difficult problem. The broken line in the learning curve for the easy group (○) indicates the transition from the easy problem to the gradually harder problem, and the final transition to the most difficult problem. (○): easy group; (●): difficult group. [*After* Sutherland, Mackintosh, and Mackintosh (1963).]

It has been known since Pavlov (see Chapters 1 and 2) that a discrimination that is difficult, if not impossible, for animals to solve can be learned if the subject is pretrained on an easier, related problem. Pavlov (1927, pp. 121–122) reported such results both for a brightness discrimination and a circle versus ellipse discrimination. Lashley (1938, p. 143) obtained similar results by varying the brightness contrast between figure and background. North (1959) has performed a more thorough experiment rather similar to Lashley's; the easy discrimination was between an erect and an inverted black triangle on a white background, while the stimuli of the difficult discrimination were broken figures. He found that a group trained initially on the easy problem and then on the difficult problem took fewer trials to learn *both* problems than the group trained on the difficult problem from the outset needed to solve the difficult problem *alone*.

These results are not perhaps intuitively surprising. Pavlov, at any rate, appears to have found no need to comment extensively upon his own findings. It was Lawrence (1955) who wrote,

> They pose an important problem for theories of discrimination learning. Usually it is assumed that a fixed amount of practice is most profitably spent if all of it is devoted to training on the specific task on which the individual ultimately will be tested. In the above experiment [Lawrence's] the converse

> of this was true: for a fixed amount of practice, better performance on the test discrimination resulted if the training had been devoted to an easier though similar task than if devoted to the test discrimination [p. 27].

The problem is that any change in the relevant stimuli normally disrupts discriminative performance (an example of this would be the breakdown of transposition in "far" tests, see Chapter 3); this generalization decrement effect would be expected to operate against subjects shifted from the easy to the hard problem. That generalization decrement does occur in experiments on transfer along a continuum is evident from Lawrence's finding that a gradual transition from black–white to gray led to better test performance than an abrupt transition. Franken (1967) has presented some evidence which suggests that habituation to stimulus change may also enhance test performance.

While the phenomenon of transfer along a continuum is readily explained by an attentional theory, it can and has been interpreted in different terms. Three specific suggestions have been advanced; of these, two can be rejected fairly confidently, but one may be a viable alternative to an attentional analysis.

### 2. *Restle's Analysis*

Restle's original model of discrimination learning was proposed as an account of additivity of cues (Restle, 1955; see Chapter 5). The explanation of additivity follows from the assumption that rate of learning is a function of the proportion of relevant to irrelevant cues (or stimulus elements). More specifically, the theory states that the proportion of relevant cues conditioned and irrelevant cues adapted on each trial is equal to the proportion of relevant cues to all cues in the situation. According to this assumption, an increase in the number of relevant stimulus elements should not only increase the overall learning rate, but should also increase the number of relevant stimulus elements conditioned and irrelevant elements adapted on each trial. Any factor, therefore, that causes an increase in the overall proportion of relevant elements should lead not only to faster current learning but also to better performance on a reduced population of relevant cues, than would have been attained by training on the reduced population from the outset. One method of increasing the proportion of relevant elements will be to increase the discriminability of the relevant stimuli; hence, Restle's theory predicts transfer along a continuum.

The problem with this explanation is that transfer from an easy to a difficult problem is not always positive. Restle's analysis predicts that any procedure that increases the proportion of relevant elements will enhance subsequent performance on a subset of those elements. One such procedure, however, has usually been found to interfere with test performance. We refer to experiments on overshadowing, discussed in Chapter 5. In these experiments, one group of animals is trained on a discrimination problem with two cues, A and B both relevant, while a second group is trained with Cue A alone relevant. After a fixed number of trials, both groups are tested for the amount learned about Cue A. While there has been no unanimity in the results of such studies, the general trend has been for the amount learned about Cue A to be *reduced* by the addition of Cue B. In an experiment by Miles (1965), for example, a graded effect was obtained, dependent on the discriminability of the overshadowing cue: the easier the discrimination based on Cue B, the less was learned about Cue A. Miles' results, therefore, directly contradict Restle's theory; over a wide range of values, an increase in the proportion of relevant stimulus elements decreased the amount learned about one set of those elements.

Since the addition of relevant stimuli in one dimension interferes with the learning of stimuli in another dimension (overshadowing), while the addition of relevant stimuli in the same dimension enhances such learning (transfer along a continuum), it appears that it is not the total number of relevant stimuli that determines performance on the test problem, but the proportion of stimuli varying along the same dimension. Unfortunately, there are no animal studies in which a direct comparison has been made between these two situations.[6] Two recent experiments with college students have, however, provided direct evidence, largely compatible with this conclusion. Marsh (1967) required his subjects to learn a very difficult hue discrimination. The control group received 70 trials on this problem, and performed at chance throughout. Two groups received 20 pretraining trials on an an easier hue discrimination. Over 50 trials on the difficult discrimination, they averaged between 75 and 80% correct responses. Two groups were given overshadowing training: for 20 trials they were presented with stimuli differing in both hue and brightness (the hue difference was the same as the difficult test problem; the addition of the brightness difference resulted in a discrimination that was as easy as the easy hue problem). These overshadowing groups, however, performed at chance when shifted for 50 trials to the difficult hue discrimination.

A similar result has been reported by Trabasso (1963) for a card-sorting (concept-identification) task. Students were required to sort cards containing stylized drawings of flowers into two categories, the basis for the classification being the angle between the leaf and the stem. For the test problem, the angular difference was relatively small (30 versus 60°). Subjects pretrained on stimuli with a wider angular difference showed perfect transfer to the test

[6] Two recent experiments correct this deficiency. Both Marsh (1969) using pigeons, and Singer, Zentall, and Riley (1969) using rats, show that training on an easy problem only facilitates subsequent learning of a difficult problem if the stimuli to be discriminated in the two problems differ along the same dimensions. See also Mackintosh and Little (1970).

problem; subjects trained on the test angle, but with the addition of a relevant color cue (the 30° angle was, say, red; the 60° angle was green) showed very little transfer to the test problem. Other groups in this experiment showed transfer effects not predictable from our analysis. For example, subjects trained on the test angle, but with *both* angles colored red (the rest of the drawing was black) showed excellent transfer to the test problem. This result (similar to one reported by Hull, 1920) is most easily interpreted in terms of observing responses: the subject looks at that part of the figure which stands out from the background, and therefore may learn about the relevant feature when it is located in that area. This observing-response analysis, however, cannot easily explain why the presence of greater angular differences should have improved test performance, and certainly cannot explain why the addition of color as a second relevant cue should have failed to improve test performance.

### 3. *Other Consequences of Differences in Problem Difficulty*

Restle's theory does not provide the only rationale for an attempt to explain transfer along a continuum by appealing to the general effects of problem difficulty. Subjects trained on a difficult problem make more errors than those trained on an easier problem; they might, therefore, become more frustrated and learn less, because a high drive interferes with discrimination learning. The results reviewed above make it reasonable to reject any such general analysis; it is also worth remembering that other results (for example, those on blocking, see Chapter 4) have been supposedly explained by the principle that learning occurs on error trials, from which it follows that animals making *fewer* errors will learn *less* about a given problem. The most decisive evidence against such speculations, however, is the demonstration that transfer along a continuum occurs in a classical conditioning situation, where the experimenter's control of reinforcement ensures that subjects trained on either easy or difficult discriminations receive identical sequences of reinforced and nonreinforced trials. Haberlandt (1968) (quoted by Wagner, 1969a), using rabbits and a differential eyelid conditioning procedure, trained one group on a difficult auditory frequency discrimination between tones of 4890 and 6480 Hz and a second group on more widely spaced tones. When both groups were tested on the difficult pair of tones, subjects initially trained on the easier discrimination showed significantly better discrimination.

### 4. *A Generalization Gradient Analysis*

Although there are good grounds for rejecting the above two analyses, a further explanation of transfer along a continuum may remain as a possible alternative to an attentional analysis. Lawrence (1955) saw that his data might be amenable to an analysis in terms of generalization gradients, and set out to show what restrictions were imposed on such an analysis by the data. The general idea underlying such an explanation is this: given a stimulus dimension A, with values A1, A2, A3, A4, transfer along a continuum will be predicted if direct training on A2 versus A3 produces a smaller net difference in response strengths to these two stimuli than is produced by training on A1 versus A4. If responses to A1 are reinforced, excitation will be built up at A1, and will generalize to other values on the dimension; similarly, nonreinforcement of responses to A4 will build up inhibition that will generalize to other stimuli. Net response strength may be assumed to be equal to the difference between excitation and inhibition. The critical question is whether there exist functions for these generalization gradients that satisfy the requirements for predicting transfer along a continuum. Figure 6.6 shows three possible shapes of generalization gradient: linear, concave upward (of the type proposed by Hull, 1952), and concave downward (of the type proposed by Spence, 1937). As Lawrence pointed out and as Bitterman (1966) and Logan (1966) have since reiterated, concave downward gradients do satisfy the requirements; in the lower panel of Figure 6.6, the

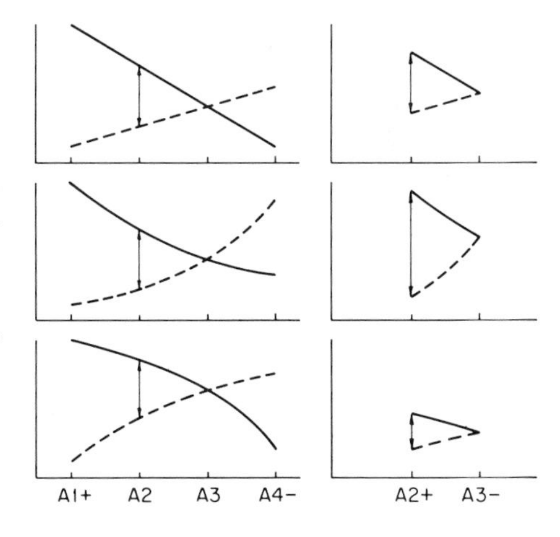


**Fig. 6.6.** Theoretical generalization gradients of excitation and inhibition. The gradients in the left-hand panels are based on training at A1+ and A4−; those in the right-hand panels are based on training at A2+ and A3−. (—): excitation; (- - -): inhibition.

difference between the net response strengths at A2 and A3 is greater when reinforcement and nonreinforcement are given at A1 and A4 than when they are given directly on A2 and A3. Since Spence (1937) showed that a general, concave downward generalization gradient would also predict transposition (see Chapter 3), there seems to be rather strong, although indirect, evidence to support the postulation of this particular theoretical function. Linear gradients, of course, predict neither transposition nor transfer along a continuum; while concave upward gradients (middle panel) predict the reverse of both phenomena.

Lawrence argued that the precise function chosen by Spence was certainly not satisfactory, since a further requirement imposed by the data is that the slope of the gradient be an increasing function of the absolute value of response strength. Spence's particular gradients were of invariant slope, but there is no general reason why a concave downward gradient cannot meet this further requirement. Logan (1966), indeed, has shown that this theoretical restriction yields a testable prediction. If the gradients of excitation and inhibition are of the same form, and if (as must be the case) the absolute amount of excitation is greater than the absolute amount of inhibition, then this restriction implies that the slope of the gradient of excitation must be steeper than the slope of the gradient of inhibition. The steeper the slope of a gradient, the greater will be the contribution of changes along that gradient to the phenomenon of transfer along a continuum. It follows, therefore, that changes in S+ should produce better transfer than changes in S−. Logan tested and confirmed this prediction.

This result would seem to provide good evidence to support an analysis of transfer along a continuum in terms of gradients of excitation and inhibition. As Logan noted, however, one feature of his results was not consistent with his analysis; the reason that a change in S− produced less efficient transfer than a change in S+ was entirely because animals in the former group started responding again in the presence of S− (Logan used a go–no-go successive training procedure). This suggests an alternative account of his results. As we pointed out above, the striking feature of transfer along a continuum is that the effect occurs despite the countereffects of generalization decrement. If this is so, then variations in the amount of generalization decrement must produce variations in the amount of transfer shown. The differences between Logan's groups could be interpreted as differences in the extent to which the stimulus change from training to testing disrupted performance. In Chapter 3, we briefly discussed the question whether discriminative performance is based on learning to approach S+, or on learning to avoid S−. One experimental method used to answer the question has been to change either S+ or S− after animals have learned a problem, and then see which operation disrupts performance more. Logan's results were that a change in S− disrupted test performance more than a change in S+, and that this disruption was manifested by an increase in responding to S−. In order to explain such results, all that we have to assume is that discrimination learning in this situation consisted largely of learning to inhibit responses to S−. From the learning curves shown, the assumption is certainly correct: all animals were pretrained to press the lever, and animals trained from the outset on the difficult discrimination responded as consistently to S+ as animals trained on easier problems; they failed to perform at the same level of accuracy solely because they continued to respond to S−. Even within the framework of Spence's theory, the prediction of Logan's results follows more properly from this analysis than from his own. This is because the use of a go–no-go discrimination procedure implies that the important theoretical value determining discriminative performance is not the relative difference between excitatory and inhibitory gradients at different points, but rather the extent to which the excitatory–inhibitory difference is above threshold at S+ or S−. Given that all animals responded at a high rate on all trials, changes in S+ would still leave the difference above threshold although producing a drop in excitation; if learning was due to the development of inhibition at S−, changes in S−, causing a drop in inhibition, would be more effective in disrupting the discrimination.

There is, therefore, no compelling evidence that supports a generalization gradient analysis of transfer along a continuum over an attentional analysis. On the other hand, there is not much evidence pointing the other way, and it is possible, of course, that both factors contribute to the effect. Some of the results of Haberlandt's (1968) experiment, however, show that differences in the amount of control exercised by irrelevant cues play an important part in transfer along a continuum and therefore support an attentional analysis. Haberlandt's rabbits were trained, as already mentioned, in a differential conditioning situation to discriminate between different auditory frequencies. The stimuli presented on positive and negative trials, however, were not simply different tones, but compounds comprising a tone-plus-light and a vibratory stimulus. The same light and vibratory stimulus appeared on both positive and negative trials, and acted therefore as constant irrelevant cues. When all subjects were finally tested on the difficult auditory discrimination, the tones were presented at either 10, 5, or 0 dB above the level of the background noise. At the two higher intensities, subjects pretrained on the easier discrimination showed better differentiation than subjects trained throughout on the difficult problem, a transfer-along-a-continuum effect. When the tones were presented at the lowest intensity, no subject showed any discrimination, and it is reasonable, therefore, to infer that responding was largely controlled by the irrelevant light and vibratory stimulus. The important point is that, on these test trials, the difficult group responded at a significantly higher level than the easy group; i.e., subjects trained throughout on the difficult problem were more likely to be controlled by irrelevant stimuli than those pretrained on the easier problem. Since there was no differential responding to the two tones on these trials, the results cannot be ascribed to differences in response strengths having been greater in the easy group and, therefore, having suppressed an equally large tendency to respond to the irrelevant stimuli.

Thus, Haberlandt's results do suggest that one factor in transfer along a continuum is that training on an easy discrimination is more effective than training on a difficult discrimination in altering the relative strengths of relevant and irrelevant analyzers. While we cannot rule out the possible contribution made by differential transfer of appropriate instrumental responses, differential transfer of analyzers must also contribute to the effect.[7]

## III. Acquired Nondistinctiveness of Cues

The evidence discussed up to this point has suggested that the relative strengths of relevant and irrelevant analyzers may be changed by appropriate pretraining procedures. We have stressed this description of the data, since it is often difficult to decide whether transfer is due to the strengthening of relevant analyzers or to the weakening of irrelevant analyzers. Before discussing this question in the final section of this chapter, we turn to a consideration of experiments that were explicitly designed to study the effects of irrelevant pretraining: does an animal that has been trained to ignore a particular cue find it harder to learn a subsequent problem with that cue relevant? We shall consider two ways in which, in principle, it might be possible to train animals to ignore particular cues.

### A. Nondifferential Training

The first method is suggested by a casual reading of the second rule of our theory. In an informal way, we have said that an analyzer is strengthened when its outputs are followed by different events of importance to the animal —in the simplest case when responses to one output are rewarded (or punished) and responses to another output are not rewarded (or not punished). From this it follows that if all outputs predict reward, the analyzer will not be strengthened. In possible contrast to the models proposed by Lovejoy (1966) and Zeaman and House (1963), for example, our own theory predicts that if an animal is rewarded when it responds both to black and to white, the brightness analyzer cannot be strengthened and, it might appear, should, in fact, be weakened.

At least one experiment has shown that "nondifferential" reward, given on both values of a cue, may retard subsequent learning with that cue relevant.

[7] A recent paper by Mackintosh and Little (1970) provides evidence against the generalization gradient analysis. They found that even *reversed* training on an easy wavelength discrimination eventually led to performance on a hard wavelength discrimination superior to that of subjects trained on the hard problem from the outset. A second experiment found no evidence that generalization gradients were of the appropriate shape to produce transfer along a continuum.

Siegel (1945) gave two groups of rats nondifferential reward in a T maze, one group for four trials, the second for 80 trials. All trials were forced, half to each goal box, and all trials were rewarded. He then gave 48 noncorrection training trials in the T maze, with choice of one goal box rewarded, and choice of the other not rewarded. The first group made an average of five errors in these 48 trials; the second group averaged nine errors. The difference was significant at the .01 level: the greater the number of trials of nondifferential pretraining, the slower was subsequent learning.

The question remains whether these results are due to the weakening of a spatial analyzer or to more general effects of nondifferential training. It is possible that Siegel's pretraining procedure would have retarded the learning of any subsequent discrimination problem, even a visual discrimination with position irrelevant (see, e.g., the results reported by Mandler, 1966, discussed on page 201). It is clear that further control groups are necessary to evaluate such a possibility.

A series of studies by Bitterman and others has relied on the following control procedure: an experimental group is given nondifferential training with the visual stimuli subsequently used in discrimination training, while a control group is given nondifferential training to blank, gray doors. In two experiments (Bitterman, Calvin, & Elam, 1953; Bitterman & Elam, 1954), the experimental group learned the test discrimination (between horizontal and vertical striations) significantly *faster* than the control group. In the second of these two experiments, moreover, different amounts of nondifferential training with the test cue were given to different subgroups, without this having any significant effect on subsequent performance. It seems that Siegel's results may as well be attributed to a general effect of nondifferential reward as to a specific effect.

On the strength of his own results, Bitterman argued that some differentiation of the relevant stimuli occurs during rewarded exposure to those stimuli, i.e., that nondifferential reward will facilitate subsequent discrimination learning. The evidence for this suggestion, however, is far from convincing. In two further experiments, there was no sign whatever that a control group trained to gray doors would take longer to learn the test problem than an experimental group previously exposed to the relevant stimuli (Billingsley, Fedderson, & Bitterman, 1954; Calvin, 1953). Moreover, a study by Bitterman, Elam, and Wortz (1953) found that nondifferential, unrewarded experience of horizontal and vertical striations tended to retard the learning of a subsequent horizontal–vertical discrimination.

A recent study by O'Malley, Arnone, and Ziegenfus (1969) used a different procedure. An experimental group was exposed to black and white plaques in their home cage while the control group was housed in ordinary cages. Both groups were then trained in a T maze with black and white arms, either on a brightness discrimination with position irrelevant or on a position discrimination with brightness irrelevant. Prior exposure to black and white plaques interfered with brightness learning but facilitated position learning. It should be noted that the brightness discrimination was easy and was learned as rapidly as the position problem.

The evidence is confusing, and it is not easy to draw any certain conclusions. A better way to test the possibility that nondifferential reinforcement selectively weakens the analyzers in question, is to provide such experience on one cue and then require subjects to learn a two-cue problem with both that and another cue relevant. Andelman and Sutherland (1970) undertook the following experiment: different groups of animals received nondifferential reward on either black and white squares, horizontal and vertical rectangles, or two gray doors; in Stage 2, all animals were trained (to criterion) on a combined brightness and orientation discrimination, and in Stage 3 were given transfer tests to discover the basis of their solution to the Stage 2 problem. The question was whether animals given nondifferential reward on black and white would learn the two-cue problem largely in terms of orientation, while animals given nondifferential reward on horizontal and vertical would learn the two-cue problem largely in terms of brightness. The results are shown in Table 6.II; they do not show any such effect. Although the differences between the brightness and orientation groups taken by themselves were (just) in this direction, none of the differences shown in Table 6.II was significant.

TABLE 6.II

**Results of Experiment by Andelman and Sutherland[a, b]**

| Groups, Stage 1, nondifferential training | Transfer tests, percentage correct | |
|---|---|---|
| | H–V | B–W |
| H–V | 83.0 | 85.4 |
| B–W | 85.1 | 78.6 |
| Control | 85.5 | 77.0 |

[a] Data obtained from Andelman and Sutherland (1970).
[b] B designates black; W, white; H, horizontal; V, vertical.

With these results available, the evidence becomes less confusing: nondifferential reward on a particular cue does not seem to have any consistent effects on subsequent learning, and certainly does not usually weaken the analyzer detecting that cue. It is conceivable that these generally negative results are to be attributed to the fact that the analyzer that was supposedly being depressed was already relatively weak; nondifferential reward might not retard the learning of an already indistinct cue, but might still have an effect on a very dominant cue. While this may be true in general, and while it would explain why significant interference has only been obtained in the experiments by Siegel and O'Malley, Arnone, and Ziegenfus, the argument is not very easily applied to Andelman and Sutherland's results. We must conclude that nondifferential training does not generally weaken attention to particular stimuli.

This conclusion, which we have arrived at empirically, is, in fact, predicted by Rule 2 of our model. Since, under conditions of nondifferential reinforcement, no analyzer makes consistently correct predictions, all analyzers should tend towards their base value which is, of course, their starting value.

It should be noted that, although nondifferential reinforcement will not change the strength of analyzers if no learning has occurred, it should reduce the strength of any analyzer that as a result of previous learning is above its base value. At first sight it might appear that a model that assumes that analyzers are strengthened by reinforcement and weakened by nonreinforcement does not make this prediction. If the relevant analyzer is used on every trial and if reward always occurs no matter which stimulus the animal selects, then the use of that analyzer will always be reinforced and hence it cannot drop in strength. This argument has in fact been put forward by Trabasso and Bower (1968, pp. 16–17). The argument is only valid if nondifferential reinforcement is introduced at a stage when the relevant analyzer is controlling behavior on every trial. If an irrelevant analyzer is used on some trials, then it will be strengthened, and if linear operators are used, its gain in strength will tend to be large because of its low starting value. If analyzer strengths sum to a constant amount, the gain in strength of irrelevant analyzers will weaken the strength of the relevant analyzer. The results of experiments introducing nondifferential reward after learning has occurred are, therefore, not likely to be very helpful in selecting the right rule to use for changes in analyzer strength. The only way to be certain that an analyzer is switched-in on every trial is if the animal is performing at 100% correct; but if it is performing at this level, it will never select the negative stimulus and, hence, it will not get any experience of the nondifferential reinforcement. This problem might be overcome to some extent by forcing choices, but such a procedure introduces additional complications and makes results difficult to interpret.

Our reasons for making the assumption that increases in analyzer strength depend upon the analyzer outputs being differentially correlated with reinforcing events were set out in Chapters 2 and 5. It is, perhaps, worth drawing attention here to some further implausible consequences that follow from the alternative assumption, namely, that analyzers are directly strengthened by reinforcement. Let us take the case of an animal trained on a black–white discrimination, with a black square positive and a white square negative. In the sense that we have used the term, the only relevant analyzer is brightness. Other analyzers, however, will detect various features of the stimuli—their size, shape, texture, etc. As soon as the animal has solved the problem, all further responses will be rewarded. If more than one analyzer can be used on a trial, and if analyzers are directly strengthened by reward, there would be nothing to stop these other analyzers all being strengthened to asymptote if sufficient overtraining were given, Now there is, in fact, excellent evidence that animals do notice some of the constant features of the stimuli they are trained to discriminate, for if changes are made in the brightness, location, or color of the stimuli, discriminative performance breaks down (Fields, 1932; Logan, 1961; Williams, 1967a), However, the assumption that all possible analyzers will be strengthened during overtraining leads to predictions much stronger than the implication that changes in contextual cues will disrupt performance. To return to the above example, during overtraining, the animal is consistently rewarded for selecting a black square. If a shape analyzer is strengthened, substantial response strength to a square will be built up, and the animal should subsequently prefer a square to other shapes. It seems plausible to suppose that the animal will indeed prefer a black square (S+) to a black diamond, but rather less plausible to imagine a preference for a white square (S−) over a white diamond. Yet, unless compound learning is to be allowed, both preferences should be equally strong. (In an unpublished experiment to be described in Chapter 7, Mackintosh, in fact, found no evidence for either preference.) Exactly the same set of implausible predictions, of course, can be be derived from theories that do not postulate any selective processes, as Spence (1936) seems to have realized. These problems can be avoided by the assumption (stated formally in Rule 2) that differential reinforcement is necessary for the strengthening of analyzers.

### B. Irrelevant Training

In the sense that we use the term, a variable irrelevant cue is one whose values are each sometimes associated with reinforcement and sometimes with nonreinforcement. Several different types of experiment have investigated whether an irrelevant cue loses distinctiveness.

The most direct experimental test of the proposition that irrelevant pretraining retards subsequent learning is provided in a study by Jeeves and North (1956). They trained two groups of rats on a horizontal–vertical discrimination, one with black–white irrelevant, the other with no brightness cue present. In the second stage of the experiment, they removed the orientation cue and trained both groups on a brightness discrimination. The results for both stages of the experiment are presented in Table 6.III. The difference in the Stage 2 scores fell short of significance, although it can be seen that it is in the expected direction. Although it was not significant, there is reason to suspect that the difference is underestimated, since a glance at the Stage 1 scores reveals that the experimental group learned this stage faster than the controls. It is highly unlikely that the presence of an irrelevant cue should actually facilitate the learning of any discrimination. Wortz and Bitterman (1953), for example, found that the presence of such a cue retarded learning. It is more probable, therefore, that Jeeves and North's two groups were not matched for learning ability in the first place; if then their experimental group contained faster learners, the difference between the groups in Stage 2 must have been underestimated.

TABLE 6.III

**Results of Experiment by Jeeves and North[a,b]**

| Groups, stage 1 training | Stage 1, errors in 128 trials | Stage 2, errors in 80 trials |
|---|---|---|
| H–V[c] | 26.7 | 15.1 |
| H–V[d] | 31.7 | 12.9 |

[a] Data obtained from Jeeves and North (1956).
[b] B designates black; W, white; H, horizontal; V, vertical.
[c] Black–white cues irrelevant.
[d] No brightness cue present.

An alternative experimental design, and one that (like Sutherland and Andelman's experiment mentioned on page 196) gets directly at the question of the specificity of the effects of irrelevant pretraining, has been used by Mackintosh (1965d) and Sutherland and Holgate (1966). In both these experiments, two groups of rats were trained on a successive brightness discrimination. For one group, orientation was present as an irrelevant cue, while for the other group it was not. In Stage 2, the two groups were trained on a simultaneous brightness and orientation discrimination, and in Stage 3, they were given transfer tests to determine how much they had learned about each cue separately. The results are shown in Table 6.IV; in both experiments, animals pretrained with orientation irrelevant performed substantially less accurately on orientation transfer tests than animals pretrained without orientation present.

TABLE 6.IV

**Results of Experiments by Mackintosh and Sutherland and Holgate** [a, b]

| Groups, stage 1 training | Stage 3, percentage correct on test trials | | | |
|---|---|---|---|---|
| | Mackintosh | | Sutherland and Holgate | |
| | B–W | H–V | B–W | H–V |
| B–W[c] | 85.9 | 83.1 | 72.0 | 71.0 |
| B–W[d] | 90.6 | 60.3 | 90.0 | 59 0 |

[a] Data obtained from Mackintosh (1965d) and Sutherland and Holgate (1966).

[b] B designates black; W, white; H. horizontal; V, vertical.

[c] No orientation cue present.

[d] Horizontal–vertical cues irrelevant.

A third demonstration that animals, trained with a particular cue irrelevant, extinguish the analyzer detecting that cue is provided by experiments by Lawrence and Mason (1955) and Goodwin and Lawrence (1955). Lawrence and Mason initially trained rats with Cue A relevant and Cue B irrelevant. In Stage 2, Cue B was made relevant and Cue A irrelevant; and in Stage 3, animals were returned to the Stage 1 discrimination (Cue A relevant), with half required simply to relearn the Stage 1 problem, but the remainder required to reverse the specific responses reinforced in Stage 1. The argument underlying the experiment was as follows: if attention to Cue A had been maintained throughout Stage 2 (when Cue A was irrelevant), then the specific preference established in Stage 1 for one value rather than the other should have been extinguished, and animals required to reverse this preference in Stage 3 would have learned as rapidly as those retrained on their original problem. Lawrence and Mason, therefore, interpreted their finding that the nonreversal was learned significantly faster than the reversal as evidence that animals had stopped attending to Cue A when it became irrelevant. This conclusion was strengthened by the results of Goodwin and Lawrence's study; here, it was found that overtraining at Stage 2 did nothing to decrease the reversal–nonreversal difference of Stage 3. These conclusions have not gone unchallenged, and the whole issue is discussed at greater length in Chapter 8. For the present, we may conclude that Lawrence's results tend to support the suggestion that attention to an irrelevant cue is weakened.

These results appear to show that irrelevant training (unlike nondifferential training) does weaken analyzers. There are several reasons, however, why such a conclusion may be premature. First, in the Goodwin and Lawrence experiment, for example, irrelevant training was given *after* differential reinforcement, i.e., after an analyzer had been strengthened and different responses attached to its outputs. According to Rule 2, such an analyzer would be weakened either by irrelevant or by nondifferential training because its responses were initially unequal and were being driven together.

Second, in all the experiments so far described, the weakening of an irrelevant analyzer has been accompanied by the strengthening of a relevant analyzer; animals being trained with a particular cue irrelevant have also been learning a problem with another cue relevant and, since analyzer strengths sum to a constant amount, this in itself should weaken the irrelevant analyzer. There have been no experiments (such as those of Bitterman and Elam or Andelman and Sutherland on nondifferential reinforcement), designed to see whether random reinforcement by itself weakens particular analyzers. There is, indeed, evidence that random reinforcement on one cue will retard learning of an entirely different cue; Mandler (1966) found that rats, given either 70 or 220 trials of random reinforcement in a Y maze without differential visual stimuli, required over 150 trials to learn a subsequent brightness discrimination. Both naive animals and animals trained on a position problem, learned the brightness problem in about 100 trials: random reinforcement on one cue, therefore, retarded learning of a discrimination problem in which that cue was irrelevant. Unfortunately, no one has directly compared the effects of random reinforcement when the visual stimuli are absent to its effects when they are present, although the results of experiments on fixation (Maier, 1949; Wilcoxon, 1952) suggest that this latter condition has extremely deleterious effects on subsequent discrimination learning.

There is one final problem about the interpretation of many of the experiments under discussion (although it applies equally to those on irrelevant and nondifferential training, and cannot account for apparent differences in outcome between the two types of study). In the experiments by Mackintosh (1965d) and Sutherland and Holgate (1966), animals pretrained on a brightness discrimination with orientation irrelevant learned less about the orientation component of a compound brightness and orientation discrimination than did animals simply pretrained on brightness alone. While this may have been because they had learned to switch-out an orientation analyzer, it may also have represented a novelty effect. For animals pretrained on brightness alone, the shift to Stage 2 involved the abrupt introduction of orientation differences; as noted in the discussion of the experiment by Turrisi, Shepp, and Eimas (1969), and as suggested by the data of Kamin (1968) discussed in Chapter 4, it may be that subjects automatically attend to novel stimuli. These results, therefore, need not represent slow learning by animals that have learned to ignore a particular cue, so much as rapid learning by animals attracted to a novel cue.

### C. Conclusions

Much of the evidence we have been discussing is somewhat ambiguous. It is possible, for example, that animals extinguish an analyzer whose outputs are associated equally with reinforcement and nonreinforcement—especially if such training is given in the context of differential reinforcement on another cue. Nevertheless there is no evidence that compels this conclusion, except in the special case where the analyzer had previously received differential reinforcement. On the other hand, there is no evidence that nondifferential reinforcement selectively suppresses particular analyzers, although both nondifferential reinforcement and random reinforcement and nonreinforcement may interfere with the learning of any new discrimination problem, possibly for reasons that lie outside the scope of our theory.

These rather tentative conclusions are consistent with Rule 2: analyzers should not be strengthened above their base value by nondifferential reinforcement and should be weakened by random reinforcement if they are above their base value. This happy agreement between data and theory is, unfortunately, complicated by some considerations to which we must now turn.

## IV. Nature of Transfer Effects

The evidence reviewed in this chapter establishes that it is possible to alter the relative distinctiveness of different cues by appropriate training techniques. An animal trained with Cue A relevant and Cue B irrelevant will learn a subsequent problem involving the same relevant and irrelevant cues more rapidly than an animal initially trained with Cue B relevant and Cue A irrelevant. While it is reasonably certain that such results are to be attributed to changes in analyzer strengths rather than to changes in response strengths or to complex orienting behavior, it is not possible to state *a priori* whether they are to be attributed more to an increase in the strength of relevant analyzers or to a decrease in the strength of irrelevant analyzers. For example, Lawrence (1950) in his second experiment on the acquired distinctiveness of cues, trained rats in Stage 1 with one cue relevant and another irrelevant. In Stage 2, both cues were relevant. Finally, he gave a series of tests to discover how the animals had solved the Stage 2 problem. They tended to solve it more in terms of their formerly relevant cue than in terms of their formerly irrelevant cue; but, in principle, this could have been either because Stage 1 training had increased attention to the first cue, or because it had decreased attention to the second.

It should be noted that on our own model, both effects necessarily occur, since analyzer strengths sum to a constant amount and no one analyzer can be strengthened without others being weakened nor vice versa. We shall present evidence in Chapter 7 (page 240) that demonstrates that training on a discrimination along a particular dimension does, in fact, weaken analyzers detecting irrelevant cues present in the situation. It is less easy to prove that the relevant analyzer is strengthened, since normally it is not possible to remove all irrelevant cues and thus demonstrate that positive transfer to a new task is based solely on the strengthening of an analyzer relevant to both tasks rather than on the weakening of irrelevant analyzers. Mackintosh (1963a) has performed an experiment that does offer direct evidence for the relevant analyzer being strengthened; this experiment is discussed in detail in Chapter 9 (page 304). For the time being, we limit ourselves to discussing the experiments quoted earlier in this chapter.

While the results of Lawrence (1950) are, fairly obviously, open to more than one interpretation, it is less obvious, but unfortunately still true, that the interpretation of several other results reviewed here is also obscured by the existence of several competing possibilities. In Lawrence's first study of acquired distinctiveness of cues (Lawrence, 1949), three groups of rats were initially trained on a simultaneous discrimination. In Stage 2 they were all shifted to a successive discrimination. Animals learned the Stage 2 problem faster if the same cue was relevant in Stages 1 and 2 than if the Stage 1 relevant cue became irrelevant or took on a neutral, intermediate value. The most obvious interpretation is that the first group had learned, in Stage 1, to switch-in the analyzer relevant in Stage 2, and, therefore, learned more rapidly than the other two groups. Even if this is accepted, a variety of other effects may also have been occurring: the last two groups also learned to switch-in analyzers in Stage 1, and such training must have had some effect on their Stage 2 performance. Furthermore, all groups had been trained in Stage 1 with position irrelevant, and should, therefore, have learned to switch-out position analyzers. In Stage 2 (a successive, i.e., conditional spatial, discrimination) position was relevant. Might not Stage 1 training have interfered with Stage 2 learning?

One seemingly obvious solution to these difficulties would be to include a control group that received no training at the first stage of the experiment. However, any results obtained from such a control group would now be difficult to interpret for a different reason. If an experimental group, trained to respond in terms of a certain cue, solves a subsequent problem involving that cue faster than such a control group, it may be either because the experimental group has specifically learned to switch in the relevant analyzer or because such general factors as habituation to the apparatus, reduction of fear, etc., have facilitated its learning. Despite the general disadvantages working against such an untreated control group, the results of two experiments with rats have shown remarkably little difference between the performance of such a group and that of a group given relevant pretraining. Mackintosh and Holgate (1967) gave two groups of rats pretraining on a successive discrimination (brightness with orientation irrelevant for one group, and orientation with brightness irrelevant for a second group) and, in Stage 2, trained these two groups, together with a naive control group, on a simultaneous (probabilistically reinforced) brightness discrimination. The performance of the two pretrained groups was significantly different, but when comparisons with the control group were made, the only significant difference was that animals trained with brightness irrelevant performed significantly less well than the controls. The group given relevant pretraining was no better than the control group. In another experiment already cited, Mackintosh (1965d) gave two groups of rats pretraining on a successive brightness discrimination, and in Stage 2, trained both groups along with a naive control group on a simultaneous brightness-plus-orientation discrimination. Although the pretrained animals made fewer errors than the controls during 60 trials of Stage 2 training (12.12 versus 15.62), the difference was small, and only just significant.

This experiment also provided for a direct test of the proposition that pretraining on, say, a successive brightness discrimination enhances subsequent learning by selectively strengthening a brightness analyzer, rather than via any more general process. In Stage 3, animals were given transfer tests on brightness and orientation alone, to determine how much they had learned about each cue. Although the pretrained animals performed significantly worse than the controls on orientation transfer tests (a result referred to in Chapter 4), they did *not* perform any better than the controls on the brightness transfer tests (the scores were 88.28 % correct for pretrained animals, and 87.81 % correct for the controls). It is true that transfer tests are relatively insensitive measures of learning, and little reliance should be placed on negative results derived from them. It is also true that the control group was performing relatively accurately on the brightness tests, thus leaving little room for other groups to show superior performance. The experiment is worth repeating with a more difficult cue. Nevertheless, the results fail to provide any evidence that relevant analyzers are selectively strengthened by prior differential reinforcement.

Although the absence of any direct evidence showing that relevant analyzers are selectively strengthened is balanced by the equal lack (noted in the preceding section) of direct evidence for the selective weakening of irrelevant analyzers, we should seriously consider the possibility that much of the data on acquired distinctiveness reflects the suppression of attention to irrelevant cues, rather than the strengthening of attention to relevant cues. In many cases, as we have pointed out, both interpretations are equally plausible. In some instances, indeed, the evidence favors the suppression interpretation: the results of Turrisi, Shepp, and Eimas (1969) suggested that EDS problems may be harder than IDS problems only when subjects have been trained in Stage 1 to ignore the cue relevant in the EDS problem. Whenever subjects have been trained with irrelevant cues that were subsequently made relevant (as in Lawrence's second acquired distinctiveness study, or in IDS–EDS studies), a suppression interpretation is entirely possible. In other cases, the interpretation is not too forced: in the case of Lawrence's experiment on transfer along a continuum, one could argue that animals trained on the easy problem learn better than those trained throughout on the difficult problem to suppress attention to the dominant cue of position. (Haberlandt, 1968, has, of course, shown that irrelevant stimuli will acquire less control in subjects trained on an easy problem, but this is a different matter.)

There are, however, cases where the argument becomes distinctly less plausible. In Lawrence's first acquired-distinctiveness experiment, animals trained on a simultaneous brightness discrimination learned a successive brightness discrimination more rapidly than those initially trained on a simultaneous texture discrimination. Since the variable irrelevant stimuli in Stage 1 were the same for both groups (namely position), we should have to make the distinctly implausible assumption that a brightness analyzer can be suppressed by training on a texture discrimination between two mid-gray alternatives. Since there is no evidence to support such an assumption, it is perhaps safer to conclude that, although often difficult to distinguish from the consequences of weakening irrelevant analyzers, selective strengthening of relevant analyzers does occur and is responsible for much of the data we have been reviewing.

## V. Summary

Chapters 4 and 5 established that the amount learned about one cue is partly a function of the nature of the other cues in the experimental situation and of the subject's prior experience with those cues. This constitutes the basic evidence demanding the incorporation of some attentional, selective processes into any theory of discrimination learning. The present chapter establishes the converse of one of these points, namely, that the amount learned about one cue is a function of the subject's prior experience with that particular cue. Three classes of experiment suggest that prior experience with a particular relevant cue enhances subsequent performance when that cue is again relevant. In all three cases, a problem has been to distinguish between the effects of stimulus distinctiveness, and those of directly appropriate response transfer. In studies of acquired distinctiveness, this problem is (hopefully) solved by varying the response requirements between training and testing situations; in studies of intra- and extradimensional shifts, the problem is solved by changing the actual values of the stimuli between training and testing; studies of transfer along a continuum have not yet got very far in separating these two possibilities.

A fourth class of experiment suggests that animals trained with a particular cue irrelevant may learn to suppress attention to that cue. This raises the possibility that many other results may be interpreted in this way. Although some of the evidence reviewed in this chapter does support such an interpretation, and none unequivocally contradicts it, we remain skeptical. It is true that there is excellent evidence that irrelevant stimuli fail to *acquire* control when presented in conjunction with more relevant stimuli, but the evidence that they are actually *suppressed* is distinctly less convincing. In Chapter 9, we shall return to the general question, when we consider whether the main effects of overtraining on a discrimination problem are to strengthen relevant analyzers or to weaken irrelevant analyzers. In this particular case, there is, in fact, some evidence pointing to the former interpretation.

CHAPTER 7

# Generalization

Most textbooks on animal learning contain a chapter or section entitled, "Generalization," preceding a chapter or section entitled, "Discrimination." The nature of the relationship between the two is implied by the order in which they are treated, as is illustrated by the following quotations from Kimble's (1961) revision of Hilgard and Marquis' *Conditioning and Learning*:

> When an organism has been conditioned to respond to a particular stimulus it can be shown that other similar stimuli will also elicit the response even though these other stimuli have not been used in training. This ability of different stimuli to evoke a conditioned response is known as stimulus generalization [p. 328].

Later he says:

> Although the tendency for a response to generalize has adaptive value in many situations, there are some circumstances which present an obvious necessity for inhibiting this tendency. . . . There must be an influence which restricts the range of generalization and restrains the organism from making the same response to all physically similar stimuli. This influence is called discrimination [*op. cit.*, p. 361].

The underlying assumption is that is there some process that causes the habit strength or excitation built up by rewarding a given response to a given stimulus to generalize (spread, irradiate) to other stimuli similar to the original; and that some active inhibiting force must be applied to suppress this generalized habit strength. Although the physiological language may have been dropped, these ideas stem directly from Pavlovian notions of the irradiation of excitation and inhibition in cortical analyzers.

An alternative view of generalization was first implied in an article by Lashley and Wade (1946) and has since been more clearly expounded by Prokasy and Hall (1963). According to this view, responses to stimuli other than the training stimulus occur not because of a process called generalization, but because the subject either cannot or does not discriminate between test and training situations. While such a view may not be entirely correct, it seems to us indisputable that it points to an important determinant of some instances of generalization. Indeed, in this chapter we shall be discussing only those aspects of generalization that may most easily be represented in these terms. We shall not discuss generalization experiments (such as those on peak shifts, e.g., Hanson, 1959) whose results imply the interaction of various *processes* of generalization.

# I. Lashley and Wade's Account of Generalization

Lashley and Wade (1946) subjected what they called the "Pavlovian theory of generalization" to a thorough and detailed criticism, and proposed in its place an entirely different explanation of the results of generalization experiments. Until recently, this paper of Lashley's received less than its fair share of attention (it is not even reprinted in the selection of Lashley papers edited by Beach, Hebb, Morgan, and Nissen, 1960), partly perhaps from a feeling that the experimental results reported in it were somewhat dubious. It is true that, as so often, Lashley's experiments were no match for Lashley's ideas; but the experiments reported by Lashley and Wade, although certainly inconclusive, are not as worthless as has sometimes been suggested.

The substance of Lashley and Wade's argument may be expressed as follows:

(1) "The phenomena of 'stimulus generalization' represent a failure of association [p. 74]." This somewhat cryptic remark may be taken to mean that an animal will respond to a generalized stimulus to the extent that the trained response has not been associated with that feature of the training stimulus which is varied in a subsequent test of generalization. The occurrence of generalized response is due not to any "spread of effect" of training, but, in the first instance, to a failure of discrimination between training and test stimuli, itself caused by the direction of the subject's attention during aquisition. The reason why, at the outset of, say, aversive conditioning, very widespread (even substantial cross-modality) generalization can be obtained, is not that the subject is incapable of discriminating a light from a buzzer, but that at this stage of training the only feature of the light about which the subject has stored information is that it produces a sudden change in the overall level of stimulation.

(2) The slope of a gradient of generalization is not determined by the physical similarity of training and test stimuli, but by their *subjective* similarity; this subjective similarity depends not only upon physical similarity, but also upon which features of the generalization stimuli are attended to during the test. Responses may, therefore, occur in a generalization test not only because the subject attended to some other feature during training, but also because attention is directed to other features during testing.

(3) Just as the occurrence of generalized responses represents a failure of discrimination (or attention), so their absence represents successful discrimination, and a gradient of generalization that has a significant slope along a given dimension implies that the subject attended to this dimension both during training and during testing.

(4) We come to that aspect of Lashley and Wade's argument that has attracted most experimental attention: "The 'dimensions' of a stimulus series are determined by comparison of two or more stimuli and do not exist for the organism until established by differential training [p. 74]." This again is an ambiguous remark, for Lashley and Wade nowhere specify what "differential training" involves.

On the one hand, it is not clear whether differential training implies discrimination training by the experimenter in the experimental situation, or whether mere experience with more than one value of a given dimension at some stage of the subject's life (provided that the different values were correlated with different reinforcing events) is sufficient to establish a dimension. The first interpretation is implied by the design of Lashley and Wade's own experiments; the second by their argument that since they are familiar with so many dimensions, human adults are unsuitable subjects with which to test the hypothesis. Moreover, even if it is accepted that mere prior experience with different values is not sufficient to establish a dimension, the nature of the discrimination training regarded as necessary is nowhere specified. As we shall see, discrimination training may be explicitly programmed by the experimenter, or it may occur more or less fortuitously—it may be along a dimension orthogonal to that varied in the generalization test, or it may be along the test dimension. Lashley and Wade never stated which of these conditions they regarded as sufficient to generate a sloping gradient of generalization.

There is a large body of published research that is directly or indirectly relevant to Lashley and Wade's fourth claim, and much of this chapter will be devoted to a discussion of these experiments. Before entering upon such a discussion, however, we should make our own position clear. To a large extent, we agree with Lashley and Wade's first three claims. In the terminology of our theory, we argue that the slope of a generalization gradient is a direct function of the strength with which the analyzer specific to the test dimension was switched-in, both during initial training and during testing. Consider a subject rewarded for responding to a red circle. If it uses only a color analyzer to discriminate the stimulus from the background, then, when given tests with orange and green circles, it will respond more slowly, i.e., show a sloping generalization gradient; but if given tests with red squares and triangles, it will respond to these at the same rate as to the training stimulus, i.e., will generalize completely to different shapes. Where we depart from the position of Lashley and Wade, however, is by arguing that the determinants of the strength with which analyzers are switched-in, and therefore of the slope of generalization gradients, are not only various types of discrimination training, but at a more fundamental level, the rules of our model—especially Rules 1 and 2. It is true, as we shall see, that discrimination training has important effects on the slope of generalization gradients; but, as we shall also see, the data on generalization gradients cannot be explained by any account in which the role of discrimination training is treated as a primitive axiom. In our model, most of the effects of discrimination training can be deduced from Rule 2 which states that analyzers are strengthened only if their outputs are correlated with differences in reinforcing events. Lashley and Wade's fourth claim will not survive detailed examination, largely because it can be given three distinct and mutually inconsistent interpretations. Evidence can be cited in support of each of these interpretations, but evidence favoring one necessarily contradicts another. The rules of our model, on the other hand, not only account for the substantial body of data agreeing with one or another version of Lashley and Wade's claim, but also provide a unified explanation of why each of the interpretations can be partly correct.

The plan of the present chapter, therefore, will be first to consider the evidence supporting the three possible interpretations of Lashley and Wade's claim about the role of discrimination training. In the light of this evidence, we shall amplify the above assertion that no single coherent interpretation of data can be provided within the framework of Lashley and Wade's account. We shall then offer, in detail, our own explanation of the role of discrimination training. The chapter will end with a discussion of a number of other phenomena of generalization interpretable within our own theory. We repeat: our treatment of generalization is not intended to be complete; we are only interested in an interpretation of those variables known to affect the slope of generalization gradients.

## II. Effects of Differential Training on the Slope of Generalization Gradients

The three versions of Lashley and Wade's claim, in ascending order of strictness are as follows:

(1) Some prior experience is necessary for the formation of generalization gradients.

(2) General prior experience is not sufficient, since some explicit discrimination training is necessary.

(3) Not all types of discrimination training are sufficient, since gradients will only be formed if subjects have received explicit tıaining along the test dimension.

## A. The Necessity of Some Prior Experience

The weakest interpretation of Lashley and Wade's claim is that a subject deprived of all experience of variation along a given dimension will show a perfectly flat generalization gradient along that dimension, i.e., that some prior experience with different values of a dimension is a necessary condition for the appearance of sloping gradients along that dimension.

Ganz and Riesen (1962) performed the first test of this hypothesis in a study of wavelength generalization in rhesus monkeys. Two groups of four infant monkeys were separated from their mothers at birth; for 10 weeks, one group was reared in total darkness, the other normally. Both groups were then transferred to the experimental situation: all animals lived in total darkness except for half an hour each day, during which they received diffuse monochromatic light projected to the right eye only. They were trained to press a key for sucrose reward in the presence of light of this wavelength, and were subsequently given a series of unreinforced generalization tests to stimuli of different wavelengths, preceded, on each day, by reinforced training in the presence of the original positive stimulus. (The procedure is adapted from that developed by Guttman and Kalish, 1956, in their work with pigeons, and is discussed in more detail below.) On the first test day, subjects reared in darkness showed very nearly complete generalization to all other colors, whereas subjects reared normally showed sloping gradients. The results are shown in Figure 7.1. During the course of successive daily tests, however, the deprived subjects showed increasingly steep gradients, until after seven day's testing, the slope for these subjects was actually steeper than that for normal subjects.

The only valid test of Lashley and Wade's hypothesis, of course, is the result of the first day's test, and here the hypothesis receives some support, although no statistical analysis of performance on this day alone is reported. After this first day, subjects received discrimination training, since the generalization stimuli were presented without reward, while the positive stimulus was presented not only with the generalization stimuli, but also during rewarded retraining trials. However, the fact that during the course of generalization testing, the deprived subjects came to produce steep gradients is of considerable

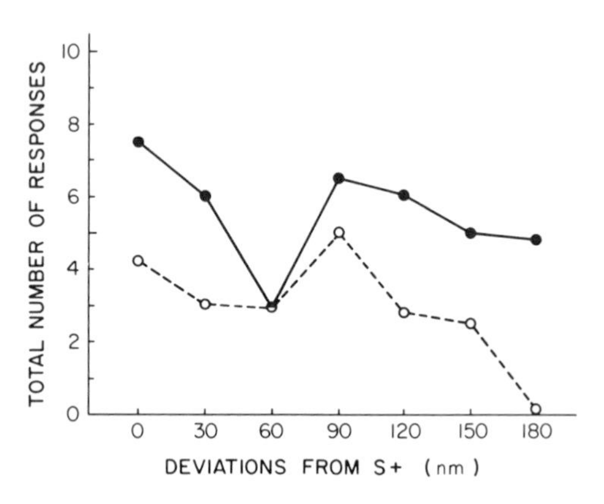


**Fig. 7.1.** Wavelength generalization in visually deprived and experienced monkeys on first day of testing. (●): dark-reared; (○): control. [*After* Ganz and Riesen (1962). Copyright by the American Psychological Association. Reproduced by permission.]

importance. It suggests that there can have been little wrong with their perceptual abilities. They were capable of discriminating one color from another, and as soon as they received differential reinforcement, they did so; their relatively complete generalization on the first day does not represent inability to distinguish different colors, but a failure to classify the positive stimulus as being of a given color, i.e., a failure to *utilize* a color analyzer.

The results of an experiment on wavelength generalization in birds, performed about the same time, seemed to confirm this conclusion. Peterson (1962) trained two groups of ducklings to peck at a key illuminated with a 589-nm sodium light and then gave generalization tests to a series of different wavelengths. One group had been reared normally, the other in individual cages with white walls illuminated with monochromatic light of 589 nm. This second group, therefore, had received no experience of variations in wavelength. While the normal birds gave a sharply sloping gradient of generalization peaking at 589 nm, the subjects reared in monochromatic light produced a flat gradient.

Peterson's results, however, have proved difficult to replicate. Rudolph, Honig, and Gerry (1969), for example, have reported an extensive series of experiments in which neither monochromatic rearing nor dark rearing succeeded in flattening wavelength generalization gradients in birds. In different experiments, they used either domestic chicks or Japanese quails and reared their experimental subjects either in total darkness, 589-nm sodium light, or light of a dominant wavelength of 530 or 630 nm. In all cases, their experimental subjects produced gradients of significant slope, indeed, in some cases, of a slope steeper than that generated by normally reared birds. While none of their studies was intended as an exact replication of Peterson's experiment, the most reasonable conclusion must be that Peterson's results represent the exception rather than the rule to be expected from such experiments with birds. This is confirmed by the results of three other studies, in none of which did deprivation conditions significantly flatten wavelength gradients in birds (Malott, 1968; Mountjoy & Malott, 1968; Tracy, 1970).

The implications for the first version of Lashley and Wade's hypothesis are, therefore, conflicting. Ganz and Riesen's results provide some support for the hypothesis, although from Figure 7.1, it is apparent both that the control subjects' gradient is only slightly steeper than the experimental subjects', and that even the experimental subjects responded more frequently to their training stimulus than to any other stimulus. In view of the small number of subjects (six) used in Peterson's experiment, and the results of Rudolph *et al.*, it is difficult to accept the conclusion that some prior experience with different values of a dimension is always (or even often) necessary for the occurrence of sloping gradients along the dimension.

### B. The Necessity of Explicit Discrimination Training

Even if prior experience of variations along a dimension is not necessary for the occurrence of a sloping gradient of generalization, it may be sufficient. In other words, we may ask whether some sort of explicit discrimination training is also necessary. If discrimination training is necessary, then we must also ask whether the training need be along the dimension to be used in the generalization test, or whether any discrimination training is sufficient. We shall first consider the weaker claim that some sort of explicit discrimination training is necessary, but that any sort is sufficient.

#### 1. *Complete Generalization in the Absence of Discrimination Training*

Presumptive evidence in favor of this claim is provided by any experiment in which no explicit discrimination training is given and in which perfectly flat generalization gradients are obtained. There are not many such experiments reported in the literature, but this may be partly because they have been regarded as yielding " negative results " without theoretical interest. Some such negative results, however, have been published, and in some cases with very large numbers of subjects. Jensen and Cotton (1961) and Thompson (1962) tested for generalization along the dimension of size; although a total of 413 rats was tested in the two experiments, no gradient was obtained in either. McCain and Garrett (1964) found no gradient of brightness generalization in 145 rats; Margolius (1963) has obtained similarly negative results in a study of brightness generalization; and Ferster (1951) after training rats to press a lever in the presence of a continuously sounded buzzer and continuously illuminated lamp, found equal rates of response in extinction whether or not the buzzer and light were present. Jenkins and Harrison (1960) found little or no evidence for a gradient of auditory frequency generalization in pigeons, while Newman and Baron (1965) found a completely flat gradient of generalization along an orientation dimension in pigeons, and Butter and Guttman (1957) also found a nearly flat gradient along this dimension.

A number of experiments normally considered under the heading of secondary reinforcement have obtained results that may be similarly interpreted. Notterman (1951) trained one group of rats for 50 trials to run down an alleyway for food. Upon entry into the goal box, a light was turned on. They were then extinguished without the light, and finally rerun with the light again on. The reintroduction of the light had no effect on running speed at all; thus the generalization gradient between "situation with light" and "situation without light" was flat. Experiments on secondary reinforcement such as this are usually interpreted as supporting the hypothesis that in order for a stimulus to become a secondary reinforcer it must be a discriminative stimulus, and, as Ferster (1951) said, "The crucial event in establishing discriminative control over behavior is the nonreinforced occurrence of the response in the absence of the positive stimulus [pp. 448–449]." This point of view has not usually been linked with the Lashley–Wade hypothesis, but since tests for secondary reinforcement depend either on demonstrating a generalization decrement in extinction (slower responding in situation without secondary reinforcer than in situation with secondary reinforcer), or on demonstrating a preference in a choice situation, it is clear that the connection exists. It simply amounts to saying that a stimulus cannot become a secondary reinforcer unless the animal attends to it.

Several of the experiments already cited in this section have gone on to provide more direct support for one version of Lashley and Wade's hypothesis by showing that although a flat generalization gradient is produced by nondiscriminative training, any explicit discrimination training produces significantly sloping gradients. Jenkins and Harrison (1960) compared generalization in two groups of birds. Both groups were originally reinforced for responding in the presence of a 1000 Hz tone, and tested for generalization to a series of tones ranging from 300 to 3500 Hz. One group received nondiscriminative training, that is to say, the positive tone was permanently on throughout each daily training session, and responses were reinforced on a variable interval (VI) schedule. The second group received discriminative training: the tone was on during only part of each session, and responses were reinforced in its presence, but not reinforced in its absence. The results are shown in Figure 7.2. Nondiscriminative training results in virtually complete generalization to all test stimuli; discrimination training between the presence and the absence of the tone results in a steep gradient to different frequencies.

Similarly, both Butter and Guttman (1957) and Newman and Baron (1965) who found flat or nearly flat gradients along an orientation dimension after nondiscriminative training, obtained significantly steeper gradients if subjects

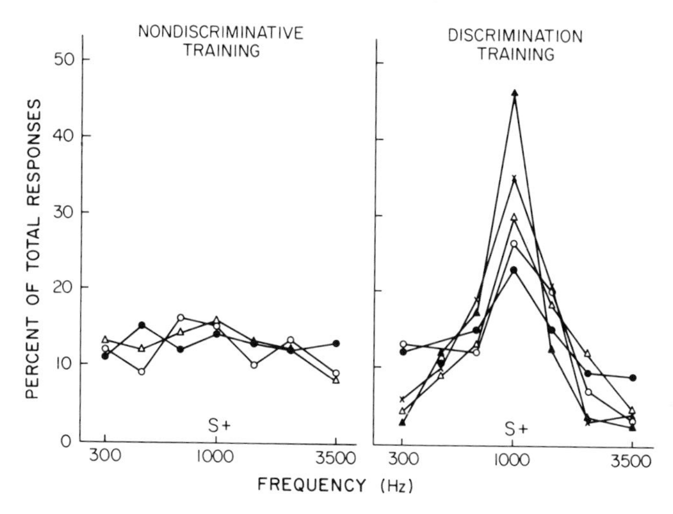


**Fig. 7.2.** Auditory frequency generalization in individual pigeons following nondiscriminative or discrimination training. [*After* Jenkins and Harrison (1960). Copyright by the American Psychological Association. Reproduced by permission.]

had been trained to discriminate the presence from the absence of the original stimulus. Notterman's study of secondary reinforcement obtained similar results: he showed that the greater the number of training trials on which his subjects were not rewarded in the absence of the light, the greater was the effect of the light on extinction running times. In other words, discrimination training produced significantly sloping generalization gradients where nondiscriminative training had produced gradients of zero slope.

### 2. *Gradients of Generalization in the Apparent Absence of Discrimination Training*

There is, therefore, a considerable amount of evidence in support of the claim that some sort of discriminative training is a necessary condition for the formation of gradients of generalization. On the other hand, there is a number of experiments whose results appear to show significantly sloping gradients in the absence of any formal training. We shall discuss here two such groups of experiments: first those by Grice (1948, 1951) which have been taken as refuting both Lashley and Wade's theoretical claims and as casting doubts on the validity of their experiments, secondly those initiated by Guttman and Kalish (1956) on wavelength generalization in pigeons. We shall argue that neither set of experiments succeeds in unequivocally disproving Lashley and Wade's position.

*a. Size Generalization in Rats.* Lashley and Wade reported a series of experiments with rats, the designs of two of which are illustrated in Table 7.I.

TABLE 7.I

**Design of Experiments by Lashley and Wade[a] and Grice[b, c]**

| | Group | Stage 1 + | Stage 1 − | Stage 2 + | Stage 2 − |
|---|---|---|---|---|---|
| Experiment I | + + | 8 | 0 | 8 | 5 |
| | + − | 5 | 0 | 8 | 5 |
| Experiment II | − − | 0 | 8 | 5 | 8 |
| | − + | 0 | 5 | 5 | 8 |

[a] Data obtained from Lashley and Wade (1946).
[b] Data obtained from Grice (1948, 1951).
[c] Here O designates blank door; 5, a 5 cm circle; 8, an 8 cm circle.

In the first experiment, the subjects were trained in a jumping stand to approach a black door with a filled-in white circle and avoid a plain black door. For one group (+ +), the circle was 8 cm in diameter, for the other (+ −), the circle was 5 cm in diameter. Both were then trained on a discrimination between the 8- and 5-cm circles, with the 8-cm circle positive for all subjects. The second experiment was of similar design, except that the subjects were originally trained to jump away from the door with a circle. It was assumed that in both experiments the subjects had learned either to approach or to avoid a single stimulus of a given size, and were then tested for size generalization. If a significantly sloping gradient of generalization were formed along the size dimension, then subjects would presumably continue to select (or avoid) the circle they had been trained with in the first stage of the experiment. In Experiment 1, the final size discrimination should be learned by Group + + faster than by Group + −; in Experiment II, Group − − should learn faster than Group + − . No such differences were found. Both these experiments were subsequently repeated by Grice (1948, 1951); in each case, Grice found that nonreversed subjects (Group + + and − −) learned faster than reversed subjects (Group + − and − +). His results have been widely accepted as providing decisive disproof of Lashley and Wade's hypothesis (e.g., by Osgood, 1953, p. 449; and by Kimble, 1961, pp. 369–371), but this enthusiastic acceptance has overlooked two striking features of his results. (1) In the first experiment, where both groups were tested with the 8-cm circle positive, the reversal group scored above chance over Trials 1–20 of the reversal. To explain this surprising result, Grice suggested that rats have a strong preference for the larger of two stimuli in a size discrimination. In his second experiment, where both groups were tested with the 5-cm circle positive, the reversal group again scored above chance over 1–20. It can hardly now be supposed that this was due to a preference for the smaller of two stimuli. (2) In his second experiment, the reversal group learned the test discrimination with an average of 11.7 errors. Naive rats learn the same discrimination in the same apparatus with an average of 14.3 errors (Grice, 1949), i.e., with more errors than rats supposedly learning a reversal.

These two findings suggest that the reversal groups in these experiments may not, after all, have been learning a reversal. Because of the nature of the design, this is not so suprising as it may sound. In the first stage of the experiments, subjects were not, in fact, trained with a single stimulus; they were trained to discriminate a white circle from a black door—a problem presumably soluble as a brightness discrimination. The Stage 2 problems—a discrimination between two white circles of different areas—could equally have been solved as a brightness discrimination. If this possibility is accepted, then inspection of the experimental designs shown in Table 7.I reveals that, in both experiments, both reversed and nonreversed groups were trained in Stage 2 in the same direction as they had been trained in Stage 1. In the first experiment, both groups, having initially learned to approach a white circle, were trained with the brighter (larger) circle positive; in the second experiment, both groups were initially trained to approach a dark door and were then trained with the darker (smaller) circle positive.

If this analysis is correct, to obtain reversal groups in such a situation, it would be necessary first to train subjects to approach either the 8- or the 5-cm circle (the brighter of two stimuli), and then to train them on the large- versus small-circle discrimination, with the small (darker) circle positive. This has been done in an experiment by Mackintosh (1965f) with results clearly supporting the present interpretation: subjects trained, e.g., with the 5-cm circle positive in the first stage, and then with the 5-cm circle positive versus the 8-cm circle negative (a nonreversal according to Grice), took four times as long to learn their test problem as the equivalent of Grice's "reversal" group. In neither Lashley and Wade's nor Grice's experiments, therefore, did subjects receive single stimulus, nondiscriminative training. The subjects learned two consecutive brightness discriminations, and Grice's results provide no evidence for the existence of sloping gradients of generalization after training with a single stimulus.

*b. Wavelength Generalization in Birds.* A second series of experiments that has been interpreted as showing that no discrimination training is necessary for the formation of gradients of generalization was initiated by Guttman and Kalish in 1956. They trained pigeons to peck at a key illuminated with light of a specific wavelength. During training, responding was reinforced on a VI schedule. On such a schedule, rewards are given for the first response occurring after a certain interval of time, but this interval, although it has an average value of, say, 1 min may range from 5 sec up to 2 min or more. This type of reinforcement schedule produces a stable and reasonably fast rate of responding, combined with great resistance to extinction. The advantage is that a long series of unrewarded generalization trials can be given without the subject ceasing to respond. In Guttman and Kalish's generalization tests, a wide range of different wavelengths was programmed to appear on the response key, and generalization was measured by calculating response rate as a function of wavelength. They obtained reliable and orderly gradients of wavelength generalization, with peaks at the training stimulus.

The number of studies of wavelength generalization in pigeons and other birds is, by now, extremely large (we have, indeed, already discussed several such examples in the series of studies by Rudolph *et al.* 1969). For our present purposes, the important feature of the experimental situation is that, at any rate according to the intentions of the experimenter, no explicit discrimination training is involved. Nevertheless significantly sloping gradients are virtually always produced and, as Guttman (1956) pointed out, this casts some doubts on Lashley and Wade's hypothesis. These doubts, however, have recently been partly allayed by an ingenious study of Heinemann and Rudolph (1963). Noting the difference between the results, on the one hand, of Jenkins and Harrison (1960), where a flat gradient was produced after training in the presence of a tone, and Ferster, where a flat gradient was produced after rats were trained to bar-press in a continuously illuminated cage, and on the other hand, those of Guttman and Kalish, where a sloping gradient was produced after training to peck at an illuminated key, Heinemann and Rudolph (1963) suggested that

> If the visual stimulus is presented on the response-key so-called nondifferential training methods actually do involve some differential training, at least in the sense of training to discriminate between the presence and absence of the stimulus. . . . Behavior such as holding the head at the appropriate height, moving the head forward, and so forth, is reinforced when it occurs near the key (when the visual stimulus is likely to be imaged on *S*'s retinas), while similar behavior goes unreinforced if it occurs in some place other than the immediate vicinity of the key (when the visual stimulus is much less likely to be imaged on the *S*'s retinas). It is obvious that such differential training could not occur if the visual stimulus completely surrounded the *S*, as in the experiment by Ferster. Furthermore, it would seem that the smaller the area of the stimulus, the smaller would be the likelihood of its being present on the *S*'s retinas during the occurrence of any given unreinforced sequence of behavior. Hence, the amount of differential training that occurs should be related to the geometric size of the stimulus [p. 654].

To test this prediction, Heinemann and Rudolph trained three groups of pigeons to peck at a response key surrounded by a small, medium, or large gray card. They were then tested for generalization to brighter and darker cards of the same size as that used for each group in training. As is shown in Figure 7.3, the prediction was confirmed. The gradient following training to the largest stimulus was completely flat, that following training to the smallest stimulus was sharply sloped, while training with the medium-sized stimulus produced gradients of intermediate slope. Although these results may be partly confounded by differences in discriminability between test stimuli under the three conditions, they do suggest first that some discriminative training (even if it is inadvertent) is necessary to produce sloping generalization gradients along a brightness dimension in pigeons. Second, and of more general importance, Heinemann and Rudolph's study shows that effective discrimination training may occur even if it is not explicitly programmed by the experimenter.

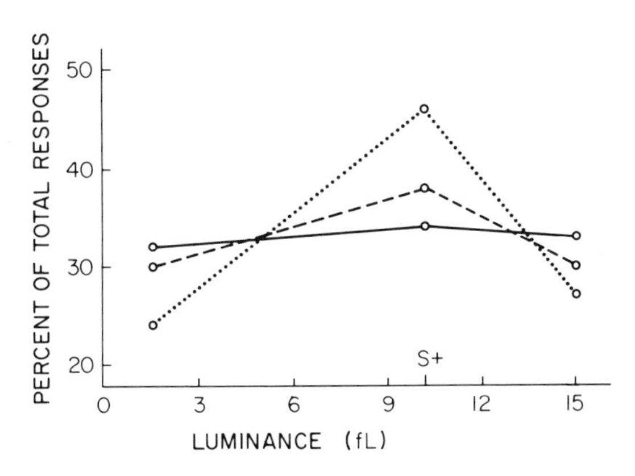


**Fig. 7.3.** Brightness generalization in pigeons following training with stimuli of different size. (—): large; (– – –): medium; (···): small. [*After* Heinemann and Rudolph (1963).]

The importance of this second conclusion will emerge when we discuss other studies that appear to have demonstrated sloping gradients of generalization after what was intended to be nondiscriminative training. Although several such studies exist, it must be admitted that the second version of Lashley and Wade's hypothesis has fared relatively well. Nondiscriminative, single-stimulus training is frequently insufficient to produce a sloping gradient of generalization, and must be supplemented with some type of discrimination training.

## C. Necessity of Relevant Discrimination Training

In the majority of the studies just discussed, discrimination training has been provided between the presence and absence of a given stimulus. With few exceptions (e.g., Ferster, 1951, Notterman, 1951), however, generalization has been measured not between the presence and absence of the stimulus, but along a specific stimulus dimension. Jenkins and Harrison (1960), for example, reinforced pigeons for responding in the presence of a 1000-Hz tone, and did not reinforce responses in the absence of the tone; they then, however, tested for generalization along the dimension of auditory frequency. The fact that this procedure produced sharp gradients of generalization suggests not only that *some* type of discrimination training is a necessary condition for the appearance of sloping gradients, but that any type is sufficient. This is equivalent to saying that specific training along the test dimension is not a necessary condition for the appearance of sloping gradients. Nevertheless, some studies have shown that specific relevant discrimination training has effects on generalization gradients over and above those ascribable to discrimination training per se; and several other studies have shown that such specific discrimination training may be a necessary condition for the formation of significantly sloping gradients of generalization.

1. *Additional Effects of Relevant Discrimination Training*

Hanson (1959) and Honig, Thomas, and Guttman (1959) compared gradients of wavelength generalization in pigeons after subjects were trained either to respond to a single wavelength (and not respond to a dark key), or after they received wavelength discrimination training. In the latter experiment, for example, a control group was reinforced for responding to a 550-nm key, while a discrimination group, in addition to being reinforced on the 550-nm key, was also not reinforced for responding to a 570-nm key. Not surprisingly, in generalization tests, the discrimination group showed a markedly slower rate of responding to wavelengths longer than S−. More importantly, the gradient for the discrimination group was sharpened not only in this direction, but also to wavelengths substantially shorter than S+. Sharpening of the gradient in the region of S− might be interpreted as the effects of extinction on response strength in this region. Sharpening in the opposite direction, however, cannot be so interpreted, especially since, in one of their experiments, the discrimination group showed a higher rate of response to stimuli immediately below the positive (the phenomenon termed "peak-shift" by Hanson). The results suggest, therefore, that relevant discrimination training sharpens the gradient by strengthening the relevant analyzers. In neither of these experiments was the discrimination training given to the control subjects very stringent; it consisted only of alternating S+ with brief periods of "time out" during which the key was not illuminated at all. Although some discrimination training must be involved here, pigeons do not normally peck at a dark key, and the procedure therefore lacks one of the features of normal discrimination training (as received by experimental subjects), namely, the occurrence of nonreinforced responses to S−. An experiment by Jenkins and Harrison (1962), however, shows that a generalization gradient that is sharpened by explicit discrimination training between the presence and absence of a stimulus is yet further sharpened by relevant discrimination training along the test dimension. In an earlier study already referred to (Jenkins & Harrison, 1960), these authors had shown that gradients of auditory frequency generalization in pigeons were completely flat after nondiscriminative training with a 1000-Hz tone, but were sharpened if subjects were trained to discriminate between the presence and absence of this tone. In the experiment under discussion, they retrained two of the subjects of the earlier study on a frequency discrimination between the 1000-Hz tone positive and a 950-Hz tone negative. The frequency generalization gradients obtained before and after this discrimination training are shown in Figure 7.4. The effect of relevant discrimination was to lower the overall rate of

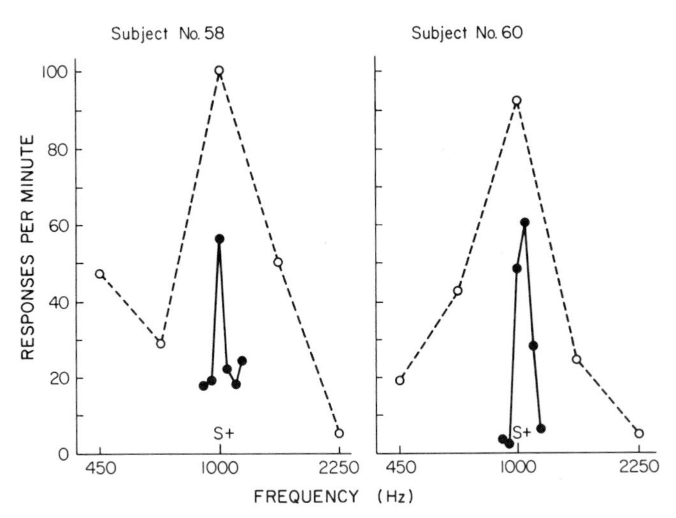


**Fig. 7.4.** Auditory frequency generalization in pigeons following discrimination training either between the presence and absence of a 1000-Hz tone, or between a 1000- and a 950-Hz tone. (○): S−, no tone; (●): S−, 950-Hz tone. [*After* Jenkins and Harrison (1962).]

responding (as the authors note, this may have been the result of the repeated conditioning and extinction sessions given to these subjects). More importantly, the gradient is considerably sharpened, both in the direction of S−, and in the opposite direction. This effect, the authors write, "cannot be reconstructed by any simple rational method of combining excitatory and inhibitory gradients [p. 440]." It directly supports the contention that discrimination training has even more marked effects on generalization gradients when given along the test dimension than when given along a less specific dimension.

### 2. *The Necessity of Relevant Discrimination Training*

Several studies have shown that nonspecific discrimination training (along any dimension) may not be sufficient to produce sloping gradients of generalization at all, and that more specific training procedures may be necessary.

TABLE 7.II

**Design of Experiment by McCaslin, Wodinsky, and Bitterman[a,b,c]**

| Groups | Stage 1 | | Stage 2 | | Stage 3 | | | |
|---|---|---|---|---|---|---|---|---|
| | | | | | Nonreversal | | Reversal | |
| | + | − | + | − | + | − | + | − |
| I | B | W | H | V | H | h | h | H |
| II | B | W | H | W | H | h | h | H |

[a] Data obtained from MacCaslin, Wodinsky, and Bitterman (1952).
[b] B designates black; W, white; V, broad vertical striations; H, broad horizontal striations; h, narrow horizontal striations.
[c] The design for only two groups is shown; other groups were trained in the opposite direction, and with different values of the relevant dimensions.

MacCaslin, Wodinsky, and Bitterman (1952) performed a somewhat complex experiment, the design of which is summarized in Table 7.II. In the first stage of the experiment, all subjects learned a discrimination between black and white cards. In the second stage, one group learned to discriminate between horizontal and vertical striations; the second group was trained with horizontal or vertical striations positive, but their negative stimulus was the black or white card that had been negative in State 1. The first group, therefore, had to learn an orientation discrimination; the second group merely continued to avoid the old negative brightness and select a stimulus of intermediate brightness (which happened to contain black and white striations). In the final stage of the experiment, all subjects were tested for generalization from their S+ of Stage 2 to another set of striations differing from S+ in width of stripe. (The generalization test was conducted by training all subjects to discriminate the two stimuli, half with their former positive remaining positive, half with their former positive now negative. Presumably, this is a very sensitive test.) MacCaslin *et al.* found that the first group showed a steeply sloping generalization gradient along the stripe-width dimension, while the second group generalized completely from one stripe width to the other.

Both groups, of course, had received discrimination training in the second stage of the experiment, with striations of a given width as S+. The authors interpret their results to mean that the difficulty of this discrimination determined whether or not a sloping generalization gradient would be produced: only if subjects were trained on a relatively difficult discrimination (as in the first group) did they notice the stripe width used in Stage 2; if they were trained on an easy problem (merely to continue avoiding their old negative) they failed to notice this feature. An alternative interpretation of their result is possible. One might argue that a single analyzer is capable of detecting differences both in the orientation and in the width of a set of striations. Therefore, the first group in Stage 2 learned to utilize an analyzer that detected differences along the test dimension, and naturally showed a sloping gradient of generalization. The second group, however, could continue to use a brightness analyzer in Stage 2; since this analyzer was already at a high strength at the outset of Stage 2, no other analyzer would be strongly switched-in, and they generalized completely across all values of the orientation–stripe-width dimensions. Using either interpretation, however, the results show that discrimination training is not a sufficient condition for the appearance of generalization gradients.

Newman and Baron (1965) tested four groups of pigeons for generalization along an orientation dimension. As already mentioned, a group given nondifferential training to a vertical line generalized completely to all other orientations. A second group reinforced in the presence of the vertical line and not reinforced in its absence, showed a sharp gradient. Thus far, the experiment adds nothing new to the studies discussed in the previous section. For our present purpose, the important group in Newman and Baron's experiment was the third. This group, like the second, was initially reinforced in the presence of the vertical line and not reinforced in its absence. During this discriminative training, however, there was another relevant cue: the vertical line (S+) was shown on a green key; S− (no line) was a red key. This group generalized completely to all orientations shown during the test session. Thus, discrimination training between the presence and absence of a line failed to result in a gradient along an orientation dimension when the discrimination could be solved by use of a (presumably dominant) color analyzer.[1]

[1] All groups in Newman and Baron's experiment were given generalization tests to lines on a green background. In a more recent study, Newman and Benefield (1968) have shown that a group given nondifferential training to a vertical line on a green background may show a sloping gradient if tested with lines on a black background. However, even if the presence of the colored background served to flatten the gradient, which might have been detected in Newman and Baron's third group had they been tested with black backgrounds, the important conclusion remains that the addition of a second relevant cue flattened the gradient shown along another dimension.

The studies mentioned so far have contrasted single-stimulus, non-differential training with some type of differential training immediately preceding the generalization test. In an experiment whose design is illustrated in Table 7.III, Mackintosh (1965e) gave three groups of rats single-stimulus

TABLE 7.III

**Design of Experiment by Mackintosh[a,b,c]**

| Groups | Stage 1 | Stage 2 | Stage 3 | | | |
|---|---|---|---|---|---|---|
| | | | Nonreversal | | Reversal | |
| | | + | + | − | + | − |
| Relevant pretraining | Successive brightness discrimination | BH | BH | WH | WH | BH |
| Irrelevant pretraining | Successive orientation discrimination | BH | BH | WH | WH | BH |
| Control | — | BH | BH | WH | WH | BH |

[a] Data obtained from Mackintosh (1965e).

[b] BH designates a black horizontal; WH, a white horizontal rectangle.

[c] The design for only three groups is shown; other groups were trained with BV, WH, or WV in Stage 2, and with appropriately different stimuli in Stage 3.

training, but for two of the groups, this stage of the experiment was preceded by discrimination training. One group was trained on a successive brightness discrimination; the second on a successive orientation discrimination; while the third (control) group received no training at this stage. All subjects then received single-stimulus training with a rectangle of given orientation and brightness. The experiment utilized a factorial design; for simplicity, we shall consider only subjects trained with a black horizontal rectangle. In the third stage of the experiment, they were tested for generalization; half of each group along a brightness dimension (from the black horizontal rectangle to a white horizontal rectangle), the other half along an orientation dimension (from the black rectangle to a black vertical rectangle). As in the experiment of MacCaslin *et al.*, generalization was measured by requiring subjects to learn a simultaneous discrimination with the single stimulus either positive or negative. Subjects that had received, in the first stage of the experiment, successive discrimination training along the same dimension that was used in the generalization test, showed steep gradients. Subjects that had received discrimination training on the dimension not used in their generalization test, showed flat (not significant) gradients. Control subjects showed intermediate gradients, significantly less steep than those of the first group and significantly steeper than those of the second. While none of the subjects received differential training during initial exposure to the single positive stimulus, it is clear that prior discrimination training has a marked effect on subsequent generalization. Most important, this prior training can act significantly to flatten the gradient if given on an irrelevant dimension. Once again, the mere occurrence of discriminative training is not sufficient to produce sloping gradients of generalization; the discrimination training must specifically involve the same dimension as that used in the test for generalization.

## III. Criticism of Lashley and Wade's Position

We have now discussed three possible interpretations of Lashley and Wade's claim that gradients of generalization only appear after discrimination training, and have found some evidence to support each of these interpretations. The very extent and diversity of this supporting evidence should give one pause: it cannot be the case that all three versions of the hypothesis are correct, for, as we have shown, the claims can be ranked in order of increasing strictness. Proof that only certain appropriate training procedures will produce gradients of generalization, while supporting the third and strongest version of the hypothesis, is incompatible with the second version, according to which any discrimination training is sufficient. Proof of this second version is equally incompatible with the weakest claim of all: either explicit discrimination training is necessary or mere experience at other times of the subject's life will suffice. Thus, the very abundance of supporting evidence that we have reviewed, since it supports mutually incompatible versions of the same hypothesis, actually serves to discredit the hypothesis. It seems more likely that discrimination training is neither a necessary nor a sufficient condition for producing generalization gradients, but that such training is often effective because it affects some other factor that is of critical importance. This possibility is increased by the existence of a number of contradictions between the evidence already cited and further experimental results.

### A. Gradients of Generalization in the Absence of Discrimination Training

We have cited a number of studies in which generalization gradients were flat unless some explicit discrimination training was given. We have also presented evidence arguing against some apparent exceptions to this rule. Yet there are other exceptions, not so far considered.

For example, although Jensen and Cotton (1961) and Thompson (1962) found complete generalization along a size dimension in rats, Grice and Salz (1950), Margolius (1955, 1963), and Muntz (1963c) all obtained sloping gradients in experiments of very similar design. While Dusek and Grice (1968), following similar, single-stimulus training to a circle of given size, measured generalization by training on a size discrimination with the training circle either positive or negative, and also obtained significant gradients. Similarly, although McCain and Garrett (1964) and Margolius (1963) found complete generalization along a brightness dimension in rats, Brown (1942) and Spence (quoted by Hull, 1947) obtained significant gradients along this dimension.

It is true that the positive results from the size generalization studies can probably be reconciled with Lashley and Wade's hypothesis by arguing, as do Heinemann and Rudolph, that many supposedly nondiscriminative situations provide implicit discrimination training. In these experiments, rats were trained to run down a brief alley and push open a door in the center of a white circle to obtain food. It is clear that the probability of reinforcement was greater when the circle impinged on the subject's retina than when it did not. In the brightness generalization experiments, however, which used either diffuse illumination or painted alleyways, the argument is much less appropriate. The results of experiments by Hearst (1960, 1962) on brightness generalization in monkeys provide particularly strong evidence against Lashley and Wade's position. Hearst trained monkeys to pull a chain for food reinforcement on a VI schedule. Throughout each training session, the test chamber was continuously illuminated with a diffuse, overhead light. Generalization tests given in extinction to other levels of illumination revealed orderly and significantly sloping gradients. Similar results were also obtained with rats, although here the gradient was only just significant and much less orderly (this is not suprising when it is considered that several experiments on brightness generalization in rats have failed to obtain significant gradients). Since in Hearst's situation, the illumination was held constant throughout training (there were no time-out periods in darkness), and since the stimulus was as little localized as the tone in Jenkins and Harrison's study of auditory generalization in pigeons, the results cannot reasonably be reconciled with the suggestion that some sort of discrimination training must be given in order to obtain reliable gradients.

Lashley and Wade's position is questioned not only by the occurrence of gradients of generalization in the absence of implicit or explicit discrimination training, but also by the variablity of the results of the studies just considered. There do not seem to have been any differences, in the extent to which discrimination training occurred, which could account for the differences between the successful and unsuccessful studies. If, for example, discrimination training was implicitly provided in the size generalization situation, why did some of those experiments *fail* to obtain significant gradients?

Such contradictions occur equally when we compare generalization along different dimensions using the same type of subject and situation. Guttman and Kalish (1956), and many others since then, have obtained steep gradients of wavelength generalization in pigeons in the absence of explicit discrimination training. We can, indeed, point to two sources of discrimination training present in most of these studies, namely the occurrence of time-outs and the fact that the stimulus was localized on the response key. Now, if one or other of these conditions is sufficient to produce the steep wavelength gradients typically observed, why do they not produce equally steep gradients along any dimension? Although gradients of orientation (line tilt) are usually significant in pigeons, they are often relatively shallow (e.g., Butter & Guttman, 1957), and in one study, when subjects were trained to respond to a white line on a green background (Newman & Baron, 1965), the gradient was flat.

The fact that the same experimental procedure, incorporating the same amount of discrimination training, has quite different effects on subsequent generalization gradients, implies that the central position accorded to the role of discrimination training in Lashley and Wade's analysis is mistaken.

### B. Variable Importance of Discrimination Training

The discordant evidence so far discussed suggests first that gradients of generalization are sometimes formed in the absence of any discrimination training, and second that whether such a result occurs depends upon (as yet unspecified) details of the experimental situation and subject that cannot be interpreted by Lashley and Wade's account. There is, however, another problem to consider: why is it that discrimination training not specifically along the test dimension is sometimes *sufficient* to produce sloping gradients and on other occasions is not? Several studies have shown that if pigeons are trained to respond in the presence of a line of specific orientation and not to respond in the absence of the line, they subsequently show a significant gradient of generalization along the dimension of orientation (Butter & Guttman, 1957; Honig, Boneau, Burstein, & Pennypacker, 1963; Newman & Baron, 1965). Other studies, however, have found that such explicit but nonspecific training may *not* be sufficient to produce a gradient along the orientation dimension. As already mentioned, Newman and Baron trained another group of pigeons to respond in the presence of a vertical line on a green background, and not to respond in the presence of a red key with no line on it. They found complete generalization to all other orientations of the line.

Boneau and Honig (1964) have also shown that when subjects are specifically required to discriminate the presence of a vertical line from its absence, they do not necessarily show a gradient of generalization along the orientation dimension. These authors trained pigeons on a conditional discrimination: if the key was illuminated with light of 550 nm, then subjects were rewarded when a vertical line was shown on the key, and not rewarded when no line was shown; but if the key was illuminated with light of 570 nm, they were rewarded when no line was shown, and not rewarded when a vertical line was on the key. In generalization tests given after subjects had learned this discrimination, lines of different orientation were shown either on the 550-nm key or on the 570-nm key. Subjects responded at the same (high) rate to all orientations of line on the 550-nm key, and at the same (very low) rate whatever the orientation of the line on the 570-nm key. Despite having learned to discriminate the presence of a vertical line from its absence, they had failed to classify the lines as *vertical*; and produced completely flat gradients of orientation generalization.

Both Boneau and Honig's and Newman and Baron's studies show that subjects trained to discriminate along one dimension (presence versus absence of a line) may generalize completely along another dimension (orientation of the line). It is certain, however, that such discrimination training (presence versus absence of a line) usually does produce a sloping gradient along the orientation dimension. Once again, it does not seem possible to reconcile this apparent contradiction within the framework of a hypothesis that seeks to explain the occurrence of gradients of generalization solely in terms of the subjects' prior discrimination experience. Just as we saw that it was impossible to predict the slope of a generalization gradient from knowledge of whether subjects had received any discrimination training at all, so here it is no more possible to do so from knowledge of the nature of this discrimination training.

## IV. The Determinants of Generalization Gradients

It will be as well at this point to recapitulate the main conclusions reached in this chapter before elaborating our own interpretation of the role of discrimination training in determining generalization gradients. The difficulties inherent in providing a single, consistent account of Lashley and Wade's hypothesis of the role of discrimination training, and one that is, at the same time, compatible with the experimental evidence, seem to us to be formidable. This makes it all the more important to distinguish between two parts of the Lashley and Wade account. Their first, theoretical, suggestion is that the slope of a generalization gradient is determined by the direction of the subject's attention during training. Their second, empirical, suggestion is that discrimination training is an essential determinant of the direction of attention. We have argued that this second suggestion cannot be coherently interpreted, but such a conclusion in no way affects the validity of the first suggestion. Indeed, the evidence reviewed so far is totally incompatible with the most obvious alternative to the Lashley and Wade position, namely, the suggestion that gradients of generalization are automatically produced along all stimulus dimensions that subjects are *capable* of discriminating. The application of our own theory to the phenomena of stimulus generalization implies, in the first place, that sloping gradients of generalization will be formed only along such stimulus dimensions as are detected by the analyzers switched-in by the subject during initial training, and this assertion corresponds to an acceptance of Lashley and Wade's theoretical position. The further implication of our own theory is that the slope of generalization gradients along any given dimension will be determined by the rules governing the strengthening and weakening of analyzers detecting that or other dimensions, together with certain assumptions about the original weights given to the various analyzers. This corresponds to a rejection of Lashley and Wade's empirical suggestion.

In terms of our theory, there are two main determinants of the slope of generalization gradients: first, the initial dominance of the various analyzers, second, the effects of Rules 1 and 2 on altering this ordering of analyzers.

### A. The Organization of Analyzers

There is no single sufficient and necessary condition for the appearance of gradients of generalization that is valid for all stimulus dimensions for all organisms in all situations. What is sufficient to produce reliable gradients of generalization along the dimension of brightness in monkeys, namely, nondifferential single-stimulus training, is not sufficient to produce gradients along the dimension of auditory frequency in pigeons. Similarly, explicit differential training is frequently necessary for the appearance of orientation gradients in birds, but is not necessary for the appearance of wavelength gradients.

It is hardly suprising that along some dimensions, sloping gradients are automatically formed, for if this were not so, it would imply that the subject had switched-in no analyzer at all during original training. It is no more suprising that this does not occur along all dimensions, for that would imply that the subject had switched-in all analyzers during original training. The actual ordering of dimensions for a given organism, in terms of the amount and nature of the discrimination training required is of more practical than theoretical importance. It must reflect the dominance order in which analyzers are arranged for that organism—what Baron (1965) has called the organism's attending hierarchy. We can distinguish three possible determinants of this ordering.

### 1. *Innate Factors*

To the extent that the slope of generalization gradients depends upon an organism's sensory capacity, it is beyond dispute that innate factors enter into the determination of generalization gradients: it will always be easier to obtain sloping gradients of wavelength generalization in pigeons than in rats. If sensory capacity is innately determined, then it is equally likely that the ordering of analyzers is also innately determined; indeed, it is difficult to distinguish empirically between a lower sensory capacity and a lower ordering of the appropriate analyzer. It is reasonable to assume that the innate ordering of analyzers in birds differs from that holding for rats; in the former, color would be high and spatial orientation low; in the latter the ordering would be reversed.

Even when all else is held constant, therefore, we should not expect to find identical conditions sufficient and necessary to produce significant gradients along different dimensions in the same class of animal, or along the same dimension in different classes. The danger with such a statement is that not only may it be itself invulnerable to disproof, it may protect a complete theoretical system from disproof. If it forms part of such a system, however, it should have implications for other parts of the system. The relative dominance of different analyzers, according to our theory, should not only affect the conditions necessary for producing gradients of generalization, it should also determine, e.g., the effects of overtraining on reversal and nonreversal shift learning (see Chapters 9 and 12 for differences between color or brightness and other dimensions in such experiments with birds).

### 2. *Extraexperimental Experience*

The first version of the Lashley–Wade hypothesis was that some prior experience of variation along a particular dimension was necessary for the appearance of a gradient of generalization along that dimension. The experiments reported by Rudolph *et al.* (1969) showed that this form of the hypothesis is incorrect, and cast severe doubts on the earlier favorable evidence reported by Peterson (1962). Equally, since even on the first day of testing, the dark-reared monkeys in Ganz and Riesen's experiment (1962) responded more frequently to their training stimulus than to any of the generalization stimuli, Ganz and Riesen's results do not unequivocally support a strong form of this hypothesis. Since we have already argued (in Chapter 3) that the analyzing mechanisms for simple physical dimensions are innately organized, such a conclusion does not suprise us. What we should expect, however, is that prior visual deprivation will result in some reordering of the normal hierarchy of analyzers. Sutherland (1959) has suggested elsewhere that earlier findings on the effects of restricted rearing are as well explained in these terms as by the more common suggestion that perceptual learning is disrupted. Other things being equal, therefore, visual deprivation (dark rearing) would be expected to flatten visual generalization gradients, and this is exactly what Ganz and Riesen found.

### 3. *The Experimental Situation*

Finally, the initial relative strengths of different analyzers will depend upon numerous features of the experimental situation. For example, a pigeon is more likely to attend to the brightness of a response key if the key differs in brightness from its background than if it does not. Heinemann and Rudolph (1963) obtained steeply sloping brightness gradients when the response key was located on a dark background, but a flat gradient when the background was approximately the same brightness as the key.

It has been known for some time that animals learn to discriminate between stereometric stimuli more rapidly than between flat, two-dimensional stimuli (e.g., Harlow & Warren, 1952). It seems reasonable to suggest that stimuli standing out from the background are more likely to be attended to than those forming part of the background. This factor may well account for much of the variability in the results of size-generalization experiments with rats. Grice and Salz (1950), Margolius (1955, 1963), and Dusek and Grice (1968) used stereometric stimuli cut from metal, projecting 2 in. out from the background; they all obtained significant gradients. Jensen and Cotton (1961) also used stereometric stimuli, but ones that only projected 3/32 in., while Thompson (1962) used stimuli painted on the background; both failed to obtain significant gradients.

If differences of this sort can affect the results of generalization experiments it is likely that other conflicting results are also due to such minor details. Many of these details will be apparently trivial, and although their consequences may be impressive, the precise initial ordering of different analyzers in any situation is of more practical than theoretical interest. Other variations in experimental conditions, such as the nature of the discrimination training given, or the number of relevant cues available will also have important consequences, and these effects will also be of greater theoretical interest. We turn now to a discussion of such variables.

## B. The Role of Discrimination Training

### 1. *The Necessity of Some Discrimination Training*

Even if an organism enters the experimental situation with certain analyzers relatively strongly switched-in, single-stimulus, nondifferential training will be the least effective method for strengthening those analyzers.

Rule 2 of our model states that analyzers are strengthened only if their outputs predict differences in reinforcing events for the subject. The only sense in which this will occur in completely nondifferential training (e.g., with diffuse bright light) is that the subject is rewarded more consistently in the experimental apparatus than outside it (a point further discussed in Chapter 10), and this may be sufficient to maintain strength of initially dominant analyzers. In general, however, nondifferential training will not be effective in selectively strengthening initially weak analyzers relevant to the experimenter's arbitrarily chosen generalization dimension. Hence, some type of discrimination training will be necessary for the formation of generalization gradients along most dimensions for most organisms.

### 2. *The Sufficiency of Nonspecific Discrimination Training*

There will often be no need for this discrimination training to be given specifically between two points on the dimension used in the generalization test; any type of discrimination training may often be sufficient to strengthen the analyzer relevant to the test dimension. The difference between the presence and absence of a vertical line, for example, can be detected by an orientation analyzer (although it is more likely to be detected by a brightness analyzer), and the difference between the presence and absence of a tone can be detected by a frequency analyzer (although it is more likely to be detected by an amplitude analyzer). Thus discrimination training between the presence and absence of a line may serve to strengthen an orientation analyzer, and discrimination training between the presence and absence of a tone may serve to strengthen a frequency analyzer. On this basis, it is possible to explain the results of Butter and Guttman (1957) and of Newman and Baron (1965), on the one hand, and those of Jenkins and Harrison (1960) on the other. A similar explanation can be offered for the other studies reviewed under this head above.

### 3. *The Importance of Relevant Discrimination Training*

We predict that discrimination training will only sharpen generalization gradients if such training strengthens the analyzer specific to the test dimension. It is clear that relevant discrimination training will be more effective in achieving this than will nonspecific discrimination training. Hence, as Jenkins and Harrison (1962) have shown, pigeons trained on a frequency discrimination will show even sharper gradients of frequency generalization than will pigeons trained to discriminate the presence of the tone from its absence. Furthermore, in a number of situations, nonspecific training should be totally ineffective in producing gradients of generalization.

First, if a subject is trained on a discrimination problem that can be solved by the use of a very dominant analyzer in the subject's repertoire, such training will have little or no effect on other, weaker analyzers. In other words, a subject trained on a combined two-cue discrimination with one obvious and one inconspicuous cue, will learn little about the latter cue. Several studies showing this result were reviewed in the discussion of overshadowing in Chapter 5. The result follows from the assumption made in Rule 2 that analyzer strengths sum to a constant amount: the effective strength of a weak analyzer will be less if a dominant cue is included in the situation. Precisely the same explanation can be provided for the results of Newman and Baron (1965): they showed that while pigeons trained to discriminate the presence of a vertical line from its absence showed sloping gradients of orientation generalization, the addition of a dominant color cue to the discrimination problem resulted in a flat gradient of orientation generalization. If the discrimination problem could be solved in terms of color, the subjects did not learn to switch-in an orientation analyzer.

A second factor that will reduce the effectiveness of nonspecific discrimination training can also be derived from Rule 1. We suggested that nonspecific training will normally serve to strengthen a number of analyzers, including that relevant in the subsequent generalization test; discrimination training between the presence and absence of a line may (almost fortuitously) strengthen an orientation analyzer. If however the discrimination problem is one involving a number of relevant cues, each of which must be attended to, the subject's analyzing capacity will be severely taxed and there will be less chance that incidental analyzers will be strengthened. In terms of Rules 1 and 2, if two analyzers are strengthened by discrimination training, less will be learned about other, incidental features of the stimulus. Boneau and Honig (1964) showed that pigeons trained on a conditional red–green, and line–no line discrimination generalized completely to all orientations of the line. Comparing this flat gradient with the steep gradient usually obtained when pigeons are trained on a line–no line discrimination, Boneau and Honig suggested that

> The attention that the animal will pay to nonessential or irrelevant features of the stimulus situation is a function of the demands made on it to process information. When these demands are not great, as in the simple discrimination, nonessential features have a chance to control behavior; but they will not do so when the animal's capacity to process information is taxed severely by relevant features of the stimulus situation [p. 92].

The results are consistent with the idea that there is a limit to the number of cues to which animals are capable of attending. When animals are only required to switch-in an analyzer detecting the presence or absence of a vertical line, they are likely also to switch-in an analyzer detecting its orientation (indeed, they are more likely to do so under these circumstances, than if given nondifferential training). However, when they are required both to switch-in an analyzer detecting the presence of a line and another analyzer to detect the color of the background, they are unlikely to switch-in an orientation analyzer strongly.

4. *The Effects of Irrelevant Discrimination Training*

Our general argument has been that the effects of discrimination training on generalization can be understood in terms of the rules governing the strengthening of analyzers. Our model obviously predicts that specific, relevant discrimination training should sharpen the slope of generalization gradients, but it is less obvious that it can explain why discrimination training between the presence and absence of a stimulus should sharpen the gradient along a particular dimension. We have suggested that such an effect may occur because this training is nonspecific, in the sense that a number of different analyzers can be used to detect the difference between, say, a tone and its absence, and that all of these analyzers may well be strengthened by such training. It would appear, however, that if specific discrimination training is given along a dimension other than that varied in the test for generalization, then our model should predict no increase in the slope of the gradient, since the analyzer strengthened is irrelevant to the dimension along which generalization is tested. We shall see below that this argument is valid only under certain conditions, but first we shall review the evidence. Not all the relevant experiments will be mentioned since there are good recent reviews by Honig (1970) and Thomas (1969).

The experiments to be reviewed fall into two main types. We shall call the dimension on which *discrimination* training is given Dimension D (or Cue D), and that on which *testing* is carried out Dimension T. In experiments of Type-1, subjects are given training on Cue D with a single value of Dimension T present with both the positive and negative stimuli of Cue D. In Type-2 experiments, discrimination training is first given on Cue D with no value of Dimension T present; single-stimulus training on some value of Cue T is then given in some cases with Cue D absent, in others with the positive value of Cue D present. In each type of experiment the final stage is of course to test for generalization along Dimension T. There are two comparatively trivial ways in which discrimination training on an irrelevant dimension might sharpen gradients along a second dimension. First, despite the experimenter's intentions, some implicit discrimination training may be given along Dimension T unless the stimuli are very carefully chosen. Second, in learning to discriminate Cue D, the subject may learn an external orienting response that facilitates acquiring information about the single value of the dimension to be tested (Dimension T).

Several of the experiments which have reported positive results using the Type-1 design seem to be open to one or other of these objections. For example, Friedman and Guttman (1965) trained pigeons to discriminate between a 550 nm positive stimulus and a negative stimulus consisting of a 550-nm key with a black cross. They reported that the gradient of wavelength generalization obtained after such discrimination training was "considerably sharper than the gradients ordinarily obtained after VI training — and would appear to indicate that discrimination along one dimension increases the sharpness of the generalization gradient on another dimension [p. 263]." It should be noted that, in two further experiments, they failed to confirm this result. Moreover, the procedure used may very well have introduced some implicit discrimination training on dimension T (wavelength). Whereas the whole of the positive key was 550 nm, half the negative key was plain black, so that there were clearly wavelength differences between the keys. At the very least, some wavelength analyzer must have been involved in detecting the difference between the plain 550-nm key and the key bearing the cross against a 550-nm background. Thus, the analyzer for Dimension T may have been strengthened directly during discrimination training.

A similar experiment by Switalski, Lyons, and Thomas (1966) (cf. also Lyons & Thomas, 1967) is open to an even more serious objection. They showed that pigeons, given discrimination training between a 555-nm key positive and a white vertical line on a black background negative, produced steeper gradients of generalization along the wavelength dimension than did control birds simply reinforced in the presence of the 555-nm key. They also showed that nondifferential reinforcement in the presence of both stimuli significantly flattened the wavelength gradient. It will be noticed that this experiment does not strictly follow the paradigm for Type-1 experiments, since one value of Dimension T (wavelength) was present on the positive stimulus of Cue D, but no value of Dimension T was present on the negative key which was achromatic. We have already seen that training with the presence versus the absence of a stimulus can sharpen generalization gradients, and it was this type of training that was administered in this experiment. Birds could solve the discrimination problem by learning to respond to green and not to respond to its absence instead of learning to respond to a plain key and not to respond to a key containing a white vertical line.

Thomas (1969) reports two further experiments with pigeons to which this critism is not applicable. In the first of these, undertaken by Klipec, discrimination training was between noise versus no noise and the key was illuminated with light of 555 nm in the presence both of the positive and of the negative stimulus. Testing was then carried out along the wavelength dimension. Although the generalization gradient was steeper after discrimination training, it was not significantly steeper than the gradient produced by controls who were given no discrimination training. Lyons conducted an experiment of the same design in which Dimension D involved a horizontal floor (positive) versus a 10° tilted floor (negative). Dimension T was again wavelength, and this time discrimination training did produce a significant increase in gradient slope. (Lyons' experiment is more fully reported by Thomas, Freeman, Svinicki, Burr, and Lyons, 1970.)

Reinhold and Perkins (1955) performed a Type-1 study that also appears to avoid implicit discrimination training on Dimension T. They trained different groups of rats to run an elevated runway painted either black or white, and with a floor texture that was either rough or smooth. In the first stage of the experiment, the control group received nondiscriminative training, being always rewarded on a runway of given brightness and texture; the discrimination group was rewarded when on a runway of one texture, but not rewarded on another runway of the same brightness but opposite texture. In the second stage, both groups were tested for generalization between the runway on which they had been rewarded and another differing from it only in brightness. The discrimination group showed a gradient of brightness generalization significantly steeper than that shown by the control group. Reinhold and Perkins themselves write,

> It seems unlikely that this finding was the result of the development of orienting or attending responses which would make the difference between the training and test stimuli more distinctive. Attending to the primary tactual cues of the runway surface would be unlikely to facilitate reception of colour cues [p. 426].

It is by no means obvious that this is true. Since the runway was elevated, it presumably had no walls and the color difference appeared on the runway floor. It is in fact uncertain that the texture difference was detected by touch rather than by vision; if it were detected by vision, the animals might well have learned a visual orienting response to the runway floor that would have facilitated learning of the brightness of the floor. Even if the texture difference was detected by touch, the rats may well have been led to look more closely at the floor. Hence, one cannot rule out the possibility that the subjects trained to discriminate between rough and smooth had learned an external orienting response that facilitated learning about the brightness value of the runway.

In one Type-1 experiment, discrimination training did not sharpen the generalization gradient. Mackintosh (unpublished) found that rats trained on a simultaneous discrimination between a white square and a black square generalized completely between the white square and a white diamond.

Similarly, although rats trained to discriminate between a white square and a white diamond showed a significant preference for the white square over a black square (i.e., did not generalize completely between the two), the preference was no greater than that shown by animals given nondifferential training to the white square alone.

In this experiment, there was no possibility that discrimination training was inadvertently provided on the test dimension. It would appear, therefore, that the one experiment to find no effect of extradimensional discrimination training on generalization gradients did not allow other factors to operate, whereas the results of all but one of those experiments that did find a significant steepening of the generalization gradient can be explained either in terms of some implicit training having been given on Dimension T or in terms of the learning of an external orienting response that facilitated learning about Dimension T.

We now turn to experiments of Type 2. Mackintosh (1965e) trained rats on a successive, conditional discrimination (either brightness or orientation), then gave single-stimulus training, and finally tested for generalization between the single training stimulus and another differing from it either in brightness or in orientation. If Stage 1 training was given along the dimension varied in the generalization test, the gradient was steep; but if discrimination training was given along the other dimension, the gradient was flatter than that of an untreated control group. Mackintosh, therefore, found that training on an irrelevant dimension (D) actually flattened generalization gradients along a second dimension (T).

Several other Type-2 experiments have, however, found that irrelevant discrimination training increases the slope of generalization gradients measured after single-stimulus training on some value of Dimension T. For example, Honig (1969) trained one group of pigeons on a wavelength discrimination (blue versus green in one experiment, white versus pink in another), gave a second ("pseudodiscrimination") group equal reinforcement in the presence of both stimuli, and gave a third group nondiscriminative training to a single, colored key. In the second stage of the experiment, all subjects were given single-stimulus, nondiscriminative training to a key with vertical lines and were finally tested for generalization along the dimension of orientation. The group given genuine discrimination training produced steeper generalization gradients than all of the other groups (see Figure 7.5). It is hardly conceivable that discrimination training can have strengthened an analyzer capable of detecting differences between the test stimuli; nor is it particularly likely that the discrimination training had facilitated an observing response that would help to detect the orientation cue (T) since the pseudodiscrimination groups must learn to orient to the key in order to peck it just as much as the discrimination group.

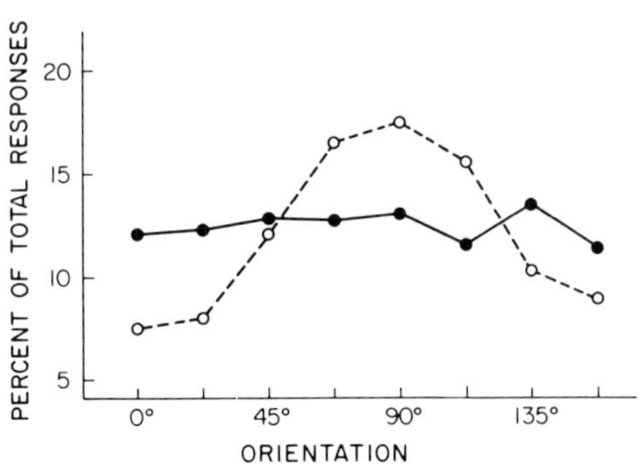


**Fig. 7.5.** Orientation generalization in pigeons following prior training either on a blue–green discrimination, or with blue and green randomly reinforced. (○): true discrimination; (●): pseudodiscrimination. [*After* Honig (1969).]

In Honig's experiment, single-stimulus training on the T dimension was given in the absence of any value of the D dimension. In Mackintosh's experiment, on the other hand, training on T was given in the presence of one value of D. We should expect this procedural difference to have an important effect, since the presence of one value of D to which the subject had already been trained should maintain the D analyzer at a high value and, therefore, prevent the strengthening of the T analyzer. Some recent studies reported by Thomas *et al.* (1970), however, suggest that the simultaneous presentation of values of the D and T dimensions is not sufficient to prevent extradimensional training from sharpening generalization gradients. In one of these studies (originally reported by Freeman, 1967, and discussed by Honig, 1970), one group of pigeons was trained on an orientation discrimination with a vertical line positive. A second, pseudodiscrimination group received equal reinforcement in the presence of lines of both orientations, while a control group received no training. In the second stage of the experiment, all animals received single-stimulus training to a 555-nm key with a vertical line superimposed, and were finally given generalization tests to stimuli varying in wavelength (with no lines present). Once again, discrimination training along one dimension sharpened the gradient along another.

This is a striking result, and one which, at first sight, directly contradicts the idea that attention to one dimension may be selectively strengthened at the expense of attention to other dimensions. It would also appear to be in conflict with the results of several experiments on blocking (see Chapter 4). In blocking experiments, one group of animals is given discrimination training with Cue A relevant, and a second group receives no such training. Both groups then receive discrimination training with two cues (A and B) relevant, and are finally tested for the amount learned about the incidental cue (Cue B). Typically, it has been found that the pretraining on Cue A reduces the amount learned about Cue B. In this design, Cue A is equivalent to Cue D in Thomas *et al.*'s study and Cue B is equivalent to Cue T; nevertheless, in contrast to the typical blocking result, they found that discrimination training increased the amount learned about Cue T as evidenced by the steeper generalization gradients obtained. The most direct contradiction is between these results and those of D. F. Johnson (1966). Johnson used exactly the same dimensions (orientation and wavelength) in a blocking experiment, and found that pretraining on orientation flattened the wavelength gradient. The most obvious procedural difference between the two studies is this: in both cases, experimental subjects were pretrained on an orientation discrimination; in Thomas *et al.*'s study the second stage of the experiment involved single-stimulus training to a compound containing the positive value of the D dimension and a value of the T dimension; in Johnson's study, the second stage involved *discrimination* training with a compound S+ and a compound S−. Mackintosh and Honig (1970) have recently shown that this is probably the critical difference. In a single experiment whose design is shown in Table 7.IV they were able to obtain a significant blocking effect when using Johnson's design, and an (insignificant) enhancement effect when using Thomas *et al.*'s design.

TABLE 7.IV

**Design and results of Experiment by Mackintosh and Honig**[a]

| | Single-stimulus groups (replication of Thomas *et al.*) | | Discrimination group (replication of Johnson) | |
|---|---|---|---|---|
| | Pretrained | Control | Pretrained | Control |
| Stage 1 | Vertical+ Horizontal− | — | Vertical+ Horizontal− | — |
| Stage 2 | Vertical on 501 nm background+ | | Vertical on 501 nm background+ Horizontal on 576 nm background− | |
| Test | Wavelength generalization test | | | |
| Results | Pretraining steepens wavelength gradient ($p < .10$) | | Pretraining flattens wavelength gradient ($p < .05$) | |

[a] Data obtained from Mackintosh and Honig (1970).

We now consider how our own model bears on these findings. Its application to the Type-1 situation is straightforward: since discrimination training should increase the analyzer strength for Cue D and since Cue D is always present in acquisition as well as Cue T, by Rule 2 attention to Cue T should be reduced, and there should be no sharpening of the generalization gradient to that cue. We have already seen that all but one of the experiments that obtained a steeper gradient can be explained in other ways.

The application of the theory to the results of Type-2 experiments is more complicated, and here we follow a theoretical suggestion first put forward by Wagner (1969a). He pointed out that if subjects are first trained to discriminate Cue D and are then reinforced on a single value of Cue T with Cue D absent, then the effect of training on Cue D might be to reduce attention to implicit background cues (e.g., the shape of the key in a pigeon experiment). When single-stimulus training is then given with Cue T, provided Cue D is absent, the analyzer for Cue T should be relatively strong since attention to the background cues has been reduced by the discrimination training. This result only follows if (as stated in Rule 2 of our model) analyzer strengths sum to a constant amount.

Wagner not only proposed this explanation for the sharpening of generalization gradients after irrelevant discrimination training, he also carried out an ingenious experiment to test it. In Chapter 5, we described an experiment (Wagner, Logan, Haberlandt, & Price, 1968) in which it was shown that when one cue predicts reinforcement with only 50% success, much less is learned about it if there is a second cue present that is consistently correlated with reinforcement than if there is no other cue that makes better predictions. Wagner (1969a) conducted a variation on this experiment using eyelid conditioning in rabbits. The design is shown in Table 7.V; it involves two groups of subjects. All the stimuli used were compounds, and for both groups one member of all the compounds was always a vibratory stimulus (V). When the vibratory stimulus was accompanied by an auditory tone (A), reinforcement (the US) always followed. Two other compounds were also used for both groups consisting of the vibratory stimulus and either a steady or a flashing light (L1 and L2). Reinforcement followed these compounds on

TABLE 7.V

**Design of Experiment by Wagner[a,b,c]**

| Groups | Stimuli | | |
|---|---|---|---|
| | AV | L1, V | L2, V |
| Correlated | 100 | 100 | 0 |
| Uncorrelated | 100 | 50 | 50 |

[a] Data obtained from Wagner (1969a).

[b] Numbers indicate the percentage of trials on which each stimulus was followed by the unconditioned stimulus.

[c] A, designates auditory stimulus; V, vibratory stimulus; L1, one light stimulus; L2, other light stimulus.

50% of trials, but for one Group (Group Correlated) one light signaled reinforcement 100% of the time and the other light was never followed by reinforcement, whereas for the second group (Group Uncorrelated) both the compounds containing a light were followed by reinforcement on 50% of presentations. After training, Wagner obtained generalization gradients to different frequencies of the tone and to the presence versus absence of the vibratory stimulus. Group Correlated showed steeper generalization gradients to the tone than Group Uncorrelated, but flatter generalization gradients to the presence versus absence of the vibratory stimulus. Wagner interprets this result as follows: when the lights make consistent predictions, their presence reduces attention to the vibratory stimulus appearing in compound with them; hence less is learned about V by Group Correlated than by Group Uncorrelated. However, if attention to V is reduced, this should allow more learning about the tone (A) appearing in the AV compound: hence Group Correlated learns more about the tone than does Group Uncorrelated.

The bearing of this result on experiments of Type 2 can be readily seen; we may consider that the discrimination between the lights is equivalent to discrimination training on Cue D, that the training on A is equivalent to single-stimulus training on Dimension T, and that the vibratory stimulus (V), which is always present, is equivalent to the background stimuli. Training on Cue D reduces attention to the background stimuli and, therefore, results in more learning about Stimulus T presented in the presence of the background stimuli but with Cue D removed. The most important feature of Wagner's experiment is that he included an explicit background stimulus (V) and hence was able to show directly that the sharpening of the gradient to Dimension T is accompanied by a flattening of the gradient to a background cue.

As Wagner points out, his results provide a simple explanation for Honig's findings, since, in Honig's experiment, single-stimulus training on a value of Dimension T was carried out without Cue D being present. According to Wagner this was a critical feature of the design, for it ensured that the decrease in attention to the background stimuli (caused by the previous discrimination training) would subsequently allow more attention to be directed to the T stimulus. The application of this analysis to the results reported by Thomas *et al.* (1970) is more difficult, and it does not, in fact, seem possible to make unambiguous predictions. In this experiment, prior training on Cue D enhanced the generalization gradient tested along Dimension T, even though initial training on T had been in the presence of the reinforced value of the D dimension.

If animals are trained on T only in the presence of D, then prior discrimination training on D should, other things being equal, reduce the amount learned about T. This, of course, is exactly the result obtained in experiments on blocking. Mackintosh and Honig's data, described above, imply that Thomas *et al.* obtained the opposite result because they were comparing the generalization gradients of two groups of subjects only one of whom had ever received any discrimination training. When pretrained and control subjects were trained on a discrimination between a compound S+ and compound S−, then prior training on one of the components blocked learning about the other; but when subjects were simply reinforced for responding to a compound S+, prior discrimination training on one component tended to enhance the gradient tested along the other dimension.

Wagner's suggestion that discrimination training weakens analyzers for background stimuli does not greatly affect the analysis of the blocking paradigm; both pretrained and control subjects receive discrimination training during the course of the experiment, and both will, therefore, learn to ignore background stimuli. In the design used by Thomas *et al.*, however, it leads to some complications. Only pretrained subjects receive discrimination training (on D); they will, therefore, be less likely than control subjects to attend to background stimuli, and for this reason should give a steeper gradient. However, they are trained on Cue T in the presence of Cue D, and prior strengthening of the D analyzer should result in the analyzer for Cue T being less strongly switched-in. In order to predict the outcome of such an experiment, we should have to know whether the strength of the analyzer for Cue T would be more greatly affected by the suppression of analyzers for background stimuli or by the strengthening of the analyzer for Cue D. With two factors pulling in opposite directions, it is impossible to make unequivocal predictions; nevertheless the results reported by Thomas *et al.* cannot be seen as unequivocal evidence against our position. What we should predict is that the outcome of experiments of this design should depend upon a variety of details. This appears to be true: Mackintosh's experiment described above employed exactly the same general design as that of Thomas *et al.*, but differed in a large number of details (subjects, type of pretraining, testing procedure); in contrast to Thomas *et al.*, Mackintosh found that extra-dimensional training flattened generalization gradients.

Two other explanations of why discrimination training on one dimension should increase attention to a second have, in fact, been put forward, but neither is particularly plausible. Sutherland and Andelman (1967) have suggested that analyzers are hierarchically arranged and that the strengthening of an analyzer within a particular modality may increase the strength of other analyzers within that modality. This explanation does not help with the Thomas *et al.* result, since they used different modalities for the stimuli to be discriminated and the dimension along which generalization was measured. Reinhold and Perkins (1955) suggested that discrimination training might establish a "set to discriminate"; if we interpret this to mean that discrimination training raises the general level of attention, then this suggestion has some initial plausibility. It is, however, contradicted by Wagner's results since he showed that, although discrimination training raises attention to Dimension T, it lowers attention to background cues.

We shall encounter a similar problem when we discuss the results of extra-dimensional shift training in Chapter 9. Such experiments are formally rather similar to those discussed here: discrimination training on one dimension is followed by training on a second and we shall see that, as in the experiments discussed in this section, such pretraining on one dimension sometimes facilitates the learning of the second problem and sometimes retards it.

## V. Further Determinants of the Slopes of Generalization Gradients

### A. Generalization as a Function of Acquisition Level

In the remainder of this chapter we shall discuss two further sets of experiments that bear on our theoretical position. Rule 3 of our model states that an initially weak analyzer will reach asymptote more slowly than will its response attachments. This implies that, in discrimination learning, the effect of overtraining will be largely confined to increasing the strength of the relevant analyzer; this implication will be spelled out in detail when we discuss the overtraining reversal effect in Chapter 8. Since the slope of a generalization gradient depends upon the strength of the appropriate analyzer, this rule also implies that if discriminative training is given and if the relevant analyzer is initially relatively weak, an increase in the number of acquisition trials given before a test for generalization will normally sharpen the generalization gradient.

#### 1. *Overtraining Steepens Generalization Gradients*

Such a result has been obtained by numerous investigators. Razran (1949), in his review of Russian studies on the generalization of classically conditioned responses in dogs, concludes,

> (a) that CR generalization increases in the very initial stages of training the CR;
> (b) with further training of the CR, it begins to decrease slowly; and
> (c) after a great number of reinforcements, the generalization may increase again [p. 348].

The first claim, that generalization initially increases, presumably represents the building up of response strength; the third claim is one that Razran himself accepts as only possibly true, and the result has not, so far as we know, been consistently repeated. The second claim, that from moderate to high levels of training generalization tends to decrease is in line with our prediction.

The interpretation of these data is complicated by the fact that Razran presents only relative amounts of generalization as a function of acquisition level. Since increased training may increase overall response rate at S+, there will be more room for a decline in response rate to generalization stimuli, and a relative generalization gradient may reflect no more than this. The only studies that can be unambiguously interpreted are those where absolute response rates show a cross-over, i.e., where overtrained subjects respond more to S+ but less to other stimuli than nonovertrained subjects. Three more recent studies satisfy this requirement. In all these studies cats were trained to avoid a shock by making a specific response upon the presentation of the CS. Generalization was then measured by extinguishing different groups of subjects either on the training stimulus or on stimuli of varying degrees of similarity to it. Thompson (1958) used a 250-Hz tone as the training stimulus and trained his subjects to a criterion of either 20, 55 or 90% avoidance response over 20 trials. The three groups were then each subdivided into six subgroups, each of which was extinguished to a single stimulus with a value ranging from 250 (the training stimulus) to 8000 Hz. For the subgroups extinguished to the training stimulus, the number of trials to extinction was an increasing function of training level. For the subgroups extinguished to generalized stimuli, the number of trials to extinction first increased, and then decreased (with increases in training). In a second study, Thompson (1959) tested for crossmodality generalization from a light to a tone, and found that, although subjects trained to a 55% criterion showed considerable generalization, subjects trained to a 90% criterion showed none at all. Finally, Hoffeld (1962) compared the effects of training either to a 50% criterion, a 90% criterion, a 90% criterion plus 60 overtraining trials, or, finally, a 90% criterion plus 120 overtraining trials, on generalization from a 250-Hz tone to 500- and 2000-Hz tones. Once again, the absolute number of generalized responses tended to decrease with increased training (and with the number of overtraining trials given here, this result casts further doubts on Razran's third claim).

### 2. *Overtraining Flattens Generalization Gradients*

At this point, we should consider some of the factors that demand qualification of our prediction that increased training will sharpen generalization gradients. The studies that have obtained such an effect have involved differential training; in the Thompson and Hoffeld experiments, electric shock only comes on in the experimental situation after presentation of a discrete CS: absence of the CS is correlated with no shock. As mentioned above, our original prediction will presumably not hold if every attempt is made to provide nondifferential training, partly because, by Rule 2, no analyzers will be strengthened by such training, and partly because the only analyzers likely to be used in such a situation are initially dominant ones which cannot be further strengthened. Friedman and Guttman (1965), in what is, so far as we know, the only investigation of wavelength generalization in birds as a function of acquisition level, reported that very extensive training appeared to produce a flatter than normal generalization gradient.[2]

A second situation in which overtraining should not serve to strengthen the originally used analyzer, is when rats are trained to run down an alleyway. In Chapter 8, we cite a number of studies that suggest that increased training in this sort of situation causes rats to switch control of the running habit from external to kinesthetic cues. If such an effect were to occur in an experiment on visual generalization the outcome would presumably be for overtraining to flatten the generalization gradient. Indeed, the one class of study that has consistently failed to find an increase in the slope of generalization gradients with increased training, is that in which rats have been the subjects. There have been three studies of size generalization in rats as a function of acquisition level. Margolius (1955) tested for generalization after various numbers of acquisition trials ranging from 4 to 164. For both absolute and relative amounts of generalization, for any one of three response measures, he found that an increase in the amount of training led to a significant decrease in the slope of the gradient. Jensen and Cotton (1961) varied acquisition between 10 and 190 trials, and obtained completely flat gradients at all levels of training. In an experiment on size and orientation generalization conducted in a similar apparatus, Muntz (1963c) found no consistent change in the slope of generalization gradients when acquisition training varied from 40 to 100 trials.

In the only study in which overtraining has been claimed to sharpen generalization gradients in rats, Bindra and Seeley (1959) trained three groups to run down an alleyway, and after either 15, 40, or 80 trials changed certain features of the runway and continued training. The only appropriate measure

[2] Hearst and Koresko (1968) have reported that *relative* gradients of orientation grow steeper with extended training in pigeons, but that the absolute number of responses to test stimuli increases. This result implies an intermediate effect of overtraining on generalization gradients, neither unequivocal sharpening (since there is no cross-over in the absolute level of responding), nor unequivocal flattening. Both Hearst and Koresko, and Friedman and Guttman used brief blackouts between stimulus presentations.

of the slope of the generalization gradient here would be the difference in running time over a certain number of trials between subjects for whom the alley has just been changed, and subjects for whom it has remained the same. The greater this difference, the steeper the generalization gradient. Unfortunately, the authors present their analysis in terms of the difference for the same subjects between prechange and postchange running times. This analysis ignores the fact that running speed would normally increase from one trial to the next and that this increase is bound to be greater early in training than later. Despite this use of a measure that is heavily biased in favor of a smaller generalization decrement (i.e., a flatter gradient) the less training subjects had received, it was only when the relative decrement was computed that such an effect was obtained. In absolute terms, the slope of the gradient did not change as a function of acquisition level.

At the very least, there is some discrepancy between these rat results and those obtained with cats in an avoidance situation. While there are doubtless many differences between the two classes of study that may be responsible for the differences in outcome, our own interpretation seems worth further investigation. Its plausibility is enhanced by the result of a further series of experiments on the way in which the slope of visual generalization gradients is affected by inferred changes in the extent to which the subject's behavior is kinesthetically controlled.

## B. Visual Generalization and Internal Control

In an important series of experiments, Hearst and his associates have shown that the slope of visual generalization gradients is affected by the training procedure used in a fashion that is most simply interpreted as being due to changes in the degree of internal control of the behavior involved. We shall say that behavior is under "internal control," not only when guided by kinesthetic feedback from responses, but also when temporal discriminations involving an "internal clock" are required. These studies have been reviewed by Hearst (1965) and we shall largely follow his account here.

### 1. *Approach and Avoidance Gradients*

The original experiments (Hearst, 1960, 1962) compared the effects of training both monkeys and rats for variable interval positive reward and fixed interval avoidable shock on the slope of brightness generalization gradients. Each monkey, for example, was trained to press a lever to postpone the onset of shock for 10 sec (thus, a rate of lever pressing greater than 1 per 10 sec avoids shock altogether), and also to pull a chain to obtain reward on a 2-min

VI schedule. Throughout training, the experimental cage was brightly (and diffusely) illuminated. Generalization tests were given with the cage illuminated by a light of varying brightness. The results for three individual monkeys are shown in Figure 7.6. A relatively sharp gradient along the brightness dimension was formed by all subjects for the appetitive, food-rewarded behavior (as noted above, this is the definitive case of sloping gradients formed after nondifferential training); but the avoidance behavior generalized completely to all other intensities of light used. Such results appear to provide a striking disproof of the view that all stimuli impinging upon an organism at the time of reinforcement are associated with the response: the intensity of the cage illumination has simultaneously become associated with one pattern of behavior, and has failed to become associated with another.

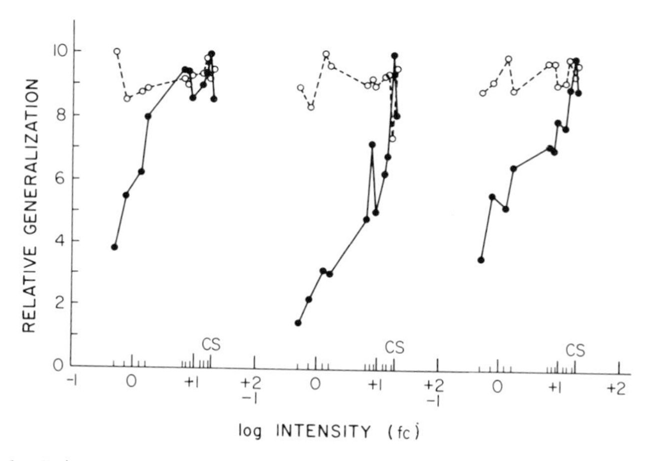


**Fig. 7.6.** Brightness generalization in monkeys concurrently trained to pull a chain for food reinforcement and press a lever to avoid shock in the continuous presence of a 28.1-fc light. (O): avoidance; (●): approach.

Some subsidiary experiments together with other results have eliminated several possible explanations of these findings; the flatter gradient for avoidance than for appetitive behavior in this situation is not likely to be due to differences in drive level, or to the direction of testing (from brighter to darker, or vice versa). Furthermore, since even those subjects whose avoidance responding extinguished relatively rapidly showed flat generalization gradients, it is unlikely that overall greater resistance to extinction of the avoidance response was the sole cause of the flatter avoidance gradient. Nor is it in general true that avoidance behavior always generalizes widely. Indeed, Miller's analysis of approach–avoidance conflict (Miller, 1959) rests upon the assumption that avoidance gradients are steeper than approach gradients. The experiments by Thompson and Hoffeld discussed in the previous section also produced steeply sloping gradients for an avoidance response. However, the avoidance situation used in those experiments differed from that used by Hearst. Thompson and Hoffeld used a discrete warning stimulus, whose onset preceded shock and whose absence was correlated with safety. Hearst used an avoidance procedure developed by Sidman (1953) in which there is no exteroceptive warning stimulus; whenever the subject is in the apparatus, electric shock occurs at fixed intervals unless the subject makes some designated response, and each occurrence of the response postpones the shock for a set period of time. It is the absence of an exteroceptive warning stimulus that Hearst (1963) sees as crucially determining the slope of the generalization gradient in this situation. He suggests that

> Internal cues play a relatively greater role in the control of many kinds of avoidance behavior than in the control of approach behavior. Many theoretical discussions have stressed the importance of proprioceptive, emotional and temporal cues in avoidance situations similar to the one studied here. This view would suggest that avoidance subjects are primarily controlled by feedback from their own behavior. For example, successful avoidance monkeys have learned that all behavior other than the correct response (lever-pressing) is punished; thus any initiation of "incorrect" behavior feeds back to the subject through his proprioceptors and provides warning stimuli for the avoidance response. . . . If it is true that this type of avoidance behavior is primarily under the control of proprioceptive and other internal cues, then variations in the intensity of the overhead light are, so to speak, "irrelevant". By this reasoning one would expect avoidance gradients for light intensity to be relatively flat [p. 493].

### 2. *Approach Gradients after Different Reinforcement Schedules*

The strongest evidence in favour of this interpretation of the difference in the generalization of appetitive and avoidance behavior in Hearst's situation comes from experiments showing that differences in the extent to which appetitive behavior itself is controlled by internal cues have similar effects on the slope of visual generalization gradients. One type of reinforcement schedule that has been assumed to produce internally controlled behavior is a differential reinforcement of low rate (DRL) schedule. In DRL training, responses are only rewarded if a fixed period of time has elapsed since the last response; e.g., on a DRL 6-sec schedule, only responses spaced at least 6 sec apart are reinforced. The cue for the subject to respond, therefore, is that it has not pressed the response key within the last 6 sec. Thus, the subject's responses are controlled by its own preceding behavior more than by any visual stimuli on the response key. To test the prediction that visual generalization gradients in pigeons would be relatively flat after such training, Hearst, Koresko, and Poppen (1964) trained two groups of pigeons to peck at a key with a line of given orientation—one group on a VI 1-min schedule, the other on a DRL 6-sec schedule. Generalization gradients along the dimension of orientation were then obtained in the usual way. The gradients were significantly flatter, on both absolute and relative measures, for the DRL group than for the VI group.

As has been noticed by Ellison (1964), a similar interpretation may be provided for Pavlov's finding that a trace conditioning procedure (in which the CS both begins and ends before the US) results in flatter generalization gradients than does a delayed conditioning procedure (in which the CS begins before the US but overlaps it). Pavlov (1927) wrote: "trace reflexes... exhibit a permanent and universal generalization involving all the analyzers [p. 113]." A trace conditioning procedure has one crucial feature in common both with a DRL schedule of reinforcement and with Sidman avoidance training, namely, that the subject is required to make a temporal discrimination. Animals sometimes make such discriminations by developing set behavior patterns and using kinesthetic feedback as the indicator of elapsed time, but even if no such "mediating behavior" occurs, timing behavior must depend in some way upon internal cues.

In a second experiment, Hearst *et al.* showed that variations even within VI schedules of reinforcement significantly affected the slope of visual generalization gradients. They trained different groups of pigeons to peck at a key with a given orientation of line on VI schedules with mean reinforcement intervals ranging from 30 sec to 4 min. The slope of both absolute and relative generalization gradients was inversely related to the length of this interval. A similar result for wavelength generalization has been obtained by Haber and Kalish (1963). Long VI schedules do not presumably demand the learning of a temporal discrimination by the subject (since the length of interval between reinforcements is random); but, just like a DRL schedule, they do produce a relatively low rate of responding, thus increasing the probability of occurrence of stereotyped behavior emerging between successive responses.[3]

Once again, however, the gradient slope cannot have been directly determined by response rate during training; it is only because the lower rate produced by the longer interreinforcement intervals of the long VI schedules encourages the development of stereotyped behavior patterns, that

[3] Given initial differences in response rates, the same problem arises here as was mentioned above in the discussion of overtraining and generalization: that a relative measure may show a steeper gradient in subjects with a higher rate at S+ simply because their rates have farther to fall. This certainly applies to Hearst *et al.*'s study on VI and DRL schedules, but in the present instance, subjects trained on the shorter VI schedules not only responded at a higher absolute rate at S+ but also at a lower absolute rate to distant generalization stimuli.

such schedules produced flatter gradients. This is shown by some results of Thomas and Switalski (1966) on the gradients obtained after VI and variable ratio (VR) training. On a VR schedule, reinforcements are obtained after a variable number of responses have been emitted by the subject. Such a schedule produces a far more rapid response rate than does a VI schedule, and it has often been suggested that rapid responding by the subject comes to serve as a discriminative cue predicting the occurrence of reward. This implies that the behavior of VR trained subjects is under kinesthetic control, and the authors predicted that such subjects would therefore show flat generalization gradients along the wavelength dimension. The prediction was confirmed: by comparison with subjects trained on a VI schedule (with the same mean interval between reinforcements) the VR subjects generalized widely to different wavelengths. It is important to note that in this experiment, in contrast to that of Hearst *et al.*, the group showing the flatter gradient responded significantly faster during training.

At the very least, the studies summarized in this section should destroy the myth that there is a single true shape and slope to the generalization gradient along any dimension for any animal group. Relatively minor differences in the schedule of reinforcement used in training, for example, have marked effects on the slope of the gradient; the gradient obtained after a VR-40 schedule is no more "correct" than that obtained after training on a VI-1 schedule. Furthermore, although these studies may individually be susceptible of alternative explanations, cumulatively they provide strong evidence for the suggestion that if behavior is kinesthetically controlled, then to that extent it will not be controlled by visual cues and visual generalization gradients will be flat. Their results, therefore, are consistent with the notion of stimulus selection expressed by Rule 1 of our model: stimuli compete for control of behavior, and the presence of a strong cue will reduce the amount learned about other cues.

## VI. Summary

If a subject has been reinforced for responding in the presence of a particular stimulus, there are many factors influencing the occurrence of responses to other stimuli differing from the training stimulus. We have not attempted an exhaustive treatment of generalization in this chapter, but only considered some of those variables, known to determine the slope of generalization gradients, whose effects suggest an attentional interpretation.

Gradients differ markedly depending on the nature of the subject, stimulus dimension, and number and nature of other stimuli present; they are affected by the amount and type of discrimination training, and by the schedule and type of reinforcement given in the presence of the training stimulus. It is clear that gradients of invariant slope are not automatically found along all stimulus dimensions that the subject is capable of detecting.

The role of discrimination training, interpreted by Hull (1950) as simply a matter of neutralizing incidental (in our terminology, irrelevant) stimuli, was elevated by Lashley and Wade to a position of crucial importance. The evidence does not bear out either of these interpretations. No single set of conditions of discrimination training can be regarded as universally sufficient and necessary to produce a gradient of generalization, and in the only study (Wagner, 1969a) to provide direct evidence that the suppression of irrelevant stimuli might sharpen subsequently obtained gradients, the suppression occurred for reasons not predicted by Hull.

Lashley and Wade's major theoretical suggestion was that the slope of a generalization gradient indicates the extent to which the subject paid attention to that feature of the training situation. When interpreted within the terms of a comprehensive theoretical system, this suggestion brings much order into what are otherwise disorderly data. We have argued in this chapter that the slope of a generalization gradient depends first on the initial strength of the analyzer detecting that dimension relative to the strength of other analyzers that can be used in the situation, and secondly on how the conditions of training alter these relative strengths.

Dimensions that can be shown independently to be dominant for a given subject (such as wavelength for birds), will yield sloping gradients under nearly all conditions of training; gradients may not be found along other dimensions unless some differential reinforcement is provided. Any type of discrimination training may sharpen gradients to a small extent, as Wagner suggests, by selectively weakening constant, irrelevant stimuli. However, differential reinforcement that actively strengthens the relevant analyzer (even if along with others) will sharpen gradients yet further. The gradient produced will depend upon the extent to which other, irrelevant analyzers are also strengthened; nonspecific discrimination training will therefore be less effective than specific, relevant discrimination training, and the presence of other strong cues, the analyzers for which may also be strengthened, will result in a relative flattening of the gradient. This decrease in slope will occur whether the other stimuli are exteroceptive as in the studies of Newman and Baron (1965) or Boneau and Honig (1964), or interoceptive and influenced by the schedule and type of reinforcer, as in the experiments of Hearst (1965) and his associates.

CHAPTER 8

# Reversal Learning

## I. Introduction

In this chapter, we discuss one of the paradoxical phenomena mentioned in Chapter 2, the so-called overtraining reversal effect (frequently abbreviated to ORE). The effect was discovered by Reid (1953). Reid trained three groups of rats on a black–white discrimination in a Y maze. One group was trained to a criterion of 9 out of 10 consecutive correct responses; the second group was trained to the same criterion, and received a further 50 overtraining trials; the third group was trained to this criterion, and received 150 overtraining trials. After completion of initial training, each group learned the reversal of the original problem. The number of trials to learn the reversal was 138.3 for the nonovertrained group, 129.0 for the group receiving 50 overtraining trials, and 70.0 for the group receiving 150 overtraining trials. Most theories of learning, whatever else they disagree on, agree in supposing that increased training serves to strengthen what is learned; if it is further assumed that the reversal of a discrimination habit simply involves the extinction of one habit and its replacement by another, then the most obvious conclusion is that overtraining should retard reversal learning. Hence, Reid's finding that overtraining significantly facilitated reversal for his rats poses something of a problem for most learning theories. Indeed, unlike several other experimental findings that pose problems for orthodox learning theories, the ORE even seems surprising from a commonsense viewpoint. This perhaps explains the enormous volume of research on the ORE undertaken in recent years; since Reid's original experiment, approximately 50 reports have appeared in the literature, some of them containing up to six experimental studies, and several lengthy reviews (Lovejoy, 1966; Mackintosh, 1965a;

Paul, 1965; Sperling, 1965) have been published. One major conclusion that has emerged from this research effort is that the ORE is a somewhat elusive phenomenon; while Reid's results have been replicated in about 20 studies, in the remainder of these experiments, overtraining has either had no effect on reversal learning, or has (more rarely) actually retarded reversal. This of course has important consequences: the problem for any learning theory has now expanded from being one of simply explaining why the ORE occurs at all, into one of not only explaining why it occurs, but also why it frequently does not occur.

The present chapter is divided into four parts. First, we show how our model is capable of predicting the ORE; second, we discuss some experimental evidence supporting our interpretation; third, we attempt to interpret some of the differences between studies obtaining an ORE and those failing to do so; and finally, we shall consider some alternative theoretical interpretations that have been offered.

## II. Theoretical Analysis of Reversal Learning

Our own explanation of the ORE involves rejecting the assumption that reversal learning is simply the extinction of one habit and its replacement by its opposite. Sutherland (1959) pointed out that an attentional two-process model should be capable of predicting the ORE, and more recently Lovejoy (1966) has proved that this is so. The prediction can follow once it is realized that to learn the reversal of a brightness discrimination, the subject must continue to attend to the originally relevant cue. A black–white discrimination and its reversal have at least one thing in common: both are brightness discriminations. Solution of both, therefore, demands utilization of a brightness analyzer. Although the response requirements of the two tasks are antagonistic, and the responses established in initial training must be extinguished before the reversal can be solved, the analyzer required for reversal is the same as that established during initial training, and if extinguished, must be reestablished.

### A. General Conditions Determining Speed of Reversal

#### 1. *Theoretical Conditions*

Our two-process theory predicts that reversal learning will be rapid to the extent to which the relevant analyzer is maintained during reversal, and slow to the extent to which it is extinguished. This follows from Rule 1, since, for reversal to take place, the strength of preexisting response attachments must be altered, and the rate of change of response attachments is, according to

Rule 1, directly related to the strength of their analyzer. In fact, of course, during the early stages of reversal training, the relevant analyzer will tend to extinguish. If an animal is trained on a black–white problem with black positive, and is then reversed to white positive, it will initially continue predominantly to select black and will not be rewarded. According to Lovejoy's model (see Chapter 2), if the animal attends to brightness, selects black, and is not rewarded, the probability of attending to brightness is decreased by a nonreward operator. According to our model, the brightness analyzer will tend to extinguish for a slightly different reason, namely that, from Rule 2, an analyzer reverts towards its base level whenever an expectancy is disconfirmed; if the strength of approaching black ($R_B$) is greater than the strength of approaching white ($R_W$), the brightness analyzer will be weakened whenever the animal selects black and is not rewarded, and also whenever it approaches white and is rewarded. On both models, therefore, attention to the relevant dimension will decrease during early reversal trials; on our model, in fact, the strength of the relevant analyzer will continue to decrease until $R_B = R_W$, i.e., until the original response attachments have been completely extinguished.

Since, from Rule 1, the rate of change of $R_B$ and $R_W$ is directly related to the strength of the brightness analyzer, it can be seen that the rate of reversal learning predicted by the model critically depends upon the extent to which the brightness analyzer has extinguished by the time that $R_B = R_W$. If, for example, an extinction parameter for analyzers were chosen that was small by comparison with the extinction parameter for responses, reversal learning would be predicted to occur rapidly, since $R_B$ would equal $R_W$ while the analyzer was still strongly switched in. If, on the other hand, parameters were chosen such that analyzers extinguished rapidly relative to responses, reversal learning would be predicted to occur slowly, since now the analyzer would be very weak by the time that $R_B$ was equal to $R_W$.

It should be obvious that this analysis does not yet provide any explanation of the ORE, unless one were to make the arbitrary assumption that overtraining somehow alters the relative magnitudes of the extinction parameters of analyzers and responses. It does, however, make explicit that one major determinant of predicted speed of reversal learning within the framework of such a model is the relative extent to which the analyzer and its responses are extinguished during early reversal trials.

Even though approximately equal (and unchanging) rates of extinction are assigned to analyzers and responses, there are several circumstances which would be expected to affect the rate of reversal learning. In the first place, reversal learning will be more rapid the greater the strength of the relevant analyzer at the outset of reversal. For given a fixed rate of extinction for the analyzer, an increase in prior strength will decrease the extent to which the analyzer is extinguished during the course of reversal. Secondly, given equal values of the relevant analyzer at the outset of reversal, differences in response strengths will yield differences in speeds of reversal. For the stronger the initial (and now inappropriate) response attachments, the more time it will take for the strengths of the two responses to equalize, and the longer the period of time in which the relevant analyzer is subject to extinction. In general, therefore, irrespective of the precise *rates* of analyzer and response extinction, we can predict that speed of reversal learning will depend upon the prior strengths of analyzers and response attachments; other things being equal, the greater the strength of the relevant analyzer relative to its response attachments, the faster a reversal will be learned.

2. *Experimental Conditions*

Some initial evidence supporting these predictions comes from an experiment by Lawrence (1950). In part of his second study on the acquired distinctiveness of cues, Lawrence trained rats in the manner outlined in Table 8.1. In Stage 1, subjects were trained on a successive black–white discrimination with the presence or absence of chains irrelevant. In Stage 2, subjects learned a simultaneous two-cue discrimination between black with chains positive, and white without chains negative. In Stage 3, the subjects were split into two subgroups; each subgroup learned a single-cue reversal problem: for one, the problem was white positive versus black negative, for the second no chains was positive and chains negative. Previous results (see Chapter 6) had shown that subjects had solved the two-cue discrimination of Stage 2 largely in terms of brightness (the cue relevant in Stage 1); the results of Stage 3 showed that animals *reversed* on brightness also learned more rapidly than those reversed on chains. These results (like the ORE) are difficult to understand in terms of a single-process theory. If subjects solved the Stage 2 problem largely in terms of brightness, then at the outset of reversal training the

TABLE 8.I

**Design of Experiment by Lawrence**[a,b]

| Stage 1[c] | Stage 2 | Stage 3 |
|---|---|---|
| B− B+ | | Group 1 W+ B− |
| | BC+ WNoC− | |
| W+ W− | | Group 2 NoC+ C− |

[a] Data obtained from Lawrence (1950).
[b] B designates black; W, white; C, chains; NoC, no chains.
[c] C/NoC irrelevant.

difference in response strengths between black and white must have been greater than the difference between chains and no chains. Yet the larger black–white difference was reversed more rapidly than the smaller chains–no chains difference. From our point of view, however, the results are readily interpretable: Stage 1 training increased the strength of the brightness analyzer, and in Stage 2 the brightness analyzer remained switched-in while new response attachments were formed. The conditions for rapid reversal learning to brightness were therefore met: the relevant analyzer had been strengthened throughout Stages 1 and 2, while its response attachments were somewhat weaker, having only been strengthened during Stage 2. The analyzer for detecting the chains cue, however, would perhaps have been weakened by Stage 1 training; by the end of Stage 2, therefore, the chains responses must have been at least as strong as the chains analyzer. On the chains cue, therefore, the conditions for somewhat slower reversal learning were met.

## B. Theoretical Conditions for Predicting the ORE

From what we have said, it can be seen that a two-process model will predict an ORE, if it can be assumed that overtraining increases attention to the relevant cue without causing a corresponding increase in response strength; this will ensure that, in our terminology, the strength of the relevant analyzer relative to its response attachments is greater for overtrained than for non-overtrained subjects. Rule 3 of the model, which states that an initially weak analyzer will reach asymptote more slowly than its response attachments, in effect incorporates this assumption. It implies that, in subjects trained only to criterion, response attachments will be nearer asymptote than the relevant analyzer, with the consequence that overtraining can strengthen the analyzer more than it can strengthen its responses. The question arises as to whether this is a reasonable assumption, or, put differently, whether Rule 3 can be derived from other properties of the model.

Lovejoy (1966) has, in fact, shown that something like our Rule 3 follows directly from some basic assumptions of a simple two-process model. We have already discussed his account in some detail in Chapter 2; for present purposes, we need only state the following: Lovejoy assumes that the probability of attending to brightness increases on any trial on which the subject so attends and makes the correct response, and decreases on any trial on which it attends and makes the incorrect response; the probability of making the correct responses conditional upon attending, however, increases on all trials on which the subject attends. From this it follows that, other things being equal, response probability will increase faster than attentional probability; it further follows that the probability of attending will only in general increase, provided that, when the subject does attend, it is more likely than not to make the correct response, i.e., once *some* response learning has already occurred. Given approximately equal starting values and equal rate parameters for attentional and response learning, the probability of making the correct response will tend to reach asymptote sooner than the probability of attending correctly, and, hence, the ORE is predicted.

Such consequences do not follow from such basic assumptions in our model. (Given the elusive nature of the ORE, this is conceivably a virtue.) The critical difference here between our model and Lovejoy's, is that the version of Rule 2 discussed in Chapter 2 states that as soon as $R_B > R_W$, the strength of a brightness analyzer will increase, not only on trials on which the subject selects black and is rewarded, but also on nonrewarded trials when white is selected. Thus, other things being equal, the brightness analyser might be expected to increase about as rapidly as correct response strength. As suggested in Chapter 2, however, one could suppose that the increment in relevant analyzer strength on a given trial is proportional to the strength of the expectancy confirmed. This would mean that the relevant analyzer was strengthened rather slowly at first, and would not increase rapidly until the correct response itself had a high strength.

However this may be, it is clearly *possible* to predict the ORE from a two-process model. Furthermore, many additional predictions derivable from such a model have been tested with some success. As we shall argue later in this chapter, this is more than can be said for most alternative accounts of the ORE that have been offered.

# III. Evidence for a Two-Process Analysis of the ORE

## A. Analysis of the Course of Reversal Learning

### 1. *Theoretical Predictions*

Although we have just been arguing that an attentional theory is capable of predicting an ORE, we wish immediately to introduce a further qualification. Lovejoy's computer simulation has, indeed, proved that overtraining can lead to faster reversal learning when attention to the relevant cue has been strengthened. As he has pointed out, however, the difference between his overtrained and nonovertrained stat-rats in trials-to-reversal-criterion scores was a relatively small one. This is not entirely suprising, since, although theoretically stronger attention should eventually speed up rate of reversal learning, there is equally good reason to predict that it should retard the apparent rate of reversal during the early stages of reversal. That is to say, from the assumptions made so far, it can be predicted that overtraining will retard initial extinction of the original discrimination habit. If, after training to criterion on a brightness discrimination, the brightness analyzer is not much stronger than the irrelevant position analyzer, then subjects reversed upon reaching criterion should soon extinguish the brightness analyzer to the point where the position analyzer again controls behavior. As soon as this happens, such subjects will be responding at the 50% level, and will appear to have extinguished the original discrimination. After overtraining, the brightness analyzer will have been further strengthened, and the position analyzer further weakened. Thus, it will take longer for the former to be weakened to the point where the latter controls behavior.

Reversal learning can, in effect, be regarded as consisting of two parts: first, the extinction of the original discrimination, and second, the acquisition of the reversal discrimination. While overtraining should speed up the second part, it should slow down the first. This leads to our further qualification: if overtraining has opposite effects on the two parts of reversal learning, it becomes difficult to determine what the effects of overtraining on an overall trials-to-criterion measure will be. This can, in fact, be rigorously decided only when the necessary rate parameters have been fixed—hence the value of quantitative models in this area. Nevertheless, even if there are some problems in deriving predictions about the effects of overtraining on an overall trials-to-criterion score, there are several detailed predictions that follow about the course of reversal learning in overtrained and nonovertrained subjects. Indeed, a good case can be made for the assertion that the theoretically most important fact about overtraining and reversal learning is not that overtrained rats often reach criterion of reversal more rapidly than nonovertrained rats, but that overtraining alters the manner in which a reversal is learned. Two predictions about the course of reversal learning can be made:

(1) Overtraining should increase the number of trials, at the outset of reversal, during which subjects continue predominantly to select their former S+.

(2) Overtraining should decrease the probability that control of behavior will be shifted from the relevant to an irrelevant analyzer during the course of reversal. In the standard simultaneous brightness discrimination, this will normally mean that nonovertrained subjects will show marked position habits and respond for many trials at the 50% level, while overtrained subjects will be less likely to do so.

### 2. *Experimental Evidence*

In Figure 8.1 are shown the daily reversal scores of two rats (the filled circle designating overtrained, and the open circle, nonovertrained) reversed on a simultaneous brightness discrimination (Mackintosh, 1963a). The data from these two subjects support both of these predictions: during the first 50

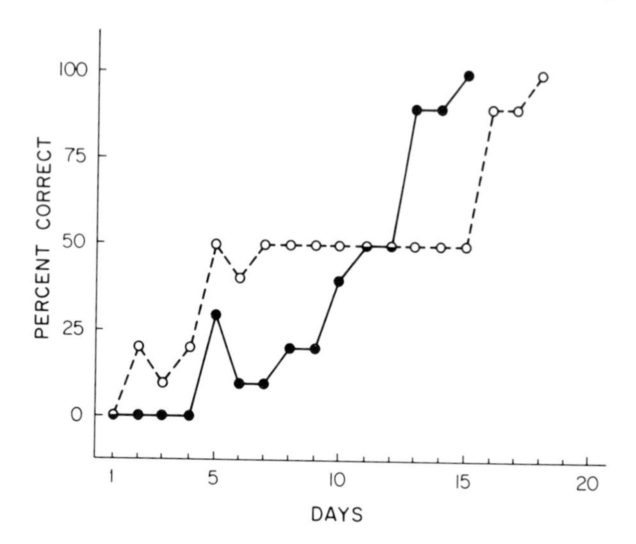


**Fig. 8.1.** Reversal learning curves for a single overtrained (●) and a single nonovertrained rat (○). [*After* Mackintosh (1963a).]

trials of reversal, the overtrained subject is making many more errors than the nonovertrained, indeed, it has not even started to extinguish the initial discrimination habit; later in reversal, however, the nonovertrained subject's score remains at the 50% level, and is overtaken by that of the overtrained subject. From Day 7 to Day 15, the nonovertrained subject responded at chance; on all trials on these days, this subject selected the left-hand stimulus.

Group mean learning curves do not give an accurate picture of the individual learning functions; nor do backward learning curves, since we are interested not only in the transition from 50 to 100% correct, but also in the transition from 0 to 50% correct. It is obviously important, however, to know whether the two subjects shown in Figure 8.1 are representative of their respective groups. In Figure 8.2, the group learning curves are plotted in a somewhat unusual manner, which serves, however, to point up the shape of the individual curves. The results confirm our predictions.

In all of our own experiments on visual reversal learning in rats, these two predictions have generally been confirmed. On three different measures (number of initial perseverative errors, number of days on which subjects select their former S+ more than any other stimulus, number of trials before reaching 50% choice of S+ and S− over 10 trials), we have found that overtraining increases resistance to extinction of choice behavior. Secondly,

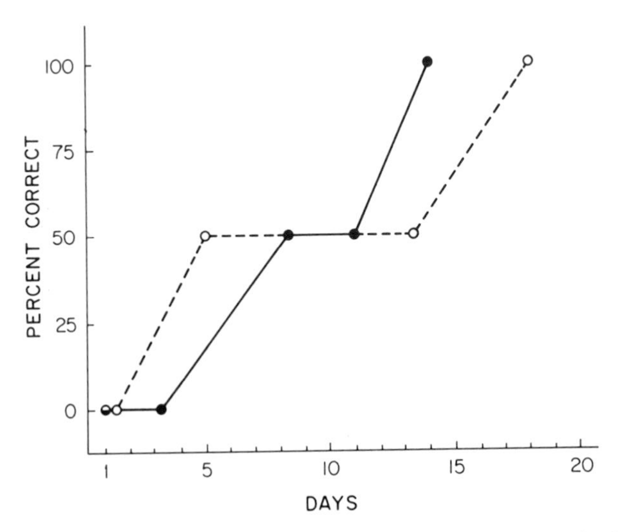


**Fig. 8.2.** Reversal learning curves for overtrained (●) and nonovertrained (○) groups. In order to represent the course of reversal in individual animals, the group curves were constructed by dividing reversal into four stages: (1) number of trials before the first correct responses; (2) number of trials before the first day with 50% correct responses; (3) number of trials before the first day with more than 50% correct responses; (4) number of trials to criterion. The group curves are constructed by joining these four points. [*After* Mackintosh (1963a).]

nonovertrained subjects have always shown a greater number of responses to irrelevant cues (usually position) during reversal than have overtrained subjects (Mackintosh, N. J., 1962, 1963a,b, 1965b). Unfortunately, despite the abundance of studies of reversal learning, few investigators have reported more than overall trial or errors-to-criterion scores. In those cases, however, where further details have been provided, these two conclusions have usually been confirmed.

In his original study, Reid (1953) reported a significant increase in perseverative errors with increased overtraining, and also that "very soon after the breakdown of the consistent response to the black [positive] card, this [overtrained] group began to consistently run to the white card .... [Nonovertrained subjects], however, began to respond in terms of position habits and to stimuli other than the cards [p. 110]." Komaki (1961) found that overtraining caused an increase in initial perseverative errors, and a decrease in subsequent position habits. Paul (1966) found, in two experiments, that overtraining had no consistent effect on initial perseverative errors, but significantly decreased the extent of position habits at a later stage of reversal. Hooper (1967) (see Mandler & Hooper, 1967), Mandler (1968), and Siegel (1967) all found that overtraining both increased initial resistance to extinction and decreased subsequent position responding (in their situations, position responding was defined in terms of the rat's initial orientation on each trial rather than its actual choice on each trial).

Even in studies which failed to show an ORE in terms of trials to criterion, a similar pattern of reversal learning has often been observed. Tighe, Brown, and Youngs (1965) found a significant increase in perseverative errors following overtraining, and also (personal communication) that overtraining somewhat (but not significantly) decreased the number of trials to reach criterion after subjects had reached a 50% level of responding. In an earlier experiment, similar in most respects to her later study, except that a small reward was used and no ORE occurred, Mandler (1966) reported both a greater number of perseverative errors and fewer position responses in overtrained than in nonovertrained subjects. In a series of experiments, none of which obtained an ORE, Eimas (1967) found that overtraining usually retarded extinction of the formerly correct response (measured as the number of trials until subjects responded at chance), but had no effect on reacquisition scores. The major exception to these results is a set of experiments by Lukaszewska (1968): in four experiments, none of which obtained an ORE, overtraining failed consistently to increase initial errors at the outset of reversal, and in only one case did it decrease responses to position.

Although, therefore, the amount of evidence on the effects of overtraining on the course of reversal learning is small relative to the amount available on overall trials-to-criterion scores, there is, to our knowledge, no published exception to the conclusion that, in simultaneous visual discrimination experiments with rats, overtraining facilitates reversal not because it facilitates initial extinction of choice behavior (for this is usually retarded), but because it decreases responses to irrelevant cues during later reversal learning. We have already argued that the predictions that follow from a two-process model concerning the course of reversal learning are less equivocal than those concerned with trials to criterion scores, so that this evidence is of considerable importance in evaluating such a theory. Furthermore, even those studies in which no ORE was obtained have often shown that overtraining has the effects on the course of reversal learning predicted by such a theory. While it therefore remains important to discover why some studies of visual reversal learning have obtained an ORE and others have not, in those cases where the relevant details have been given, many of the negative results still provide some support for an attentional theory.

### B. Salience of the Relevant Cue

The discussion so far has been concerned with the reversal of simultaneous visual discriminations by rats. The majority of such experiments can be regarded as providing the subject with a discrimination problem containing a single relevant cue and a single, rather dominant, irrelevant cue—namely, position. When overtraining does facilitate reversal learning in these experiments, it does so by decreasing the probability of position habits, and we interpret this by saying that in nonovertrained subjects the relevant analyzer is extinguished well below the point where the irrelevant position analyzer comes to control behavior, while in overtrained subjects the relevant analyzer does not extinguish so far.

### 1. *Importance of Irrelevant Cues*

From these considerations, it seems reasonable to infer that the extent to which overtraining will facilitate reversal should depend partly upon the number and prepotence of irrelevant cues. If there were no other cues in the experimental situation which could come to control the subject's behavior (a state of affairs presumably not perfectly realizable) nonovertrained subjects should not be at such a great disadvantage *vis à vis* overtrained subjects, whereas if there were several such cues present, the rate of reversal learning in nonovertrained subjects should be severely retarded, while overtrained subjects, maintaining attention to the relevant cue at a higher level, should be less affected. One discrimination situation in which there are few obvious irrelevant cues (certainly none deliberately manipulated by the experimenter) is the standard spatial discrimination in a T or Y maze. It is possible that this is one reason for the general failure to find the ORE in such situations. (Reversal learning in spatial problems is discussed in detail in Section IVB.)

We have shown in two experiments that the magnitude of the ORE in visual problems varies directly with the number of irrelevant cues. Mackintosh (1963a) trained three groups of rats on a brightness discrimination and reversal. For Group B–W, the stimuli were black and white squares; for Group Bhv–Whv, the stimuli were black and white rectangles, shown either horizontally or vertically, the orientation of the rectangles acting as an additional irrelevant cue. Reversal learning scores for these two groups, as a function of overtraining, are shown in Figure 8.3; the magnitude of the ORE was greater in Group Bhv–Whv than in Group B–W. Descriptively, the reason for this result is that the extra irrelevant cue retarded reversal in nonovertrained subjects, but had no significant effect on overtrained subjects. This is exactly what we should expect.

The fact that the overtrained subjects of Group Bhv–Whv reversed as rapidly as those of Group B–W could be interpreted either by saying that both had learned to attend to brightness, or that the former, by being trained with orientation irrelevant had learned to ignore this cue. To test between these two possibilities, a third group of subjects was run. Group BW–Bhv–Whv was initially trained with black and white squares, but was reversed to black and

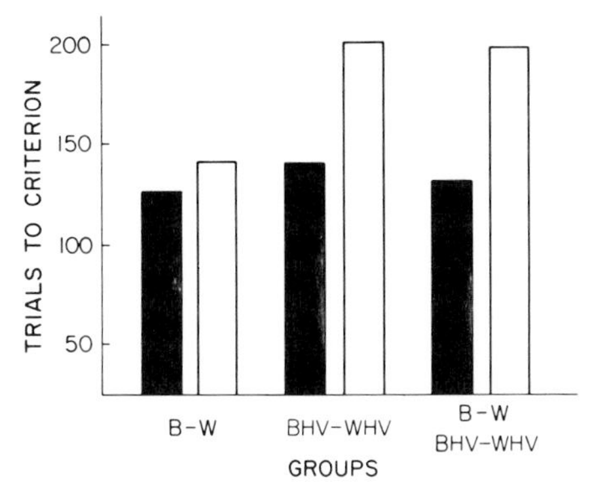


**Fig. 8.3.** Reversal learning scores for overtrained (■) and nonovertrained (□) animals as a function of number of irrelevant cues. [*After* Mackintosh (1963a).]

white rectangles; for this group therefore the extra irrelevant cue was not introduced until after original training, Nevertheless, overtrained subjects in this group reversed as rapidly as those in other groups. In this situation, therefore, with the particular stimuli used, rate of reversal learning was determined more by what analyzers had been strengthened by original training than by what analyzers had been weakened. This conclusion will be further discussed in Chapter 9, where we take up this distinction in greater detail.

An experiment with octopuses (Mackintosh & Mackintosh, 1963) succeeded in replicating several of the results of the above rat experiment. Here again, the effect of overtraining on reversal learning was significantly affected by the number of irrelevant cues; and here again, the presence of irrelevant cues retarded reversal learning in nonovertrained subjects without having any significant effect on overtrained subjects. This experiment is discussed in greater detail in Chapter 12 on comparative psychology. Although some other results to be discussed later (notably those of Clayton, 1963b, on the effects of irrelevant cues on spatial reversal learning) do not agree with these theoretical predictions, it is certain that our interpretation of the ORE has received some experimental support.

### 2. *Problem Difficulty*

Lovejoy (1966) has shown that his own model predicts an ORE only when the initial probability of attending to the relevant cue is moderately low—below about .7. With this probability set at 1.0, indeed, Lovejoy found overtraining to retard reversal. He argues that differences in the initial probability of attending to the relevant cue can be regarded as representing differences in the discriminability of the relevant stimuli: an easy discrimination is one where subjects are very likely to attend correctly, while a difficult discrimination is one where subjects do not automatically attend to the relevant cue. (It is also, of course, likely that the number and salience of irrelevant cues will affect the initial probability of attending to the relevant cue, but it still seems worth distinguishing between relevant and irrelevant cue salience.) This identification seems a reasonable one, and leads to the prediction that the more difficult the discrimination the greater should be the chances of observing an ORE. Exactly the same prediction will follow from our own theory, for an analyzer that is already strong cannot be strengthened (more than its response attachments) during overtraining.

In two experiments with birds, problem difficulty has been directly varied, with results consistent with this analysis. Mackintosh (1965b) trained chicks on a very easy black–white discrimination, and found that overtraining retarded reversal. However, when chicks were trained on a relatively difficult orientation discrimination (with brightness irrelevant), overtraining facilitated reversal. Schade and Bitterman (1965) found that pigeons showed a reverse ORE when trained on a color problem with shape irrelevant, but no effect of overtraining on reversal when trained on a (much harder) shape problem with color irrelevant. Although both experiments suggest that the discriminability of the relevant stimuli is an important variable, in each case either the number or the salience of irrelevant cues was varied concomitantly, and it is not possible, therefore, to separate the two effects.

A further experiment by Mackintosh (1969a) provides unequivocal evidence. Rats were trained in a Grice box either on a black–white discrimination or on a dark-gray–light-gray discrimination. In both cases, the only irrelevant cue was position, and all conditions except the discriminability of the brightness stimuli were held constant. There was a significant interaction between the effects of overtraining and problem difficulty on reversal scores; and separate analyses revealed that overtraining facilitated reversal only for subjects trained on the difficult problem.

These direct comparisons suggest that problem difficulty has a critical effect on the occurrence of the ORE; as we shall see below, further comparisons between successful and unsuccessful experiments support this conclusion. The results are entirely consistent with a two-process analysis, and, so far as we know, do not follow from any other theoretical account of the ORE.

### C. Extinction of Analyzers and Responses

Our explanation of the ORE assumes that the effect of overtraining is to strengthen the relevant analyzer relative to its response attachments. During early reversal trials, both the relevant analyzer and its response attachments are weakened; the analyzer continues to weaken until the two antagonistic choice responses are of equal strength, and from then on, the relevant analyzer will be strengthened again. The assumption that overtraining strengthens the analyzer relative to its response attachments implies that, after overtraining, the analyzer will not be as greatly extinguished during reversal as it will be in animals trained only to criterion. We have attempted to test this hypothesis in a number of studies designed to separate out the effects of analyzer and response extinction. Once again, it is not possible to make any exact predictions without detailed specification of certain rate parameters. Although we can assert that the extent to which the analyzer will be extinguished should be decreased by overtraining, we cannot predict the absolute extent to which the analyzer will be extinguished either with or without overtraining. We can predict the direction of effects, but not their absolute magnitude.

Several studies have provided results which suggest that if animals are trained on a discrimination problem only to criterion, and if this discrimination is then extinguished, the effect of these extinction trials is to weaken the relevant analyzer to a greater extent than its response attachments. We can now predict that these effects should be decreased by prior overtraining, and we have conducted several experiments of this sort.

### 1. *Extinction of a Discrimination Habit Learned to Criterion*

*a. Lawrence's Experiments.* Lawrence and Mason (1955) trained rats on a discrimination problem with one cue relevant and another irrelevant; after animals had reached criterion on this problem, they were next trained with the originally relevant cue irrelevant and the originally irrelevant cue relevant. In order to solve the second problem, animals had to select each value of the originally relevant cue equally often, so that in this stage, the original discrimination was being extinguished. In a third stage of the experiment, the animals were shifted back to the original discrimination: half relearned this problem, the other half learned the reversal. Lawrence and Mason found that Stage 2 training had clearly not succeeded in extinguishing the original discrimination, since subjects retrained with their former S+ still positive learned very much more rapidly than subjects trained with their former S+ negative. They suggested that such results would follow if one assumed that, during Stage 2, subjects stopped attending to the originally relevant cue before they had fully extinguished their choice responses to the values of that cue. In our terminology, they assumed that if animals are trained to criterion on a given discrimination, then during extinction trials, the relevant analyzer reaches any given level of extinction more rapidly than its response attachments.

As Lawrence and Mason recognized, their results do not compel this interpretation, since, even on a single-stage continuity view of discrimination learning, one might not expect complete equalization of response strengths to the Stage 1 positive and negative stimuli during Stage 2 training. In a subsequent experiment, however, Goodwin and Lawrence (1955) presented results that pointed more strongly to their attentional interpretation. Even if continuity theory does not necessarily predict that animals trained to criterion in Stage 2 should have equalized response strengths to the original positive and negative, it must predict that further training in Stage 2 should further equalize these response strengths. Goodwin and Lawrence, however, found that even if rats received overtraining after reaching criterion in Stage 2, they still learned more slowly in Stage 3 when reversed than when not reversed, and that this reversal–nonreversal difference was not affected by the amount of Stage 2 training given. The results suggest that in Stage 2 animals rapidly extinguished the originally relevant analyzer without fully extinguishing its original response attachments, so that once the analyzer was extinguished little further change in those response strengths occurred.

*b. Alternative Interpretations.* This interpretation has been questioned by Ross, and in two studies (Ross, 1962a,b) he has presented evidence suggesting that further factors may be operating to preserve original response strengths in addition to the possibility that they are preserved because subjects are no longer attending to that cue. In his first experiment, rats were trained on a simultaneous black–white discrimination, and were then given a series of equally reinforced trials to the black and white stimuli presented singly. When subjects were returned to the simultaneous discrimination, he found that nonreversed subjects relearned more rapidly than reversed subjects. He suggested that the preservation of differential response strengths to black and white through the intervening stage of equal reinforcement could not be explained by supposing that subjects had stopped attending to brightness, since they were not required to attend to some new cue as they had been in Lawrence and Mason's experiment. This does not seem entirely convincing; although we should expect that the presence of a relevant cue would hasten extinction of the brightness analyzer (cf. the discussion of Wagner *et al.*, 1968, in Chapter 5), it does not follow that in the absence of a relevant cue no extinction of the brightness analyzer would occur. However, Ross also recorded running speeds to black and white during the intervening equal-reward period, and found that, although speed to the former negative increased during this stage, it remained significantly less than speed to the former positive. Since this latency measure revealed differential responding to black and white throughout the equal reward period, subjects cannot have completely extinguished attention to brightness.

In Ross' second experiment, rats were trained first on a simultaneous brightness discrimination with position irrelevant, then on a position discrimination with brightness irrelevant, and finally returned to the brightness discrimination, with half relearning the initial problem and half learning the reversal. Once again, the nonreversal was learned more rapidly than the reversal, and once again, the latency difference to black and white remained significant throughout the second stage of the experiment.

Ross has argued that the preservation of a discrimination through a period of equal reinforcement cannot be entirely explained in terms of extinction of attention to the relevant cue during this period, since the latency data indicate differential responding and therefore attention to the relevant cue. He has suggested that inhibition built up by nonreward of a given stimulus is partially maintained even when that stimulus is now rewarded. While the latency data certainly indicate that the original analyzer cannot have been fully extinguished during the period of equal reinforcement (a position which is not necessary to the argument), the question remains whether the maintenance of different running speeds to S+ and S− throughout the period of equal reinforcement is necessary for the occurrence of reversal–nonreversal differences in the test phase. A recent experiment by Stettner (1965) suggests that it is not, and therefore that Ross' notion of the preservation of the inhibitory potential of a nonreinforced stimulus does not provide a complete explanation of the observed reversal–nonreversal differences. Stettner trained, and then reversed rats on a go–no-go brightness discrimination. All subjects were then given a period of equal reinforcement in the black and white runways, and running speeds were recorded. Running speeds to black and white rapidly converged, and over the last 72 of the 80 trials of equal reinforcement given, there was no sign of any difference in speed. Despite this failure to find Ross' evidence of persistence of inhibition to the more recently negative stimulus, when the animals were retrained on the successive discrimination, Stettner observed a significant difference in speed of learning between those who relearned their second problem and those who were reversed. This implies that the differences in response strength revealed by the test phase were indeed latent during the period of equal reinforcement, and supports the argument that the reversal–nonreversal difference must be partially attributed to a more rapid extinction of the relevant analyzer than of its response attachments.

*c. The "Dip Effect."* A further finding, first reported by D'Amato and Jagoda (1960), also suggests that in rats trained only to criterion on a brightness discrimination, extinction trials will produce more complete extinction of the brightness analyzer than of its response attachments. In this experiment rats learned a brightness discrimination to an 18 out of 20 criterion, and then received 60 extinction trials in which choice of neither alternative was rewarded. They were then trained on the reversal of the original problem, i.e., rewarded for choice of their former S−. Paradoxically, they immediately and significantly increased their choice of their former *positive* stimulus; the reversal learning curve after extinction showed an initial dip in probability of selection of the new positive, a phenomenon which, for lack of a better name, we shall call the "dip effect." We have replicated this result on a number of occasions (Mackintosh, 1963b, 1965b). In the first experiment, subjects having been trained to criterion on a brightness discrimination were extinguished to a criterion of equal choice of positive and negative over 10 trials. Latencies were recorded during extinction, and by the end of extinction, subjects were responding at the same speed to both stimuli. As in Stettner's experiment, therefore, there was no evidence that differential response strength was preserved during extinction. Nevertheless, as soon as reward was reintroduced into the experimental situation, it was evident that differential response strength had, in fact, been partially preserved, since subjects showed a significant preference for their former positive stimulus (see Figure 8.4, page 271).

The dip effect can be explained by the argument presented so far, with the addition of a further assumption. During extinction trials, the relevant analyzer is extinguished to the point where it no longer controls behavior and subjects respond at chance with respect to the relevant cue. The responses attached to the outputs of the analyzer, however, have not been equalized, although this is not apparent from the subject's choice behavior since his behavior is now predominantly controlled by some other analyzer. If the reintroduction of reward into the situation at the outset of reversal training causes the initially relevant analyzer to be switched-in again, the subject's behavior will again be determined by the strengths of these response attachments, and above chance choice of the former positive stimulus will occur.

The assumption introduced here is that an analyzer established in original training, and weakened during a series of extinction trials, will immediately control behavior again upon the reintroduction of reward. The implication is that the analyzer (regarded as a response) is conditioned to situational stimuli, —which include the presence of reward in the goal box. In part then, the weakening of the analyzer during extinction trials is regarded as being due to a generalization decrement factor (a change, by the removal of reward, in the stimulus complex to which the analyzer was originally conditioned), so that a reinstatement of that stimulus complex will partially reinstate the analyzer. The notion that analyzers are conditioned to such stimuli is not farfetched: a number of studies of interhemispheric transfer, for example, have shown that a discrimination trained in one hemisphere shows no transfer whatsoever to the other until the original reinforcement is reintroduced, whereupon, substantial transfer effects appear (Levine, 1945; Myers, 1961). Considerable evidence will be discussed in Chapter 10 showing that responses can be conditioned to the memory of reward or nonreward, and if responses can be thus controlled by memory stimuli, it is possible that analyzers can also be conditioned to them. The real question, in fact, is whether it is the reinstatement of the relevant analyzer, or simply of the original response (as originally suggested by D'Amato and Jagoda, 1960) that is the cause of the dip. This can be answered by looking at the effects of overtraining.

### 2. *Extinction of Overlearned Discrimination Habits*

If the preservation of a discrimination habit through a series of extinction trials (as evidenced by faster nonreversal than reversal learning or by the dip effect) occurs in nonovertrained animals because of a rapid extinction of the relevant analyzer, then our assumption that overtraining strengthens the relevant analyzer relative to its response attachments predicts that these effects should be decreased by prior overtraining. In nonovertrained animals differential response strength is preserved because the relevant analyzer rapidly reaches a level of extinction where further changes in those response strengths are small. After overtraining, the relevant analyzer has been further strengthened and will therefore take longer to extinguish to the point where response strengths change only slowly: if these response strengths have not been equally increased by overtraining, it follows that they will be more completely extinguished during the course of extinction trials. To reiterate a point made earlier, we cannot assert that responses will be entirely extinguished before the analyzer is extinguished in overtrained animals, since this depends critically upon the precise rate parameters assumed; we can only predict that extinction of responses should be carried further following overtraining.

Tests of this prediction simply involve repeating the experimental designs used by Lawrence and Mason and by D'Amato and Jagoda, incorporating animals trained either to criterion or overtrained on the initial problem. Mackintosh (1963b) trained two such groups of rats on a discrimination between black and white, horizontal and vertical rectangles, with brightness relevant and orientation irrelevant. In Stage 2 of the experiment, all subjects received a fixed number of trials on the horizontal–vertical discrimination with brightness irrelevant. In Stage 3, the subjects were divided into two subgroups, half relearning their original brightness discrimination, half learning the reversal of their original brightness discrimination. The results are shown in Table 8.II. The reversal–nonreversal difference in Stage 3 was significantly

TABLE 8.II

**Results of Experiment by Mackintosh**[a]

| Groups | Nonreversal | Reversal |
|---|---|---|
| Overtrained | 63.75 | 90.25 |
| Nonovertrained | 77.25 | 172.50 |

[a] Data obtained from Mackintosh (1963b).

greater in subjects trained to criterion in Stage 1 than in subjects overtrained in Stage 1. It is worth noting that there still appears to be a substantial difference between nonreversed and reversed overtrained subjects, although, since only a small number of subjects was used, the difference was not significant ($.10 > p > .05$). The important point is that this difference was substantially greater in nonovertrained subjects.

One feature of the results makes them particularly difficult to account for within the terms of a single-stage model. Ross' explanation of Lawrence and Mason's results depended upon showing that the differential response strength to the original S+ and S−, evidenced by the reversal–nonreversal difference in Stage 3, was also shown by latency differences in the equal reinforcement period of Stage 2. Since Mackintosh's results showed that the difference in response strengths in Stage 3 was greater for nonovertrained than for overtrained subjects, it should follow that a similar difference should have appeared in Stage 2. Although latency measures were not taken in this experiment, all subjects received an equal number (150) of Stage 2 trials; nonovertrained subjects actually learned Stage 2 significantly more rapidly than overtrained subjects. This implies that by the end of Stage 2, the difference in response strengths between the original S+ and S− should have been *less* in nonovertrained than in overtrained subjects, and yet the Stage 3 results indicate precisely the opposite.

We have also tested our prediction in a situation similar to that used by D'Amato and Jagoda, i.e., we investigated the dip effect as a function of overtraining (Mackintosh, 1963b, 1965b). Rats were trained either to criterion or overtrained on a brightness discrimination, and then extinguished to a criterion of equal choice of black and white over 10 trials. In the final stage of the experiment, all subjects learned the reversal of their original brightness discrimination. The reversal learning curves obtained from the first experiment are shown in Figure 8.4. (The results obtained in the second experiment were essentially similar.) As mentioned previously, nonovertrained subjects showed a significant dip in choice of the S+ upon reversal ($p < .001$), while it is clear that overtrained subjects did not show the effect (in fact, 2 out of 8 subjects selected the new S+ on 4 out of the first 10 reversal trials, and all other subjects scored at least at chance throughout reversal training).

These results provide some of the most compelling evidence to support our interpretation of the ORE. Although only indirectly, these experimental designs do allow one to get at the crucial theoretical assumption, namely, that overtraining affects the relative extents to which analyzers and responses are extinguished during early reversal trials. We have also utilized such designs with favorable results in our comparative research on the effects of overtraining in octopuses, goldfish, and domestic chicks. These studies, discussed in detail in Chapter 12, extend our theoretical conclusions by correlating the effects of overtraining on reversal learning across different groups with the effects of overtraining on the extinction of analyzers and responses across these groups.

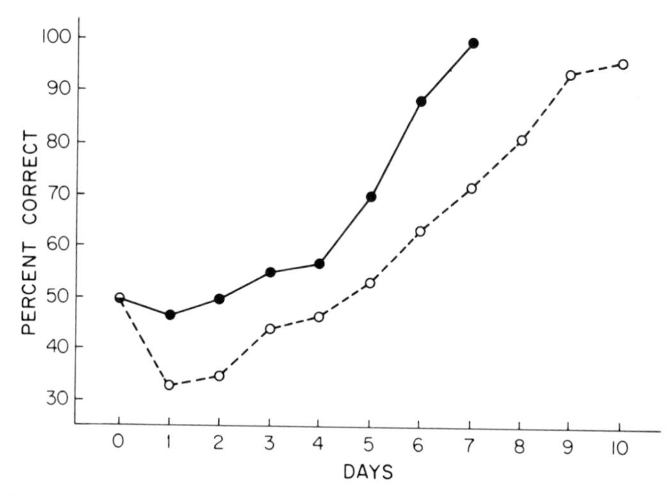


**Fig. 8.4.** Reversal learning by overtrained (●) and nonovertrained (○) animals following extinction. Day 0 was the last day of extinction. [*After* Mackintosh (1963b). Copyright by the American Psychological Association. Reproduced by permission.]

# IV. The Elusive Nature of the ORE

Much of what we have said so far in this chapter might lead the unwary to suppose that the ORE is a reliable, easily reproducible phenomenon. Anyone familiar with the literature on the subject will know that this is not the case; indeed, our initial argument in this chapter, to the effect that the ORE is a phenomenon that can be explained by, but does not necessarily follow from, a two-process model, may have been recognized as a device for hedging our bets. Although it can hardly be doubted that the ORE does occur,[1] the number of experiments in which the effect has been found (about 20) is substantially outweighed by the number in which it has not been found (over 40). There is good reason to believe that several factors are responsible for such a degree of variability. We shall first consider those factors directly suggested by the arguments presented up to this point, and then show how the importance of another factor (reward size) can be handled by a two-process analysis. For reasons that will become apparent, it is convenient to discuss visual and spatial experiments separately.

## A. Visual Studies

In Table 8.III, we list the majority of studies on overtraining and simultaneous visual discrimination reversal, divided into those in which an ORE

[1] It has, of course, been doubted, e.g., by Gardner (1966). His argument was that the ORE has only been obtained in early, relatively careless studies; fortunately this has been invalidated by the recent appearance of several large-scale studies in which positive results were obtained (e.g., Hooper, 1967; Paul, 1966; Seigel, 1967).

TABLE 8.III

**Results of Visual Overtraining Reversal Experiments**

| Overtraining facilitates reversal | Overtraining does not facilitate reversal |
| --- | --- |
| *Rats* | *Rats* |
| Birnbaum (1967, difficult problem, extended overtraining) | Birnbaum (1967, easy problem) |
| Brookshire, Warren, and Ball (1961) | D'Amato and Schiff (1965, eight experiments) |
| Capaldi and Stevenson (1957) | Eimas (1967, two experiments) |
| D'Amato and Jagoda (1961) | Erlebacher (1963) |
| Hooper (1967, large reward) | Hooper (1967, small reward) |
| Komaki (1961) | Lukaszewska (1968, four experiments) |
| Mackintosh (1962) | Mackintosh (1969a, easy problem or small reward) |
| Mackintosh (1963a) | Mandler (1966) |
| Mackintosh (1965b) | Paul and Havlena (1965) |
| Mackintosh (1969a, Experiment II, Experiment III, difficult problem, large reward) | Tighe, Brown, and Youngs (1965) |
| Mandler (1968) | Weyant (1966) |
| North and Clayton (1959) | |
| Paul (1966) | |
| Pubols (1956) | |
| Reid (1953) | |
| Siegel (1967) | |
| Sperling (1970) | |
| *Other animals* | *Other animals* |
| Mackintosh (1965b, chicks, difficult problem) | Beck, Warren, and Sterner (1966, cats, rhesus monkeys) |
| Mackintosh and Mackintosh (1963, octopuses) | Boyer and Cross (1965, stump-tailed monkeys) |
| Williams (1967b, pigeons) | Brookshire, Warren, and Ball (1961, chicks) |
| | Cross and Brown (1965, squirrel monkeys) |
| | Cross, Fickling, Carpenter, and Brown (1964, squirrel monkeys) |
| | D'Amato (1965, capuchin monkeys) |
| | Hirayoshi and Warren (1967, cats) |
| | Mackintosh (1965b, chicks, easy problem) |
| | Mackintosh, Mackintosh, Safriel-Jorne, and Sutherland (1966, goldfish) |
| | Schade and Bitterman (1965, pigeons) |
| | Tighe (1965, rhesus monkeys) |
| | Warren (1960a, paradise fish) |

was found, and those in which overtraining had no effect on reversal (included in this category are three studies with birds in which overtraining significantly retarded reversal). We shall not discuss here the results of the bird, fish, and octopus experiments, since they are treated at length in Chapter 12, where we attempt to explain certain similarities and differences between rats and these classes of animals. The consistently negative results obtained with monkeys and cats (no ORE has been found in either animal) might suggest a difference in learning mechanisms between these animals and rats. (We shall also discuss this in Chapter 12.) We must confess at this point, however, that we can offer no good suggestion as to the nature of this difference between rats and other mammals.

As we have already pointed out, in some studies in which overtraining did not facilitate reversal, the details of the course of reversal learning tend to support our analysis: Mandler (1966) and Tighe *et al.* (1965), for example, both found that overtraining increased initial perservative errors, and tended to facilitate the reacquisition phase of reversal. There is, however, no uniformity about the rat results, for even were we to assume that, in the remaining negative cases, an analysis of daily reversal scores would reveal the pattern obtained in other published studies,[2] we would not have answered the question as to why some studies have yielded an effect of overtraining on trials to criterion scores and others have not. From the discussion so far presented in this chapter, two possible suggestions emerge that may be relevant to understanding some of the discrepancies.

### 1. *Problem Difficulty*

As we have already argued, a two-process theory predicts that the effect of overtraining on reversal will depend upon the initial strength of the relevant analyzer, and this will depend on the discriminability of the relevant stimuli. At least one experiment has provided direct evidence of the importance of this variable: Mackintosh (1969a) found an ORE when rats were trained on a dark-gray–light-gray discrimination, but no effect when they were trained on a black–white problem.

It is likely, therefore, that some of the between-experiment variability in the occurrence of the ORE will be attributable to differences in problem difficulty. This possibility was suggested by Lovejoy (1966), who noted that in both Erlebacher's and D'Amato and Schiff's experiments, the relevant brightness cue may have been more obvious (corresponding to a greater initial

[2] In the case of D'Amato and Schiff's (1965) results, this is not an unreasonable assumption, since they noted "a tendency for the reversal curve of the overtrained group to be slightly steeper than the controls [p. 380]."

strength for that analyzer) than in most other studies of brightness reversal learning in rats. Erlebacher used a T maze whose goal arms were completely painted either black or white; D'Amato and Schiff (1965) used an automatic Y maze, one of whose arms was diffusely illuminated, while the other was dark; in this apparatus only one of nine experiments (that of D'Amato and Jagoda, 1961) yielded an ORE—and that was of only marginal significance. Erlebacher's rats learned initially in less than 30 trials; in the majority of successful studies with rats, initial learning has taken at least 50 trials.

Two more recent studies are consistent with this hypothesis. Eimas (1967) failed to obtain an ORE in rats trained on a very easy black–white discrimination in a Grice box (whether a large or a small reward was used). Lukaszewska (1968) failed to find a significant ORE in four separate experiments: in three of the experiments, indeed, overtraining tended to retard reversal; only in the fourth (the most difficult problem) did overtraining even slightly facilitate reversal (and the effect, although not significant, was relatively large: the nonovertrained group reversing in 267 trials, the overtrained in 217 trials).

### 2. *Salience of Irrelevant Cues*

The initial strength of the relevant analyzer should be affected not only by the discriminability of the relevant stimuli, but also, as we have argued, by the salience and number of irrelevant cues. Direct evidence of the role of irrelevant cues in increasing the magnitude of the ORE has already been presented; it remains to be considered whether any of the variability between experiments can be accounted for in this way.

In D'Amato's Y maze, it is arguable that the irrelevant spatial cue is less salient than in most other situations. The start arm for any given trial is that arm entered by the subject on the preceding trial. The spatial difference between the two goal arms over a series of trials, therefore, is defined only in terms of left and right with reference to the subject, not in terms of absolute position in space. In the language of the place versus response controversy of T maze learning (Restle, 1957) the spatial cue in D'Amato's situation is solely a response-defined cue; it is not, as in the ordinary Y or T maze or jumping stand, defined both in terms of place and response. It is not surprising, therefore, that rats take a long time to learn a simple spatial discrimination in this apparatus (over 50 trials to criterion in the experiments of D'Amato and Jagoda, 1962); nor is it unreasonable to assume that when rats are trained on a simultaneous brightness discrimination in this apparatus, the salience of the irrelevant spatial cue will be lower than it is in other types of apparatus.

Tighe *et al.* (1965) trained rats either on a horizontal–vertical discrimination, or on a flat–raised discrimination. For each group, the second visual cue was irrelevant. Although this would, at first sight, appear to have been an optimal situation for obtaining the ORE, they may have reduced the chances of finding a significant effect, since during reversal training, the two values of the irrelevant cue were no longer simultaneously present on each trial. This would presumably reduce the chances of that cue coming to control behavior during reversal. Furthermore, the apparatus used was a type of shuttle box, in which the subject started from one end on a given trial, ran to the other end to make a choice, and then started from that end at the beginning of the next trial. As in D'Amato's Y maze, therefore, position was defined only in terms of response cues, and despite initial appearance to the contrary, the salience of irrelevant cues may have been quite low.

### 3. *Reward Size*

Although the arguments of the preceding sections follow naturally from the assumptions of a two-process theory as stated so far, they are insufficient to account for all the variability in the visual ORE data. It is certain that one further condition must be satisfied: an ORE does not usually occur with simultaneous visual discriminations in rats unless a relatively large reward is used. Two studies (Hooper, 1967; Mackintosh, 1969a) have directly compared the effects of overtraining on visual reversal with large (250 or 380 mg) or small (37 or 45 mg) rewards. In both, a significant interaction between the effects of overtraining and reward size was observed. In Mackintosh's experiment, overtraining had no effect at all on reversal when a small reward was used. In Hooper's experiment, overtraining slightly (but significantly) facilitated reversal with a small reward and a correction training procedure, but had no effect with a small reward and a noncorrection procedure. In both experiments, a large reward resulted in a substantial ORE. Further relatively direct evidence comes from two studies by Mandler (1966, 1968). In the first, Mandler used a 45-mg reward and found no ORE; in the second, with similar apparatus and procedures but a 250-mg reward, she found that overtraining significantly facilitated reversal.

The suggestion that a large reward is a necessary condition for the occurrence of the ORE is not only directly supported by these results, it also accounts for a relatively large proportion of the variance in visual ORE experiments. Many of the experiments listed in Table 8.III as yielding no ORE employed quite small rewards: Eimas (1967, Experiment I) and Paul and Havlena (1965) used a single 45-mg pellet; Erlebacher (1963) used two 45-mg pellets, and Tighe *et al.* (1965) a single 94-mg pellet. In seven of their eight experiments, D'Amato and Schiff (1965) gave 2-sec access to water; in their eighth they gave 5-sec access to a .25-ml dipper. Equally impressive is the fact that, so far as can be ascertained, there is only one exception (D'Amato & Jagoda, 1961) to the generalization that *all* studies obtaining an ORE have employed moderately large rewards; e.g., Mackintosh (1962, 1963a) gave 20-sec access to food; Paul (1966) used three 94-mg pellets, and Siegel (1967) used eight 45-mg pellets. The smallest rewards appear to have been 5-sec access to food (North & Clayton, 1959), or 150 mg (Brookshire, Warren, & Ball, 1961[3]); in each of these two experiments the ORE was quite small, and only significant at the .05 level.

While there is, therefore, good evidence that a large reward is a necessary condition for the appearance of an ORE, there is equally good evidence that it is not a sufficient condition; with very easy visual problems, no ORE occurs whatever the size of reward. Eimas (1967, Experiment II), training rats on an easy black–white discrimination in a Grice box, found no ORE whether the reward was 45 or 450 mg. In Mackintosh's (1969a) experiment, already referred to, a large reward yielded an ORE only when the discrimination was relatively difficult. Four groups of rats were trained on brightness discriminations in a Grice box, two groups on black–white, two on dark-gray–light-gray. Within each pair of groups, one received a 45-mg reward, the other a 380-mg reward. The reversal learning scores of overtrained and nonovertrained subjects in all four groups are shown in Figure 8.5. Only in the group trained

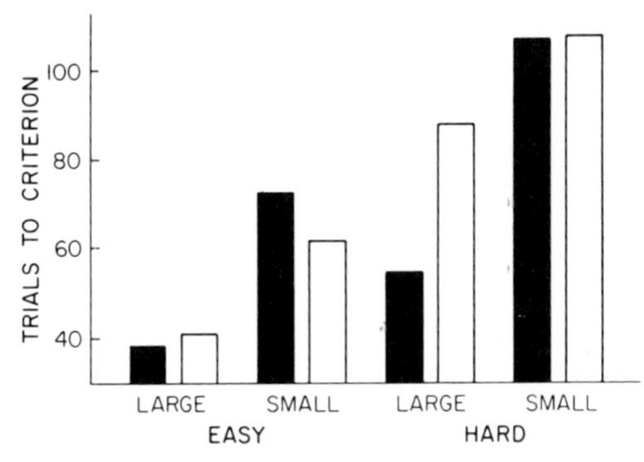


**Fig. 8.5.** Reversal learning scores for overtrained (■) and nonovertrained (□) animals as a function of problem difficulty and reward size. (Large and Small are reward sizes. [*After* Mackintosh (1969a). Copyright by the American Psychological Association. Reproduced by permission.]

with a large reward on the harder problem, did overtraining have any significant effect on reversal.

The importance of reward size as a determinant of the ORE was first suggested by Theios and Blosser (1965a,b), and attributed by them to its effect on extinction. Their analysis in terms of frustration theory will be discussed later in this chapter. Although it seems to us indisputable that a frustration analysis is not sufficient to account for all ORE data (for it appears

[3] We are grateful to K. H. Brookshire for providing us with this information.

to have nothing to say about the role of problem difficulty), it is tempting to suggest that it be brought in to supplement the account provided by an attentional theory. Of the two conditions jointly sufficient to produce an ORE, a two-process theory can account for one (problem difficulty), while frustration theory can account for the other (reward size). The argument is appealing; however, as we attempt to show below, it is also suspect, and it is therefore worth showing that it is unnecessary. That is to say, while a two-process theory certainly cannot be said to *predict* the effects of reward size on the ORE, it can be modified easily enough so as to be able, *post hoc*, to account for such effects.

A large reward leads to more rapid visual discrimination learning than a small reward (e.g., Hooper, 1967; Mackintosh, 1969a; Waller, 1968). Any model of discrimination learning must represent changes in reward size in such a way as to predict this fact. Most plausibly, changes in reward size will be represented as changes in the effectiveness of reward (or possibly non-reward) operators. The critical question, therefore, is whether appropriate changes in these operators can predict not only the effects of reward size on simple acquisition, but also its effects on the ORE. As we have argued above, a two-process model predicts an ORE if and only if overtraining strengthens the relevant analyzer more than its response attachments, and this will most easily happen if, at criterion, the relevant analyzer is weaker than its responses, and both tend to asymptote during overtraining. The question, therefore, becomes whether changes in reward operators might affect the relative strengths of analyzers and responses at criterion.

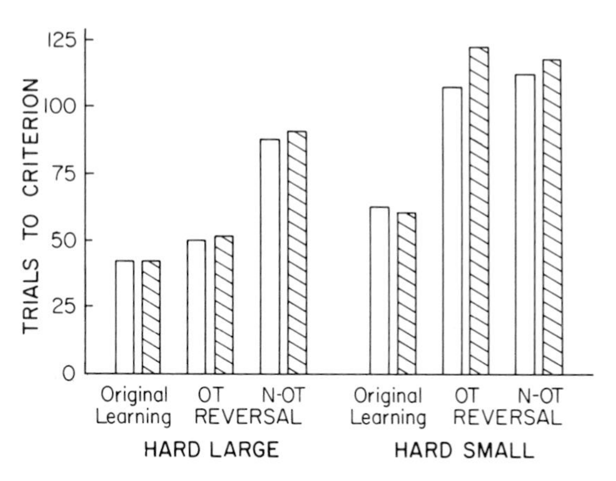


**Fig. 8.6.** Obtained (□) and predicted (▧) original and reversal learning scores for overtrained and nonovertrained animals trained on a difficult problem with a large or small reward. [*After* Mackintosh (1969a). Copyright by the American Psychological Association. Reproduced by permission.]

Stated thus, the question can obviously be answered in the affirmative. The relative strengths of analyzers and responses at criterion must partly depend (other things being equal) on the relative effectiveness of their respective reward operators. By assuming that increases in reward size have a greater effect on response operators than on analyzer operators, we will ensure that the discrepancy between response strength and analyzer strength at criterion (which underlies our explanation of the ORE) varies directly with the magnitude of reward. By computer simulation of Lovejoy's model, Mackintosh (1969a) has shown that these (admittedly arbitrary) assumptions successfully predict the effects of reward size on the ORE. Figure 8.6 shows the learning scores of two of the four groups of Mackintosh's experiment, together with the scores of two groups of stat-rats run with operators simulating large and small rewards. The fit of theory to data is close.

3. *Conclusions*

Before concluding that we have brought order into what was chaos, we must reluctantly point to the existence of a small number of studies that resist our analysis. Birnbaum (1967), Lukaszewska (1968), and Weyant (1966) have all trained rats on relatively difficult problems with relatively large rewards, and still failed to obtain an ORE. We have, of course, deliberately failed to specify the exact level of difficulty or reward size that might be regarded as necessary to produce the effect; however an ORE has been obtained with a reward of 150 mg (Brookshire, Warren, & Ball, 1961) and with a black–white discrimination learned in just under 40 trials (Mackintosh, 1969a), and in all these studies the reward was larger and the problem more difficult. If, however, we ignore absolute levels of problem difficulty, it is worth pointing out that Weyant used a problem that was learned with fewer than 20 errors, that Birnbaum did obtain an ORE for one of her groups trained on the more difficult problem, and that the one case where Lukaszewska's overtrained rats reversed faster than her nonovertrained rats was with her most difficult discrimination.

One or two further observations are in order. In three of her four experiments, Lukaszewska used some type of correction training procedure. This might be expected to interfere with the ORE by minimizing the importance of position habits (Hooper, 1967, however, found that the ORE was more readily obtained with correction than with noncorrection training—from our point of view, an extremely puzzling result). Birnbaum found an ORE after 400, but not after 200 overtraining trials. It is not only a two-process model that has difficulty in predicting this result.[4] Weyant reported his results in

[4] Sperling (1970) has recently reported that the number of overtraining trials necessary to produce an ORE may be as small as 20. In her experiment, overtraining beyond this point had no further effect on reversal learning.

terms of errors to reversal criterion. Since overtraining often increases perseverative errors at the outset of reversal, such a measure may obscure a genuine ORE. For example, in the experiment whose results were shown in Figures 8.1 and 8.2 (Mackintosh, 1963a), overtraining facilitated reversal in terms of trials to criterion (133 versus 180 trials), but had virtually no effect on the number of errors to criterion (90 versus 96 errors).

### B. Spatial Studies

In Table 8.IV is presented a relatively complete list of studies of overtraining and reversal learning in which position was a relevant cue. In some of these studies, another (visual) cue was also relevant; in some there was an irrelevant cue; but in most cases a simple, undifferentiated T or Y maze was used. It can be seen that in relatively few instances has overtraining been found to facilitate spatial reversal, and that it has at least as frequently retarded reversal learning. In this section, we shall discuss three suggestions that may help to answer the two main questions posed by these results, namely, first, why the general trend of results from studies of spatial reversal should differ from the general trend for visual reversal experiments, and second, why there should be such variability in the results of different spatial studies.

#### 1. *Dominance of Spatial Cues*

It seems a reasonable presumption that rats, like other mammals that naturally live underground and are mainly active at night, should neither have particularly good eyesight, nor rely on visual cues in the way that birds, for example, might be presumed to. In an environment such as an underground burrow, where spatial orientation would seem to be important, visual cues, when relied upon at all, are likely to be used as indicators of spatial location rather than as indicators of events of interest regardless of spatial location. If these arguments fail to convince the reader of the plausibility of assuming that a simple spatial discrimination in a T maze is a relatively natural problem for the rat, in a way in which a simultaneous visual discrimination is not, we may simply point to the differences in the rates of learning typically observed in each type of problem. It is not easy to devise a simple, two-choice spatial problem for the rat that will take more than 20 to 40 trials to solve; it is equally not particularly easy to devise a visual problem soluble in much less than 40 trials. What we wish to infer from these observations, of course, is the conclusion that in the simple spatial problem the relevant analyzer will normally start at a higher strength than it will in the more difficult visual problem; and that just as it is not easy to obtain an ORE when rats are trained

TABLE 8.IV

**Results of Spatial Overtraining Reversal Experiments with Rats**

| Overtraining facilitates reversal | Overtraining has no effect on reversal | Overtraining retards reversal |
| --- | --- | --- |
| Brookshire, Warren, and Ball (1961) | Clayton (1963a) | Clayton (1965) |
| Bruner, Mandler, O'Dowd, and Wallach (1958) | Clayton (1963b, two experiments) | Eimas (1967, Experiment II) |
| Capaldi (1963) | D'Amato and Jagoda (1962, four experiments) | Galanter and Bush (1959) |
| Ison and Birch (1961) | D'Amato and Schiff (1964, two experiments) | Hill and Spear (1963a) |
| Pubols (1956) | Eimas (1967, Experiment III) | Hill, Spear, and Clayton (1962, Experiment III) |
| Theios and Blosser (1965b, large reward) | Fidell and Birch (1967) | Krechevsky and Honzik (1932) |
| | Hill, Spear, and Clayton (1962, Experiments I and II) | Mackintosh (1965c, Experiment I) |
| | Kendler and Kimm (1964) | |
| | Kendler and Kimm (1967) | |
| | Komaki (1962) | |
| | Mackintosh (1965c, Experiment II) | |
| | Mackintosh (1969a, Experiment I) | |
| | Mandler (1966) | |
| | Paul and Kesner (1963) | |
| | Theios (1965) | |
| | Theios and Blosser (1965b, small reward) | |

on relatively obvious visual problems, so it is never easy to obtain an ORE when rats are trained on spatial problems.

While this argument, originally advanced by Lovejoy (1966), seems eminently reasonable, it is not one for which it is easy to provide direct support. The most compelling evidence, in fact, comes from experiments with chicks. If the reason that rats are more likely to show an ORE in visual than in spatial experiments lies in their natural preference for spatial cues, then an animal in which it is reasonable to postulate a preference for visual over spatial cues should be more likely to show an ORE in spatial than in visual experiments. Most birds would seem to come into this category, and Brookshire *et al.* (1961) found that in chicks, the direction of the effect of overtraining on reversal was as we should predict; while overtraining actually retarded the reversal of a simple brightness discrimination, it had no effect on the reversal of a spatial problem.

### 2. *Weakness of Irrelevant Cues*

In straightforward studies of spatial learning in a T maze, the experimenter normally intends that there should be no obvious irrelevant cues. Unless more than usually careful, he may well not fully succeed; but unless he is more than usually careless and unlucky, he is not likely fortuitously to introduce an irrelevant cue as potent as that automatically present in experiments on simultaneous visual discrimination learning. From what we have already said about the effects of irrelevant cues on reversal, this factor provides a second reason for not expecting the ORE to occur in spatial studies. There was, furthermore, some early evidence which suggested that spatial studies in which an irrelevant visual cue was present were more likely to find an ORE than studies in which (by intention at least) no irrelevant cues were present. Both Pubols (1956) and Brookshire *et al.* (1961) had an irrelevant brightness cue present throughout original learning and reversal, and both studies are listed under the positive results. However, Pubols' groups were not well matched for original learning speeds; and the position discrimination used by Brookshire *et al.* was a difficult one, being based only on response, not on place cues, and their results may be more properly attributable to this factor rather than to the presence of the irrelevant brightness cue. More recently, Clayton (1963b) has performed two experiments directly comparing groups trained with and without irrelevant cues. In the first, overtraining had no effect on reversal learning whether or not an irrelevant cue was present; but as he pointed out, since the irrelevant visual cue was not sufficiently salient to affect initial learning, the test may not have been entirely fair. In the second experiment, with a more salient irrelevant cue, no ORE was obtained, and the irrelevant cue did not interact with overtraining to affect trials to criterion scores. Indeed, over early reversal trials, the presence of the irrelevant cue retarded learning more in overtrained than in nonovertrained subjects. Clayton's results, to say the least, provide little encouragement for this analysis of the difference between visual and spatial studies.

### 3. *Effect of Overtraining on Stimulus Control of Spatial Discriminations*

It is obvious that the preceding analyses cannot account for all the variance in the results of spatial studies, and although the argument appealing to the dominance of spatial cues for the rat may provide a valid account of one major difference between visual and spatial studies, the variability of the spatial studies poses a problem for explanation at least equally important. A possible explanation of some of this variance stems from yet a third analysis of the difference between visual and spatial experiments.

*a. Reversal Learning and Transfer to Proprioceptive Control.* Krechevsky and Honzik (1932), in one of the first studies of overtraining and reversal, found that a group trained to criterion made seven errors in learning the reversal of a spatial problem, while a group that had received 80 overtraining trials made 28 errors. An inverse ORE of this order of magnitude has not often been obtained since. Krechevsky and Honzik's subjects were initially trained to run down a straight alley, *past* a first turning. In an attempt to see whether it was this feature of their study that accounted for the inverse ORE, Mackintosh (1965c) trained rats in the maze illustrated in Figure 8.7. If subjects initially learned to go to the far goal box, overtraining significantly retarded reversal; while if subjects were initially trained to the near, left-hand

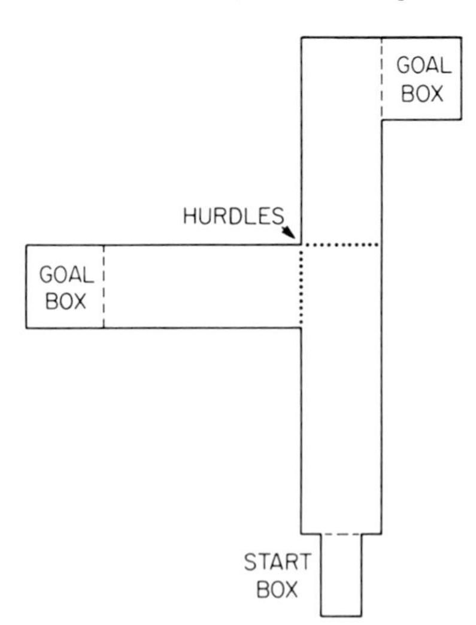


**Fig. 8.7.** Floor plan of maze. [*After* Mackintosh (1965c).]

goal box, overtraining had no effect on reversal. This suggested that rats that had been trained well beyond criterion to run straight past the first turning in the maze, had (in some sense or other) ceased to notice the turning. A second experiment, in which hurdles were placed at the choice point along the dotted lines, tended to support such a suggestion: the hurdles forced the rats to stop at the choice point, climb and squeeze through a narrow gap; their presence produced slightly faster reversal learning in all groups, but had its major effect on subjects overtrained with the far goal box positive. These subjects now reversed as rapidly as nonovertrained subjects.

Our interpretation of these results rests basically on a suggestion made by by the early behaviorists, and stated most explicitly by Carr (1917) who suggested

> . . . that the white rat learns the standard type of maze primarily in tactual and kinaesthetic terms, that during the course of learning the control is gradually transferred from contact to kinaesthesis, and that after the problem is thoroughly mastered the act is to be regarded as a kinaesthetic-motor coordination with an occasional reliance upon contact in times of emergency [p. 259].

A similar suggestion had been made earlier by Vincent (1915) and has more recently been revived by Restle (1957) and Watson (1959). The application of this suggestion to reversal learning is worked out as follows: animals initially guide their maze-running behavior in terms of external (visual, olfactory, etc.) stimuli. With increased practice it becomes possible for them to run off a "response-chain" either programmed centrally (as suggested by Lashley, 1929) or as a kinesthetic sequence; overtraining, therefore is likely to lead to a shift from external control of behavior. If one assumes that in order for reversal learning to occur, external control must be reestablished (a reasonable assumption, since the type of response chain we envisage would automatically produce the originally correct response and no other), it follows that the more complete this shift in stimulus control, the slower (other things being equal) reversal learning will be.

*b. Application to Different Results.* We are thus provided with a *general reason* for supposing that overtraining will not benefit reversal learning in a maze, and also an account sufficiently flexible to explain at least some of the diversity of spatial reversal results. For if speed of spatial reversal learning is determined by the extent to which the prereversal habit is under external control, since there is no reason to suppose this will always be the same in all varieties of experimental situation, there is a *prima facie* reason for expecting different results from different experiments.

In an ordinary T maze, where the subject must make a definite turn at a definite choice point, this automatization may well occur, but presumably less rapidly and less completely than in a straight alleyway. In Mackintosh's experiment, the group trained with the near, left-hand goal box positive was learning a problem very similar to that presented by the standard T maze. On the other hand, the group trained with the far goal box positive, had to run straight to the end of the alley before turning into their goal box, and this would seem to be remarkably similar to an alleyway situation. Therefore we should expect this latter group to reverse more slowly after overtraining, and the former to show little effect of overtraining, The placing of hurdles would presumably prevent the formation of smoothly run-off response chains, and thus abolish the effect of overtraining for the first group.

If we were to suppose that the longer the start arm of a maze, the greater the stability of running patterns at the choice point, it would follow that kinesthetic stimulation would also be more stable in long-arm mazes, and therefore (following Restle, 1957) that the maze habit would more readily come under kinesthetic control. Such a suggestion is supported by Adams and McCulloch's finding that reversal learning was significantly slower with a long start arm than with a short arm (Adams & McCulloch, 1934); and would also explain why Hill, Spear, and Clayton (1962) and Hill and Spear (1963a) have always found overtraining to retard reversal (in two out of four cases significantly), when using a maze with a 4-ft start arm, while few others have used such a long-armed maze, or found overtraining so consistently to retard reversal.

This account may also help to explain some of the successful demonstrations of an ORE in spatial studies. At least three of these successes may perhaps be explained by invoking differences in the probability or effects of response chaining:

(1) Capaldi (1963) is the only investigator of T-maze reversal to have used an *elevated* maze; it is intuitively obvious that rats are less likely to forego the benefits of reliance on visual cues when placed on a thin strip of wood several feet above the ground, than when safely inside an enclosed maze. Not surprisingly, such experimental evidence as there is, points to the vastly greater role of external cues in controlling the maze habit on elevated as opposed to enclosed mazes (Tsang, 1934). (Restle, 1957, in his review of place versus response learning, shows that response learning is more likely to be found in enclosed than in elevated mazes.)

(2) Ison and Birch (1961) may have succeeded in obtaining an ORE, because they precluded the possibility of any response chaining occurring; they trained their rats by placing them in the positive and negative goal boxes for either 50 or 200 trials before running them on the reversal. Since it is always unsatisfactory to make comparison between experiments, it is worth mentioning that within one and the same experiment it has been shown that

overtraining will benefit spatial reversal if subjects are placed in the goal boxes for training, but retard reversal if they are run in the normal fashion (D. A. Riley, personal communication). This seems very strong evidence in favor of our suggestion.

(3) Bruner, Mandler, O'Dowd, and Wallach (1958) trained rats in a four-unit maze to take an alternating path to food, and found that overtraining on a left–right–left–right pattern would normally benefit reversal to right–left–right–left. It is conceivable that in such a situation response chaining (if programmed as ⟨turn in the opposite direction from that taken previously⟩) might benefit reversal, since now the subjects could learn the reversal simply by altering their choice at the first choice-point, instead of having to reverse all four choices. There is evidence that this is, indeed, how rats learn reversals in multiple-unit mazes. Dabrowska (1963) has shown that a complete reversal at all choice-points of such a maze is learned more rapidly than a reversal at a single choice-point.

*c. Independent Evidence.* The suggestion that rats may transfer control of the maze-running habit from external to internal stimuli, serves as an explanatory principle of some value. It provides a clear differentiation between visual and spatial studies, and also serves to resolve a number of apparent inconsistencies in the evidence from spatial studies. It remains to be shown that there is good, independent evidence to support the basic assertion.

Carr and Watson (1908) trained rats to run down an alley and turn left or right at the end. They then shortened the alley, and found that subjects bumped into the end wall. This suggests that the subjects' behavior was not controlled by the maze stimuli but formed a kinesthetic chain. Carr and Watson did not, however, compare the incidence of such behavior after varying degrees of original training, so that their results do not provide particularly strong evidence in favor of our interpretation of the effects of overtraining. Macfarlane (1930) did compare the effects of different degrees of training in his experiment. He trained rats either in a water maze, through which they had to swim, or in the same maze with a false floor inserted, over which they could run. After different amounts of training, he transferred them to the alternative condition and found that the degree of disruption produced increased with the amount of prior training. All that is changed by transfer from running to swimming the maze is the response pattern; in terms of stimulus control, the animal must still approach the same external cues in both conditions. Thus increasing disruption of performance would indicate increasing importance of response chaining. Still more recently Kendler and Gasser (1948) trained rats in a T maze, and then tested for spatial orientation to the goal by blocking the arm of the T half way down and allowing the subjects to choose between a number of alternative paths radiating from this point. (This is the test of spatial orientation first devised by Tolman, Ritchie, and Kalish, 1946.) They found better orientation after 24 trials of training than after 100; this suggests that spatial orientation depends upon external control, and that this is better after a smaller number of trials.

These experiments have been discussed both in chronological order and in order of increasing importance as tests of our suggestion. None, however, is as direct and simple a test as that first undertaken by Aderman (1958) with negative results. It is simple to put external place and internal response cues into opposition and discover which is controlling the animal's behavior: if a rat is trained in a T maze, and, after learning, is tested with the start arm on the opposite side of the maze, then it can either approach the positive goal box or make the same turn at the choice point as it did during training: it cannot do both. This method of testing "place versus response" tendencies was first used by Ritchie, Aeschliman, and Pierce, (1950), and a similar method was used by Aderman in his experiment. He found that overtraining had no significant effect on the proportion of response choices. Since, however, response choices were predominant at the lowest level of training given, it is clear that this situation provided little opportunity for the *development* of response choices. As Restle's review clearly indicates, the probability of place learning varies enormously with variation in experimental conditions, and the prediction we are making is that where there is some *initial* preponderance of place choices this will give way to a subsequent preponderance of response choices. Furthermore, it is not at all certain that in Aderman's experiment, a response choice was really independent of external cues. The experiment was run in a homogeneous "dome" with a light disk at the end of the positive goal arm, as the only major directional cue. For the test trials, the only change was to move this disk to the other goal arm, and a place choice consisted of choosing the lighted arm. It seems rash to suppose however that there were *no* other external (e.g., intramaze) cues that could have controlled subjects' behavior, and for this reason, the results are hardly conclusive.

Two more recent experiments (Hicks, 1964; Mackintosh, 1965c) have shown unequivocally that the incidence of response choices increases with overtraining. Mackintosh, for example, trained rats for either 20 or 200 trials in an enclosed T maze, before giving a single test trial with the start arm on the opposite side of the maze. Whereas all the subjects in the 20-trial group made a place choice on the test trial, 75% of the 200-trial group made a response choice. In Hicks' experiment, 70% of subjects trained to criterion in a cross maze made a place choice on the test trial, while only 30% of subjects receiving 100 overtraining trials made a place choice. We can conclude that overtraining does often lead to a transfer from external to internal control, and the application of this finding to studies of reversal learning becomes correspondingly more reasonable.

## V. Alternative Explanations of the ORE

We have now presented our own analysis of the ORE and much of the evidence relevant to evaluating it, although we shall reserve for Chapter 9 a discussion of reversal and nonreversal shifts. In view of the large amount of research undertaken in this area, with different experimenters utilizing a wide variety of apparatus and procedures, it is not entirely surprising that the picture presented by the evidence is far from simple. While we certainly do not claim that we can account for all the data collected, it does seem that certain conclusions can be stated:

(1) The ORE is more likely to occur with visual than with spatial discriminations.

(2) The ORE is more likely to occur with difficult than with easy visual discriminations.

(3) Even with difficult visual problems, the ORE is more likely to occur when a large reward rather than a small reward is used.

(4) When the ORE does occur in visual studies (and sometimes when it does not), overtraining tends to increase the number of errors early in reversal, but leads to more rapid learning later in reversal.

(5) The crossover in the learning curves of overtrained and nonovertrained subjects occurs because the latter develop more marked position habits.

Since all but the third of these conclusions follow directly from a two-process model, and since it is possible to account for the third within such a framework, the evidence for our analysis of the ORE is considerable. Our confidence in this assertion is increased by the failure of any alternative hypothesis to account for all of the facts. Several such alternative explanations have been offered; some of these, which stress the development of appropriate strategies or the weakening of irrelevant strategies, are relatively similar to our own explanation and are discussed in the following chapter. In the remainder of this chapter, we consider several explanations which seek to account for the ORE in entirely different terms.

### A. Overtraining and Resistance to Extinction

At the outset of this chapter, we pointed out that the paradoxical nature of the ORE arises from its confrontation of two commonly held views: that overtraining strengthens whatever is learned, and that reversal involves extinguishing what has been learned. One obvious way to avoid the paradox is to assert that overtraining, far from increasing resistance to extinction, actually decreases it. The suggestion has the merit of simplicity and seemingly the further advantage of actually being true: when rats are trained in a runway situation, it has frequently been found that increasing the number of acquisition trials leads to more rapid extinction of the running response (a finding sometimes referred to as the "overtraining extinction effect"; for a review, see Sperling, 1965). The suggestion does, however, appear to replace one paradox with another, and immediately invites further speculation as to why overtraining should decrease resistance to extinction.

Several suggestions have been advanced to explain such a relationship. One possibility is to assume that overtraining increases the discriminability of the transition from acquisition to extinction conditions, and then to invoke the discrimination hypothesis of extinction (see Chapter 10). An alternative explanation in terms of frustration theory (again, see Chapter 10) has been advanced by Amsel (1962), Birch (1961), and Theios and Brelsford (1964): resistance to extinction is partly determined by the amount of frustration produced by an unrewarded trial; this, in turn, is determined by the magnitude or vigor of fractional anticipatory goal responses; and this, in turn, is determined by the number of rewards received in the goal box. We shall concentrate on this second explanation, since implicit in it is the possibility of accounting for some of the variability of ORE results in terms of variations in size of reward. As we saw above, there is excellent evidence that problem difficulty is not the sole determinant of the ORE, for even when rats are trained on relatively difficult visual discriminations, no ORE occurs when a small reward is used. If, as has been argued by Ison and Cook (1964) and by Theios and Blosser (1965a,b), the effect of overtraining on extinction is dependent on reward size, it is possible that a frustration analysis may contribute to a complete account of the ORE.

### 1. *Overtraining and Successive Discrimination Reversal*

The most direct evidence to support the basic assertion that the ORE can be accounted for in terms of resistance to extinction comes from an experiment by Birch, Ison, and Sperling (1960). They trained rats on a go–no-go brightness discrimination in a runway and found that overtraining significantly facilitated reversal. In this situation they were able to record running speeds separately to S+ and S− during reversal, and found that the effects of overtraining were confined to producing more rapid extinction of running to the original S+. At the least, it seems economical to ascribe the ORE obtained to the observed effects of overtraining on extinction of running to S+.

Until recently, there was some doubt about the reliability of this effect, since several studies failed to obtain an ORE in the go–no-go situation. North (1962) failed to find any difference in speed of reversal between subjects receiving 96 and 192 trials on a brightness discrimination in a runway, although many of the details of his procedure (apparatus, intertrial interval, reward size) were identical to those present in an earlier experiment where overtraining led to more rapid extinction (North & Stimmel, 1960). Uhl (1964) found that overtraining retarded the reversal of a go–no-go discrimination in a Skinner box. Sperling (1970), however, found a significant ORE in a successive discrimination in a runway and, recording latencies to S+ and S− separately, again observed that the principal effect of overtraining was to produce more rapid extinction of responses to the former S+.

### 2. *Applicability of an Extinction Analysis to the ORE*

Although it may be reasonable to account for the ORE observed by Birch *et al.* and by Sperling in terms of resistance to extinction, there are several assumptions involved in transferring this explanation from the go–no-go situation to the more usually studied simultaneous discrimination situation. The evidence suggests that few of these assumptions are justified.

*a. Overtraining and Choice Extinction.* The simplest application of the extinction analysis would be to assume that overtraining has a like effect on speed and choice measures of extinction. If overtraining did decrease resistance to extinction of choice behavior and running speed equally, then it would be natural to suggest that the ORE occurs in simultaneous discriminations because overtrained animals give up consistent selection of their former S+ sooner than nonovertrained animals. As we have already pointed out, however, all the available evidence from reversal learning studies in two-choice situations suggests just the opposite: if extinction is measured in terms of initial perseverative errors, or the number of days on which subjects select the original S+ in preference to any other stimulus, then overtraining facilitates reversal not because it reduces resistance to extinction, but despite the fact that it tends to increase resistance to extinction.

In order to provide a more extensive measure of the effects of overtraining on resistance to extinction of choice behavior, Mackintosh (1963b) trained rats on a brightness discrimination in a jumping stand, and after either 0 or 150 overtraining trials, extinguished each group to a criterion of equal choice of S+ and S− over 10 trials. Overtraining significantly increased the number of trials required to reach this extinction criterion (a finding which we have since replicated in a different apparatus, Mackintosh, 1965b). When subjects were trained on the reversal problem after reaching the extinction criterion, overtrained subjects learned significantly more rapidly, suggesting that it is the reacquisition phase of reversal that benefits from overtraining. In the first experiment, latencies were recorded during extinction, and on this measure there was evidence that overtraining did decrease resistance to extinction, just as it does in some runway studies. Whatever may be the explanation of this discrepancy between choice and latency measures, it is, to say the least, implausible to ascribe the ORE observed in simultaneous visual discriminations to the effects of overtraining on extinction of choice responses.

It is still possible, however, to support a weaker conclusion: even if overtraining does not facilitate reversal by facilitating choice extinction, it may only facilitate reversal provided that it does not too greatly increase resistance to extinction of choice behavior. Specifically, it would seem reasonable to suggest that overtraining might slightly increase choice extinction scores when a large reward is used, but have a much greater effect when a small reward is used.

There is very little evidence to support this suggestion. Several experiments have examined the effects of reward size on the ORE without obtaining any effect either on reversal or on extinction. Eimas (1967, Experiment II) trained rats on either an easy visual or on a spatial problem, giving either one or ten 45-mg pellets as reward. Overtraining retarded both reversal and extinction, irrespective of the size of reward. In a third experiment, Eimas found no significant effect of overtraining and no interaction with reward size on either score. Mackintosh (1969a, Experiment I) also found no significant effects of either overtraining or reward size (and again no interaction) on either extinction or reversal of a position discrimination.

In the two visual experiments where the ORE has directly been shown to be a function of reward size (Hooper, 1967; Mackintosh, 1969a, Experiment III), there was still no evidence that reward size affected choice extinction. In both experiments (Hooper's data have been analyzed by Mandler and Hooper, 1967), overtraining significantly increased perseverative errors, and the effect was entirely independent of reward size. It is only in two studies by Kendler and Kimm (1964; 1967) that there has been any evidence of the looked for interaction between the effects of overtraining and reward size on perseverative errors in reversal learning. These studies are discussed below.

*b. Specificity of Overtraining Extinction Effect.* If there is little evidence from choice extinction data to support a frustration analysis of the ORE, there is even less from latency data. Even if we leave aside the problem of translating an effect of overtraining on speed of responding into an effect on choice reversal, the data suggest that the former effect is far too general to account for the latter. Even in runway situations, the often repeated generalization that overtraining decreases resistance to extinction only with large rewards is distinctly suspect. Such a result has been reported by Ison and Cook (1964), but there are at least two studies (Likely & Schnitzer, 1968; Madison, 1964) in which the effect was obtained with small rewards; and a third small reward experiment (Hill & Spear, 1963b) in which, although overtrained rats did not reach a criterion of extinction faster than nonovertrained rats, there was a strong suggestion that the former showed a greater decline in running speed during extinction.

Studies reporting speed measures during reversal have also suggested that the effect of overtraining on speed of responding in extinction or reversal is more general than its effects on choice reversal. Birnbaum (1967) found a significant ORE in only one of the four groups she studied; but when four comparable groups were given extinction trials to their former S+, all four extinguished more rapidly after overtraining. Mackintosh (1969a, Experiment III), also found a significant ORE in only one of four groups studied, but when speed of responding to the former S+ was measured over early reversal trials, overtraining led to more rapid extinction in all groups.

The effect of overtraining on speed measures of extinction, although it presumably accounts for the ORE observed in successive discriminations, appears to be more general than the effect of overtraining on simultaneous discrimination reversal. It does not seem possible to explain one effect by an appeal to the other, and, unpalatable as the conclusion may seem, the present state of the evidence suggests that the two effects may be unrelated.

### 3. *Reward Size and Position Reversal*

The main argument advanced by Theios and Blosser (1965a,b) for a frustration analysis of the ORE, was that reward size was a critical determinant of the effect of overtraining on reversal learning. They claimed that an analysis of earlier studies supported their argument and performed two separate studies both of which suggested that T-maze reversal was retarded by overtraining when a small reward was used, while one of the studies obtained a significant ORE when a very large (930 gm) reward was given.

However, Clayton (1965) has argued that Theios and Blosser's results may have been confounded by changes in drive level. After a preliminary experiment in which he found that overtraining significantly retarded spatial reversal even when a large reward was used, Clayton noted that the total amount of food provided daily for animals in the large reward groups of Theios and Blosser's experiments should have been sufficient to cause them to gain weight rapidly. A further study definitely proved this, and suggested therefore that a marked decrease in drive level must have occurred during the course of overtraining in Theios and Blosser's experiment. There is independent evidence (e.g., Bruner *et al.*, 1958; Kendler & Lachman, 1958) that rate of reversal learning is inversely related to current level of hunger. Thus Theios and Blosser may have found that overtraining facilitated reversal not because they used a large reward, but because they allowed their large-reward subjects to gain weight during overtraining and hence enter reversal under a low drive. In a final experiment, Clayton was able to replicate Theios and Blosser's results, but only when sufficient food outside the maze was allowed so that subjects increased in weight.

Kendler and Kimm (1964, 1967) have also investigated the effects of reward size and overtraining on position reversal. Although neither of their studies yielded a significant ORE, they did observe some interaction between reward size and overtraining: overtraining tended to increase perseverative errors and, hence, retard reversal when subjects were trained with a small reward, but had no effect on reversal performance when subjects were trained with a large reward. Although these results appear consistent with a frustration analysis, they too are open to criticism. Kendler and Kimm's small reward consisted of two 45-mg pellets; their large reward consisted of 8-sec access to food. Since they gave no rewarded pretraining in the apparatus, and since rats commonly take some time to learn to eat in a novel environment (e.g., Zimbardo & Montgomery, 1957), it is likely that throughout original learning (completed in 20 to 30 trials) subjects nominally receiving a large reward consumed no more food than small-reward subjects. This conjecture is supported by the observation that the two groups did not differ in original learning scores; large rewards normally facilitate the learning of spatial problems (e.g., Eimas, 1967; Theios & Blosser, 1965b; Waller, 1968). With continued training the animals should have learned to eat more rapidly, and indeed the authors observed precisely such an effect. They do not, however, draw the conclusion suggested by this observation, namely, that overtraining and reward size may have been partially confounded, and that the only animals to recieve a large reward at the outset of reversal may have been those that were overtrained with a large reward. Since large reward directly affects speed of position reversal (e.g., Eimas, 1967), this confounding may be sufficient to explain Kendler and Kimm's results.

Three further experimental studies of overtraining and position reversal have directly compared the effects of different sizes of reward. Eimas (1967, Experiment II) found that 100 overtraining trials retarded position reversal whether the reward was one or ten 45-mg pellets. In a subsequent experiment, Eimas found that 200 overtraining trials tended (not significantly) to facilitate position reversal, but again the effect was independent of reward size. Finally, Mackintosh (1969a, Experiment I) found no significant effect of either overtraining, or reward size, or their interaction, on position reversal. Taken in conjunction with Clayton's results, these studies imply very strongly that size of reward is not the critical variable determining the effect of overtraining on position reversal.

### 4. *Conclusions*

There seems to us to be very little evidence to support an analysis of the ORE in terms of the effect of overtraining on extinction, and therefore little to be gained by distinguishing between different possible explanations. We have concentrated on the analysis proposed by frustration theorists partly because it has been the most widely accepted, but also because it has suggested one variable that might differentiate the successful from the unsuccessful studies. Our objections are twofold. First, the ORE does not seem to occur because of anything happening during early reversal trials (i.e., during extinction). Overtraining, although usually decreasing running speed during extinction, increases perseverative choices of the former positive stimulus. Each of these effects, moreover, occurs both when overtraining subsequently facilitates reversal, and when it does not. Secondly, reward size does not appear to have any effect on the occurrence of the ORE in spatial problems. In view of the conclusion reached earlier in this chapter, namely, that reward size is a critical determinant of the ORE in visual situations, this may seem surprising. We argued there that it is possible for an attentional theory to account for the effect of reward size on the visual ORE, and this argument is supported by the failure of reward size to have any effect either on easy visual problems, or on spatial problems. It is also supported by the findings reported both by Mackintosh (1969a) and Mandler and Hooper (1967), that the *cause* of the effect of reward size on the visual ORE is nothing to do with its effects on perseverative errors. In both experiments, an ORE occurred for large reward groups more than for small reward groups, not because overtraining increased perseverative errors to a greater extent when a small reward was used, but because it decreased position habits to a greater extent when a large reward was used. This result is exactly what would be expected if the effect of increases in reward size was to increase the extent to which overtraining strengthened the relevant analyzer more than its attached responses.

### B. Overtraining and Avoidance of S−

In the course of overtraining on a simultaneous discrimination problem, rats normally make very few errors. At the outset of reversal, therefore, overtrained animals have had little recent experience of nonreinforcement in the presence of S−, while nonovertrained animals have usually been responding to S+ and S− equally often until a relatively short time before reversal. A nonreinforced choice of (or experience with) a stimulus may be presumed to increase some inhibitory (aversive, frustrative) tendency which reduces the probability of subsequent selection of that stimulus. Taking these considerations into account, D'Amato and Jagoda (1960) suggested a simple explanation of the ORE: other things being equal, overtrained animals should find S− less aversive than nonovertrained animals, and should therefore be more likely to select it. Although this suggestion cannot reasonably be regarded as providing a complete account of the data discussed in this chapter, and although it runs counter to Ross' evidence of the persistence of inhibitory tendencies (see above), it does yield some testable predictions.

1. *Evidence*

*a. Forced Experience with* S−. In order to test their analysis, D'Amato and Jagoda (1961) gave one group of overtrained rats a number of forced trials to S− during the course of overtraining. They found that this procedure retarded reversal, and abolished the small ORE observed with the control animals. Several studies of position reversal have reported a similar effect (D'Amato & Jagoda, 1962; Hill, Spear, & Clayton, 1962; Komaki, 1962); in these experiments, where no ORE was observed for control subjects, overtraining with forced trials to S− produced a reverse ORE. While such results are consistent with the general hypothesis (if we ignore the problem of distinguishing between visual and spatial experiments), they are also consistent with the well-known fact that rats tend to avoid any alternative to which they have been forced; e.g., Denny (1957) found that the greater the number of forced *rewarded* trials rats were given to one arm of a T maze, the more likely they were to choose the other arm on a free trial.

*b. Successive Discriminations.* D'Amato and Jagoda predicted that no ORE should occur in a successive, go–no-go discrimination situation. Their argument was that during overtraining on a simultaneous discrimination, subjects select S+ on every trial (thereby removing themselves from S−), whereas in a successive discrimination they must continue to inhibit responses to S−. This argument reveals some ambiguity in their hypothesis, for in the successive situation subjects make few unrewarded *responses to* S−, just as in the simultaneous situation they make few unrewarded *choices of* S−. There is, moreover, some evidence to suggest that extensive overtraining on a successive discrimination does indeed reduce the aversiveness of S− (Terrace, 1966). Nevertheless, there is a sense in which experience with S− is reduced more by overtraining on a simultaneous problem than by overtraining on a successive problem, and the prediction is probably legitimate. If it is, the hypothesis receives little support: as we have mentioned, both Birch, Ison, and Sperling (1960,) and Sperling (1970) have obtained an ORE in a successive situation.

*c. Speed of Learning to Approach* S−. A successive discrimination situation permits independent measurement of speed of responding to S+ and S− during reversal learning. D'Amato and Jagoda's hypothesis predicts that the main effect of overtraining should be on acquisition of responding to the former S−. The evidence here is largely negative. Birch *et al.* (1960) found that overtraining caused more rapid extinction of responses to the former S+, but had no effect on acquisition of responses to the former S−. Sperling (1970) repeated these observations, both when initial training had been given in a successive situation, and when it had been given in a simultaneous situation. Birnbaum (1967) gave single-stimulus training following simultaneous acquisition, and, regardless of whether independent groups did or did not show an ORE, found that overtraining always facilitated extinction to the former S+, and always tended to retard acquisition to the former S−.

The only study yielding favorable evidence employed octopuses as subjects, and involved punishment for errors. Mackintosh and Mackintosh (1963) trained octopuses on a successive brightness discrimination, overtrained half, and then presented S− alone and rewarded all responses. Overtrained animals learned to approach S− significantly more rapidly.

*d. Preference Tests.* As we have already argued, there is no guarantee that latency measures correlate sufficiently closely with choice measures to tell us much about the causes of the ORE. It would therefore be better to test D'Amato and Jagoda's hypothesis by a choice measure, and in an ingenious experiment Deutsch and Biederman (1965) have shown how this may be done. They trained rats concurrently on two simultaneous visual discriminations, but gave a larger proportion of each day's trials on one of the problems than on the other. After a number of days of such training, test trials were given on which the negative stimuli from the two problems were presented together, and their relative aversiveness was determined by subjects' choices of one over the other. Deutsch and Biederman found that after 9 days of concurrent training, subjects showed a significant preference for the more frequently presented negative stimulus, a result which tends to support D'Amato and Jagoda's hypothesis (for a similar result with monkeys, see Behar, 1962). A second group of rats, however, received 12 days of concurrent training and showed no preference between the two negatives.

While these results are interesting, and while the experimental procedure potentially provides the most direct possible test of D'Amato and Jagoda's hypothesis, Deutsch and Biederman unfortunately neither correlated relative preferences between the two negative stimuli with speed of learning the reversal of each discrimination nor even gave sufficient training to ensure that one discrimination had been learned to criterion and the other had been overlearned. Indeed the fact that the preference for the more frequently experienced S− disappeared at the higher level of training (itself still insufficient for the less frequent problem to have been fully learned) suggests that the hypothesis is probably incorrect.

### 2. *Conclusions*

Although the basic contention of D'Amato and Jagoda's hypothesis seems reasonable in its own right, there is no very convincing evidence to support it. Furthermore, there is some ambiguity about the way in which a decrease in aversion towards S− is supposed to facilitate reversal. On the most natural interpretation, it should result in a more rapid extinction of the original discrimination habit, i.e., a faster appearance of a chance level of performance. If this interpretation is accepted, then the objections raised against the frustration–extinction explanation of the ORE apply as well to this hypothesis. Overtraining facilitates reversal in spite of, not because of, its effects on extinction. In the case of D'Amato and Jagoda's hypothesis, there is one experimental result which directly suggests that the causes of the ORE are to be found at a later stage of reversal. Mackintosh and Mackintosh (1963) in an experiment already referred to, interposed a series of trials between the end of acquisition and beginning of reversal, in which all subjects were trained (to the same high criterion) to approach their former S−. Although this procedure should have largely abolished any differences in aversion towards S− before reversal began, a substantial ORE occurred.

## VI. Summary

There is more than one requirement for a satisfactory account of the ORE. Not only must the theory explain the occurrence of the effect, it must also explain the conditions under which the effect does and does not occur. Perhaps even more important, it must explain why overtraining may not only decrease overall trials to reversal criterion, but also increase initial errors and decrease position habits and other determinants of chance-level performance. Judged by these standards, neither of the two hypotheses we have been considering can be regarded as at all satisfactory. D'Amato and Jagoda have done no more than suggest a reason for the occurrence of the effect; the frustration–extinction account relies on the insufficient notion of reward size to explain the variability of the effect. Neither offers a satisfactory account of the course of reversal learning.

A two-process theory of discrimination learning does satisfy these requirements. According to such a theory, speed of reversal partly depends upon the relative strengths of analyzers and responses at the outset of reversal, and an ORE is predicted by the assumption that overtraining may strengthen the relevant analyzer more than its response attachments. Of the two conditions that appear to be necessary for the ORE to occur, one (that a moderately difficult, nonspatial problem be used) follows naturally from such a theory, while the other (that a large reward be used) *can* be explained by the theory. The theory correctly predicts the course of reversal learning following training to criterion and overtraining. Finally, it is worth mentioning that these predictions can all be rigorously derived from formal models.

The present chapter, therefore, has presented the evidence that seems to compel acceptance of a two-stage analysis of reversal learning. In the following chapter, we shall consider whether the particular form of two-process theory that we propose is better able than the variety of others that has been proposed to account for the facts of reversal learning.

CHAPTER 9

# Reversal and Nonreversal Shifts, Serial Reversal Learning

## I. Alternative Analyses of the ORE

The analysis of reversal learning presented in Chapter 8 rested upon two main assumptions: first, discrimination learning involves two processes, one of which is appropriate to reversal, while the other is not; second, overtraining sometimes strengthens the first of these processes more than the second. This explanation was contrasted with others that postulated effects of overtraining on extinction to the former positive stimulus or on readiness to approach the former negative stimulus. In this chapter, we compare our analysis with others rather more similar to it.

The attentional analysis is a special case of a more general class of explanation; any theory that assumes that discrimination learning involves more than the acquisition of approach and avoidance responses to the positive and negative stimuli, and which assumes that the additional processes involved are appropriate to reversal learning, can handle the ORE in the manner outlined above. One of these additional processes may be selective attention; others are possible. They may range from something as general as habituation to the experimental situation to something as particular as the conditional reorientation response proposed by Siegel (1967). We have already (particularly in Chapters 4 and 5) discussed some of the possibilities. Two main suggestions demand most consideration. It is possible that in order to solve a discrimination problem, animals must learn to make appropriate observing or orienting responses. A second possibility is that more stress should be placed on the suppression of attention to irrelevant cues and less on the strengthening of attention to relevant cues.

In this chapter, we shall first discuss the application of these ideas to overtraining and reversal learning, and then discuss one experimental area (comparison of reversal and nonreversal shift learning) that is particularly suited to distinguishing general from more specific effects of overtraining. The final section of the chapter provides an interpretation of serial reversal learning which, although it also raises other issues, is partly concerned with similar problems.

## A. Observing-Response Theories

As mentioned before, the logical structure of observing-response theories and some attentional theories is identical: both assume that, in addition to the learning of appropriate instrumental responses, discrimination learning involves the isolation of the relevant stimuli from other, irrelevant stimuli. The two types of theory differ over what is regarded as necessary for such isolation to occur. Observing-response theories have traditionally emphasized the appropriate orientation of the subject's head and receptors; attention theories have traditionally asserted that such orientation, although necessary, is not sufficient. To a considerable extent, of course, such theoretical approaches are not mutually incompatible. It is self-evident that if a rat in a jumping stand keeps his eyes shut, or looks elsewhere when jumping at the stimuli, it cannot start learning the discrimination. We agree that correct orientation is necessary. The question is whether it is sufficient. We have already discussed this problem in places where it was appropriate. Since an observing-response explanation was the first to be offered of the ORE (Reid, 1953; Pubols, 1956), it will be appropriate to discuss it in this connection also.

### 1. *Specific Orientation Responses*

Two rather specific versions of observing-response theory have recently been advanced, one by Mandler (1966; 1968), the other by Siegel (1967). Mandler trained rats in a Y maze, with the discriminanda positioned some distance from the choice point. As mentioned in Chapter 8, she employed a retrace correction procedure with an error being counted only when the subject actually touched the negative card—entrance into the negative arm alone did not count as an error. Mandler observed that, at the time rats reached criterion, they were still entering the same side on almost all trials and retracing when the negative stimulus was on that side. Only during overtraining did subjects come to make their choice from the choice point, i.e., without even entering the negative arm. Mandler's suggestion is simply that this choice-point strategy is more efficient than the retrace strategy (simultaneous scanning of the positive and negative stimuli is more effective than successive scanning), and that overtraining can therefore be thought to establish an overt set of orienting responses appropriate to reversal learning. In support of this, she found that in subjects trained to criterion only, changes in the negative stimulus disrupted performance more than changes in the positive stimulus; while in overtrained subjects, the reverse relationship held (Mandler, 1968). This implies that nonovertrained animals were more likely to base their decision simply on retracing in the presence of the negative stimulus than overtrained animals.

In a study, part of which has already been described in Chapter 6, Siegel (1967) trained rats in a T maze with swing doors at the choice point. Subjects had to push open these doors in order to perceive the discriminanda in the goal arms. Because only one side could be viewed at a time, Siegel's rats could not, of course, ever adopt the kind of strategy observed by Mandler in overtraining. Siegel suggested instead that overtraining strengthened the retracing tendency, which he called a "conditional reorientation."

The nature of Siegel's apparatus clearly demanded of his rats that they learn to reorientate when they perceived the negative stimulus. Given that only one stimulus could be seen at a time and that it appeared unpredictably on either side, it is not surprising that the subjects' solution consisted of selecting one side to look at and, if it contained the negative, going to the other side. Evidence that his rats adopted this solution was provided by Siegel's analysis of the transfer from a simultaneous to a successive problem (see Chapter 6). In order to apply this idea to the analysis of the ORE, Siegel has to alter the emphasis of the argument. For if the occurrence of the reorientation response were conditional upon the sight of the negative stimulus, it would hinder rather than facilitate reversal learning. A reorientation response conditioned to the particular negative stimulus used in original learning is not a response of sufficient generality to be appropriate to reversal learning; indeed, it is simply a rather involved avoidance response to the negative stimulus and as such no more appropriate to reversal than an approach response to the positive stimulus. Only if the reorientation response were somehow partially independent of the actual stimulus to which it was originally conditioned, could strengthening it be expected to have facilitated reversal. On this interpretation, Siegel is offering a variant of an observing-response theory.

### B. Suppression of Irrelevant Cues

As stated in the preceding chapter, our analysis stresses that overtraining strengthens analyzers relevant to reversal. Within the context of our theory, however, and of any other in which analyzers sum to a constant value, strengthening of relevant analyzers is necessarily accompanied by weakening of irrelevant analyzers. We could, therefore, say that overtraining facilitates the reversal of a simultaneous brightness discrimination both because it strengthens the brightness analyzer and because it weakens the position analyzer. In this sense, the possibility that overtraining facilitates reversal by suppression of irrelevant cues is not inconsistent with our theory. Analyses of discrimination learning have, however, been proposed which lay exclusive emphasis on the weakening of control by irrelevant stimuli; such analyses can be and have been applied to the ORE, and demand consideration in this context.

### 1. *Spence's Explanation*

Two possible explanations of this sort can be distinguished. The first was suggested by Spence (quoted by Reid, 1953). It will be recalled from Chapter 4 that Spence analyzed the learning of a simultaneous visual discrimination into the building up of response strength to four component stimuli: the two visual stimuli and the two positions. With suitable parameter values, Spence was able to predict that subjects will develop position habits during original training, will eventually break their position habit when the negative stimulus is on the preferred side, and will then rapidly attain criterion. According to the model, the subject will remain in a position habit until the difference in response strengths between S+ and S− is greater than the difference in response strengths to the two positions. At the point of breaking the position habit, this latter difference may well be quite large; and Spence's explanation of the ORE rests upon supposing that it will remain significant even when the criterion run has been completed, but will be reduced nearly to zero during overtraining. If this were so, then nonovertrained subjects would start reversal with a greater difference in response strengths to the two positions than would overtrained subjects; therefore, they would be more likely to fall into a position habit during reversal; and, finally, they would take longer to learn, since reversal cannot occur until the difference in response strengths to the new S+ and S− is greater than the difference in response strengths to the two positions.

Wolford and Bower (1969) have recently simulated Spence's model by computer, and for some parameter settings were able to obtain an ORE. In an extension of this work, Turner and Mackintosh (1970) found that the basis for the ORE in these cases was indeed that nonovertrained subjects reverted to their original position habit during reversal, while overtrained subjects did not develop a position habit. They also found, however, that using Spence's operators it was not possible to get an ORE at all, unless starting values, which represented a marked *initial* position bias for response strengths, were chosen. In other words, if subjects did not enter upon *original* learning with a strong position habit, they would not show an ORE. Since Pubols (1956) has shown that a pretraining procedure that abolishes initial position biases does not prevent the ORE from occurring, there is considerable question whether Spence's model can be said to provide a satisfactory analysis of the ORE. We discuss below a further objection to Spence's account, which arises from the fact that it must necessarily predict that non-overtrained subjects will revert to their original position habit during reversal.

#### 2. *Suppression of Attention to Position*

Other analyses are not constrained in the same way as Spence's theory is; overtraining might directly reduce attention to position, or a tendency to respond in terms of position, without this being defined as the reduction of a bias to one position over another. Harlow (1959) has talked of the tendency of subjects to respond in terms of position when solving visual discrimination problems as an "error factor," and has suggested that the cause of the ORE might be the reduction of this error factor.

### C. Initial Evidence

Although we believe that some of these alternative accounts must enter into a complete explanation of the ORE, before discussing the evidence that compels this conclusion, it is worth presenting some reasons for refusing to believe that they can completely supplant our own explanation.

#### 1. *Observing Responses*

The role of observing or orienting responses in discrimination learning is never easy to disentangle from that of attention, although the more explicit the type of observing response postulated, the easier it is to see whether it is applicable to other situations. We have already, for example, given reasons for supposing that the reorienting response learned by Siegel's rats was forced upon them by the particular apparatus used, and probably does not occur to such a marked extent in other types of apparatus (see Chapter 6). We also believe that the change from a retrace to a choice-point strategy observed by Mandler in her apparatus is, to some extent, specific to that type of apparatus. It is, for example, unusual to find that changes in the negative stimulus disrupt rats' performance more than changes in the positive stimulus at any stage of learning a simultaneous discrimination (e.g., Gardner & Coate, 1965).

Short of actually measuring the orientation of a subject's receptors, just about the only way of distinguishing between the effects of observing responses and the effects of attention is to use an apparatus and relevant stimulus dimension that guarantee reception of the relevant stimuli from the outset of training. It is difficult to prove unequivocally that this has been achieved; but one type of apparatus that would be likely to minimize the role of observing response is that shown in Figure 9.1. In this discrimination box (as we have

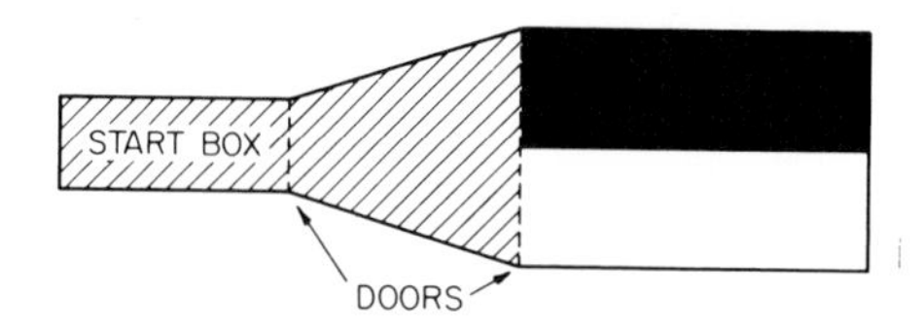


**Fig. 9.1.** Floor plan of discrimination apparatus (Grice box) designed to minimize importance of orienting responses.

used it), the subject is placed in the start box at the start of a trial, detained there for 5 sec behind a transparent door, and then, when the door is raised, the subject chooses which of the two goal boxes to enter. The floor and back and side walls of both goal boxes are painted (black and white). The relevant stimuli, therefore, cover a wide area, and appear side by side directly in front of the subject waiting in the start box; they are preeminently of a type such that according to Spence (1945), "It is practically impossible for the subject to respond and not receive discriminably different retinal stimulations [p. 256]." At the very least, this sort of apparatus does not seem to require the type of orienting responses observed by Mandler and Siegel.

Therefore if an ORE were obtained in this apparatus, this would be one case where the effect could not easily be ascribed to any type of observing response. There is, however, an excellent reason why, on our own account, no ORE would be expected: with black and white goal-boxes, the problem is too easy, the initial strength of the relevant analyzer is relatively high, and overtraining is unlikely to strengthen the analyzer any more than it strengthens the correct response. As all accounts predict, the apparatus when used with black and white goal-boxes, does indeed produce either no ORE or a very small one (Mackintosh, 1969a). We have, however, in the same study obtained a large and highly significant ORE by using dark gray and light gray instead of black and white goal-boxes. This change would have decreased the initial strength of attention to the relevant cue, but cannot reasonably be said to have decreased the initial probability of observing it. This result therefore suggests that the ORE can be obtained when no receptor orientation or response strategy is necessary, although it does not mean that these factors never contribute to the effect.

### 2. *Suppression of Irrelevant Cues*

The second possibility was that overtraining facilitated reversal by weakening irrelevant response tendencies. It is quite certain that it does not do so for the reasons derivable from Spence's theory. Quite apart from the impossibility of simulating an ORE from Spence's model as it stands without incorporating initial position biases, the theory can only predict that overtraining reduces position habits in reversal by reducing the discrepancy in response strengths to the two positions that still exist when animals have reached criterion in original learning. The explanation rests upon the assumption that nonovertrained animals enter reversal learning with their original position habit still largely intact, and therefore predicts that such animals will revert to their original position habit during reversal. If a given subject had a left position habit during initial training, then he should have a left position habit during reversal. This correlation between initial and reversal position habits should be reduced by overtraining.

To test these predictions, we have analyzed the position habits (defined as 20 consecutive trials of consistent choice of the same position) of 99 rats trained and reversed on a brightness discrimination in our jumping stand. In all the experiments analyzed, overtraining both facilitated reversal and decreased the number of position responses during reversal. The results of the analysis are shown in Table 9.1. They provide no support for Spence's

TABLE 9.I

**Position Habits in Acquisition and Reversal**

| Groups | Number of subjects | Percentage of subjects | |
|---|---|---|---|
| | | With same position habit in acquisition and reversal | With opposite position habit in acquisition and reversal |
| Overtrained | 48 | 58.3 | 41.7 |
| Nonovertrained | 51 | 56.9 | 43.1 |

account. Less than 60% of nonovertrained subjects had the same position habit during acquisition and reversal. The proportion is not significantly different from chance (by a one-tail binomial test, $p = .16$), and is marginally *less* than the proportion of overtrained subjects that returned to the same position habit. Others who have examined data obtained in other types of apparatus, have reported a similarly low correlation between original and reversal position habits (E. Lovejoy and S. E. Sperling, personal communication). It is not entirely surprising that the correlation should be so low, since, as we saw in Chapter 4, a similar prediction derivable from Spence's theory (that any errors during the criterion run of initial learning should be to the preferred position) was also disconfirmed. Position habits look stable, but once broken, any biases disappear rapidly, and no explanation that relies on their permanence is likely to be correct.

Even if overtraining does not benefit reversal by eliminating position biases, it may do so by directly weakening attention to position or position error factors; the above evidence is not critical for this theoretical possibility. An experiment by Mackintosh (1963a) does not, in fact, offer much support even for this idea. In this brightness reversal experiment, the magnitude of the ORE was found to be a direct function of the number of irrelevant cues; the addition of horizontal–vertical as an irrelevant cue retarded reversal for nonovertrained but not for overtrained animals. This of course is quite consistent with the idea that overtraining weakens attention to irrelevant cues; but the results of a third group for whom the irrelevant horizontal and vertical stimuli were not introduced until reversal began, are not consistent with the hypothesis. In this group also (where overtraining cannot have weakened attention to orientation), nonovertrained animals reversed slowly, but overtrained animals were not affected by the presence of the irrelevant cue. In this study, overtraining benefited reversal more because it strengthened attention to brightness than because it weakened attention to orientation. Therefore, it is reasonable to insist that the weakening of irrelevant analyzers cannot be the sole cause of the ORE. No stronger conclusion is, however, possible; overtraining may still facilitate reversal partly because it weakens position analyzers. The fact that it did not weaken attention to an already relatively weak cue (orientation) which failed to control subjects' behavior during original learning, does not mean that it fails to weaken attention to a cue like position which overridingly controls behavior until shortly before criterion is reached.

## II. Reversal and Nonreversal Shifts

The evidence just discussed seems to us to show that the ORE must be partly attributed to the effect of overtraining on strengthening attention to the relevant cue. It does not, however, rule out the possibility that other factors contribute to the effect; to demonstrate that an effect can occur without a given factor having contributed to it, is not to prove that the factor is always unimportant. Experiments that have compared the effects of overtraining on reversal and nonreversal shifts seem to show equally strongly that overtraining may not only strengthen the relevant analyzer, but must also have more general effects. Unfortunately the evidence is both confusing (for this is an area with a number of conflicting results), and too imprecise to determine which of the possibilities that we have considered so far (if indeed any of them) is most important.

The term, nonreversal shift, has been used to describe several different transfer tasks. We have already distinguished between two broad classes (Chapter 6): EDS, where the subject is required to learn a second task whose relevant cue differs from that used for the first task, and IDS, where the relevant cue for the second task is the same as that used for the first. In this chapter, we are concerned with only a single type of IDS, namely, reversal learning, but with a number of different EDS tasks. Our main interest lies in a comparison of the effects of overtraining on EDS and reversals, but before discussing the variety of experiments designed to make this comparison, we shall briefly discuss experiments that have directly compared the relative difficulty of the two types of task.

## A. Relative Difficulty of Reversal and Nonreversal Shifts

### 1. *Problems of Interpretation*

We can deal very briefly with these studies, since their outcome is not particularly critical for evaluating the virtues of any two-stage learning models, let alone for assessing the relative contributions of different processes. For our own type of theory, of course, the important difference between reversal and nonreversal shifts is that the same analyzer remains relevant for a reversal problem but a new analyzer becomes relevant for a nonreversal shift (as we saw in Chapter 6, this is also the interesting difference between EDS and IDS). This has led a number of writers (e.g., Kendler & Kendler, 1962; Schade & Bitterman, 1965) to assert that a model of this class inevitably predicts that a reversal should be easier to learn than a nonreversal shift, and it is presumably this belief that has motivated the relatively large number of studies making this particular comparison. By contrast the IDS–EDS comparison (theoretically simpler, and therefore more important) has only been made in two or three animal studies.

In fact, comparisons of the relative ease of reversal and nonreversal shifts may provide very little theoretical information. This is simply because no two-stage model can make unequivocal predictions about the outcome of any such comparison until several parameters have been specified; and no single-stage model can make predictions about the outcome of one of the more commonly made comparisons. As we argued at the outset of Chapter 8, the reversal learning situation is one of considerable theoretical complexity for a two-stage model such as ours. We say that the analyzer strengthened by initial training is appropriate for reversal, but the responses so strengthened are not. In general, therefore, a reversal will be learned rapidly if the analyzer extinguishes slowly, while the response attachments extinguish rapidly. Equally, the reversal will be learned slowly if the responses extinguish slowly but the analyzer extinguishes rapidly (once the analyzer has been extensively weakened, rate of response change will slow down yet further). Which of these two extremes will occur must depend on the values of the parameters used in the model. It seems clear that this will be true of any two-stage model, and that very large changes in predicted speed of reversal will be produced by quite small changes in the relative rates of extinction of analyzers and responses. Now the rate of nonreversal shift learning predicted by such a model will also depend, among other things, upon the rate of extinction of analyzers: other things being equal, the faster the originally relevant analyzer is extinguished, the more rapidly the subject will learn about the new relevant cue. Thus, while one set of parameter values (including rapid analyzer extinction) might predict more rapid nonreversal than reversal learning, another set (including slow analyzer extinction) might predict the exact opposite. Without specification of these parameter values, no two-stage model can make unequivocal predictions about the relative ease of the two problems, and therefore without some independent means of estimating these parameters, the experimental outcome is essentially uncritical.

### 2. *Experimental Designs*

*a. Stage-1 Cue Present and Irrelevant.* Prediction of the relative difficulty of reversal and EDS is not always easy even for a simple, single-stage model (see Eimas, 1965). Ever since the initial study demonstrating that college students tend to learn the reversal more rapidly (Buss, 1953), it has been assumed that one particular design could not provide a critical test of the adequacy of a single-stage model. Several ways of making the reversal–nonreversal comparison have been used; that originally employed by Buss is schematically illustrated as Design A in Figure 9.2. Subjects are initially

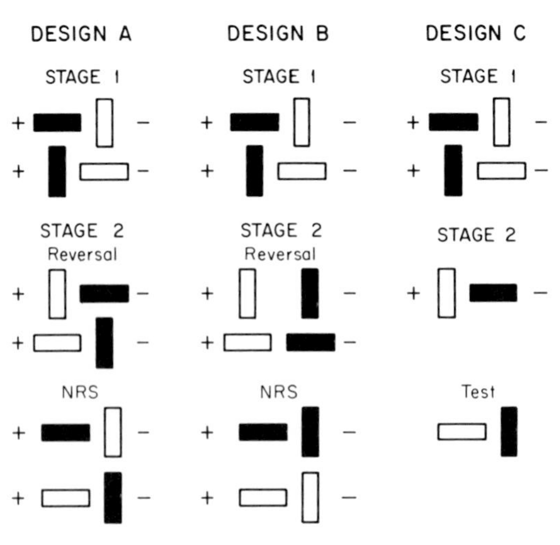


**Fig. 9.2.** Schematic illustration of three experimental designs used to compare reversal and nonreversal EDS learning.

trained on a black–white discrimination with orientation irrelevant. In Stage 2, they either learn a reversal or a nonreversal shift. For both groups, the same stimulus pairs that were presented in original learning are presented for the shift problem; for the reversal subjects, white is now positive, for nonreversal subjects horizontal is now positive. The problem created by this design is that nonreversal subjects can continue to make the response correct in Stage 1, and are rewarded on 50% of trials for doing so. This partial reinforcement (PR) (which does not, of course, occur for reversal subjects) may interfere with the course of learning the nonreversal shift. If it does so, then the fact that the reversal is learned more rapidly is not critical for any theory.

*b. Irrelevant Cue Held Constant.* There are two main ways, which have been used to evade this problem. The first is shown as Design B in Figure 9.2. Here the Stage-1 problem is the same as in Design A, and subjects are again shifted to a reversal (white positive) or nonreversal (horizontal positive) in Stage 2. The difference between the two is that with Design B, the irrelevant cue in Stage 2 (orientation for reversal subjects, brightness for nonreversal subjects) is only permitted to vary between trials, not within trials. Thus the nonreversal subject cannot continue to make the response correct in Stage 1, and PR for this response cannot occur.

It has usually been assumed that Design B is a more appropriate test of the relative difficulty of reversal and nonreversal shifts than Design A. Certainly the two designs do not appear to produce the same results. All animal studies using Design B have found that reversals are learned more slowly than nonreversals: Kelleher (1956) and Tighe *et al.* (1965) found this in rats, and Tighe (1964, 1965) in monkeys. Although some studies using Design A have also found a nonreversal to be learned more rapidly both by rats (Kelleher, 1956) and by cats (Coutant & Warren, 1966), several have found no difference in rate of learning the two kinds of problem. Monkeys (Coutant & Warren, 1966), rats (Brookshire *et al.*, 1961), and pigeons (Schade & Bitterman, 1965) have all been found sometimes to learn both problems at the same rate.

It is possible that Design A produces results biased in favor of reversal rather than nonreversal learning because it permits PR of the formerly correct response to occur during nonreversal learning. It is also possible, however, that Design B produces results biased in favor of nonreversal learning by eliminating another potential source of interference. In our discussion of IDS and EDS in Chapter 6, we pointed out that the difficulty of an EDS problem should be markedly affected by the presence of the Stage-1 relevant cue. When two values of this cue are present but irrelevant, learning should be harder than when they are not present (and this would hold, irrespective of the actual values of the cue employed, i.e., irrespective of the possibility of PR for the formerly correct response). If this cue is present, then in order to learn the shift problem a subject must extinguish the analyzer reinforced in Stage 1; if the cue is not present, the Stage-1 analyzer can no longer detect differences between the discriminanda and becomes nonfunctional. One difference between Designs A and B is that for the nonreversal subjects Design A continues with simultaneous presentation of both values of the Stage-1 relevant cue and Design B does not.

*c. Optional Shift.* Figure 9.2 shows a third type of design that has been used to study the relative ease of the two types of shift. In Design C (first introduced with college students by Kendler & D'Amato, 1955; and Buss, 1956), all subjects are trained on the same Stage-2 problem. The Stage-2 problem can be solved either as a reversal of the Stage-1 problem or as a nonreversal. In the former case, a subject would be expected to select the white horizontal rectangle in the test; in the latter case, he should select the black vertical rectangle. All animals tested on this optional shift problem (as it is usually called) have made predominantly nonreversal choices in the final test phase: Kendler, Kendler, and Silfan (1964), Tighe and Tighe (1966), and Sutherland and Mackintosh (1966) have all found that rats show negligibly little sign of solving the optional shift as a reversal.

*d. Experiments with Human Subjects.* This pattern of results from animal studies (especially the results from optional shift experiments) stands in marked contrast to that obtained with human subjects. While animals never learn reversals more rapidly than nonreversals, and usually find nonreversals significantly easier, human subjects, whether adult or children, usually find reversals at least as easy as nonreversals. About the only exception to this rule is provided by a study of Kendler, Kendler, and Wells (1960) with 4-year-old children using Design B; but as we have argued, Design B may bias the results in favor of the nonreversal shift. In no optional shift experiment with children has the nonreversal option been selected significantly more frequently than the reversal option (when absolute preferences for different stimulus dimensions have been controlled for). This is true even for a study (Kendler, Kendler, & Learnard, 1962) sometimes cited as showing such a difference. Kendler *et al.* (1962) did find a decrease in the proportion of subjects giving reversal responses on test trials as age decreased, but this decrease in reversal responses was not accompanied by an increase in nonreversal responses, only in an increase in random choices. However, the significance of this age trend is somewhat doubtful, since it disappears entirely with a seemingly minor change in procedure (Jeffrey, 1965).

It is (fortunately for us) beyond the scope of this book to enter into a detailed discussion of the causes of these differences between animals and children. Although we have argued above that the relative ease of reversal and nonreversal shifts is not a critical test of any two-stage model of discrimination learning, we should certainly agree with Tighe and Tighe (1965) that the differences obtained with different classes of subject are matters of considerable interest, and must (eventually) throw some light on the nature of the "mediating processes" possibly available to children but not to animals.

### B. Effects of Overtraining on Nonreversal Shifts

To argue, as we have done, that a comparison of the relative difficulty of reversal and nonreversal shifts in animals does not provide a critical test of the adequacy of any two-stage model, is not to deny the importance of the reversal–nonreversal comparison in another context. For, as we have also pointed out, the critical difference between the two tasks from the point of view of our theory is that one demands continued use of an old analyzer, while the other demands a change to a new analyzer. Consequently, we might expect that any variable that affected the difficulty of reversal learning by changing the strength of the relevant analyzer, should have an opposite effect on the difficulty of nonreversal shift learning.

One such variable should be overtraining (others are discussed elsewhere —serial reversal training later in this chapter, inconsistent reinforcement in Chapter 11). If the main effect of overtraining is to strengthen the relevant analyzer, then, while overtraining should facilitate reversal learning, it should not facilitate (and, under some circumstances, should retard) nonreversal shift learning. To say that this is so is not to say that, after overtraining, a reversal should be learned faster than a nonreversal (as is implied, e.g., by Schade and Bitterman, 1965). The most that follows is that overtraining should have opposite effects on the two problems: the nonreversal, although made harder, may still be easier than the reversal, although made easier.

This argument is only valid if the Stage-1 cue is still present (but irrelevant) at Stage 2. If the original cue is omitted at this stage, then overtraining at Stage 1 might actually result in faster learning of an EDS, since it will depress analyzers for background cues such as position, and hence, the new relevant cue will have greater relative strength. This argument is spelled out in more detail below (page 312). The other possible theoretical accounts considered above, in general, predict that overtraining will either facilitate or have no effect on nonreversal shift learning. As in the case of our own model, the exact prediction made depends largely upon the experimental design employed, so that, as before in this chapter, we must distinguish between the possible designs, specifically between nonreversal shifts in which the Stage-1 relevant cue is present and irrelevant (Design A), nonreversal shifts in which the Stage-1 relevant cue either takes a single value on each trial or is altogether absent (Design B), and optional shifts (Design C). We discuss these in reverse order.

1. *Optional Shifts* (*Design C*)

Two experiments, both with rats, have compared optional shift performance following training to criterion or overtraining (Tighe & Tighe, 1966; Sutherland & Mackintosh, 1966). Both obtained essentially similar results: overtraining neither increased nor decreased either reversal or nonreversal scores. In both experiments, the proportion of reversal choices was very small after all levels of training. Tighe and Tighe used a situation (specifically, a very small reward) that does not produce an ORE. There is perhaps, therefore, nothing particularly surprising about their results. In our own experiment, however, we used an apparatus (jumping stand) and a reward (20-sec access to food) that have produced an ORE in past experiments. The failure of overtraining to have any effect on reversal scores becomes, therefore, somewhat surprising. We can suggest two possibilities. In our experiment, both overtrained and nonovertrained subjects solved the optional shift problem (in approximately the same number of trials), almost entirely as a nonreversal shift. The failure of overtraining to increase reversal scores may conceivably have been simply because (even after overtraining) the nonreversal was so much easier than the reversal. In other words, because lower animals tend to find a reversal harder than a nonreversal, the optional shift procedures may be too insensitive to detect the effects of variables (such as overtraining) presumed to have some effect on the relative difficulty of the two types of problem.

It is also possible that choices on test trials after optional shift learning do not always reveal relative amounts of reversal and nonreversal learning. This point has been dramatically made in a study by Jeffrey (1965) to which we have already alluded. Part of the design of Jeffrey's experiments is shown schematically in Figure 9.3. In Panel A is shown the standard optional shift design: a subject first learns a black–white discrimination with size irrelevant, in Stage 2 the subject is presented with a small white square positive and large black square negative, and in Stage 3, he is tested for the basis of Stage-2 solution. Choice of the small black square indicates a nonreversal solution;

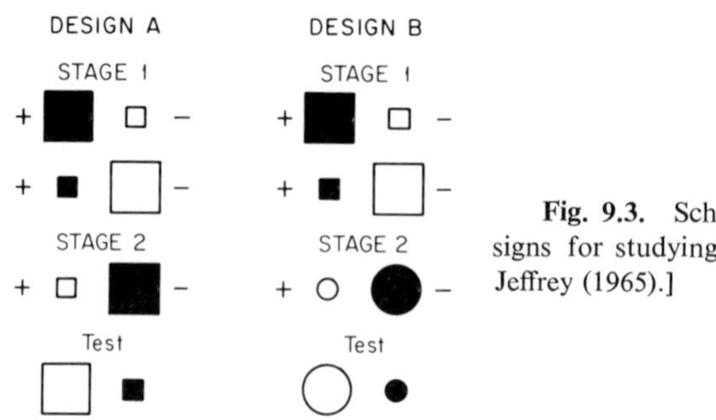


**Fig. 9.3.** Schematic illustration of two designs for studying optional shift learning. [*After* Jeffrey (1965).]

choice of the large white square indicates a reversal solution. Using this design, Jeffrey obtained, as did Kendler *et al.* (1962) a significant increase in reversal choices with age. When tested on the design shown in Panel B, however, 4-year-old children made 76% reversal choices on test trials (compared with 75% made by college students). The only difference between the two designs, of course, is that in the second the test stimuli are circles not squares. Jeffrey points out that with the standard optional shift design, the test pair of stimuli in fact consists of one of the pairs of stimuli used in original training. If the subject chooses on test trials that stimulus compound he was rewarded for choosing in original learning (the small black square), he is counted as having made a nonreversal choice. In order to make a reversal choice, he must select the nonrewarded compound from Stage 1, even though that reward contingency has not been explicitly reversed. When this conflict is not imposed on young children, as in the design shown in Panel B, they show that they have learned Stage-2 as a reversal at least as well as have older subjects.

The problem is the same as one we have come across before (in Chapter 5), namely, that under some not easily specified circumstances, subjects may respond to compounds rather than to the component stimuli making up the compounds. To the extent to which this happens in animals as well as younger children, the standard optional shift design may be an unsuitable one to use with such subjects. At the least it may be as well to treat the available evidence with some caution.

2. *Irrelevant Cue Held Constant* (*Design B*)

*a. Experimental Results.* This experimental design has yielded all possible results. Mandler (1966, 1968) and Komaki (1961) have found that overtraining on a black–white discrimination in rats facilitates the learning of a horizontal–vertical striations discrimination. Sutherland and Andelman (1969) obtained a similar result using a jumping stand. Tighe (1965) with monkeys, and Tighe *et al.* (1965) with rats, have found that overtraining has no significant effect on nonreversal shift learning (nor, in their situation, on reversal learning). Turner (1968) trained rats on a black–white discrimination in a jumping stand, and looked at the effect of overtraining on the shift to a horizontal–vertical discrimination; he found no significant effect. Finally, Mackintosh (1962) gave rats three levels of training on a black–white discrimination and then trained them to discriminate horizontal and vertical striations. In this experiment, overtraining significantly retarded nonreversal shift learning.

*b. Theoretical Predictions.* Before attempting to resolve these apparent discrepancies, we should consider what sorts of predictions follow from different theoretical positions. In all these experiments, at Stage 2, the Stage-1 relevant cue either is absent, takes on an intermediate value, or takes on one training value on some trials and the other training value on other trials. The analyzer strengthened in Stage 1, therefore, can no longer detect differences between the two stimuli shown within a trial in Stage 2. We must, of course, assume that analyzers can be utilized even though no within-trial differences are present, for otherwise, animals would not be able to solve successive conditional discriminations. We must also assume that a brightness analyzer strengthened in simultaneous training will remain relatively strong when brightness differences occur between, not within, trials, for otherwise we could not account for Lawrence's results on the acquired distinctiveness of cues. Nevertheless, it is reasonable to suggest that the absence of both within- and between-trial differences in a previously relevant cue will cause a substantial drop in the strength of that analyzer, and may be sufficient to obscure differences brought about by overtraining. If this interpretation is accepted (it should be noted that Mackintosh, 1962, did not advance it), then it becomes extremely difficult to make predictions about the effect of overtraining on nonreversal shift learning. If the Stage-1 analyzer is inoperative at Stage 2, our model predicts that overtraining will facilitate EDS learning. The argument is the same as that advanced in Chapter 7 (page 240) to explain why training animals to discriminate stimuli differing on one dimension sharpens generalization gradients along a second dimension. To recapitulate the argument, one effect of Stage-1 training will be to depress the analyzers for irrelevant cues; hence, at the start of Stage-2 training, the base value of the new relevant analyzer will be higher than for nonovertrained animals since (under Rule 2 of our model) analyzer strengths sum to a constant amount. This argument of course holds good only if the Stage-2 cue is not included as an irrelevant cue during Stage 1. If it is included at that stage, its analyzer strength will be reduced by the same proportion as the analyzer strengths for irrelevant cues and hence there will be no effect of overtraining on the EDS learning. In the experiments with which we are dealing, the extent to which the analyzer for the Stage-1 cue is operative at Stage 2 cannot be determined. Hence, on our model the results of such experiments are determined by two factors pulling in opposite directions. On the one hand, overtraining at Stage 1 should facilitate EDS learning by weakening analyzers for background cues; on the other hand, it should impair EDS learning by strengthening an irrelevant analyzer which will be operative to an undetermined extent during Stage 2. The fact that contradictory experimental results have been obtained is therefore scarcely surprising on our model.

*c. Theoretical Implications of the Evidence.* Mandler (1966, 1968), Komaki (1961), and Sutherland and Andelman (1969) all found that overtraining facilitated the subsequent learning of an EDS problem. Turner, and

Tighe *et al.* found overtraining had no effect, and Mackintosh found it facilitated subsequent EDS learning. These contradictory results may be partly, but are not entirely due to apparatus differences, since the three studies by Mackintosh, Turner, and Sutherland and Andelman were all conducted in a jumping stand. Although our own theory can predict all three possible results depending on how far the analyzer relevant to the Stage-1 problem is actually switched-in at Stage 2, it is impossible to identify the controlling variable from the above results. The main difference between these three jumping-stand studies is that Sutherland and Andelman used white horizontal and vertical rectangles for the nonreversal shift, Turner used gray horizontal and vertical rectangles, and Mackintosh a series of gray blocks forming either horizontal or vertical striations on a hardboard background. These stimuli are increasingly difficult to discriminate both from each other and from the background. Why this difference in discriminability should have an effect, we cannot say, but it appears to be the main variable differentiating the three experiments.

Disregarding these differences within the jumping-stand studies, the overall trend of their results contrasts with that of the maze studies of Mandler and Komaki; in general, overtraining is less likely to facilitate shift learning in the jumping stand than in a maze. If we assume that this contrast is largely due to differences in apparatus, then it seems to us to be most plausibly ascribed to the fact that the jumping stand requires less in the way of elaborate observing or orienting response learning. This is not to insist that a rat in a jumping stand automatically fixates the relevant stimuli from the outset of training; but it is to say that appropriate orientation is something that is learned relatively rapidly, relatively early in training, and does not need or get substantial strengthening during overtraining. If we assume that this does not hold for most mazes, perhaps accepting Mandler's interpretation of her observed changes in response strategies during overtraining, then overtraining on a visual problem in a maze may establish habits appropriate for any new visual problem.

### 3. *Stage-1 Cue Present and Irrelevant* (*Design A*)

*a. Experimental Results.* The studies reviewed in the preceding section yielded every possible result: overtraining either facilitated, had no effect on, or retarded nonreversal shift learning when the Stage-1 cue was absent or held constant. When the nonreversal is learned with the Stage-1 cue still present but now irrelevant, this variability decreases by one step. In some experiments, overtraining has retarded such nonreversal shift learning (Goodwin & Lawrence, 1955; Mackintosh, 1963b; Mackintosh & Holgate, 1965; Schade & Bitterman, 1965); in others, overtraining has had no effect on the shift (Brookshire *et al.*, 1961; Schade & Bitterman, 1965; Mandler, 1966; Uhl, Parker, & Wooton, 1967; Siegel, 1967); in no experiment, has overtraining facilitated shift learning.

*b. Theoretical Predictions.* The theoretical issues posed by these studies depend on how the effect of overtraining on the nonreversal compares with its effect on reversal learning. According to our own theory, if overtraining in a given situation has no effect on reversal learning (presumably because it does not strengthen the relevant analyzer), then it would also be expected to have little effect on nonreversal shift learning.[1] A similar conclusion would hold for an observing-response model. If, however, overtraining facilitates reversal, then an attentional theory and an observing-response theory may make diametrically opposed predictions. On our own theory, if overtraining benefits reversal by strengthening the relevant analyzer, it should retard extinction of that analyzer, and thereby interfere with nonreversal shift learning. When we include consideration of the effects of overtraining on irrelevant analyzers, the argument becomes more complicated, but the conclusion probably stands. On the one hand, if both the original problem and the nonreversal shift are simultaneous discriminations, then position is an irrelevant cue in both and any weakening of position analyzers should benefit nonreversal learning. On the other hand, in all the experiments considered here, the relevant cue of the nonreversal shift problem was present and irrelevant in Stage 1; if Stage 1 overtraining weakened attention to that cue, it should have retarded nonreversal learning. We have argued above that overtraining at Stage 1 should weaken both irrelevant cues (position and the Stage-2 cue) by the same proportionate amount, leaving their relative strengths the same. If this is correct, then the only important effect of overtraining will be to strengthen the analyzer for the cue relevant at Stage 1, and this should retard EDS learning.

If, however, overtraining benefits reversal by strengthening general observing responses or by establishing some appropriate strategy of orientation—either of which should be relevant to the learning of any new visual discrimination in the same apparatus—then it should equally benefit nonreversal shift learning. The predictions derivable from the idea that overtraining facilitates reversal entirely by weakening attention to irrelevant cues are equivocal.

[1] We ignore the problem posed by the possible effects of overtraining on the strength of the correct response. Overtraining must be supposed either to strengthen or to have no effect on this response. Whether strengthening of the Stage 1 response will be expected to add to the interfering effects of overtraining on Stage 2 learning depends on the exact nature of the model. On our model, the effect would probably be negligible. Response strengths become a matter of much greater importance when we come to discuss serial reversal training and nonreversal shift learning.

*c. Theoretical Implications of the Evidence.* Since, in all experiments of this design, overtraining failed to facilitate nonreversal shift learning, we must reject explanations of the ORE based solely either on the weakening of attention to position or on the strengthening of a generally appropriate orienting response. The only possible observing-response analysis would be one that assumed different observing responses for different visual stimuli. Since in some cases the stimuli used have not subtended very small visual angles (Goodwin & Lawrence, 1955; Siegel, 1967), the explanation is strained. In some of the experiments, it is true, a reversal group trained in the same situation failed to show an ORE. Uhl *et al.* (1967) trained rats on a go–no-go discrimination with light and noise as the two cues. Overtraining had no effect on either reversal or nonreversal shift. There are several possible explanations for this: a small reward was used; no irrelevant position cue was present; the relevant stimuli were such as to require no observing response (the explanation favored by the authors); the relevant stimuli were sufficiently obvious to be attended to at the outset of training. With this array of explanations, the results are not critical for any of the positions under discussion. Schade and Bitterman (1965) found that overtraining pigeons on a red–green discrimination retarded reversal but had no effect on a shift to a shape discrimination. We shall discuss these results in conjunction with the other half of their experiment in more detail below. Finally, Mandler (1966) trained or overtrained rats on a position discrimination in a Y maze, then shifted them to a visual problem, and found no effect of overtraining on speed of learning the shift. Although no reversal groups were run, it is highly unlikely that position overtraining would have facilitated reversal.

In two further experiments, however, overtraining simultaneously facilitated reversal and had no effect on shift learning. Brookshire *et al.* (1961) trained rats in a cross maze on a place (black versus white goal-arms) or response discrimination. Overtraining facilitated reversal (with only marginal significance in the response problem), but had no effect on either shift. Siegel (1967) trained rats on brightness or texture problems, obtained a large ORE, and no effect of overtraining on the shifts.

In attempting to explain these results in terms of his conditional reorientation response, Siegel argued that although this response is appropriate both to the reversal and to the shift, and although it is strengthened by overtraining, such strengthening may only be detected in the reversal problem, because only in reversal does reinforcement drop from 100 to 0%. When shifted to the nonreversal problem, animals initially received 50% reinforcement and this may have been sufficient to maintain the conditional reorientation response in nonovertrained animals, weakly established though it was. The argument implies something of an all-or-none view of the probability of occurrence of such conditional reorientation; for on any other view, a more weakly established response should have been extinguished to a lower level than a more strongly established response, and some difference in speed of learning should have been detected. In fact, such difference as there was favored the nonovertrained subjects.

Furthermore, four studies have actually found that overtraining may retard nonreversal shift learning. We have trained rats (Mackintosh, 1963b) and octopuses (Mackintosh & Holgate, 1965) on a brightness discrimination, and found that overtraining retarded the learning of an orientation discrimination with brightness irrelevant. Although reversal groups were not run simultaneously, both rats (Mackintosh, 1963a) and octopuses (Mackintosh & Mackintosh, 1963) have shown an ORE in exactly the same situation. In an experiment directed largely at a different set of questions, Goodwin and Lawrence (1955) trained rats on a series of black versus white and high versus low hurdle discriminations. For example, one representative group was first trained on brightness (black positive, white negative) with hurdles irrelevant, and then on the hurdles discrimination with brightness irrelevant. After reaching criterion, different subgroups received varying amounts of overtraining on the hurdles problem. In Stage 3, all subjects learned the shift back to brightness, half with black, half with white positive. Regardless of whether subjects were trained on their old brightness problem or its reversal, 100 overtraining trials on the hurdles problem significantly retarded this shift learning.

Finally, in Schade and Bitterman's pigeon experiment, in addition to the group trained on color and either reversed or shifted to shape, a second group was trained on shape and either reversed or shifted to color. In the first group, as already mentioned, overtraining on color retarded reversal, but had no effect on the nonreversal shift; in the second group, overtraining on the shape problem had no effect on reversal learning, but retarded the shift to color. Much of this complex interaction of overtraining, reversal or nonreversal, and color or shape, follows nicely from an attentional analysis. As Schade and Bitterman (1965) say,

> On the assumption that the animal comes to the situation with a relatively strong set for color, overtraining on color contributes little to the set and therefore does not retard the shift to shape but it retards reversal because it strengthens response-attachments. The fact that overtraining on shape retards shift to color follows from the assumption that the initial set for shape is relatively weak. By the same token, overtraining on shape should facilitate reversal, but there was no ORE in the main experiment [p. 284].

We attempt, in Chapter 12, to account for some of the differences between rats, birds, and fish that have emerged from studies of (among other things) overtraining and reversal. Despite this problem, Schade and Bitterman's somewhat intricate results seem to offer more support for our own point of view than for any alternative, and add weight to the general conclusion of this section. There is sufficient evidence demonstrating an interaction of the effect of overtraining with the type of shift imposed on the animal to suggest that at least part of the effect of overtraining must be to strengthen specific analyzers.

## C. Conclusions

We should not, however, conclude this discussion of reversal and nonreversal shifts by exaggerating the extent to which an attentional theory is supported by the evidence. In view of the variability of the results, indeed, it would be surprising if any single theory were sufficient to account for them. Our own theory makes two relatively unequivocal predictions: overtraining should facilitate the learning of a nonreversal shift when the Stage-1 relevant cue is absent at Stage 2 and the Stage-2 relevant cue is absent at Stage 1; it should retard the learning of a shift when the old relevant cue is present and irrelevant. There are enough results contradicting this position to render our theory unsatisfactory as a complete account of the effects of overtraining. It is likely that overtraining has other, more generally beneficial effects. The evidence is not sufficient to decide unequivocally what these effects are. It is, however, suggestive. It is possible to attribute some of the differences in experimental outcome to differences in apparatus and to relate these to the different requirements that various types of apparatus impose on animals to learn specific orienting responses in order to ensure efficient discrimination. Several of the results contradicting our position (e.g., Mandler's and Siegel's) were obtained in mazes with visual discriminanda that may not always have been easily seen by the subjects, whereas many of the results supporting our position were obtained in a jumping stand, or Grice box (Goodwin & Lawrence), or key-pecking apparatus for pigeons (Schade & Bitterman)—all situations that may reasonably be regarded as decreasing the importance of overt observing responses. Unfortunately, no direct attempts have been made to manipulate the probability of appropriate orienting behavior. Nevertheless the general trend of all results, and the pattern of the differences that has appeared, seem consistent with the following propositions: while the effects of overtraining on reversal learning cannot be understood solely in terms of its effects on attention, they equally cannot be understood by appeal only to overt orienting behavior. Both factors enter into most situations, with the relative contributions of the two depending on the exact nature of the apparatus.

# III. Serial Reversal Learning

## A. Preliminary Analysis of Serial Reversal Improvement

In the remainder of this chapter, we shall present an analysis of serial reversal learning. The evidence on which the analysis is based, is, unfortunately, relatively sparse; there have been few experiments designed to investigate the effects of serial reversal training on attention. In part, of course, we hope that the following discussion may stimulate more theoretically directed experiments; but the major justification for the discussion will not become apparent until Chapter 12, where we attempt to show that serial reversal learning is one of several experimental situations in which different vertebrates behave differently, and that some of these behavioral differences can be explained by an attentional theory.

Training an animal on a brightness discrimination with black positive and white negative establishes one habit ⟨attend to brightness⟩ appropriate to learning the reversal and another ⟨approach black and avoid white⟩ inappropriate to reversal. Other things being equal, strengthening the first will facilitate reversal, strengthening the second will interfere with reversal. Our analysis of the ORE assumed that overtraining would, under certain circumstances, strengthen attention more than responses; but it is not surprising that the ORE is an elusive phenomenon, since a variable such as overtraining is likely to strengthen both attention and responses, and its effect on reversal in any given situation will critically depend on the balance struck between these two antagonistic effects. A variable that strengthened analyzers without strengthening response attachments would be expected to facilitate reversal much more reliably.

One variable that certainly has a substantial and unequivocal effect on speed of reversal is serial reversal training. Since the early experiments of Buytendijk (1930) on serial position reversal, and Krechevsky (1932b) on serial brightness reversal, there have been at least two dozen demonstrations of the basic phenomenon of serial reversal improvement. When rats are trained on a simultaneous discrimination problem, and the positive and negative stimuli are repeatedly reversed (either after each problem has been learned to criterion or after a fixed number of trials has been given), early reversals are usually learned relatively slowly, but later reversals are learned more and more rapidly. With both spatial problems (e.g., Dufort, Guttman, & Kimble, 1954), and easy visual problems (Mackintosh, McGonigle, Holgate, & Vanderver, 1968), animals may eventually make only a single error in learning a reversal. Almost invariably, later reversals are learned at least as fast as the original discrimination, and in most cases, they are learned substantially faster (in the following, unsystematically chosen studies, fewer than 10 reversals were required before each reversal was learned more rapidly than the original problem: Gatling, 1952; Dufort *et al.*, 1954; Lawrence & Mason, 1955; Pubols, 1962; Stretch, McGonigle, & Rodger, 1963; Gonzalez, Roberts, & Bitterman, 1964; Mackintosh *et al.*, 1968; Mackintosh & Holgate, 1969).

There are, of course, some exceptions to this general trend of the evidence: e.g., fish and the majority of invertebrates so far tested show little sign of serial reversal improvement (see Chapter 12); but in rats, it must rank as one of the more robust of learning phenomena. One early study failed to find any improvement (Fritz, 1930); but Fritz used only four rats, and trained them on only 2, 4, 6, or 13 reversals, so that the failure is not entirely surprising. In two studies in which animals were trained at a rate of one trial per day, no improvement occurred (Estes & Lauer, 1955; Clayton, 1962). It is probable that such extreme spacing of trials caused interference from a factor that normally operates to facilitate reversal. Finally, a more recent study by Weyant (1966) found improvement over 10 spatial reversals, but none over 10 brightness reversals. It is conceivable that the rats in this study, allowed $2\frac{1}{2}$ hr free access to food each day, became progressively less motivated during the course of the experiment. With these exceptions, however, all published studies of serial reversal learning in rats have found that later reversals are learned with significantly fewer errors than early reversals.

We shall argue that serial reversal training has at least two beneficial effects on later reversals: it both strengthens the correct analyzer, and reduces interference from the previously correct response. It is for this reason that improved performance over a series of reversals is found so consistently. The remainder of this chapter is devoted to a discussion of the evidence bearing on this argument. We first discuss evidence which suggests that changes in the strength of interfering responses cause changes in speed of reversal, but although it is probable that this is one factor responsible for serial reversal improvement, it is quite certain it is not the only factor. Whether further improvement is caused by changes in the strength of the relevant analyzer, or whether more general factors, e.g., habituation to the experimental situation or the suppression of irrelevant cues, are responsible for improvement, will occupy our attention in the final part of the chapter.

### B. Interference from the Previously Correct Response

It will simplify the discussion in this section if we assume that we are dealing with simple, spatial discriminations with no irrelevant cues, where the relevant cue is so dominant that the relevant analyzer remains in complete control of behavior throughout the experiment. Even if this is an ideal case, unlikely to occur in practice, introduction of the concept of attention will only complicate the picture. The ideas under discussion are not incompatible with a two-stage theory of discrimination learning, and can easily be incorporated into any such theory. We are concerned, however, solely with changes in response strengths over time and changes in rate of change of response strength, and are in no way concerned with the validity of two-stage models. Simply as a matter of convenience, therefore, we shall talk as if the only thing that an animal must do in order to learn a reversal is to substitute new responses for old.

If we accept that it takes longer to learn the first reversal of a problem than to learn the original problem, because the previously established response interferes with subsequent reversal, then it follows that the weaker the preference carried over from the previous problem, the easier reversal learning will be. Illustration of this point is provided by several experiments in which rats have learned spatial reversals with a fixed number of trials being given on each reversal. The usual result of this procedure is that groups receiving fewer trials per reversal make fewer errors in early reversals (although the effect disappears quite rapidly: North, 1950; Pubols, 1962; Gonzalez, Berger, & Bitterman, 1966). With only a few trials per problem, animals will learn little by the end of each problem; consequently there will be little to be reversed in the next problem, and performance at the outset of each problem will not start substantially below chance.

These considerations suggest two possible sources of serial reversal improvement. With repeated reversals, the preference established within each problem, although still sufficient to ensure accurate performance, might become less extreme. Second, irrespective of the strength of the preference established by the end of each problem, after several reversals the preference *carried over* to the next problem might become progressively weaker.

### 1. *Changes in Operators*

When trained to criterion on each problem, rats typically make runs of repetitive errors at the outset of early reversals and then continue responding below chance for many more trials. In other words, many trials are required to cancel the preference established in the preceding problem. This preference must have been built up during those (predominantly rewarded) trials when performance was improving from chance to the criterion level. If the strength of the preference established during these trials were to decrease with repeated reversals, the number of initial errors might decrease.

In fact, this requirement is satisfied by most response strength models, although this does not mean that the desired conclusion necessarily follows. After a large number of position reversals, response strengths to left and right will tend to asymptote. Assuming a negatively accelerated growth of response strength, the effect of the rewarded criterion run of trials will decrease as response strengths approach asymptote, i.e., as reversals continue. This in itself will not help reversal, for if the strength of the established preference decreases because of a decrease in the rate of change of response strength, by the same token, it will take longer to reverse a preference of given magnitude. The two effects will exactly cancel each other, leaving speed of reversal unaffected. This, however, is to ignore the effects of nonreward. On Spence's model, for example, the decremental effect of a nonreinforced trial increases linearly as the absolute magnitude of response strength increases. Even the weaker assumption, that decrements in response strength remain constant, predicts that with repeated reversals the weaker preference established in the preceding problem will be eliminated more and more rapidly.

### 2. *Forgetting*

These arguments are somewhat speculative; no attempt has been made to see whether Spence's model actually would generate serial reversal improvement, nor is there any empirical evidence available to test these ideas. A second, better documented source of serial reversal improvement is that whatever may happen to the strength of the preference at the *end* of each problem, after several reversals, the strength of the (incorrect) preference at the *outset* of each new problem decreases.

In many serial reversal experiments, animals are trained for a fixed number of trials on each problem, and are reversed every day. When animals are reversed between days in this manner, it has been noticed several times that with repeated reversals the probability of an error on Trial 1 of each problem decreases from about 1.0 to about 0.5 (Stretch, McGonigle, & Morton, 1964; Gonzalez *et al.*, 1966; Mackintosh *et al.*, 1968). Eventually, the preference established by the end of the preceding day appears to have no effect on performance at the beginning of the following day.

There are at least two possible explanations for this effect. First, the animals may be capable of anticipating reversals. The stimulus complex distinguishing Trial 1 of each day from other trials may become associated with nonreinforcement of the previously reinforced response, and animals may learn accordingly to reverse responses. Two points argue against this possibility. The probability of a Trial-1 error has never been observed to drop consistently below chance: if animals learn to anticipate reversals, there is no reason why they should not learn to do so more than half the time. Second, although rhesus monkeys are capable of using a conditional cue as a sign of reversal (Riopelle & Copelan, 1954), neither cebus nor cynamolgus monkeys (Crawford, 1962), nor cats (Warren, 1960b) have shown reliable evidence of such an ability.

A second possible explanation (first suggested by Gonzalez *et al.*, 1966) is that serial reversal training may generate considerable interference between the two opposing responses. The theory of proactive interference, as developed in the study of human verbal learning (Underwood, 1957), supposes that when different responses are learned to the same stimulus, establishment of the second response causes extinction of the first, but that eventual spontaneous recovery of the first causes forgetting of the second. It is easy to see how the serial reversal situation could result in such proactive interference effects, with the consequence that animals eventually come to forget from one day to the next which alternative was last correct.

The following evidence supports this interpretation.

(1) Gonzalez, Behrend, and Bitterman (1967) trained pigeons on a series of red–green reversals, with reversals occurring only every second day. Trial 1 of each day was not, therefore, consistently associated with nonreinforcement of the previously correct response. After performance had stabilized, animals were given a series of preference tests at different intervals after each problem. After 20 min, 72% of choices were of the last rewarded alternative; but after 24 or 48 hr, this had dropped to 48 and 51%, respectively.

(2) Gonzalez and Bitterman (1968) trained pigeons on a series of red–green reversals, with reversals occurring randomly every two or three days. They thus had a measure of reversal learning and of nonreversal relearning over the course of the experiment. Over the early part of the experiment, they found that as errors on reversal days decreased, so errors on nonreversal days increased.

(3) Mackintosh *et al.* (1968) trained rats on a series of position reversals. All animals received 20 training trials each day and were reversed every 20 trials. However, some subjects (Group 20) were reversed on Trial 1 of each day; while others (Group 10-10) were reversed on Trial 11, and were *not* reversed on Trial 1. Figure 9.4 shows the trial-by-trial performance of the two

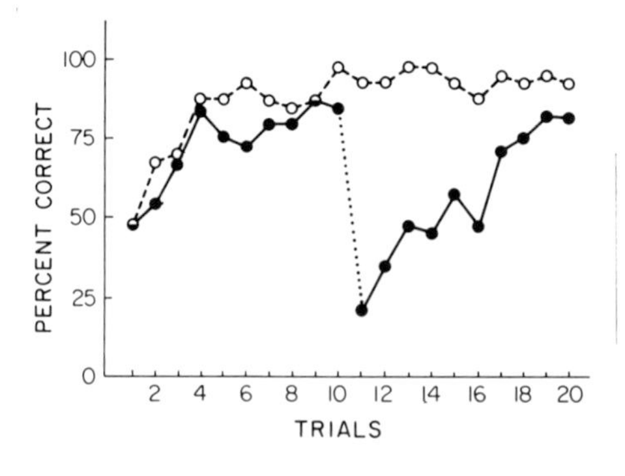


**Fig. 9.4.** Within-problem learning curves of animals reversed either on Trial 1 of each day (Group 20) (○) or on Trial 11 (Group 10-10)(●). [*After* Mackintosh *et al.* (1968).]

groups averaged over the last five reversals of the experiment. Both groups performed at chance (48.6% correct) on Trial 1 of each day.

This reversion to random choice of the two alternatives at the beginning of the day, even when subjects are not being reversed, and even when (as in our own experiment) they have *never* been reversed on Trial 1, suggests very strongly that they are incapable of retaining the information necessary for above chance performance, i.e., knowledge of which alternative was last correct.

The fact that animals may come to forget from one reversal to the next the direction in which they were most recently trained, can only reduce the number of errors in later reversals. Between-problem forgetting will reduce between-problem interference. Forgetting may also occur *within* a single problem, for if animals forget completely between successive reversals which response was last rewarded, some forgetting may occur between successive trials of a single reversal. This can only hinder learning. That interference from original learning may cause forgetting of a reversal, has been demonstrated by Maier and Gleitman (1967). The idea has been used by Clayton (1966) to account for the effects of intertrial intervals on reversal learning. Several studies have shown that performance both on a single reversal and on a series of reversals is less efficient the longer the interval between trials (North, 1959; Stretch *et al.*, 1964; Clayton, 1966); and as mentioned above, in two studies no serial reversal improvement occurred at all when trials were spaced 24 hr apart (Estes & Lauer, 1955; Clayton, 1962).

### 3. *Insufficiency of an Interference Explanation of Serial Reversal Improvement*

The two ideas discussed in this section—that the magnitude of preference established within a reversal and the preference carried over from one reversal to the next both decrease with repeated reversals—are essentially suggestions as to how negative transfer might be reduced. Early reversals are learned more slowly than the original problem, at least partly because of the interfering effects of these preferences. Disappearance of these effects, therefore, will bring the number of errors per reversal down to the level attained in original training, but no farther. In the case of the forgetting factor, even this degree of error reduction will only occur if the interval between successive reversals is long relative to the interval between trials within each reversal. It is therefore certain that other factors must contribute to serial reversal improvement; later reversals may be learned even more rapidly than the original problem, and this degree of improvement may occur when many reversals are given in a single session (Theios, 1965).

## C. Attention and Serial Reversal Learning

### 1. *Evidence for Changes in Rate of Learning Successive Reversals*

Further proof that factors other than changes in the strength of the incorrect response contribute to serial reversal improvement is provided by an examination of the course of within-problem learning in early and late reversals. While part of the reduction in errors occurring over the course of a series of reversals can be attributed to a decline in the persistence of below-chance responding at the outset of each reversal, another part is attributable to a decline in the number of errors occurring after a chance level of accuracy has been attained but before criterion is reached. To illustrate this, we can examine the performance of 20 rats trained on a series of black–white reversals in a Grice box—an apparatus that produces extremely rapid reversal learning (half the animals, from Experiment I by Mackintosh *et al.*, 1968, learned 8 reversals; the remainder, from Experiment II of Mackintosh and Holgate, 1969, learned 10). When learning their first reversal, these animals averaged 8.2 errors from the time they started scoring at chance (arbitrarily defined as 2 correct responses in 4 consecutive trials), to the time they reached criterion (18 correct responses in 20 consecutive trials). The *total* number of errors made in the final reversal, however, averaged only 2.4 per subject. The difference is significant at better than the .001 level. It is certain that the rate of within-problem learning may increase from early to late reversals, and that this change in rate cannot be attributed to decreases in response interference from one reversal to the next.

A change in rate of reversal learning would, of course, be produced by a progressive strengthening of the relevant analyzers; if animals stop attending to the relevant cue during early reversals they will learn slowly, but if they maintain attention during later reversals, they will reverse rapidly. Without, for the moment, considering how such changes in attention might come about, we can consider whether there is any evidence to implicate such changes in attention in serial reversal improvement.

### 2. *Changes in Responses to Irrelevant Cues*

In several experiments on serial visual reversals, it has been noticed that the increase in speed of reversal learning has been accompanied by a decrease in the number of responses to irrelevant cues (Lawrence & Mason, 1955; Mackintosh & Mackintosh, 1964; Mackintosh *et al.*, 1968; Mackintosh & Holgate, 1969). Such an effect would certainly be expected if the relevant analyzer were extinguished during early reversals but maintained during later reversals. It is, nevertheless, far from critical evidence; even if it could be shown that the reduction in responses to irrelevant cues was the cause of faster reversal learning, rather than the other way round, such a reduction is as well explained by a suppression of attention to irrelevant cues as by an enhancement of attention to the relevant cue. There is, however, some evidence to distinguish between these last two possibilities. Mackintosh and Mackintosh (1964) trained octopuses on a series of brightness reversals. For all animals the stimuli were black and white rectangles whose orientation was irrelevant to solution. For one group the rectangles were oriented horizontally and vertically, for the other at 45 and 135°. Octopuses are essentially unable to learn a discrimination between two oblique rectangles, but find horizontal–vertical extremely easy (Sutherland, 1957a). Not surprisingly, therefore, the animals trained with horizontal–vertical irrelevant learned the early reversals more slowly, and were the only ones to show systematic responses to orientation. After eight reversals, however, both groups were reversing equally fast and the horizontal–vertical animals had stopped responding to orientation. At this point, those trained with obliques irrelevant were shifted to horizontal–vertical irrelevant, but showed no increase in reversal scores and no signs of responding to the irrelevant cue. This implies that a decrease in responses to an irrelevant cue need not depend upon the opportunity to learn that the cue is irrelevant.

We need, however, much stronger evidence to justify the claim that changes in rate of reversal learning are partly dependent upon changes in attention, and are not entirely attributable to general factors. For in addition to the general factors that we have discussed earlier in this chapter, such as habituation to handling and the apparatus, or the development of appropriate observing responses or orienting behavior, serial reversal training has often been supposed to develop a "reversal learning set" or a "win–stay, lose–shift" strategy. The former notion may be incurably vague, but the latter is quite precise, implying only that stimuli associated with reinforcement and nonreinforcement come to control the subject's choices. Formal theories incorporating such an idea have been put forward by Restle (1962) and Levine (1965), and, as we shall see later (Chapter 12), there is excellent evidence that rhesus monkeys and chimpanzees do develop rather generally applicable strategies of this sort. Even if, as we shall argue, there is little direct evidence for the development of such strategies in rats, there is still a problem of distingushing between changes in attention and changes in *some* general factors as causes of serial reversal improvement.

### 3. *Nonreversal Shifts*

The type of comparison that ought to distinguish between these two broad possibilities has already been discussed at length in this chapter. If serial reversal training strengthens attention to the relevant cue, then such an effect should be revealed by a comparison of serial reversal (or IDS) and nonreversal (EDS) training. In an experiment already mentioned in Chapter 6, Shepp and Schrier (1969) trained one group of monkeys on a series of three IDS problems (either a series of color discriminations or a series of shape discriminations), and a second group on a series of EDS (from color to shape and back). The intradimensional animals showed substantial improvement over this short series, learning their final problem with fewer than 10 errors; the extradimensional animals showed no improvement, and learned their final problem with more than 30 errors. In this situation, therefore, the improvement can be attributed entirely to changes in the strength of the relevant analyzer.

A second way of distinguishing between general and specific effects of serial reversal training is to investigate the effect of such training on a final EDS. If, for example, rats trained on a series of black–white reversals have learned to switch-in a brightness analyzer very strongly, then, other things being equal, they should take longer to learn a shift to position (with brightness irrelevant) than a group trained on only a single brightness problem before the shift. It is certain, however, that other things will not be equal. In order to solve the shift problem, animals must stop consistently selecting their old positive stimulus (black or white), and one common effect of serial reversal training is to weaken the preference for the old positive stimulus that is carried over from one problem to the next. Proactive interference effects will (like any generally appropriate habit) act to facilitate shift learning following reversal training. We have little reason, therefore, to expect that serial reversal training will actually increase the number of trials required to learn a nonreversal shift. Nevertheless, two weaker predictions may be derived from an attentional analysis. First, if serial reversal training were to benefit shift learning *less* than it benefits reversal learning, we should have *prima facie* evidence that one of its effects was specific to the originally relevant dimension, i.e., was to strengthen the analyzer relevant to the reversal problem. Second, a closer examination of the course of nonreversal shift learning after training on a single problem or on a series of reversals, should enable one to distinguish between the effects of proactive interference and any possible attentional effects. Proactive interference should reduce the number of trials during which animals consistently select their last positive stimulus; any attentional effects of reversal training should increase the number of trials required to reach criterion once animals have stopped consistently selecting this stimulus.

In an experiment involving three main groups, Mackintosh and Holgate (1969) obtained evidence generally consistent with these weaker predictions. They trained rats in a Grice box, either on brightness shifting to position or on position shifting to brightness. In each case, one subgroup was trained on a single problem and then shifted, the other was trained on a series of 10 reversals before being shifted. In all three groups, serial reversal training facilitated reversal learning, but had no significant effect on the total number of trials required to learn the nonreversal shift problem. This confirms the first prediction. When the nonreversal shift scores were divided into trials of consistent response to the former positive stimulus and residual trials to criterion, then in all three groups reversal training reduced these initial errors and increased the residual trials to criterion. In only one of the three groups, however, was this last effect significant. The results for this group are shown in Figure 9.5.

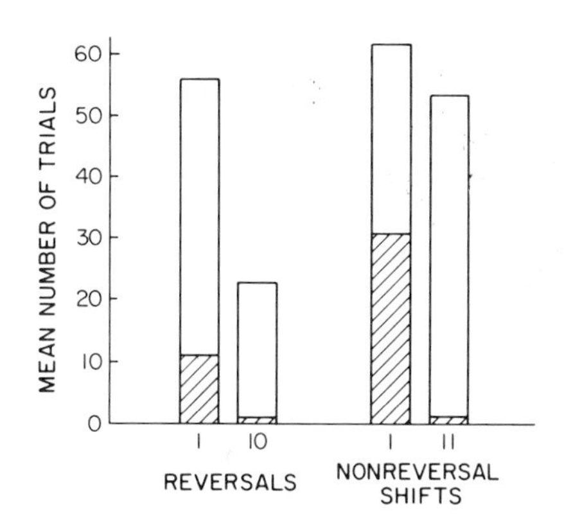


**Fig. 9.5.** Performance of rats on their first (R1) and tenth (R10) brightness reversal, and on a nonreversal shift to position learned after 1 brightness problem (NRS 1) or after 10 brightness reversals (NRS 11). The total height of the bars represents total trials to criterion; the shading represents the number of trials for which subjects consistently selected their former positive stimulus. [*After* Mackintosh and Holgate (1969).]

The second prediction is thus partially confirmed. In general, serial reversal training is more likely to slow down than to speed up the *rate* of nonreversal shift learning. To this extent, the effects of reversal training on rate of reversal learning are most reasonably ascribed to a strengthening of the relevant analyzer. However, in two of the three groups, serial reversal training did not significantly interfere with the acquisition of the shift problem, and this suggests that specific changes in attention are accompanied by changes in more general factors. We have argued above that a similar sort of conclusion is suggested by the results of overtraining and nonreversal shift experiments. The two possible general factors suggested there, however, cannot plausibly be put forward to account for the present results. It is unlikely that any elaborate orienting behavior is required to solve brightness or spatial problems in the Grice box; and suppression of attention to an irrelevant cue would have retarded shift learning (to that cue), not facilitated it.

4. *Insufficiency of an Attentional Analysis*

There is other evidence pointing to the importance of some unspecified general factor in serial reversal learning. Schade and Bitterman (1966), for example, trained pigeons concurrently on color and position problems. Specifically, the birds were trained in two-key boxes with red and green stimuli for 40 trials per day, a new problem being presented each day. Any one of the four stimuli (red, green, left, or right) might be positive on any day. For a subject learning Red+ on Day 1, the Day 2 problem might be either a color reversal (Green+) or an EDS to position (Left+ or Right+). If left were correct on Day 2, three types of problem might be presented on Day 3: a position reversal (Right+), a nonreversal EDS (Red+) back to the last positive color, or a reversal EDS (Green+), back to the other dimension but with the reinforcement assignments reversed. Error scores for all types of problem on both dimensions declined over the course of the experiment. That is to say, animals improved simultaneously on *both* color and position reversals and on EDS problems.

Much of this decline in errors can almost certainly be attributed to proactive interference. With all four stimuli equally often positive, the situation seems designed to maximize such interference effects. Performance on later reversals and reversal shifts may, therefore, have improved partly owing to a reduction of negative transfer. Unfortunately, the total error scores per day (the only data presented) do not allow one to distinguish between a decline in negative transfer and an increase in *rate* of learning. It is probable that some such increase occurred, for nonreversal EDS errors also declined over the course of the experiment (although much less than errors on the reversal and reversal shift problems), and the contribution of forgetting cannot have been solely beneficial. It is true that the learning of an extradimensional nonreversal (to Red+) will be interfered with by the competing tendency (established on the preceding day) to select, say, left; to this extent, forgetting might benefit nonreversal learning. It is equally true, however, that nonreversal learning will benefit from the retention of the most recent day of color discrimination training (Red+); and to this extent, forgetting should decrease a positive transfer effect. If the development of proactive interference is unlikely to have caused the reduction in extradimensional nonreversal errors, so also is an appeal to the concept of attention inappropriate in this instance; it is unreasonable to attribute a change in rate of learning solely to the simultaneous strengthening of attention to both color and position. Although therefore, the results do not (as the authors argue) show that changes in the strength of attention contribute nothing to serial reversal improvement, they do suggest that other factors may also increase rate of learning. Also, (just as in Mackintosh and Holgate's experiment discussed above) this other factor could have been neither a reduction in responses to irrelevant cues (the only cues obviously present were both equally often relevant) nor any general observing response; it is most improbable that pigeons ever fail to see the color of a key they peck.

5. *Problems for an Attentional Analysis*

Although other factors may contribute to serial reversal learning, it is worth repeating that the evidence for the contribution of attention is relatively strong. In visual problems, reversal improvement is accompanied by a reduction in responses to irrelevant cues; the presence of irrelevant cues retards early reversals, but animals that have learned several reversals ignore an irrelevant cue even when it is first introduced; finally, the beneficial effects of serial reversal training are largely confined to reversal learning—they do not extend to nonreversal shift learning.

While all this implies that in most serial reversal experiments, animals maintain attention to the relevant cue better in later reversals than in earlier reversals, it does not tell us how such an effect comes about. Much less interest has been been shown in this question than in the question of the effects of overtraining on attention, so that what follows is extremely tentative.

It is not too difficult to see how the relevant analyzer might be strengthened by a series of reversals when its initial strength is extremely low. The criterion of original learning will be reached with the relevant analyzer still substantially below asymptote. During the first reversal the analyzer will be weakened until response strengths are equalized, and then be strengthened again. Provided the lowest level reached during reversal is higher than the starting value at the outset of initial training, criterion on the first reversal will be reached with the analyzer somewhat stronger than it was at the end of the first problem. The second reversal will therefore be learned faster than the first, and because the analyzer started at a higher value, it will not reach the same level of extinction as before, and will end up higher than ever. The process is (for a few problems) self-reinforcing, but, in order to get off the ground, depends upon supposing that the initial strength of the relevant analyzer is less than the lowest point reached during the first reversal. While this analysis, therefore, may apply to difficult problems, it is clearly irrelevant to an understanding of serial reversal experiments involving a very dominant cue.

In fact, serial reversal improvement occurs with problems such as position discriminations in a T maze, brightness in a Grice box, and color with pigeons, where the initial strength of the relevant analyzer must be very high. If, in these situations, the analyzer is close to asymptote at the end of original learning, reversal training cannot result in higher terminal analyzer strengths, and we must look not to changes in rate of acquisition, but to changes in rate of extinction, to account for serial reversal improvement. Since the analyzer will be extinguished during each reversal until such time as response strengths to the two stimuli are equal, those factors such as progressive forgetting that reduce the strength of the incorrect preference carried over from one reversal to the next, will automatically ensure that the relevant analyzer is maintained at a higher level with successive reversals. In this indirect way the relevant analyzer may be subject to less extinction in later reversals, without any change in *rate* of extinction having occurred. It is unlikely, however, that this will serve to account for all the available data; e.g., some such change in rate must probably be postulated to account for the slower EDS learning observed by Mackintosh and Holgate.

As we shall discuss in the next chapter, a rat that has sometimes been rewarded and sometimes not rewarded for running down an alleyway, will continue to run for a long time in extinction. One cause of this effect of PR on extinction is that the running response becomes conditioned to the stimulus consequences of nonreinforcement (conceived in the widest possible sense). One might argue that when rats are trained on a series of black–white reversals, the brightness analyzer is subject to a complex PR schedule. It is conceivable therefore (although difficult to reconcile with our suggestion that analyzers are strengthened by consistency of reinforcement), that rats may learn to maintain a relevant analyzer in the face of nonreinforcement just as they learn to maintain running.[2] The idea is, to say the least, speculative, but not untestable; if correct, it would certainly account for a progressive strengthening of the relevant analyzer with repeated reversals.

## D. Serial Nonreversal Shift Learning

The evidence suggests that changes in rate of reversal learning depend partly on the development of attention to the relevant cue, and partly on some more general factor. The possibility that needs most discussion in this connection is that animals develop an appropriate win–stay, lose–shift strategy. Before dealing with this question, however, we should briefly discuss a somewhat different experimental situation, serial nonreversal shift learning, both because it suggests yet another cause of serial reversal improvement, and because it throws some light on the possibility that rats are capable of basing their current choice on whether their last choice was rewarded.

### 1. *Data*

In a complex experiment already cited, Lawrence and Mason (1955) were the first to study serial nonreversal shift learning. One of their groups of rats was trained first on a brightness problem with hurdles irrelevant, then on a hurdles problem with brightness irrelevant, then on the brightness problem

[2] In Chapter 8, we argued that the dip effect was evidence that analyzers were conditioned to the memory of reinforcement. If this is true, then the present suggestion follows easily.

again, etc. On both hurdles and brightness problems, the same stimulus (within each dimension) always remained positive. Serial nonreversal shift learning was extremely rapid, and over the first few problems was considerably faster than serial reversal learning with the same stimuli. Mackintosh (1969b) has also shown that rats will learn a series of brightness and position nonreversal shifts very rapidly.

### 2. *Interpretation*

It is tempting to assume that just as serial reversal improvement may depend upon learning to maintain attention following nonreinforcement, so serial nonreversal improvement depends upon learning to shift attention following nonreinforcement. The two cases, however, are not strictly comparable. We have already discussed how strengthening of the relevant analyzer over a series of reversals might occur: the most that would seem to be required is that the analyzer become conditioned to the stimulus consequences of no reward. In order to learn to shift analyzers, however, an animal must store information not only about the consequences of each trial (i.e., use reinforcement and nonreinforcement as stimuli to control current behavior), he must also remember which analyzer he used on the previous trial. When an animal is trained on brightness and position shifts, for example, use of a brightness analyzer on a given trial is appropriate both when a brightness analyzer was used on the last trial and the animal was rewarded, and when a position analyzer was used on the last trial and the animal was not rewarded. There is no guarantee that rats are capable of storing the two bits of information necessary for the development of this strategy. Nor is it clear that serial nonreversal shift improvement, in fact, requires such a strategy.

At least two other factors will reduce the errors made over a series of, say, brightness and position shifts. Let us assume that at the outset of original learning, both analyzers are of equal strength, and no preferences exist for either value of either dimension. If the first problem is Black+, its solution requires strengthening of the brightness analyzer and the response of selecting black. Solution of the second problem (Left+), will now be relatively slow; the relevant analyzer is the weaker of the two, and the correct response strength is only .5. Solution of the third problem will be somewhat faster, since training on position with black–white irrelevant will not have eliminated the original response strength to black (for evidence of this, see Chapter 8). If the third problem is learned faster than the second, there will have been fewer trials of random reinforcement on the preceding relevant cue, and appropriate response strengths will have survived more nearly intact. Consequently, the fourth problem will be even easier than the third. Both in

Lawrence and Mason's and in Mackintosh's study, asymptotic nonreversal shift performance was, in fact, reached within four problems.

It is probable that a second source of improvement exists. Just as serial reversal training causes animals to forget which response was last correct, so serial nonreversal shift training may cause some forgetting of which dimension was last relevant. If this happened, then eventually animals would be as likely to select the correct as the incorrect dimension at the outset of each day, and, since there has been no interference between the two values of each dimension, may solve some problems with no errors at all. In Mackintosh's experiment, there was some evidence for forgetting. If an animal is trained with black positive on brightness problems, and left positive on position problems, then only on trials when black is on the right is it possible to tell which dimension is controlling behavior. An examination of performance on the first such opposition trial of each shift problem revealed a decrease in the probability of an error occurring on this trial over the course of 10 nonreversal shifts. On Shift 1, the probability of an error was .94; on Shift 10, the probability of an error was only .62.

Given these two sources of improvement, it is not clear that it is necessary to postulate the development of a nonreversal shift strategy. Indeed, the improvement shown in the two experiments we have discussed was less than might have been expected. Asymptotic performance was reached within four problems, but certainly not because of any ceiling effect. (In Mackintosh's experiment, for example, animals averaged 3.0 errors per problem over Problems 5–10.) While it may be possible for rats to learn such problems with only a single, informative error, in neither study was there any sign of this happening.

### 3. *Optional Reversal and Nonreversal Shift Learning*

As has already been discussed in this chapter, a nonreversal shift is usually learned more rapidly than a reversal; even though asymptotic performance in serial nonreversal shift experiments may be less than perfect, over early problems many fewer errors are made than in comparable serial reversal experiments. At a theoretical level, this implies something about the relative rates of analyzer and response extinction; at a practical level, it implies that, given the option, a rat will solve a series of problems as a series of nonreversal shifts rather than as a series of reversals.

Any serial reversal experiment, where the discrimination problem contains more than one relevant cue, provides the subject with this option. We have already seen that if rats are trained on a two-cue discrimination (between a white horizontal and a black vertical rectangle), the majority will solve the problem predominantly in terms of one or other cue alone. Suppose a group of rats solved the problem by learning to approach white. If now they are ostensibly reversed, they can either learn to reverse black and white, or they can shift to orientation and learn to approach vertical. If the first reversal is learned by shifting to orientation, then the second can equally be solved by shifting back to brightness and reselecting white. The series of reversals is equally a series of nonreversal shifts.

Sutherland (1966a) trained rats on a series of eight such reversals, and by giving transfer tests to the component cues in isolation after the seventh and eighth reversals, was able to determine the extent to which nonreversal shift learning occurred. Of 24 rats, only 5 performed more accurately on the same cue on the two successive sets of transfer tests; 5 performed equally accurately on both; and 14 performed more accurately on brightness tests after one reversal and more accurately on orientation tests after the other reversal. The majority of animals, therefore, shifted the cue they learned more about from one problem to the next. These results may well have relevance to many serial reversal experiments. Even where the experimenter assumes that a single cue is relevant, there are excellent reasons for supposing that more than one analyzer is capable of detecting the difference between positive and negative stimuli. Position, for example, is a cue that almost certainly combines a large number of component cues.

Sutherland's data, however, confirm that serial nonreversal shift learning is never particularly rapid. Improvement occurred over the first four problems only; over the final five problems animals continued to average nearly 100 trials to criterion per problem. While it is certain that the number of errors required to learn successive nonreversal shifts does decrease, the data do not appear to require any new explanatory principles. In particular, the data do not require the conclusion that rats learn to base their current choice of analyzer on whether the analyzer used on a preceding trial was reinforced.

## E. Response Strategies

### 1. *General Reversal Learning Set*

We have seen that not all the increase in learning rate occurring in serial reversal experiments can be attributed to changes in attention. While it is possible that such very general factors as habituation to handling and to the apparatus increase an animal's efficiency, it has also been assumed that serial reversal training must establish a set of habits describable as a "reversal learning set" or "reversal strategy." At its most general, the idea implies that training on any series of reversals will facilitate any subsequent discrimination or reversal learning. This appears to be true in the case of chimpanzees and rhesus monkeys. Schusterman (1962) found that chimpanzees trained on a series of visual reversals showed immediate positive transfer when presented with a series of discrimination learning set problems involving entirely different stimuli. This effect is conceivably attributable to appropriate changes in attentiveness to visual (relevant) and spatial (irrelevant) cues. Warren (1966), however, found that rhesus monkeys trained on a series of position reversals also showed virtually immediate positive transfer to visual discrimination learning set problems. This is a striking result; in higher primates, the response strategy acquired from training on a series of problems is a much more important determinant of transfer effects than is any change in the relevant or irrelevant cues. We shall return to these experiments in Chapter 12. For the moment, however, they need not particularly concern us; whatever may be the case with primates, it is reasonably certain that strategies of this degree of generality are not easily established in lower animals. In the study just mentioned, Warren also trained cats on either visual or spatial reversals to test for transfer to visual discrimination learning set problems. Neither group showed positive transfer relative to naive control animals; and, as one might expect, there was some (albeit not significant) tendency for spatial reversal training to interfere with visual discrimination learning. In rats, spatial reversal training undoubtedly interferes with subsequent visual reversal learning. This has been shown in two experiments (Bitterman, Wodinsky, & Candland, 1958; Mackintosh *et al.*, 1968); in the latter, we found that the cause of the interference was an increase in the number of systematic responses to the (now irrelevant) cue of position. In nonprimates, therefore, it seems that if any response strategies are acquired, they are restricted to the cues the animal has been trained on; there is no evidence of strategies that generalize across stimulus dimensions.

### 2. *Win–Stay, Lose–Shift Strategy*

This does not rule out the possibility that within a given dimension rats may learn to base their current choice on whether or not their last choice was rewarded. As we have already indirectly mentioned, it is certain that rats can learn to control their running speed on a given trial by the memory of whether or not they were rewarded on the preceding trial; e.g., rats given a sufficient number of alternately rewarded and nonrewarded trials in a runway will eventually run rapidly on rewarded trials and slowly on nonrewarded trials (Capaldi, 1958). As we also argued, however, control of choice behavior by the reinforcement conditions of the preceding trial is not the same thing as control of running speed by memory of reinforcement. The latter requires use of only one bit of information: whether or not the last trial ended in reward; the former requires the use of two bits of information: both whether the last choice was rewarded and what the last choice was. This may appear to be a trivial addition to the requirements imposed on the rat's memory; nevertheless there is no satisfactory evidence that rats will learn to control choice behavior in this way.

Nor is there any evidence to the contrary; there is simply no evidence either way. Indirect evidence that reversal learning depends upon a win–stay, lose-shift strategy is provided by the already mentioned finding that reversal efficiency decreases with an increasing intertrial interval. Patterned running in an alley also varies inversely with the intertrial interval (e.g., Surridge & Amsel, 1965a; Katz, Woods, & Carrithers, 1966); and it is reasonable to attribute both effects to a weakening of the memory trace from the preceding trial which is used to determine performance on the current trial. Although both effects are most probably evidence of forgetting, it is equally possible to provide a different underlying rationale. In the case of reversal learning, the joint assumptions that the establishment of a response to the positive stimulus depends partly upon the extinction of responses to the negative stimulus, and that an extinguished response shows spontaneous recovery, will suffice to explain the results obtained.

This illustrates the general difficulty of establishing the reality of any strategy in rats; an effect that may appear to be based on a win–stay, lose–shift strategy (i.e., on the stimulus consequences of reward and nonreward) is equally well explained as a consequence of the *reinforcing* effects of reward and nonreward. For example, when trained by a correction or guidance procedure on a simultaneous visual discrimination problem, rats do in fact, before they solve the problem, typically respond on each trial to the position where they were rewarded on the previous trial. While it is true that this behavior does not depend on remembering both which choice they made on the last trial and whether or not they were rewarded (for they are always rewarded), it does look like at least part of the strategy we are seeking. It is obvious, however, that given approximately equal response strengths to the two sides, rewarding a response to the left on one trial will (via the reinforcing effect of reward) make a left choice more probable than a right choice on the following trial. There is no need to appeal to a win–stay, lose–shift strategy when the simple operations of reinforcement and nonreinforcement are sufficient to explain the data. To pursue the present example, it turns out that rats may continue to show some tendency to select the last rewarded position over several hundred trials when given guidance training on simultaneous visual discrimination problems (Mackintosh, 1970). Such a tendency obviously only occurs when performance is less than perfect; but, as we discuss in Chapter 11, on a probabilistic reinforcement schedule, rats rarely attain an asymptote of better than 90% correct responding, and most errors are traceable to a habit of selecting the last rewarded position. Now this tendency persists in spite of the fact that the reinforcement schedule marginally favors the opposite strategy. In most Gellermann sequences, the probability that the positive stimulus will stay on the same side is less than the probability that it will change from one side to the other. This is also true of the sequences used in the probability learning experiments. In practice, this meant that on only 45% of trials were animals actually rewarded for selecting the last rewarded position. In spite of this small difference operating in favor of a win–shift, lose–stay strategy, our rats continued with the opposite habit. Such insensitivity to a reinforcement schedule is hardly the sign of a flexible strategy.

While rats (and other nonprimates capable of serial reversal improvement) may be capable of learning response strategies appropriate to the serial reversal situation, there does not seem to be sufficient evidence to prove it. It is relatively certain that if such strategies are available to these animals, they are much less readily generalized across stimulus dimensions than they are in advanced primates. It may be unnecessarily conservative to conclude that the only general habit that a rat learns appropriate to reversal learning is not to be afraid of the experimenter or apparatus, but such a possibility should be borne in mind.

## IV. Summary

The discussion in this chapter has presupposed that discrimination learning involves more than the establishment of different response strengths to the positive and negative stimuli. This is not a novel suggestion; MacCorquodale and Meehl (1954), for example, in a review of experiments on latent learning, concluded that maze learning must involve more than the establishment of appropriate habit strengths or expectancies. The question at issue in the present chapter has been: what is the nature of these additional learning processes? Our own main interest is to distinguish between the suggestion that animals learn to attend to the relevant dimension and a variety of other possibilities that include the suggestion that rats only gradually become habituated to the experimental situation, that they must learn to orient towards the relevant stimuli, or that they acquire a response strategy based on the memory of their last choice and its consequences. While there are obvious and substantial differences among this list of alternative possibilities, they all differ from an attentional analysis in the specificity of the effects predicted.

The two findings discussed in this chapter, which have been taken as showing that some further processes occur during discrimination learning, are first that overtraining sometimes facilitates reversal learning, and secondly that serial reversal training nearly always facilitates reversal learning. Neither of these results can be completely explained in terms of the acquisition and extinction of choice responses. We reviewed in Chapter 8 the suggestion that the ORE is a consequence of the effect of overtraining on extinction, and concluded that it received little experimental support. In the present chapter, we saw that although one cause of serial reversal improvement is almost certainly the effect of forgetting in reducing (response-based) negative transfer between problems, such a process cannot account for the increase in positive transfer observed in serial reversal experiments (e.g., the increase in the *rate* of within-problem learning). That *one* of the further processes responsible for these effects is the strengthening of the relevant analyzer is suggested by the finding that the beneficial effects of overtraining and serial reversal training are (at least in part) specific to reversal learning, i.e., to a problem involving the same relevant stimuli, and do not extend equally to EDS problems.

The evidence, however, cannot support any stronger conclusion than this: it is most probable that there are other, more generally beneficial effects of overtraining and serial reversal training. The results of studies of overtraining and shift learning have been inconsistent: when the formerly relevant cue is still present during the nonreversal shift, overtraining has sometimes been found to retard shift learning (as is predicted by an attentional theory), but has sometimes been found to have no effect; while when the formerly relevant cue is removed from the situation, overtraining has sometimes been found to have no effect on shift learning, and has sometimes been found to facilitate shift learning. It is possible that these differences in outcome are due to differences in the importance of specific patterns of orienting responses in different types of apparatus, and therefore these results suggest that such orienting responses form an important part of what must be learned (in certain types of apparatus) during the course of discrimination training.

In the case of serial reversal learning, the evidence that changes in the strength of the relevant analyzer contribute to faster reversal consists largely of the finding that reversal training certainly does not facilitate shift learning when the old cue is still present, and may (marginally) interfere with such learning. This is strong evidence only for the weak conclusion that changes in analyzers are *one* consequence of serial reversal training. It would be surprising if such training did not in some more general way contribute to greater behavioral flexibility, but at the present time, there is little evidence on the nature of such additional effects.

CHAPTER 10

# Partial Reinforcement and Extinction

## I. Introduction

It is well known that if an instrumental response is reinforced on some of the occasions on which it is made and not reinforced on others, the response is often much more resistant to extinction than if it is reinforced on every occasion. The phenomenon is known as the partial reinforcement effect (PRE). Although Pavlov (1927) reports an experiment involving training under partial reinforcement, he does not appear to have investigated the effects of partial reinforcement on extinction. The PRE is, therefore, along with some of the effects associated with reversal training just discussed, one of the few major phenomena of animal learning discovered outside Russia. As we shall see below, it is in fact debatable whether a PRE can be obtained using a classically conditioned response.

Skinner (1934) appears to have been the first to demonstrate the PRE. He showed that when a rat is trained to press a bar for food, many more responses are given in extinction when the rat is rewarded during training for only a proportion of its bar-presses than when it is always rewarded. This experiment involves a free-operant procedure, since the animal stays in the situation continuously during training sessions, and it has complete control over the times at which responses are emitted. An alternative training situation is that involving discrete trials: here, the organism is either removed from the situation between each response or the responses are controlled by presenting a stimulus to signal the start of each trial. Humphreys (1939) was the first to demonstrate a PRE in a discrete trials situation, though, as we shall see (page 397), there is some doubt about the interpretation of his results, since he used classical conditioning.

The PRE, like many of the other phenomena with which we have been concerned, is a paradoxical effect. Rats trained in a runway and given 100 reinforced trials extinguish more quickly than rats given 50 reinforced and 50 nonreinforced trials, despite the fact that the consistently reinforced animals have received twice as many reinforcements. Either the strength of a habit is not directly related to the number of reinforcements received or the number of trials to extinction is not directly related to habit strength. Either assumption is against the general trend of S–R theorizing in the thirties as exemplified for instance by Hull. The phenomenon is also at variance with one of the assumptions usually made by statistical learning theorists, namely, "the independence of pathways" assumption (cf. Sternberg, 1963). The PRE is, therefore, comparable to the overtraining reversal effect discussed in Chapter 8; both effects pose problems for theories that regard learning as a single, monotonically increasing process.

A great deal of research has been undertaken on the PRE, much of it in recent years, and several very ingenious theories have been suggested to account for the effect. It turns out that our two-process model itself provides some explanation of the PRE, although the theory was not originally proposed with this effect in mind. In this chapter we shall first put forward our own explanation of the effect and then discuss at some length other explanations. In the course of describing each suggested explanation, we shall discuss the experiments to which that explanation has given rise and evaluate the different theories against the findings. Because of the large number of experiments performed on problems to do with PRE, this will involve us in a mass of detail. However, the question is an interesting and important one and to evaluate how successfully our own hypothesis accounts for the findings it is necessary both to list the findings and to see how well other theories account for them. Moreover, so much recent experimental work has been done that the most recent reviews of the subject (Lewis, 1960; Lawrence & Festinger, 1962) are already out of date and there is no very recent appraisal of the merits and demerits of the different theories put forward.

In what follows, we shall use the abbreviations PR and CR to stand for partial and consistent reinforcement, respectively. For convenience, we shall number sections giving experimental results on each of the many phenomena connected with the PRE [e.g., (E-1), (E-2), etc.].

The main question to be answered in this chapter is why PR retards extinction measured in terms of *latencies*. In the next chapter, we discuss a related phenomenon—namely, the effects on *choice* behavior of rewarding discriminatory responses with different schedules of PR, a situation known as probability learning. For convenience, discussion of the effects of PR on choice behavior in extinction will also be postponed until the next chapter.

# II. The Two-Process Model

## A. Introduction

In describing our own explanation of the PRE, we shall use for illustrative purposes the runway training situation. It will be remembered that analyzers are strengthened when their outputs consistently predict reinforcement. When PR training is given, no analyzer can consistently predict reinforcement by the very nature of partial reinforcement. During CR training, however, any analyzer that yields an output signaling that the animal is in the runway will consistently predict reinforcement given that the animal runs. Other outputs from the same analyzer will signal that the animal is not in the runway situation and will predict that the animal will not be reinforced. As we saw in Chapter 5, when multiple cues are present and CR is used, there will be a tendency for the response to be attached only to the dominant cues, since the strengthening of the dominant analyzers leads to other analyzers being depressed during learning. Under PR training, however, the tendency for the strength of the most dominant relevant analyzer to "run away" should be much reduced. Because it will be weakened when nonreinforcement occurs, it can never reach asymptote. Hence, under PR, animals should learn more readily to attach the correct responses to nondominant relevant analyzers.

We might expect, therefore, that a CR animal would learn to attach the response only or mainly to the outputs from the dominant analyzers, whereas the PR animal would attach the response to the outputs from many different analyzers. During extinction, the series of unrewarded trials will lead to the weakening of the responses attached to any analyzer that is switched-in, and any analyzers above their base value will also be weakened. If response strengths have an asymptote, it is possible that the response attached to the dominant analyzers will be weakened to a level at which the response will no longer be performed at a time when the dominant analyzers are still controlling behavior. At that point, an animal trained under CR will stop running. However, PR animals will have the response attached to the outputs from many different analyzers; since no one analyzer is firmly switched-in, the response will be weakened slowly to each of them. Moreover, if the dominant analyzers drop below the level of other analyzers, the animals will still continue to run, since the response will still be attached to the analyzers that now control behavior.

It must be admitted that these predictions do not necessarily follow from the rather loose rules of our model. The explanation may be recast in anthropomorphic terms as follows: if a given feature of the environment consistently predicts reward, an animal may learn to make a response to that feature and to no other. If no feature of the environment consistently predicts reward, but reward nevertheless occurs on some trials, the animal may continue to search through different features in order to find one that does consistently predict reward. During this search, the animal will learn that running in the presence of a large number of different features is sometimes followed by reward. It will, therefore, attach the response of running to many different features of the environment. If, in extinction, animals only notice a limited number of features on any one trial, PR animals will take longer to extinguish, since they must again search through all the different features to which the response has been attached before they can learn to stop running. In order to stop running, CR animals only have to learn that the few features of the environment to which they paid attention during training no longer predict reward.

Although the explanation as it stands is vague, it is possible to undertake some tests of it and it is perhaps advisable to test it in its loose form before trying to make it more rigorous. We have, in fact, derived and tested three different kinds of predictions:

(1) Animals should learn about more cues under PR than under CR.

(2) The model predicts that when CR training is given before PR training, the CR training will reduce resistance to extinction more than when PR is followed by CR.

(3) We would expect the difference between trials to extinction after PR and CR training to increase as the number of possible discriminatory cues in the environment is increased.

In the remainder of this section we shall explain the rationale behind these predictions and set out the experiments undertaken to test them.

### B. (E-1) Breadth of Learning

The most obvious way of testing our theory of the PRE is to train animals in a situation where there are several relevant cues that can be controlled by the experimenter and to discover by giving tests after training whether animals trained under PR really do learn about more cues than animals trained under CR.

Sutherland (1966b) undertook an experiment of this kind using rats in a jumping stand. Five visual cues were always present and relevant: they are shown in Figure 10.1, and were horizontal versus vertical stripes, narrow versus broad stripes, white versus black central shape, square versus diamond, and low circles versus high circles. Half the animals were trained to jump to A (in Figure 10.1) and half to B. Two cues from other modalities were also present and relevant. One arm of the Y-shaped jumping stand was covered with a metal grid, the other with rubber, and animals had to learn to follow either the metal arm or the rubber arm; we cannot, of course, be sure whether this particular cue was learned visually or by touch. Finally, a bell was sounded behind and to one side of the rat while it was on the jumping platform; all animals had to learn to jump directly away from the bell, i.e., to select the stimulus on the other side of the apparatus from the bell.

All subjects were given 200 training trials. Eighteen were trained with CR; they were always rewarded for correct jumps, never rewarded for incorrect jumps. A second group of 17 subjects (the 18th became ill in the course of the experiment) were trained under PR; when they jumped correctly they were rewarded 50% of the time, when they jumped incorrectly they were never rewarded. By the end of training all subjects had learned the discrimination. As might be expected the PR group learned more slowly than the CR group. At the next stage of the experiment, all animals received transfer tests on which only one of the seven cues was present. Each subject received 20 transfer tests on each of the seven single cues and the transfer tests were alternated with140 retraining trials on which the original training stimuli were present and the reinforcement schedule was the same as during training. On the transfer tests, animals were reinforced, whichever stimulus they chose.

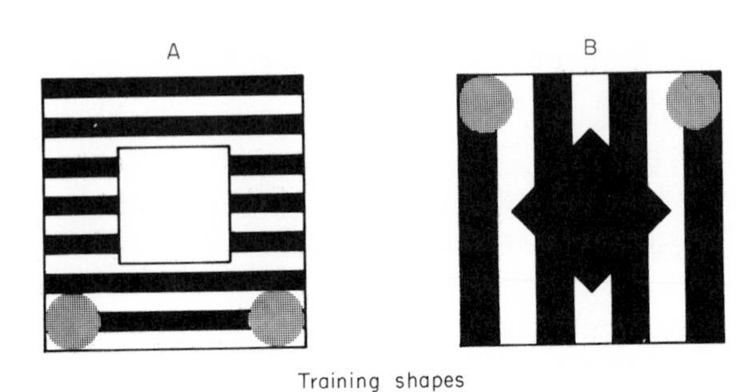


**Fig. 10.1.** Multidimensional stimuli used by Sutherland (1966b).

Table 10.I shows animals' performance during transfer tests. It will be seen that although the CR group performed better on the retraining trials, the PR group performed better on every set of transfer trials with the individual cues. Although, as a group, animals performed best in transfer tests on the horizontal–vertical cue, individual animals gave their best performance on different cues. Thus, within the partial group the best score made on any cue by individual animals was distributed over all seven cues. To overcome the variation in which cue is best learned from animal to animal, the results were retabulated by taking the average score on each animal's best cue, the average score on each animal's second best cue and so on down to the average

TABLE 10.I

**Transfer Tests**

| Cue | Group PR | Group CR | Cue | Group PR[a] | Group CR[a] |
|---|---|---|---|---|---|
| Retraining trials | 94 | 99 | | | |
| Horizontal–vertical | 71 | 67 | Cue 1 | 76 | 70 |
| Metal–rubber | 69 | 54 | Cue 2 | 67 | 59 |
| Narrow–broad | 64 | 51 | Cue 3 | 66 | 54 |
| White–black | 59 | 48 | Cue 4 | 64 | 51 |
| Circles | 59 | 47 | Cue 5 | 59 | 49 |
| Square–diamond | 58 | 48 | Cue 6 | 55 | 44 |
| Bell | 57 | 47 | Cue 7 | 49 | 39 |

[a] These columns' scores are tabulated by taking the average of the highest scores on any single cue by each subject (Cue 1), the average of the second highest scores (Cue 2), etc.

score on each animal's worst cue. The results tabulated in this way are shown on the right-hand side of Table 10.I. Once again, Group PR scored better on every single cue than Group CR. The CR animals, in fact, give evidence of having learned only about one cue. They scored 59% on their second best cue; although this is 9% above chance, they actually scored 11% below chance on their worst cue. The below-chance scores for the CR group on Cues 5, 6, and 7 are, of course, determined by the method of selecting these cues as being the ones on which each animal performed most poorly. On the other hand, the PR animals give evidence of having learned six different cues.

The results, therefore, strongly suggest that PR animals learn about more cues than do CR animals during training. We might, however, have expected the CR group to have learned more about their best single cue than the PR group, since the most dominant analyzer should be more strongly switched-in for the CR group than for the PR group. It is true that the difference between the scores of the two groups is least on Cue 1 and that it is the only difference between scores that is not significant. Nevertheless, Group PR did perform better than Group CR on this cue. This difference could again be a statistical artifact. Since Group PR performed at 60% or better on four cues a chance high score on any one of these four cues could have spuriously raised performance level on their best cue; Group CR, on the other hand, only performed above 60% on *one* cue, and it is, therefore, less likely that chance high scores on other cues could contribute toward their best cue performance.

Despite the fact that they performed better on transfer tests, PR subjects formed position habits more readily during transfer than CR subjects; if we consider only those series of transfer shapes on which individual animals performed at or below chance (50%), on 67% of each series of 20 trials PR subjects selected the same side on every trial, whereas only on 39% of series did CR subjects do so. Nor can the greater breadth of learning of the PR animals be explained by the difference in the number of reinforcements received by the two groups, since we saw in Chapter 5 that increasing the number of training trials tends to increase the breadth of learning when two relevant cues are present. The animals trained under PR with seven relevant cues, therefore, appear to have learned about more individual cues than animals trained under CR, despite the fact that the PR animals were performing less well with all seven cues present, were more prone to form position habits, and had received fewer reinforced trials during training.

There is, however, one feature of the results that was not predicted. The performance of CR animals on their best cues does not predict performance on the conjoined cues. If we apply the formula given in Chapter 5,

$$P = (p_1 \cdot p_2)/[p_1 \cdot p_2 + (1 - p_1)(1 - p_2)],$$

to the scores on the best two cues (70 and 59%) a performance of only 77% is predicted for all cues combined, whereas the obtained performance was 99%. It may be that although Group CR animals had not learned to attach the jumping response to the correct values of most of the seven cues, they had learned to use them as situational cues, i.e., they had learned to switch-in the analyzers detecting their best cues only when some or all of the remaining seven cues were present. This possibility was discussed in Chapter 5, and it makes it difficult to apply formulas to predict results of this type of experiment with any certainty.

It should also be added that on completing tests with single cues, Sutherland tried giving tests with two cues paired using all the 21 pairings that can be obtained from 7 different cues. The two groups now gave results that were not significantly different from one another. However, the PR animals actually performed worse with two cues present than they had performed with only one cue present and it seems likely that during the long series of transfer tests on single cues they had learned that they could obtain 100% reinforcement whichever way they jumped during transfer tests. They might well be more likely to learn this than animals trained under CR, since during re-training trials PR animals received less than 50% reinforcement for jumping but they obtained 100% reinforcement on transfer tests, whereas CR animals were rewarded for 99% of their responses to training shapes and for 100% of their responses to transfer shapes. The marked difference between training and transfer schedules for the PR animals may have enabled them to learn more readily than the CR animals that when transfer tests were presented no cue was correlated with reward and nonreward. To test this possibility, it will be necessary to repeat the experiment giving tests with pairs of cues at the outset of transfer to see whether the PR animals now do better on paired cues as well as on individual cues than the CR animals.

Unfortunately, the above experiment is open to two criticisms. First, it could be argued that the extra breadth of learning under PR was produced not by PR *per se*, but because animals under PR training made more *errors* than animals under CR. If discrimination learning occurs mainly on unrewarded trials, the PR group would have more opportunity to learn. We feel this is an unlikely explanation, since it would not explain why PR animals did not learn significantly more about their most dominant cue than did CR animals. Second, early in training, the latencies of PR animals were longer than those of the CR group and, therefore, PR animals had more time to inspect the stimuli—this might have resulted in their learning about more cues. Once again, this would not explain why the additional learning was limited to subsidiary cues. Nevertheless, these criticisms do have some force. They can be met by repeating the experiment using classical conditioning (as suggested by Wagner, 1969a). With this procedure, both the reinforcement contingencies and the duration of the stimulus on each trial are brought completely under the experimenter's control and cannot be influenced by the animals' own responses.

McFarland (1966) has also produced evidence that animals learn about more cues under PR than under CR. He trained ring doves to criterion on a black–white discrimination. Half were trained under CR, and half under PR: there was almost no difference in the number of trials taken to criterion by the two groups (CR, 87.5 trials; PR, 90.8 trials). He then gave further training with the black–white cue still relevant but adding a second relevant cue (horizontal–vertical); all birds were retrained with the same reinforcement schedule that they had received during initial training. Finally, he gave tests to find out how much each subject had learned about the added orientation cue and discovered that the PR birds had learned it much better than those trained throughout under CR. Moreover, when animals were extinguished on the conjoint cues, there was a positive correlation within CR birds between the amount they had learned about the orientation cue and their resistance to extinction. This experiment, therefore, shows not only that breadth of learning is increased under PR but that resistance to extinction is correlated with breadth of learning. Since all animals were trained to the same criterion at the first stage, there is no reason to suppose that the PR group differed from the CR group in terms of errors or latencies at the time when the second cue was introduced. This experiment, therefore, seems to be exempt from the two criticisms made of Sutherland's experiment. Nevertheless, it does not prove that the correlation between breadth of learning and trials to extinction is a causal one; this point will be discussed further later.

In a second experiment, McFarland introduced a different test for breadth of learning. He trained ring doves to run for water under CR or PR. He then introduced a distracting stimulus into the runway (a red flashing light) and found that, as measured by increases in running times, it interfered far more with the performance of the PR birds than with that of the CR birds. The inference McFarland draws is that CR birds had narrowed their range of attention to one or two dominant cues and were not therefore distracted by the light; since PR birds, on the other hand, were using many cues, they registered the presence of the flashing light, and consequently, their performance was disrupted by it.

A further prediction related to breadth of learning can be made from our theory. We have already seen in Chapter 5 that when animals are trained with two relevant cues under CR, the more an individual animal learns about one of the cues, the less it learns about the other. This should not be true of animals trained under PR with two relevant cues; since neither cue predicts reinforcement with complete consistency, it will be more difficult to strengthen one analyzer at the expense of the other. Accordingly, Sutherland and Holgate (1966) trained animals to discriminate a white horizontal rectangle from a black vertical rectangle. Half the animals were rewarded 100% of the time for jumps to the positive stimulus, and 0% for jumps to the negative; the other group (PR) were rewarded only 50% of the time for jumps to the positive and 0% for jumps to the negative stimulus. All animals were again given 200 training trials. They were subsequently run with no reward present, and at this stage received in random order 20 trials with the original two-cue shapes (the first pair shown in Figure 5.5, page 154), and 10 trials with each of the four transfer pairs shown in the same figure.

During extinction, there was a correlation of $-.86$ between the number of correct jumps made by individual CR animals when the orientation transfer cue was present on its own, and the number of correct jumps made when the brightness transfer cue was present on its own; the corresponding correlation coefficient for PR animals had a value of $+.80$. This result was corroborated by the results of the experiment using seven cues; correlation coefficients were calculated on animals' scores on the two most dominant cues (orientation and rubber versus metal), and once again there was a tendency for CR animals to give negative correlations while PR animals gave positive correlations.

## C. (E-2) Sequential Effects of Giving Both PR and CR

A second area within which our theory is capable of making predictions relates to the effects of giving a block of CR training trials either before or after PR training. Consider first the consequences of giving some CR trials after PR training in a runway situation. If, during PR, animals learn to attach the running response to many different cues, subsequent CR training will not alter these response attachments. It may lead to one or two dominant analyzers being strengthened, but when these are weakened in extinction the running response will be maintained because it is attached to other analyzers and these will control behavior when the dominant analyzers are weakened. We would, therefore, expect that the effects of PR on resistance to extinction will not be washed out by giving a block of CR trials between PR training and extinction. This prediction agrees with a result obtained by Theios (1962) using rats in a runway. He found that animals given 70 PR trials followed by 70 CR trials were almost as resistant to extinction as animals given just the 70 PR trials and then extinguished. Jenkins (1962) obtained a similar effect using pigeons.

The situation is very different when PR trials are *preceded by* CR training. With this experimental paradigm, we must predict that the CR training will considerably weaken the effects of PR on resistance to extinction. If CR training strengthens one or two dominant analyzers, then, when PR training begins, the analyzers detecting other cues will be comparatively weak and, hence, it will be difficult for the animal to attach the response to features of the situation not detected during CR training. It is true that if sufficient PR trials were given after CR, the dominant analyzers would be weakened, since they would now fail to predict reinforcement consistently; this would allow response attachments to be formed to less dominant analyzers. However, we can at least predict that animals given CR trials before PR will extinguish faster than animals given CR after PR. This prediction is again to some extent paradoxical, since it states that the further removed in time the CR training is from extinction the more CR training will reduce the PRE. Most other theories would make the opposite prediction. For example, on the discrimination hypothesis (see page 354), it would be expected that giving PR training after CR would make it difficult for an animal to detect the change in reinforcement schedules at the beginning of extinction, since it then moves from 50 to 0% reinforcement, whereas if PR is given before CR, the animal changes from 100 to 0% reinforcement at the start of extinction, and this change should be easier to detect, leading to faster extinction.

In order to test the prediction, Sutherland, Mackintosh, and Wolfe (1965) trained five groups of rats on a runway with different schedules of PR and CR. The schedules for the different groups were as follows: P, 60 PR trials; P–C, 60 PR trials followed by 100 CR trials; C–P, 100 CR trials followed by 60 PR trials; C-60, 60 CR trials; C-160, 160 CR trials. After being trained on these schedules, each animal was extinguished to a criterion of three successive trials on which it failed to reach the goal box within 2 min. The results are shown in Table 10.II. It is clear that in comparison with the extinction scores of the group given no CR training, giving CR after PR had very little effect

TABLE 10.II

**Trials to Extinction**

| Group | Mean trials to extinction | *SD* |
|---|---|---|
| P | 82 | 41 |
| P–C | 77 | 36 |
| C–P | 33 | 17 |
| C–60 | 17 | 6 |
| C–160 | 17 | 7 |

on the number of trials to extinction, whereas, giving CR before PR led to a marked reduction in resistance to extinction. A further point of interest is that the difference in number of trials to extinction between Group P given PR training only and groups C-60 and C-160 given CR training only is abnormally large—82 trials as against 17 trials. The corresponding figures in Theios' experiment were 61 and 31 trials. The reason that we obtained such a large PRE is almost certainly that none of our groups received any "pre-training" trials. In most experiments on PR, the experimenter gives all groups "pretraining trials"; thus Theios gave all his groups 30 pretraining trials on which animals were rewarded 100% of the time. Such pretraining trials are, in fact, CR training so that many experiments that claim to compare the effects of PR with the effects of CR are really comparing the effects of CR followed by PR with the effects of CR training throughout. Although some PRE may be obtained with this design (compare our Groups C–P and C-60), the size of the PRE will be less than if no pretraining trials on a CR schedule are given.

The finding that animals are more resistant to extinction after training with PR followed by CR than after training with CR followed by PR had in fact already been obtained by Jenkins (1962) in pigeons. He found that birds given three sessions of PR followed by ten of CR averaged 197 trials to meet a criterion of extinction, whereas when ten sessions of CR preceded three of PR, the animals extinguished in 118 trials (as compared to 100 with 13 sessions of CR and no PR training).

The finding that giving CR training before PR reduces resistance to extinction was confirmed in a parametric study by Hothersall (1966). He trained rats in a free-operant situation and found that giving CR before PR greatly weakened the PRE; e.g., a group given 50 reinforcements under CR and then given 50 reinforcements under PR made 292 responses in extinction, whereas a group given no prior CR training but receiving the same number of reinforcements under PR extinguished after 581 responses. Similar results were obtained when 100 or 200 PR trials were given and the effect of prior CR training on number of responses to extinction was studied.

A discordant note has, however, been sounded by Theios and McGinnis (1967). They point out that in our own experiment terminal running speed was faster in the P–C group than in the C–P group, and that this fact alone could account for the former group being slower to extinguish. They repeated part of our experiment and found that if extinction is measured by plotting the ratio of running speed in extinction to speed at the end of acquisition then extinction actually proceeds at a faster rate for the P–C group than for the C–P group. However, Theios and McGinnis used a low criterion of extinction (three successive failures to run in 30 sec) and they only gave a total of 40 extinction trials, whereas our Group P–C required an average of 77 trials to meet our criterion of extinction. Our own results cannot be explained by differences in rates of extinction, since running times after 20 extinction trials were the same for Groups P–C and C–P but diverged from there on. Our hypothesis, in fact, predicts that the main effect of giving CR before PR will be late in extinction: the CR training will strengthen one or two dominant analyzers so that at first extinction will be controlled by the same factors in both Groups C–P and P–C; a difference between the two groups should only be manifested later in extinction when the dominant analyzers are switched off. At that stage Group P–C should continue running because the response has been attached to other analyzers, whereas Group C–P will stop running since the response is only attached to the dominant analyzers. Theios and McGinnis may have failed to give enough extinction trials for this difference to show.

Two further studies should be noted briefly. Perry and Moore (1965) using the conditioned eye-blink response in human subjects found that a PRE was sustained through a block of CR trials given after the PR training. Fitzgerald, Vardaris, and Teyler (1966), however, failed to find any PRE under these conditions; they used classically conditioned heart rate in dogs. These two studies will be further referred to in our discussion of the PRE in classical conditioning.

### D. (E-3) Effects of Varying Number of Relevant Cues

The predictions discussed so far have all provided rather indirect tests of our explanation of the PRE. For example, although we have established that animals do learn about more cues under PR than under CR (E-1), this does not prove that the greater resistance to extinction of PR animals is actually caused by their having learned more cues. The most direct test of our explanation would be to show that a PRE occurs only when there are several possible cues in the environment. If the situation is such that behavior can only be controlled by one cue, then PR animals cannot learn about more cues than CR animals and there should be no PRE. For reasons discussed in Chapter 2, it is extremely difficult to be sure that from the animal's viewpoint only one cue is relevant. However, a weaker form of the same prediction can be tested. One might expect that the PRE would be stronger when many cues are present in an environment than when very few cues are present.

In our first attempt to test this prediction we used rats in two different runways: one runway was made as simple as possible, the other was made very complex. The simple alley was placed in a dark room and had a cheesecloth roof and dim interior lighting so that no extramaze cues were present; it was painted plain gray throughout. The complex alley had no roof so that extramaze cues were present, it also contained different visual and tactile cues at different points along the alley, and animals' arrival at two separate points on the runway produced a light and buzzer, respectively. One PR group and one CR group were given 100 training trials in each runway, and our prediction was that the PRE would be larger in the complex runway than in the simple.

At the end of training there were no significant differences in starting or running times between the groups, though the PR animals tended to have slower starting times than those receiving CR. Each rat was run in extinction until it had failed to reach the goal box in less than 3 min on three trials (not necessarily consecutive). By this criterion, the number of trials to extinction was as follows: simple runway CR, 41.4; simple runway PR, 41.5; complex runway CR, 43.8; complex runway PR, 63.4. The difference in trials to extinction between the complex runway Group PR and each of the other groups was significant at the .02 level of confidence. The results confirm our prediction to some extent, since the complex runway groups show a larger PRE than the simple runway groups. However, we are unable to account for the simple runway groups showing no PRE at all, since, even in our simple runway, several different cues must have been present.

There are other findings in the literature that are similar to the result just described. McNamara and Wike (1958) trained rats in a runway and then extinguished them. One of their groups was trained and tested in exactly the same runway throughout. A second group was trained in a runway that provided variable stimulation during training; on different trials, the experimenter inserted hurdles, and altered the color of the runway and the goal box. This group was then extinguished in the same constant runway as the first. The group that learned under changing stimulus conditions took longer to extinguish than the group that learned under constant stimulus conditions. We might expect that animals trained under constantly changing stimulus conditions would attach the response to the outputs from more different analyzers than animals trained under constant conditions and would, therefore, take longer to extinguish. Where stimulus conditions are constantly changing, it will be more difficult for the outputs from any one analyzer to predict reward, and so, several different analyzers would be brought into play. Brown and Bass (1958) investigated the effect on extinction of manipulating variability of the environment during both training and extinction. They found that animals extinguished more slowly when they were both trained and extinguished in a variable environment than when the environment was constant throughout. However, varying the environment only in training had no effect on resistance to extinction. Unfortunately, in both experiments both stimuli and responses were changed for the variable groups, since hurdles were included in the runway, thus altering the response the animal had to make. A further experiment (Wike, Kintsch, & Gutekinst, 1959) found no effect when environmental variability was changed merely by including hurdles, and this suggests that the results of McNamara and Wike may have been due to stimulus changes rather than to response changes.

It is difficult for the experimenter to be sure what cues are controlling the animals' behavior in a straightway. A better method of testing the same prediction would appear to be the use of an explicit discrimination situation in which it is easier to vary the number of possible cues that can control the animals' responses. McFarland and McGonigle (1967) report one such experiment. They trained rats in a Grice box. All animals were trained first on a black–white discrimination problem until they met a criterion of 18 out of 20 successive correct responses. At the next stage, half the animals were given additional training trials on the black–white discrimination and half were given the same number of trials on a new problem, horizontal versus vertical. Within each group, half the animals received CR training at both stages and the other half were given PR training (on a 50 : 0 schedule). At the final stage, all subjects were extinguished; rats trained on black–white throughout were extinguished on their training stimuli but rats trained with two different cues were extinguished with both cues (black–white and horizontal–vertical) present simultaneously. In terms of trials to a latency criterion of extinction there was a large difference between the CR and PR groups, but there was no difference between the PR group trained with one cue and that trained with two cues. Nor was there any difference between these two groups in terms of the percent correct responses made during extinction. McFarland and McGonigle argue that, on our own model, we should predict a difference in the percentage of correct responses made during extinction between the two PR groups: the group that had learned both cues should make more correct choices, since more analyzers should be controlling the response. The design used by McFarland and McGonigle is, however, subject to two criticisms: since the two-cue groups were extinguished with stimuli they had never seen before (both cues present instead of only one at a time) they should suffer some generalization decrement and this would militate against the effects on extinction of the two-cue training. Second, it may be that the difference between one and two cues is not sufficient for an effect to show.

Sutherland and Andelman (unpublished) performed a similar experiment with rats on a jumping stand. Rats were trained with either only one cue relevant or with six cues simultaneously present and relevant. All animals received 200 training trials and half of each group was trained with CR, half with PR (60% reward for correct jumps, no reward for incorrect jumps). Finally, all animals were extinguished with the stimuli upon which they had been trained. No interaction was found between the number of cues present and the size of the PRE, whether measured by latencies during extinction or by percentage of correct choices.

This design may, however, not be such a good test of our own theory as at first appears. The difficulty is that although the six-cue PR Group will learn about more of the relevant cues than the CR group, they should also depress the position analyzer *less* than the CR group. The CR group can obtain 100% success by using a relevant analyzer, and 50% success using the position analyzer, whereas the PR group can obtain 60% success with any of the relevant analyzers and 30% success using the position analyzer. That the position analyzer does, in fact, remain stronger in the PR than in the CR groups with this design is adequately attested by the results of Sutherland (1966b) set out under (E-1). Now if the main reason for incorrect jumps being made during extinction is that other analyzers are weakened below the level of the position analyzer, we cannot safely predict differences in performance between the one-cue and six-cue PR groups, since it may take just as long to weaken six weak analyzers below the level of the originally dominant position analyzer as to weaken one strong analyzer to this level. The point can be put in a slightly different way: in a runway situation, both PR and CR animals will attach the running response to the dominant analyzer; PR subjects will continue running longer because when this analyzer is weakened, the response will still be evoked by other cues. In the jumping stand, however, with visual cues relevant, neither group will attach the response to the dominant analyzer, and the number of correct responses made will depend upon how long it takes to weaken other analyzers below the level of the dominant analyzer. It is clear that in order to test our own theory in this sort of situation, it would be necessary to make sure that the dominant cues are included among the relevant cues.

## E. Conclusions

Our own theory, as we shall see below, is the only one that predicts the results given under (E-1) to (E-3). We shall, however, show that there are some findings on the PRE that our theory cannot explain. It seems to us likely that the effect is "overdetermined" and that there are really several different factors that contribute towards producing it. We shall consider five other theories that have been proposed to account for the effect; whether right or wrong, these theories have led to a wealth of interesting experiments. We shall try to show that two of these theories (the discrimination hypothesis and frustration theory) must be at least partially correct, and that all the known phenomena of PR can be explained by our own theory taken in conjunction with these two theories.

## III. The Response Unit Hypothesis

### A. Introduction

Mowrer and Jones (1945) suggested that under PR the response unit being reinforced might be not just one response, but a chain of responses; Skinner (1934) made a similar suggestion. Thus, for a rat reinforced each time it presses a lever, the response learned would be a single press, whereas for a rat reinforced on average once every two presses, the response learned would be a run of presses. During extinction, the CR animals must extinguish a response of one press, PR animals must extinguish a response consisting of several presses in series. Animals that have learned a reponse unit of several presses will make more presses in extinction than animals that have learned a response unit of one press.

### B. (E-4) Response Units

Mowrer and Jones showed that the lower the ratio of reinforcements to bar presses during training, the more responses were emitted to a criterion of extinction. However, when the results were replotted in response units (on the assumption that a response unit is one press for animals reinforced every response, four presses for animals reinforced every fourth response and so on), animals under PR actually extinguished faster; animals reinforced for every fourth response did not emit four times as many responses in extinction.

Mowrer and Jones' suggestion is plausible in the context of a situation in which an animal can emit continuous responses (e.g., under free-operant training in a Skinner box). It is not plausible in a discrete-trials situation such as the runway, particularly as it is known that a PRE can be obtained with intertrial intervals of 24 hr (Weinstock, 1954); it is hard to believe that two or more responses each separated by a 24-hr interval could come to constitute one response unit for the animal. Nevertheless, it is almost certain that response chaining does occur in a Skinner box. Thus, Denny, Wells, and

Maatsch (1957) trained animals to bar-press for food delivered after every fifth bar-press; when the animals were extinguished, they visited the food cup only after an average of 3.4 bar-presses, not after every press.

### C. Conclusions

It is, therefore, likely that response chaining is at least one factor in determining PRE in a free-responding situation. A simple test of this assumption could be made. If PRE in a free-responding situation is partly determined by the response unit learned, animals trained on a given ratio should be more resistant to extinction if they are allowed to emit continuous presses than animals trained on the same ratio but only allowed to emit one response at a time (e.g., if the bar is withdrawn for an interval after each response). Moreover, animals allowed to emit continuous responses during training should extinguish more quickly if they are only allowed to make one response at a time during extinction than if they are allowed to emit continuous responses. So far as we know, systematic experiments of this type have not been undertaken.

The response unit hypothesis can clearly not explain any of our own results, nor can it explain many of the results we shall quote below. Although, therefore, the learning of response units may be a factor in a free-operant situation, it is unlikely to contribute to the PRE with a discrete-trials paradigm, and we shall not discuss the hypothesis further.

## IV. The Discrimination Hypothesis and Generalization Decrement

### A. Introduction

Mowrer and Jones (1945) suggested an alternative hypothesis to explain the PRE. Extinction may be speeded up by any factor that makes it easy for the animal to discriminate extinction conditions from training conditions, and retarded by any factor that makes such a discrimination difficult. On being switched from CR to extinction there is a big difference between the schedules, and this difference should be easy for the animal to discriminate. A switch from PR to extinction produces a smaller difference between schedules and it will take the animal longer to discriminate this difference.

It has also been proposed that the PRE can be explained in terms of generalization decrement. The suggestion was inspired by Hull, but was first published by Sheffield (1949, 1950). Sheffield suggested that stimulus traces of reinforcement may be carried forward from one trial to the next. Under

CR, animals will learn to attach the running response to the stimulus traces resulting from reinforcement on their previous trial (except for the first trial of each day) as well as to external stimuli. Under PR, the stimulus traces will be absent following nonreinforced trials so the animal will learn to run both in the presence and in the absence of these stimulus traces. Since stimulus traces of reinforcement are absent in extinction, PR animals will suffer less generalization decrement than animals trained under CR, and hence continue to run longer in extinction. The theory is not very precise about what constitutes a stimulus trace; it could be either a stimulus due to the aftereffects of eating (like food particles still in the mouth or gullet) actually affecting receptors at the start of each trial or it could be a fading stimulus trace surviving from the stimuli associated with eating in the goal box on the previous trial but not mediated by current stimulation of receptors at the start of the next trial. Hull, following Pavlov, was forced to postulate such fading traces to account for trace conditioning.

In its most general form, the generalization decrement hypothesis becomes practically identical to the discrimination hypothesis. The one hypothesis states that the PRE occurs because there is less *generalization decrement* between training and extinction conditions after PR than after CR, whereas the other states that animals can *discriminate* between acquisition and extinction conditions less readily after PR training than after CR. The most fully elaborated version of these two hypotheses is that developed by Capaldi, and we shall consider his theory in detail below, together with some of the tests to which it has led. First, however, we will dispose of the limited form of the generalization decrement hypothesis that postulates that the variable producing the PRE is whether or not the response has been conditioned to stimuli arising from the actual presence of food in the mouth or gullet.

## B. (E-5) PRE WITH LONG INTERTRIAL INTERVALS

The clearest prediction from this theory is that no PRE should occur if trials are separated by long intervals, since no trace of previous reinforcement will then survive until the next trial. Sheffield (1949) found no PRE with an intertrial interval of 15 min. However, Weinstock (1954) obtained a clear-cut effect with intertrial intervals of 24 hr, and Wilson, Weiss, and Amsel (1955) also found that the PRE was little influenced by trial separation. Several more recent investigations have also obtained a large PRE with 24-hr intertrial intervals (e.g., Wagner, 1961). The reason for this discrepancy in experimental findings is not known, but it is safer to trust the results showing a positive effect. Since these results show that under at least some conditions a PRE can be obtained with large intertrial intervals, the form of the generalization decrement hypothesis that attributes the effect to ongoing stimulation of receptors produced by the aftereffects of food reinforcement must be rejected. Capaldi and his associates have undertaken many other experiments that demonstrate that this limited form of the generalization decrement hypothesis is inadequate. They have shown that speed of extinction depends on many different aspects of similarity between acquisition and extinction conditions. These experiments will be reviewed below.

### C. Capaldi's Version of Generalization Decrement

We now turn to a statement of Capaldi's (1966, 1967) version of the generalization decrement hypothesis. He suggests that the memory of a recent event associated with the training situation can act as a stimulus. In particular, he suggests that the memory of an unrewarded trial can act as a stimulus and that a different stimulus is provided by different numbers of previously nonreinforced trials. We shall refer to stimuli produced by memories of previous events as "memory stimuli"; it is in fact somewhat Pickwickian to call such memories stimuli at all: the temptation to do so arises from the S–R tradition in which it was thought that all learning consists of attaching responses to stimuli. If it is postulated that responses can be attached to the memory of a stimulus, this is a large departure from S–R theory and should be acknowledged as such. Capaldi himself, following Sheffield, refers to memory stimuli as "stimulus after-effects."

He suggests that "memory stimuli" can be ordered along a dimension of similarity between the memory stimuli arising from a trial with a very large reinforcement and those arising from a long sequence of trials with zero reinforcement. The smaller the reinforcement, the nearer the memory stimulus lies to that for nonreinforcement. One can also order along this dimension the memory stimuli resulting from delay of reward; the longer the delay of reward, the more similar the memory stimuli present on the next trial is to the memory stimulus for nonreward. As we shall see below, there are various other variables (e.g., goal-box confinement time on nonrewarded trials), which affect the position of a memory stimulus on this dimension. The dimension is illustrated in Figure 10.2; the notation used in that figure (e.g., R for a reinforced trial, N for a nonreinforced trial) will be used throughout the remainder of this chapter. The memory stimuli represented along the abscissa as one goes from left to right become increasingly similar to the memory stimuli produced by large numbers of consecutive nonreinforcements. Although small rewards must always lie to the right of large, and delayed reward to the right of no delay, the relative positions of RSI and RLD can clearly only be established from empirical data. It is important to bear in mind that a single reinforced trial replaces the memory stimuli for nonreinforced trials with the memory stimulus of a reinforced trial of a particular type. It will be noticed that in the figure the spacing between $N_{50}$ and $N_{10}$ is the same as between $N_{10}$ and $N_1$. This follows a hypothesis of Capaldi that increasing the number of nonreinforced trials has decreasingly small effects in changing the memory stimulus as the number of preceding nonreinforced trials increases; the same applies to runs of delayed trials, or trials with small rewards.

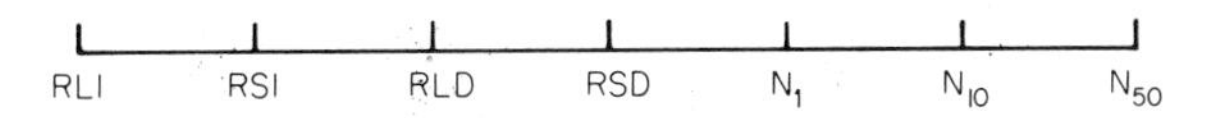


**Fig. 10.2.** Dimensionality of stimulus aftereffects. R: Reinforced trial; N: nonreinforced trial; L: large reinforcement; S: small reinforcement; I: immediate reinforcement; D: delayed reinforcement. (Numbers indicate number of nonreinforced trials in a row, i.e., " N-length.")

It should now be clear how Capaldi's theory explains the basic PRE. If, during training, the animal experiences a run of nonreinforced trials and then receives a reinforced trial, the response of running will be conditioned on the reinforced trial to the memory stimulus produced by the run of nonreinforced trials. This memory stimulus will be present in extinction and, hence, the PR animal will continue to respond longer in extinction than the CR animal that has never learned to run in the presence of a memory stimulus produced by a run of nonreinforced trials. The main predictions made by Capaldi's theory have to do with the effects of giving different sequences of reinforced and nonreinforced trials during training, the effect of varying goal-box confinement times, and goal-box similarity, the effects of interspersing R and N trials with placement trials on which the animal is placed in the goal box and fed without actually having made the instrumental reponse, the effects of varying reward size, and the effects of varying the delay of reinforcement. Experiments on the first three topics will be reviewed in this section; since other theories make specific predictions about the remaining variables, the experiments on these other topics will be reviewed later in the chapter.

### D. (E-6) Sequence of Reinforced and Nonreinforced Trials

Both the discrimination hypothesis, in general, and Capaldi's theory, in particular, predict that where the number of reinforcements and nonreinforcements given in acquisition remains the same, animals that have experienced long runs of nonreinforced trials during training will extinguish more slowly than animals that have only received short runs of nonreinforced trials. This follows from Capaldi's theory because the memory stimulus provided by long runs of nonreinforced trials will be more similar to the memory stimulus in extinction (at least after the first few trials of extinction) than will the memory stimulus provided by short runs of reinforced trials. This prediction has been tested and verified by Gonzalez and Bitterman (1964). Bloom and Capaldi (1961), Capaldi (1958), Spear, Hill, and O'Sullivan (1965) and Tyler, Wortz, and Bitterman (1953) have shown that, provided enough trials are given, regular alternation of N and R trials (minimizing N-length) leads to less resistance to extinction than giving 50% random partial reinforcement. Here and below we use "N-length" to mean a sequence of nonreinforced trials. When reinforcement is random, a high percentage of reinforced trials will result in shorter runs of nonreinforced trials than will a low percentage of reinforcement; hence, the hypothesis predicts that with random reinforcement, animals should be more resistant to extinction after PR training with low percentages of reward than with high. This finding has been obtained by Weinstock (1954, 1958). In a further experiment Capaldi and Stanley (1965) demonstrated that it is not percentage of reinforcement *per se* that influences resistance to extinction: they showed that animals with a high percentage of reinforcement but with sequences involving long runs of nonreinforced trials extinguish slower than animals trained with a low percentage of reinforcement with sequences containing only short runs of nonreinforced trials.

The above deductions on the effects of varying N-length only follow from Capaldi's model if a reasonably large number of training trials is given. Consider the two sequences of trials, NRNRNR and RNNNRR. The second sequence contains one N-length of 3, the first, three N-lengths of 1. It is clear that the memory stimulus $N_1$ will acquire more habit strength than will the memory stimulus $N_3$ since the former is reinforced three times, the latter only once. This extra habit strength could outweigh the fact that $N_3$ lies nearer to the memory stimulus prevailing in extinction than does $N_1$. Hence we might expect that with a very small number of trials resistance to extinction will be determined more by the number of N-lengths followed by a reinforced trial than by the length of reinforced N-lengths. This prediction has again been confirmed in a number of experiments. For example, Capaldi and Hart (1962) found that, when only a few trials were given, alternating N and R trials lead to greater resistance to extinction than giving a random sequence of 50% N and 50% R trials: alternation of reinforcement of course maximizes the number of N-lengths followed by a reinforced trial but minimizes the size of N-length. Spivey (1967) showed that with small numbers of trials it was the number of N–R transitions that determined resistance to extinction rather than the percentage of reinforced trials. Spivey and Hess (1968) and Spivey, Hess, and Black (1968) gave rats either the daily sequence NNRR or the sequence RRNN for three days. The first sequence involves three N–R transitions (one a day), the latter does not contain any N–R transitions; as predicted by Capaldi's theory the sequence containing N–R transitions produced a PRE, whereas the other sequence did not.

There is one further phenomenon of interest connected with the sequence of N and R trials. When reward and nonreward are alternated, rats come to run fast on rewarded trials and slowly on nonrewarded trials (Capaldi, 1958). Capaldi explains the development of slow running on nonrewarded trials by assuming that inhibition is conditioned to the memory stimulus of an R trial. Capaldi and Stanley (1963) found that differential running times on N and R trials developed on an alternating trial sequence with intertrial intervals of 20 min. This is a very remarkable finding, since it means that the rats' behavior is being controlled by an event that happened 20 min previously and that it is discriminated from an event happening 20 min before that. That the behavior was not based merely on temporal discrimination is shown by two further experiments (Capaldi, Veatch, & Stefaniak, 1966; Bowen & Strickert, 1966) in which reward and nonreward were again alternated but the intertrial interval was varied from one trial to another; rats again gave evidence of learning the alternating pattern of reinforcement.

Capaldi has since claimed (Capaldi & Spivey, 1964; Capaldi & Lynch, 1966) that rats can actually learn to respond differentially on rewarded and nonrewarded trials with a 24 hr intertrial interval; this finding has, however, been disputed by Surridge and Amsel (1965a, 1965b) who argue that Capaldi's results may have been due to the unintended presence of environmental cues signalling the presence versus absence of reward. Capaldi and Lynch (1966) in turn argued that Amsel and Surridge's failure to replicate their results on patterning with a 24-hr intertrial interval was due to their using a short goal-box confinement time on nonrewarded trials. On Capaldi's theory, the memory stimulus for a nonrewarded trial with a short confinement time would be more similar to the memory stimulus for an R trial than would the stimulus from an N trial with long confinement; hence, with short confinement, the memory stimuli might not be sufficiently different for inhibition to be differentially conditioned to the memory stimulus of an R trial. This is not a particularly plausible suggestion since a clear-cut PRE develops with short confinement times and, hence, on Capaldi's own theory, the memory stimulus of N with short confinement must be discriminably different from the memory stimulus for R. Moreover, Surridge and Amsel (1968) failed to obtain a patterning effect with a 24-hr intertrial interval using a long goal-box confinement time and taking care to eliminate situational cues that might inform the animal whether the current trial was rewarded or nonrewarded. Katz, Woods, and Carrithers (1966) failed to obtain a patterning effect of running speeds with alternating reinforcement with a 20-min trial interval.

Using short intertrial intervals, Harris and Thomas (1966) have shown that patterning is not solely due to situational cues; they ran rats in a runway with alternating reinforcement and allowed the rat to circle back from goal box to start without being handled and found that patterned running developed.

McHose (1967) and Ludvigson and Sytsma (1967) (compare also McHose & Ludvigson, 1966 and Spear & Spitzner, 1966) have suggested that some patterning effects may be due to animals being run in squads. For example, if animals are run on an RNRN schedule and if all animals complete the first daily trial in turn and then commence the second daily trial, on rewarded trials all but the first animal will be running immediately after another animal that has experienced reward, whereas on nonrewarded trials all but the first animal will be running after an animal that has experienced nonreward. McHose and Ludvigson argue that rats may leave a different odor in the goal box depending on whether they are rewarded or not. When animals are run in squads, the odor left by the last rat could serve as a signal to the next rat to inform it whether or not food is present. Perhaps the most convincing demonstration of this effect is the experiment performed by Ludvigson and Sytsma (1967) reported in a paper entitled "The Sweet Smell of Success." They ran rats for eight trials a day on a double alternation schedule (RRNNRRNN). One group of animals was run in a squad in such a way that (except for the first animal in the squad) an R trial was always administered following an R trial to the preceding subject, and similarly an N trial always followed an N trial to the preceding subject. In the other group, different animals received different orders of trials (e.g., some received NNRRNNRR and others RNNRRNNR) and this meant that, in this group, trial outcome could not be predicted by the outcome of the trial given to the last animal run. The first group came to run significantly faster on R trials than on N, whereas the second group showed no differences in running speed on R and N trials. McHose (1967) ran rats on an RNN pattern. He found that if rats were run in squads, they came to run fast–slow–slow on the three successive trials. If one rat completed its three trials before the next animal was run they developed a fast–slow–fast running pattern. With this procedure, animals experience reward only on trials after the previous animal has experienced nonreward; they never experience reward after a previously rewarded trial; hence, they develop a fast–slow–fast pattern. Although both experiments can be criticized (Ludvigson's because individual subjects in his second group experienced different reinforcement patterns from those used for the first group, and McHose's because he failed to equate intertrial intervals for the two groups), the results at least establish a *prima facie* case for the importance of odors deposited by the previous animal in determining patterning effects.

In the light of the evidence, it is unsafe to conclude that patterning with long intertrial intervals is based on internal memory cues from the preceding trial rather than on external cues present on the current trial. Hence, such experiments cannot be used to support Capaldi's position that the instrumental response is conditioned to memory cues even with long intertrial intervals.

The findings reported at the beginning of this section do lend strong support to the discrimination hypothesis. Neither our own theory nor the two theories to be reviewed below can adequately account for why with sufficient training the PRE should be increased as N-length increases, nor for the fact that, with a small number of trials, it is the number of N–R transitions that determines the size of the PRE. Moreover, as we shall see, Capaldi's theory predicts similar sequential effects of varying reward size, amount of delay, and goal-box confinement time and the predictions from his version of the discrimination hypothesis are also borne out in these cases.

### E. (E-7) Intertrial Feeding

It will be remembered that one assumption made by Capaldi (1967) is that when an animal feeds in the goal box, the memory cue for the previous sequence of nonreinforced trials is replaced by the memory cue of R. An ingenious prediction follows from this assumption. If, in the intertrial interval following each N trial, animals are placed in the goal box and fed, then the instrumental response will never be rewarded in the presence of the memory stimulus of a nonreinforced trial or trials. Although nonreinforced trials are succeeded by feeding, the feeding occurs without the instrumental response having been made and, therefore, the instrumental response is not rewarded in the presence of the memory stimulus of an N trial. The instrumental response is only rewarded in the presence of the memory stimulus of R. It follows that when reinforced placements are given between each N-length and the succeeding R trial, the PRE should be eliminated or substantially reduced. This prediction has been confirmed in several experiments involving relatively small numbers of trials (Capaldi, Hart, & Stanley, 1963; Capaldi & Spivey, 1963; Katz, 1957; McCain, 1966; Spence, Platt, & Matsumoto, 1965). Moreover, Capaldi (1964), Spivey, Hess, and Aponte (1968) and Spivey, Hess, and Klemic (1968) have used intertrial reinforced placements to vary the N-length that precedes R without an intervening feeding. For example, if the sequence RNNRNN repeating is given, there is an effective N-length of two if rewarded placements are given after the R trials, of one if they are given after the N trial following an R trial and of none if given between the second N and the succeeding R trials. As predicted by Capaldi, the longer the effective N-length, the larger is the PRE.

When a large number of training trials is given, it has been found that interposing reinforced placements between N and R trials does not reduce the PRE (Black & Spence, 1965; Spence *et al.*, 1965). Capaldi (1967) suggests that this is because the memory stimulus from a placement trial may eventually come to be discriminated from the memory stimulus of a rewarded trial; it is not very clear why, even if this does happen, the memory stimulus for N is not replaced by the memory stimulus for a rewarded placement. Perhaps the memory stimulus for a placement trial is eventually not evoked by the stimuli at the start of the apparatus since the animal never experiences the stimuli in the runway in connection with placement.[1] However this may be, some support is lent to Capaldi's notion by the finding that when animals receive rewarded placement trials in a goal box that differs from the goal box in which the instrumental response is reinforced, then the placements produce a smaller reduction in the PRE than do placements in the same goal box (Capaldi & Spivey, 1963, 1964).

### F. (E-8) Similarity of Acquisition and Extinction Conditions

It is well known that where stimulus conditions are changed between acquisition and extinction, extinction is faster due to generalization decrement. Hulicka, Capehart, and Viney (1960) (cf. Melching, 1954) demonstrated this effect directly by conditioning animals to a compound stimulus that had five different components and showing that the fewer the number of components present in extinction, the faster extinction occurred. What the discrimination hypothesis, in general, and Capaldi, in particular, have added to the straightforward notion of generalization decrement is that a memory stimulus can control responses and that the memory stimuli present in extinction may differ from those present in acquisition. So far, we have dealt only with the memory stimuli produced by reinforced and nonreinforced trials and by intertrial reinforced placements. It is possible, however, that animals' behavior is controlled by memory stimuli for other aspects of the situation that do not form part of the physical stimulus impinging on the animals' receptors at the time the instrumental response is made. We review here some experiments undertaken to test this possibility.

At the end of the previous section, we saw that the effects of reinforced placing depend upon the similarity of the goal box in which the animal is placed to the goal box in which the instrumental response is reinforced. This suggests that animals may learn to run not merely in the presence of the memory stimulus for N but in the presence of the memory stimulus for N occurring in a particular goal box. If this is correct, we might expect that if animals are run under PR and are rewarded in one goal box, and receive nonreward in a discriminably different goal box then extinction might actually take longer if they are run in extinction to the goal box associated with nonreward than if they are run to that associated with reward. Running has been conditioned to the memory stimulus of reward in Goal Box A, and non-

[1] The Hullian tradition does not allow anything to be learned other than a S–R bond, and stimulus aftereffects are stimuli not responses. Since the idea that all learning is the learning of S–R bonds is false, there seems no reason why memories should not themselves be conditioned to an experimental situation and serve as a " stimulus " to control behavior only in that situation.

reward in Goal Box B; it has not been conditioned to the memory stimulus of nonreward in Goal Box A. Although Capaldi himself does not seem to have made exactly this deduction, it appears to follow naturally from his theory. Bitterman, Fedderson, and Tyler (1953) and Elam, Tyler, and Bitterman (1954) performed experiments of this type in order to test the discrimination hypothesis and obtained the predicted result.

Unfortunately, the result has proved difficult to repeat. Lawrence and Festinger (1962, p. 106) report two attempts at replication. In both of their experiments, animals were equally resistant to extinction whether they were extinguished with the original positive goal box present or with the original negative goal box. Similar findings have been obtained by Freides (1957), Marx (1960), and Paige and McNamara (1963). However, even the finding that rate of extinction is not influenced by which goal box is present in extinction is a striking result, since the positive goal box should acquire the properties of a secondary reinforcer, and, on these grounds, we might expect animals to be slower to extinguish when run to the goal box that had been positive during training than when run to the negative goal box. In a more recent study, Black (1965) trained animals to run using both different alleys and different goal boxes: on half the training trials Alley A and Goal Box A were always used and rats were always reinforced for running; on the remaining training trials Alley B and Goal Box B were used and animals were not reinforced for running. Black then extinguished different groups of animals on all the four possible combinations of alley and goal box and found that whether the positive or negative alley was used, animals extinguished more slowly with the positive goal box present than with the negative. These results appear to contradict those of Bitterman *et al.*, since, with alley held constant, animals were more resistant to extinction with the positive goal box present than with the negative. This is the result that would be predicted if extinction were determined by the secondary reinforcing value of the goal box. However, it must be remembered that Black's rats could predict during training which goal box would be present whereas those of Bitterman *et al.* could not. Black also found that animals extinguished on the positive alley and goal box were less resistant to extinction than those extinguished on the negative alley and positive goal box and it is hard for any theory to predict this result.

In summary, the evidence suggests that when animals are trained with a goal box in which they are always rewarded and one in which they are never rewarded, and then tested with one or other goal box present, they take as long to extinguish with the negative goal box as with the positive. This finding is in line with the discrimination hypothesis if we assume that the secondary reinforcing power of the positive goal box is roughly canceled by the fact that there is a greater change in the situation when the positive goal box is used without food than when the negative goal box is present without food.

Capaldi and Minkoff (1966) have suggested that there may be a memory stimulus for the time interval between trials. They performed the following experiment. Rats were trained on a runway with the following daily sequences of trials: RNRNR. One group received a 30-sec interval between N and R trials and 8 min between R and N trials; the time intervals were reversed for a second group. After 30 training trials, both groups were extinguished; half of the animals in each group were extinguished with a 30-sec intertrial interval, and half with an 8-min intertrial interval. If there is a memory stimulus for a time interval, then, since the memory stimulus in extinction is for an N trial, it would be expected that animals would be more resistant to extinction when extinguished with an intertrial interval that was the same as the N–R intertrial interval than when extinguished with one that was the same as the R–N interval. This prediction was confirmed.

Capaldi (1967) has also demonstrated that animals are initially more resistant to extinction when extinguished with the same goal-box confinement time as used in acquisition (60 or 10 sec) than when a different goal-box confinement time is used. Late in extinction, the animals shifted to 60-sec confinement ran more slowly than animals shifted to 10-sec confinement regardless of the confinement time used in acquisition. Bloom (reported in Capaldi, 1967) performed a further experiment on the effects of the similarity of goal-box confinement times during training and acquisition. He gave one group of animals the following daily sequence of trials:

N60, R60, N10, R10,

where the numbers indicate the number of seconds spent in the goal box. A second group received the sequence,

N60, R10, N10, R60.

Both groups were then divided into two and extinguished either with 60-sec confinement, or with 10 sec. As might be expected animals extinguished faster with 60-sec confinement in extinction than with 10-sec confinement. However, Bloom argues that if goal-box confinement time is part of the memory stimulus then running should be more strongly conditioned to the N goal-box confinement time that was succeeded by an R trial with large reward than to the confinement time that was followed by a small reward. (Size of reward was determined by the time spent in the goal box on rewarded trials.) Hence, when 60-sec confinement is used in extinction, the first group should extinguish more slowly than the second, but when 10-sec confinement is used, the second group should extinguish more slowly than the first. These predictions were confirmed.

The experiments reviewed in this section provide some evidence that animals develop memory stimuli for various aspects of the situation and that rate of extinction depends to some extent on the similarity of the memory stimuli present in extinction and those that were present in training at the start of trials on which the instrumental response was reinforced. In Sections (E-10) and (E-16), we shall review additional experiments showing that rate of extinction is influenced by memory stimuli for size of reward and for delay of reinforcement.

### G. Conclusions

The discrimination hypothesis (or generalization decrement), particularly as manipulated by Capaldi, has suggested many interesting experiments. The original hypothesis was somewhat vague, but by introducing the notion of the memory of a previous event serving as a stimulus, Capaldi made the whole theory very much more specific. That the memory of a previous event can control behavior is well attested by the experiments on patterning, and we argued in Chapter 9 that it is necessary to postulate control of behavior by the memory of a previous event in order to explain such phenomena as one-trial reversals. The actual time interval over which the memory for a *single* event can be preserved and exert behavioral control in the rat is still in doubt. The experiments reviewed in Section (E-8) suggest that the memory for *repeated* events (such as goal-box confinement times or intertrial intervals) may exert behavioral control over a period of many hours. Some of these experiments should, perhaps, be treated with caution until they have been replicated. Despite this, most of the results reviewed in this section cannot readily be explained by any other current hypothesis of the PRE including our own. We conclude, therefore, that generalization decrement is certainly one of the factors at work in producing the PRE. However, generalization decrement cannot explain the results set out under (E-1)–(E-3) and, indeed, it seems to predict a greater PRE when CR training is followed by PR than when PR is followed by CR.

As we pointed out earlier, the introduction of the concept of a memory cue is a far cry from straightforward S–R theories. It would, in fact, be possible to recast Capaldi's theory in terms of expectancies without radically altering the predictions it makes (cf. Chapter 3). If it is assumed that whether or not an animal makes an instrumental response depends upon the expectancies aroused by a situation, then altering such features as goal-box confinement times will operate by disconfirming animals' expectancies rather than by changing the memory cue.

It seems likely that the extent to which a response is controlled by a memory stimulus may vary from species to species and in general lower species may be more "stimulus bound" than higher and learn responses less readily to a memory cue. If this is correct, we might expect the strength of the PRE in a species to covary with the ability of the species to learn delayed reaction tasks, alternation tasks, and various other tasks that involve a response being controlled by a recent memory. Gonzalez, Longo, and Bitterman (1961) have shown that mouthbreeders show no PRE if the numbers of training trials is equated between PR and CR animals. Gonzalez, Eskin, and Bitterman (1963) found that when the number of reinforcements is equated, mouthbreeders show only a small PRE. Sutherland and Mackintosh (unpublished) have found that in goldfish also PRE are weak. It seems clear that at least in these two species of fishes the PRE is weaker than in rats. Gonzalez *et al.* (1962) found that mouthbreeders gave no evidence of learning a pattern of alternating reinforcement and nonreinforcement and that resistance to extinction was the same under random and alternating reinforcement conditions. Moreover, Behrend, Domesick, and Bitterman (1965) have shown that goldfish do not learn to reverse more quickly if given a series of reversals in succession. Such tasks almost certainly depend on the animals' ability to control responses by a memory cue. It is known that monkeys are able to solve alternation and delayed reaction tasks readily, and also form reversal learning sets readily; unfortunately, there appear to be no systematic studies on PRE in subhuman primates. There is some evidence therefore that control of behavior by a memory cue and the size of the PRE vary concomitantly across species, and this again supports Capaldi's version of the discrimination hypothesis.

Although the extent to which behavior can be controlled by a memory stimulus is likely to vary across species, it is not the only variable at work. Gonzalez and Bitterman (1967) have, in fact, shown that some of the failures to obtain a PRE in fish mentioned above were due to animals' receiving very small rewards. The whole issue of interspecies differences will be discussed at greater length in Chapter 12.

## V. The Frustration Hypothesis

### A. Introduction

Amsel has developed an explanation of PRE in terms of the concept of frustration. The theory makes the following assumptions:

(1) Frustration acts as a negative drive in the same sort of way as pain; i.e., animals will learn to avoid and to escape from situations that have previously given rise to frustration.

(2) Since frustration is a drive, it will energize behavior; if both frustration drive and hunger are operating, responses will be more vigorous than if hunger alone is operating.

(3) Responses may be conditioned to frustration in the same sort of way that they can be conditioned to the drive stimuli associated with hunger and thirst. Amsel assumes that this conditioning occurs to response-produced stimuli associated with frustration, though the prediction to be made from the theory would be exactly the same if responses were conditioned directly to the central drive.

(4) One of the situations that gives rise to frustration is when an animal has built up an expectancy of a reinforcing event and the event fails to occur.

(5) Frustration can itself be conditioned to a stimulus immediately preceding frustration in the same way in which fear or anxiety conditioning occurs.

Although the word "frustration" has mentalistic overtones, the theory is a mechanistic one; Amsel has attempted to state the ground rules precisely, and the theory would not suffer if we substituted the phrase "factor X" for frustration. Moreover, there is considerable evidence that it is necessary to postulate an intervening process corresponding to frustration; not all this evidence can be reviewed here but good recent reviews are available (Amsel, 1962; Amsel 1967; Wagner, 1966). We shall first give some evidence in support of the general postulates of frustration theory and then see how the theory applies to PR.

The clearest demonstration of Assumptions (1) and (5) is provided by an experiment performed by Wagner (1963b) on rats. He ran rats in an alley under PR; one group of animals experienced an interrupted noise and a flashing light on trials on which no reinforcement was present in the alley. The other group never experienced these stimuli in the apparatus but were exposed to them in their home cage. Subsequently, both groups were tested in a situation in which they were exposed to the flashing light and interrupted noise in an apparatus in which these stimuli were turned off if the rats jumped a hurdle. The animals that had experienced the light and noise in association with nonreinforcement in the alley gave significantly more hurdle-jumping responses than those that had experienced these stimuli only in their home cage. It appears then that when the conditioned stimuli were paired with frustrative nonreward, frustration was learned as a response to the conditioned stimuli [Assumption (5)] and subsequently, animals learned to jump a hurdle in the presence of the conditioned stimuli in order to reduce the conditioned frustration [Assumption (1)]. Adelman and Maatsch (1956) present evidence that if a rat has learned to find food in a box, and is then allowed to run to the same box with no food present thus creating frustration, it will learn to jump out of the box onto a ledge in order to reduce the frustration evoked by the box, and this also confirms Assumption (1).

Amsel and Roussel (1952) performed an experiment to confirm Assumption (2), that frustration energizes behavior. They ran rats in a double runway with a goal box at the midpoint and a goal box at the end (see Figure 10.3). Animals received PR in Goal Box A and CR in Goal Box B. It

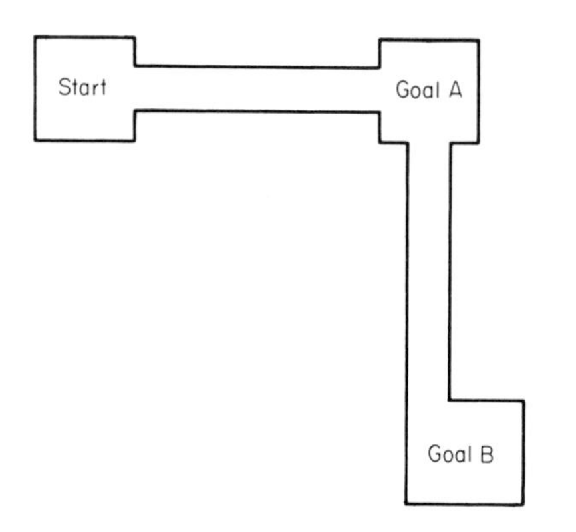


**Fig. 10.3.** Amsel–Roussel double runway.

was found that, as learning proceeded, animals came to run faster in the second section of the runway on trials on which they received no reward at A than on trials on which they were rewarded at A. Amsel and Roussel attribute this difference in running times to frustration experienced at A on nonreinforced trials energizing the running response to B.

Numerous experimental changes have been rung on this situation. Seward, Pereboom, Butler, and Jones (1957) suggested that the slower running times after feeding might be accounted for by the digestive processes interfering with the running response; they showed that prefeeding animals slowed down running times even when no frustration was involved if animals were not prefed. That this is not the whole explanation of the Amsel and Roussel finding is shown by a further experiment of Wagner (1959). He trained rats on the double runway; one group received PR at A, the other was never reinforced at A. Both groups were given CR at B. The group receiving PR at A ran at the same speed as animals receiving no reinforcement at A on trials when they were reinforced but they ran faster after nonreinforced trials. Since one group received no reinforcement at A on any trial, this difference cannot be attributed to the interfering effects of the ingestion of food. The idea that frustration may energize ongoing behavior also explains why early in extinction the learned response is usually made faster and more vigorously than at the end of acquisition (Trotter, 1956;

Bernstein, 1957) and we shall see below how it helps to explain some of the phenomena associated with PR.

All the experiments so far quoted support Assumption (4) that frustration occurs when an animal expects reward in a situation and fails to obtain it. As we shall see, several experiments suggest that the bigger the reward expected the greater the frustration if the reward does not materialize. The frustration hypothesis could also be applied to experiments showing that if animals are trained with a large reward and then shifted to a small reward they run more slowly when the small reward is first introduced than animals trained with the smaller reward from the outset of training. This result was first discovered by Crespi (1942), and has been confirmed by Zeaman (1949), Gonzalez, Gleitman, and Bitterman (1962), and Bower (1961); for a review of this subject see Black (1968). These findings can be explained if we assume that frustration is produced when the animal finds a smaller reward in the goal box than that to which it is accustomed, and this either evokes some avoidance responses in the runway that slow down running, or the frustration itself gets conditioned to runway cues and interferes with the running response.

The main weakness of frustration theory is that it does not specify very clearly under what circumstances frustration will lead to avoidance and under what circumstances it will lead to the energizing of ongoing behavior. Amsel thinks that when frustration is first elicited, it will reduce the vigor of responses immediately preceding its occurrence and increase the vigor of succeeding responses. Provided the responses preceding the occurrence of frustration continue to be made because some reward is still present in the situation, the frustration will be conditioned to stimuli preceding the point of its original occurrence: once the instrumental response is itself conditioned to the cue of frustration, frustration acting as a drive will energize responses made prior to the point of the original occurrence of frustration.

The postulated similarity between frustration drive and the fear drive has led to further experiments showing first that certain drugs have the same effect on both drives and second that avoidance gradients are steeper than approach gradients both when there is a conflict between frustration and hunger and when there is a conflict between fear and hunger. These experiments are reviewed in Wagner (1966); some of them will be further discussed later (see page 381).

We now turn to the application of frustration theory to partial reinforcement. The theory assumes that extinction is largly due to the development of frustration leading the animal to avoid the situation in which it was frustrated and hence not to make the instrumental response.

There are three possible reasons why according to frustration theory, animals should be more resistant to extinction when trained under PR than when trained under CR.

(1) During PR, frustration will become conditioned to runway cues. Since animals are rewarded on reinforcement trials in the presence of frustration, they will learn to run to the stimulus of frustration during training and, hence, will continue to run longer in extinction than CR animals; since the latter have never been rewarded for running in the presence of frustration, when frustration is conditioned to runway cues during extinction, there should be considerable generalization decrement due to changed stimulus conditions during extinction for CR animals.

(2) Since frustration is itself a response, when it occurs in the runway responses will be evoked that compete with the running response. Partially reinforced animals extinguish such competing responses during training, since reward does not depend upon their occurrence; CR animals have no opportunity to extinguish these competing responses, since no frustration is experienced during training.

(3) It may well be that PR animals do not build up such a strong expectation of reward as CR animals; hence during extinction frustration should be stronger for CR animals than for PR, and lead to faster extinction.

Of these three ways in which frustration theory can be applied to explain the PRE, Amsel himself has stressed only the first. Although generalization decrement is, therefore, involved in both Amsel's and Capaldi's theories, they do, in fact, make some different predictions, as we shall see. The three different explanations of the PRE that can be given in terms of frustration usually make the same predictions about the effect of PR on extinction, although they do make some differential predictions about performance during acquisition. For example, the first explanation would predict faster running times under PR than under CR after a sufficient number of acquisition trials, because of the energizing function of frustration once ongoing responses have been conditioned to it. This prediction is borne out by the data, and is not made by either of the other two explanations. Unless otherwise stated, in what follows, we shall be deriving predictions from the first form of the explanation.

It should be noted in passing that Spence (1960) has suggested an explanation of the PRE somewhat similar to Explanation (3). In addition, Weinstock (1954) and Estes (1959) have suggested explanations in terms of the adapting out of competing responses; these explanations have much in common with Explanation (2), although they are not couched in terms of frustration. Weinstock's explanation has not received any experimental support; e.g., Longstreth (1964) found that trials to extinction do not correlate with the number of interfering responses made such as backing up from the goal box. Logan (1960) has also suggested a hypothesis that resembles, in some respects, the frustration theory account of PR. Since these other suggestions do not usually lead to predictions different from those made by frustration theory, they will not be further discussed here.

The frustration hypothesis fits well with most of the experiments so far described, except for our own results. It can explain why the PRE occurs even with long intertrial intervals (E-5); we would expect a reduction in the PRE where reinforcement and nonreinforcement are regularly alternated during training, since once the animal has learned this pattern, little frustration will occur during training as the animal no longer expects reward on nonrewarded trials (E-6). The theory can also explain, to some extent, the puzzling result (E-8) of Bitterman *et al.* (1953). Animals that have been reinforced in one goal box and not in another may continue to run longer if extinguished with the goal box *not* associated with reinforcement during training, since they do not have a strong expectation of reward in this goal box and, hence, frustration will be slow to build up. The theory, however, cannot adequately explain the effects of manipulating other aspects of similarity between the training and extinction situation (E-8) nor can it satisfactorily explain the reduction in the PRE when a small number of training trials are used with intertrial feeding given after each run of nonreinforced trials (E-7). Frustration theory has been fruitful in making new predictions, which we will now list and discuss in the light of the experimental evidence.

### B. (E-9) Running Speed in Acquisition

During PR training in a runway situation, frustration should be conditioned to the runway stimuli: once the animal has learned to run in the presence of frustration as a result of being rewarded for so doing, the frustration in the runway should increase the running speed because of its energizing action on behavior. Several experimenters have, in fact, found that, over the first 20 or 30 trials of acquisition, PR animals run more slowly than those trained under CR, but if sufficient training is given, they eventually come to run faster (Weinstock, 1958; Goodrich, 1959; Wagner, 1961; Surridge, Boenhert, & Amsel, 1966). While there have been several failures to obtain such an effect even with extended training (e.g., Bacon, 1962; Brown & Wagner, 1964; Amsel, Rashotte, & MacKinnon, 1966), the weight of the evidence suggests that it is real enough. Several of the above-mentioned positive results were obtained when trials were given at a rate of one per day, thus probably ruling out any explanation in terms of difference in drive level, and certainly ruling out the possibility that CR animals run slowly because they are still eating at the start of a trial (cf. Seward *et al.*'s explanation of the frustration effect). A further important feature of the results is that PR animals usually only run faster than CR animals in the first half of the runway. As they approach the goal box, they actually run more slowly. Amsel (1967) has suggested that frustration may be stronger near the goal box and, whereas when it is weak it energizes ongoing behavior, when it is strong the aversive component may predominate. Robbins and Weinstock (1967) suggested that the slower running in the goal box may be due to animals being able to see whether food is present or not; when they performed an experiment in which no food was visible even on reinforced trials, they found that PR animals gave faster running times in the goal box as well as in the alley. This suggestion is rendered less plausible by the finding of Becker and Bruning (1966) that partial reinforcement may invigorate early members of an operant chain but depress the final member. There can have been no question of subjects seeing whether the water dipper was about to operate or not in this situation. Moreover, if Robbins and Weinstock's suggestion were correct, PR animals should run faster in the goal box on reinforced trials than on nonreinforced; Goodrich (1959) reported that no such effect occurred.

It is established, then, that at least in the early part of the alley, PR training can lead to faster times at asymptote than does CR training and of the theories discussed so far, only frustration theory clearly predicts this phenomenon. Our own model could account for the faster running times but only some with difficulty. According to us one of the main differences between PR and CR animals is that the former learn about many cues, the latter only about a few; this might mean that CR animals would be delayed on some trials on which the wrong analyzer was initially switched-in, since running would not start until the analyzer with the running response attached was found. This is not very convincing and we much prefer the explanation given by frustration theory.

### C. (E-10) Effects of Reward Size

Wagner (1961) and Amsel (1962) argue that the frustration hypothesis predicts that the PRE should be greater after training with large rewards. The larger the rewards given on rewarded trials, the bigger the frustration will be on nonrewarded trials; the larger the frustration is when reward is not given, the bigger will be the difference in resistance to extinction between animals that have had a chance to learn to run in the presence of frustration (PR animals) and animals that have not learned to run in the presence of frustration. Wagner (1961) and Hulse (1958) have both obtained this effect. In addition Hulse found that increasing the size of reward increased resistance to extinction in PR animals and reduced it in CR animals; the latter finding has also been obtained by Armus (1959) and Ison and Cook (1964). Gonzalez and Bitterman (1967) have shown that a PRE only appears in fish if large rewards are given. Although these results can be explained by frustration theory, the predictions cannot in fact be rigorously derived.

The theory has to assume not only that the stronger frustration conditioned to the runway cues continues to energize running for longer than weak frustration but this energizing effect will outweigh any aversive effect of the extra frustration experienced in the goal box itself during extinction. The difficulty in making this prediction as a strict deduction from the theory lies in the problem to which we have previously referred: it is not always clear when the energizing effects of conditioned frustration will outweigh its aversive effects. In summary, frustration theory must assume that, after PR training, large reward leads to increased resistance to extinction because of the additional strength of frustration, whereas, after CR, it leads to less resistance because of the stronger frustration; this is perhaps not wholly unreasonable since only in PR training will the running response be conditioned to frustration.

Capaldi's theory of the PRE encounters similar difficulties to Amsel's in providing a rigorous explanation of the opposite effects of reward size on extinction after PR and after CR training. It will be remembered that Capaldi very reasonably postulates that the memory stimulus for a small reward is more similar to that for nonreward than is the memory stimulus for a large reward. It follows that, after CR training with small reward, the memory stimulus developed is more similar to that present in extinction than after training with large reward; hence, animals trained with small reward will be more resistant to extinction. To derive this prediction, Capaldi must assume that the extra similarity of the memory stimuli in acquisition and extinction with small reward outweigh the effects of increased habit strength (or incentive motivation) with large reward. The situation is as depicted by the solid lines in Figure 10.4. It can be seen that the difference between the strengths of the generalized memory stimuli for small and large rewards at $N_1$ is more than enough to overcome the difference in the strength with which the running response is conditioned to the actual stimuli present under large and small rewards.

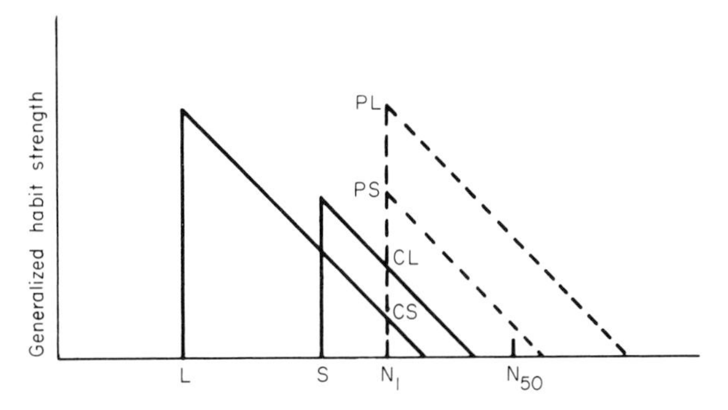


**Fig. 10.4.** Illustration of effect of size of reward on Capaldi's theory.

Capaldi can also explain why large rewards should increase resistance to extinction under PR. The situation is now as represented by the dashed lines in Figure 10.4. If small rewards are used the instrumental response will be less strongly conditioned to the memory stimulus for a given N-length. As the graph shows, this effect may outweigh the difference between the strengths of the memory stimulus for large and small rewards at $N_1$ and for all other N lengths; in Figure 10.4, the distance between PL and PS is greater than the distance between CL and CS, and this results in resistance to extinction after PR being greater with large rewards than with small reward. It is clear, however, that the graph could have been drawn in such a way that the distance between CL and CS was greater than the distance between PL and PS. For example, this result will be obtained if the point L is displaced to the left. It is for this reason that Capaldi cannot unequivocally predict that large rewards will increase resistance to extinction after PR training. The prediction follows only with a limited range of parameter settings.

It is most important to note that Capaldi's explanation depends upon assuming that changes in reward size affect the asymptote of habit strength, an assumption counter to Hull's (1952) later theoretical position. As we shall see below, this assumption makes it difficult for Capaldi to explain the effects of shifts in reward size. It was, of course, mainly data on the effects of such shifts that led Hull to abandon the idea that reward size directly affected habit strength.

Leonard (1969) performed an ingenious experiment to test the idea that the main effect of large reward in PR training is to increase the strength of conditioning of the instrumental response to the memory stimulus of nonreward. He trained rats in a runway with three trials a day in the sequence RNR: he ran four groups, using the four possible combinations of large and small reward on the two daily rewarded trials. Using L for large reward and S for small reward, this yields the following trial sequences for different groups:

SNL, LNS, LNL, SNS.

He argued that if the main effect of large reward was to strengthen conditioning of the running response to the memory stimulus of the N trial, then groups with large reward following this trial should be more resistant to extinction than groups with a small reward following this trial. The prediction was confirmed: groups LNL and SNL both ran faster than groups LNS and SNS. Moreover, his results confirmed the suggestion that the memory stimulus for a small reward lies nearer to the stimulus for nonreward than does that for large reward, since the effects of varying the size of reward on the first trial of each day were exactly the opposite to those of varying the size of reward on the third trial. Group SNL extinguished more slowly than Group LNL and Group SNS extinguished more slowly than Group LNS. These results are further confirmed in the experiment of Bloom's cited in (E-8).

Although Capaldi's theory is capable of explaining the effects of reward size on extinction after PR, the explanation raises the following difficulty: both Wagner (1961) and Hulse (1958) found that large rewards increased resistance to extinction after PR training when only one trial a day was given in acquisition. To explain the occurrence of the effect with a 24-hr intertrial interval, Capaldi must assume that stimulus aftereffects (or the memory stimulus) persist for 24 hr, but we have already presented evidence (on trial patterning) that calls this assumption in question. It may be that although the memory stimulus for the last event does not survive for 24 hr, a memory stimulus for repeated events built up gradually may last this long (cf. the arguments on page 365).

Like Amsel's theory, Capaldi's can handle the effects of shifting CR trained animals from large to small reward; when shifted, they run more slowly than animals trained throughout on small reward. Capaldi (1967) attributes this to generalization decrement, Amsel to frustration. If non-reward is similar to small reward, this effect should not occur when PR trained animals are shifted from large to small reward. Mikulka, Lehr, and Pavlik (1967) confirmed this prediction. Amsel's theory makes the same prediction, since, during PR, animals will learn to run to the cue of frustration; hence, they should not drop in running speed when trained under PR and shifted from large to small rewards. One further correct prediction made by both Capaldi and Amsel is that varying reward size during CR training will increase resistance to extinction (Yamaguchi, 1961).

Capaldi's theory, however, has some difficulty in explaining the results obtained when animals are shifted from small to large reward. This results in a rapid increment in performance, at least as measured by running speed in a runway (Crespi, 1942). In fact, when animals trained on small reward are shifted to large reward, their running speeds increase at a faster rate than those of animals trained throughout on large reward. Capaldi has difficulty explaining this finding. This is illustrated in Figure 10.5. On Capaldi's theory, the distance L should be the same as the distance S–L. Indeed, animals shifted from small to large rewards might be expected to attain asymptote more slowly than animals trained throughout on large reward, since the shift will produce some generalization decrement. Capaldi (1967) attempts to account for the difference in terms of incentive, but since this factor operates equally for both groups, his explanation is unsatisfactory.

Although, therefore, the general effects of reward size on extinction after PR and CR training can be explained by both frustration theory and Capaldi's version of generalization decrement, only Capaldi's theory can

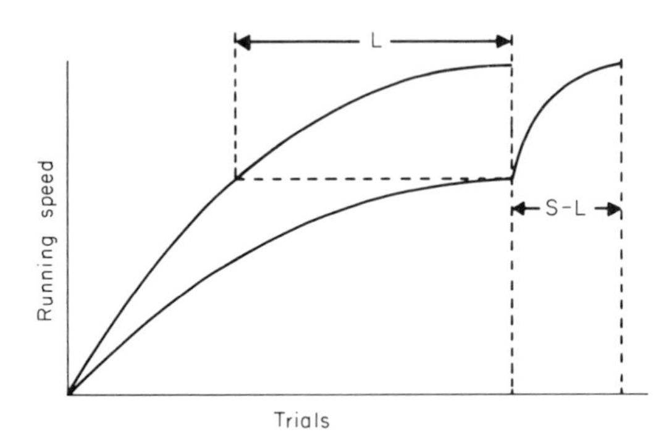


**Fig. 10.5.** Theoretical curves illustrating effect of switch from small to large reward on running speed.

explain why resistance to extinction is only greater after large rewards if the large reward *follows* nonreinforcement trials rather than precedes them.

In Chapter 8, it was argued that, on our own model, the role of large reward might be to increase the size of the operators governing response learning. This would explain why, with CR training, large reward decreases resistance to extinction: with large rewards, the response attachments to the dominant analyzer will be very rapidly learned and this in turn will boost the strength of the dominant analyzer at the expense of others, so that the breadth of learning will be decreased. It is more difficult for us to explain why large rewards should increase resistance to extinction after PR: perhaps when no one analyzer can "run away" in strength, the main effect of the large rewards will be to produce a higher asymptote in the strengths of the responses attached to the weaker analyzer and this will result in greater resistance to extinction.

## D. (E-11) Number of Training Trials

A further prediction made by frustration theory is that giving a very large number of CR training trials will actually decrease resistance to extinction: the stronger the expectation of food, the stronger will be frustration when food is not found in the goal box. Several investigators have found that if sufficient trials are given, resistance to extinction is decreased, particularly under conditions of large reward (North & Stimmel, 1960; Ison, 1962; Ison & Cook, 1964; Siegel & Wagner, 1963). However, Wagner (1961) and McCain, Lee, and Powell (1962) have shown that resistance to extinction after PR is also decreased by giving additional training trials, and as Wagner points out this is not predicted by frustration theory. If increasing conditioned frustration by giving large rewards during PR increases resistance to extinction, then increasing conditioned frustration by giving a large number of training trials should also increase resistance to extinction in PR animals.

It should be noted that the results of some other experiments varying the number of PR acquisition trials do not agree with those quoted above; e.g., Bacon (1962) found that running speed in extinction increased with number of acquisition trials after both PR and CR training; terminal acquisition speeds were higher after 300 trials than after 30 and if this factor is balanced out, *rate* of extinction was not affected by number of trials. Wilson (1964) and Uhl and Young (1967) found that overtraining increased resistance to extinction following either CR or PR acquisition; in both cases, however, the effect was more marked after PR training.

Part of the problem concerns the definition of extinction. Hill and Spear (1963), for example, found that overall speed of running during extinction was increased by overtraining after both PR and CR acquisition. However, after CR training, overtraining did result in a more abrupt decline in running speed, while after PR training, overtrained animals maintained their faster running speed throughout extinction. Their results, therefore (like those of Wilson and of Uhl and Young) suggest some interaction of reinforcement schedule and overtraining.

Although there is a possibility of some interaction between the effects of overtraining and training schedule, the case is far from proven. In fact, all the above studies have one thing in common: although the direction of the effect of overtraining on resistance to extinction varies from study to study, within each study the direction of the effect was the same for both PR and CR animals.

Our own theory can readily explain why, under some circumstances, increasing the number of training trials with CR may decrease resistance to extinction. The explanation is along the same lines as that given for the effects of large rewards. Giving a large number of training trials should greatly strengthen the one dominant analyzer; if that analyzer remains switched in after the running response has been extinguished to it, the animals should stop running. With fewer training trials, the dominant analyzer will be less strong and may get switched-out before the response attachments to it are extinguished. If the analyzer is switched-out, other analyzers to which the running response is attached will control behavior and, hence, the animal will continue to run. There is some direct evidence for this explanation: Mackintosh (1965c) showed that increasing the number of training trials in a T maze leads to an increase in proprioceptive control of running (see Chapter 8). Likely and Schnitzer (1968) provide confirmatory evidence for this interpretation. Following Mackintosh, they argue that running would be less likely to come under control of proprioception in a runway containing hurdles than in an ordinary alley without obstacles. They found as predicted that, whereas increasing the number of CR training trials from 30 to 120 in a plain alley decreased resistance to extinction, it increased resistance to extinction when hurdles were present during both acquisition and extinction.

The generalization decrement hypothesis may also be able to handle the effects of giving large numbers of trials under CR training. Provided habit strength reaches asympotote after a moderate number of trials, the main effect of additional trials may be to make the memory stimulus for the series of rewarded trials different from that for a series of nonrewarded trials and, hence, increase generalization decrement in extinction. There would appear to be no particular reason to expect any change in resistance to extinction on this theory due to giving a large number of trials under PR.

A further result on PR and length of training is important for frustration theory. Both Capaldi and Deutsch (1967) and McCain (1966) have shown that a PRE may occur with very small numbers of trials (between two and five). It can occur with the sequence NR where no previous reinforcement has been given. It is impossible for frustration theory to explain this result since presumably no frustration would be set up on a nonreinforced trial not preceded by any reinforced trials. It is, in fact, hard to explain this result on any theory other than Capaldi's version of generalization decrement.

### E. (E-12) Generality of PRE

A further type of experiment to which frustration theory has given rise concerns the way in which the PRE will generalize from one situation to another. The argument is as follows: the PRE is said to be due to the ongoing response being conditioned to the cue of frustration. Now, provided such conditioning takes place, it will not matter whether the response is conditioned to frustration produced by PRE in the goal box or whether it is conditioned as a result of frustration being produced in some other way or in some other situation.

It would be predicted that if an animal experiences frustration in the *start* box, then, since the frustration will be present when it starts running, the running response should be increased even with CR in the goal box. Longstreth (1964) and McCain and Power (1966) tested and confirmed this prediction; they frustrated animals during training by giving them food on a PR schedule in the start box.

Ross (1964) performed a more elaborate experiment the results of which confirmed similar predictions from frustration theory. The experiment was conducted in three stages. At Stage I, animals were trained in a black box either to run, or to jump a gap, or to climb onto a ledge in order to obtain food. Within each of these three groups, half the animals were trained under PR and half under CR. At Stage II, all animals were made thirsty and were taught to run a white straightway for water reward under CR. At Stage III, they were all extinguished in the white runway while thirsty. Ross argued that

PR animals trained at Stage I to run or jump should have conditioned running and jumping to the cue of frustration and should, therefore, be slower to extinguish at Stage III than the corresponding CR groups, since the responses of running and jumping would maintain behavior in the runway. This prediction was confirmed. Since climbing is incompatible with running, PR animals taught to climb at Stage I should extinguish faster at Stage III than the corresponding CR group; again, this prediction was confirmed. Moreover, he found that PR animals trained to jump gave very many more jumping responses at Stage III than any other group. This experiment is a very powerful demonstration of the way in which a PRE will generalize from one situation to another even when animals are under different drives and a different reward is used.

A further experimental paradigm has been used to test the generality of the PRE. Rats are trained with CR on one alley and PR on a very different alley. They are then extinguished on the CR alley and their rate of extinction compared with that of a group that has received training only on the CR alley. Katz (1957), Amsel, Rashotte, and McKinnon (1966), and Amsel (1967) report experiments of this kind showing that a large PRE is obtained. Moreover, the size of the PRE is as great when animals are extinguished on the CR alley as when they are extinguished on the PR alley. Amsel argues that the transfer of the PRE from the PR to the CR alley is mediated by the cue of frustration to which the running response has been attached. It is not merely due to a failure to discriminate between the two alleys since differential running times in the two alleys do develop towards the end of training.

The results reviewed above are very difficult for any other theory of PR to explain and we regard them as convincing proof that frustration does play a role in producing the PRE.

### F. (E-13) The Effects of Pretraining on Subsequent Discrimination Learning

Amsel (1962) and Amsel and Ward (1965) have developed yet another series of predictions from frustration theory; these predictions relate to the different times at which different conditioning processes will occur. Before any frustration can be conditioned to runway cues, it is clear that an expectation of reward has to be built up. Early in PR training, therefore, the animal will condition $R_G$ (the fractional anticipatory goal response to food) to runway cues and will condition approach responses to the associated stimulus ($S_G$). At a later stage in training $R_F$ (frustration) will be conditioned, but its associated stimulus ($S_F$) will not have been conditioned to the approach response. Late in training both $R_G$ and $R_F$ will have been conditioned and approach responses will be conditioned to both $S_F$ and $S_G$.

Suppose now that PR training is given on one of two stimuli to which the animal is later to be taught to make discriminatory responses. If only a few PR trials are administered, animals should be helped more on the subsequent discrimination task when the PR training is given on the positive stimulus than when it is given on the negative stimulus, since the animal will have learned only to approach in the presence of $S_G$. If rather more PR trials are given, $R_F$ will be conditioned but not the approach response to $S_F$; therefore animals will be more helped on the discrimination problem if PR training is given to the negative stimulus to which they must develop $R_F$ in order to solve the discrimination task. After many training trials, however, approach responses will be conditioned to $S_F$ and now the animal will be impeded by training on the negative stimulus since it must learn not to approach in the presence of $S_F$ in order to solve the discrimination problem.

Amsel has tested these and similar predictions and found that usually, although not invariably, they hold up. The problem of testing the predictions is not a simple one, since there is nothing in the theory itself that specifies exactly how many trials are needed to reach a particular stage in the conditioning of $R_G$ and $R_F$. Moreover, the derivation of the predictions is not completely rigorous, since the extent to which different amounts of PR training on the positive and negative stimuli will affect subsequent discrimination learning will depend upon how the two responses generalize from one stimulus to another and also on the relative rates at which different responses are learned and extinguished. For example, if the approach response to $S_F$ extinguished relatively fast while $R_F$ itself is learned slowly, giving a large number of PR trials on the negative stimulus might well produce faster subsequent learning of the discrimination than giving many PR trials on the positive stimulus. We shall therefore not go into more detail on this line of experiments; Amsel's reasoning is highly ingenious but because of the number of different processes involved and the possibility that they have different time courses both during learning and during extinction, it would need a very large experimental program to be certain about what was going on.

## G. (E-14) Effects of Punishment during Acquisition

Brown and Wagner (1964) devised the following ingenious argument: if the PRE results from PR animals having learned to run to stimuli associated with frustration during training, then it is possible that if animals could be trained in the presence of a different aversive drive during training (such as fear), then the stimuli resulting from the two drives would be sufficiently similar to mediate transfer of the running habit from the fear experienced during training to the frustration experienced during extinction. To test this possibility, they trained rats in an alley with food reward present on every trial, but the animals were punished with an electric shock in the goal box on 50% of trials. The animals experiencing punishment proved more resistant to extinction than animals trained without punishment. A similar result was obtained by Uhl (1967) using a Skinner box. Uhl also found that rats punished on every trial were more resistant to extinction than animals punished only on 50% of trials. Although these experiments were undertaken to test a prediction from frustration theory, the results can also be explained by Capaldi's version of aftereffects theory, provided the assumption is made that the memory stimulus for a punished trial is more similar to that for a nonreinforced trial than is the memory stimulus for an unpunished reinforced trial.

### H. (E-15) Physiological Effects

There are two further ways in which the frustration theory of extinction has been tested: first by investigating the effects of administering barbiturate drugs on the phenomena produced by frustration, and second, by investigating and manipulating the physiological correlates of frustration.

It has been known for some time that barbiturate administration impairs the learning of a passive avoidance habit (Miller, 1959). The most obvious interpretation of this result is that such drugs reduce the fear response. If they reduce fear, they might also reduce frustration, since both fear and frustration are aversive drives. In what follows, we shall describe a number of studies that suggest that frustration is reduced by barbiturates. There are, however, two results that indicate that this is not the whole story. These will be discussed subsequently.

If barbiturates reduce frustration, and if extinction is at least in part due to the buildup of frustration then administration of amobarbital sodium during extinction should increase resistance to extinction. This result has been obtained by Barry, Wagner, and Miller (1962), Gray (1969), and Ison and Pennes (1969). It can also be predicted that administration of barbiturates will reduce the latency of response to the negative stimulus in differentation training and this prediction was confirmed by Ison and Rosen (1967). These results establish a *prima facie* case for supposing that frustration is reduced by barbiturates and it is now possible to draw some specific predictions about the effects of barbiturates on PR.

If the reason why PR trained rats run faster at the end of acquisition than CR animals is that frustration is energizing their behavior, then giving sodium amytal during training should abolish this effect. This prediction has been confirmed by Wagner (1963a) and by Gray (1969). Moreover, if the PRE itself is due to running being conditioned to frustration during training, then we might expect that animals trained under amobarbital and extinguished without the drug would show a reduced PRE, since they should not condition the running response to frustration during training. This prediction has again been confirmed (Gray, 1969, Ison & Pennes, 1969).

It will be remembered that Adelman and Maatsch (1956) found that rats trained under CR and then extinguished would learn to jump onto a ledge in the goal box during extinction if they were then removed from the apparatus by the experimenter. Adelman and Maatsch argued that this demonstrated that in extinction the goal became aversive as a result of the frustration experienced there. Gray (1969) found that the latency of jumping under extinction in the goal box was much higher under amobarbital than in animals not given a drug; once again, this would be expected if the drug had reduced frustration.

There are two other possible interpretations of some of the above results. It could be that barbiturates simply impair performance and, hence, abolish faster running times in acquisition under PRE and slow down the latency of jumping in the Adelman and Maatsch situation. This interpretation is implausible, since Gray found that if animals were given a food reward for jumping onto a ledge, drugged subjects actually jumped with shorter latencies than controls. It could also be argued that the reduction in the PRE found in rats trained under the drug and extinguished without it is due to generalization decrement. The explanation is again unlikely since animals trained without the drug and extinguished with it take longer to extinguish than animals trained and extinguished with no drug. These experiments, therefore, offer good evidence both that frustration is, to some extent, reduced by barbiturates and that conditioned frustration is implicated in the PRE.

There are, however, two discordant results. If the effect of administering barbiturates were to reduce frustration as a primary response to a frustrating situation, then we might expect that in the double runway situation invented by Amsel and Roussel (see page 368), rats under barbiturates would no longer run faster in the second half of the runway as a result of the frustration experienced due to food being omitted in the goal box at the end of the first half. Three experiments have found that barbiturates do not reduce this effect (Ison, Daly, & Glass, 1967; Ludvigson, 1967; Gray, 1969). Moreover, it might be expected that if animals are both trained and extinguished under barbiturates then the PRE should be abolished or at least considerably reduced. Both Gray (1969) and Ison and Pennes (1969) found that rats trained and tested under the drug showed just as big a PRE as animals trained and tested without the drug. This result is in marked contrast to what happens when animals are trained under the drug and tested without it.

The first of these two discordant results could be explained if barbiturates affected conditioned frustration (in the runway), but not the primary frustration experienced in the goal box. This explanation cannot be correct, however, since barbiturates decrease the tendency to jump out of a goal box during extinction and this is a response to primary frustration. Moreover, if the drug affected only conditioned frustration, it should still abolish the PRE when animals are both trained and tested under the drug. Gray has suggested that the reason for the occurrence of a PRE when animals are both trained and extinguished under barbiturates is that the drug does not completely block frustration, but only attenuates it; if some frustration is present during both training and extinction, running should be conditioned to a weak $S_F$ in training and since the same weak $S_F$ will be present in extinction, resistance to extinction will be increased. This suggestion is not very satisfactory, since, if barbiturates attenuate but do not abolish frustration, a PRE should be obtained when animals are trained without the drug and extinguished under it. Consequently CR animals should extinguish faster than PR animals, because they will experience *some* frustration in extinction, and running has not been conditioned to frustration.

Although we do not yet have a full understanding of the effects of barbiturates on frustration, the experiments quoted (with the exception of the last two) suggest that frustration is implicated in the PRE and therefore offer support to the frustration theory account of PRE.

Further evidence that frustration is implicated in the PRE comes from a series of ingenious experiments conducted by Gray (1970). In the rat, the theta rhythm in the hippocampus varies between 6 and 10 Hz. Gray found that a rhythm of about 8 Hz appears when the rat is placed in a novel (fear producing) environment and when the animal experiences primary frustration. In other circumstances, the rhythm is either lower (during consummatory behavior) or higher (e.g., when running under CR to food). The theta rhythm in the hippocampus can be driven by stimulating the medial septal area. Gray found that when animals were given sodium amobarbital, the current threshold necessary for driving the theta rhythm by stimulating the septal area was greatly increased for the 8-Hz rhythm, but that thresholds for rhythms above and below this were not altered. These findings suggest that the 8-Hz theta rhythm is a physiological correlate of frustration and fear. The conclusion is strengthened by the fact that lesions to the medial septal area have behavioral consequences similar to the administration of barbiturates; e.g., such lesions impair passive avoidance and increase resistance to extinction.

Gray performed a series of experiments in which he attempted to manipulate frustration by producing the 8-Hz theta rhythm through electrodes implanted in the medial septal area. Rats trained under CR and extinguished under the artifically produced 8-Hz theta rhythm extinguished faster than control animals. Rats run under CR and given stimulation in the goal box on 50% of trials, extinguished more slowly than controls; by stimulating in the goal box, Gray had presumably produced artificial "frustration" and the running response had become conditioned to this during extinction thus producing an analog of the PRE. In a futher experiment, Gray ran rats under PR and stimulated at 200 Hz in the goal box on nonrewarded trials; such stimulation prevents the 8-Hz theta rhythm from occurring, and the animals showed no PRE, presumably because frustration could no longer be conditioned to running during acquisition. Gray also found that, under CR training, there was a negative correlation between the occurrence of the 8-Hz rhythm in extinction and the number of trials to extinction, whereas after PR training there was a positive correlation between the occurrence of the 8-Hz theta rhythm at the end of acquisition and the number of trials to extinction. Finally lesions to the medial septal area whether made before acquisition or before extinction abolished the PRE.

This ingenious series of experiments offers rather convincing evidence that the 8-Hz theta rhythm is a physiological correlate of frustration and that frustration is implicated in the PRE.

Before leaving these experiments on drugs and the physiological correlates of frustration, it is worth saying something about their relevance to our own account of the PRE. Gray points out that the theta rhythm appears when rats are placed in a novel environment and it would be expected that in such an environment animals are rapidly switching attention from one cue to another. He further notes that lesions to the medial septal area result in "a pronounced inability to make normal use of the vibrissae," and Komisaruk (1968) has observed a close correlation between the 8-Hz theta rhythm and vibrissae movements in the normal rat. Movements of the vibrissae are, of course, strongly associated with exploratory behavior in the rat, and again we might expect them to occur most where the animal was seeking new information and switching attention from one cue to another. These observations suggest that the 8-Hz theta rhythm may be implicated in the processing of novel information and in switches of attention. It may be that the greater breadth of learning that is exhibited under PR results from switching of attention produced by frustration. McGonigle, McFarland, and Collier (1967) ran the following experiment to test this possibility. They trained rats on a black–white discrimination rewarding choices of the positive stimulus 50% of the time, and giving no reward for choices of the negative stimulus (Stage 1). Keeping animals on the same reward schedule, they next introduced horizontal–vertical as an incidental cue presented simultaneously with the already learned black–white cue (Stage 2). During this stage half the animals were run under sodium amytal. They then tested how much had been learned about the incidental cue (Stage 3), and found that the animals that had been run under the drug had learned very much less about this cue than the controls. This is of course exactly what would be expected if sodium amytal reduced frustration and if frustration increases switching of attention. In the fourth and final stage of the experiment, animals were extinguished with both cues present and it was found that animals run under the drug at Stage 2 extinguished faster than controls. Unfortunately, both the above results, although predicted by frustration theory, could also have been due to generalization decrement between the drugged and nondrugged states. However, in the nondrug groups there was a positive correlation between trials to extinction and how much each animal had learned about the incidental cue and this result is some direct support for our own theory. As the authors point out, it is not possible to decide from this experiment how far resistance to extinction was determined by the amount of frustration present during Stage 2 and how far it was determined by the amount learned about the additional cue. It seems likely that switches of attention are closely linked to frustration and it may be that the PRE is determined both by performance being conditioned to $S_F$ during acquisition and by the extra breadth of learning brought about by switches of attention induced by frustration.

### I. Conclusions

The frustration theory explanation of the PRE appears to stand up reasonably well. It has been fruitful in yielding predictions and it is the only theory that handles the evidence on the generality of the PRE and on response latencies during PR acquisition. Moreover, the role of frustration in learning is well supported by studies not directly connected with the PRE itself.

There are, however, several findings on the PRE that it is difficult for frustration theory to encompass. The first is our own result (confirmed by Hothersall, 1966) that the PRE is reduced if a block of CR training is given before PR training (E-2). Frustration theory should make the opposite prediction: animals that are used to a CR schedule should experience very strong frustration when PR is introduced and hence the frustration response should become strongly conditioned to the runway cues. This should lead to rapid learning of an approach response to $S_F$ and hence to great resistance to extinction. Wagner in a personal communication has queried this argument. He has previously argued (1961),

> $R_F$–$S_F$ may be expected to become conditioned to the approach response when the $S_F$ cues are introduced initially at a weak value, and hence with a negligible tendency to elicit competing responses, and are then increased gradually at the same time that the approach response is being strengthened.

It seems to us that this argument is only valid in the case where initial frustration is so strong that it would prevent the animals from running altogether. For example, Brown and Wagner (1964) introduced shock initially at low intensities in the goal box and gradually increased its intensity; had they not done so, presumably animals might have ceased running completely and then there would have been no possibility of conditioning the running response to the cue of learned fear. However, if animals continue to run when frustration first occurs and are rewarded on some trials, then there seems to be no postulate in frustration theory that would prevent the running response being conditioned to frustration no matter how strong the frustration. On the assumption that strong frustration is evoked in extinction, the stronger the frustration during learning the more similar will be $S_F$ in training and extinction and therefore the longer it should continue to mediate running in extinction.

A second result that is not predicted by frustration theory is the greater breadth of learning under PR than under CR (E-1). If frustration acts as a drive, breadth of learning should be less under PR than under CR since there is evidence that high drive reduces the breadth of learning. For example, Maier (1949) found that rats trained on a jumping stand under conditions of extreme frustration often develop very strong position habits and are unable to learn a discrimination based on a less salient visual cue. The effects of drive on breadth of learning have been discussed in more detail in Chapter 4.

Frustration theory also cannot account for many of the findings of Capaldi and his associates on the effects of varying the sequence of rewarded and nonrewarded trials. For example, the PRE increases as N-length increases, but only if long N lengths are followed by a rewarded trial and not by an intertrial reinforcement [with small number of trials, see (E-8)]. Similarly, frustration theory cannot explain Leonard's results on varying the sequence of trials on which large and small rewards are given (E-10). It also has difficulty in explaining why any PRE should occur with very small numbers of trials (E-11). Despite these difficulties, the fact that there are two phenomena associated with PR that appear to be predicted by frustration theory and no other leads us to suppose that frustration may be at least a contributory factor to the effect.

There is, of course, no incompatibility between our own two-process theory and frustration theory, and it may be that both frustration and selective attention play a role in producing the PRE. It must be admitted, however, that the evidence suggests that frustration contributes more strongly to producing the effect than does the direction of attention. Our own theory is incapable of explaining the very striking results on the generality of the PRE.

It is perhaps worth concluding this section by comparing frustration theory with Capaldi's generalization decrement hypothesis. Amsel relies heavily on generalization decrement to explain the PRE: frustration is always present in extinction—because it is present for animals trained under PR in acquisition but not for CR animals, PR animals are more resistant to extinction that are CR animals. The two theories differ mainly in their specification of what are the stimuli that change from CR to extinction and hence produce generalization decrement. It is therefore not suprising that many of the results predicted by frustration theory can also be handled as we have seen above by Capaldi's version of generalization decrement.

## VI. The Cognitive Dissonance Hypothesis

### A. Introduction

A very ingenious but intuitively implausible theory of PR has recently been put forward by Lawrence and Festinger (1962). To borrow a phrase from the philosophers, it might be called the "Boo–Hurrah theory" of PR. These authors suggest that when an organism expends effort in seeking a reinforcement and fails to obtain a reinforcement, cognitive dissonance is set up. In anthropomorphic terms it is as though, when a rat runs to a goal box expecting food and finds that there is no food there, it feels that it has wasted its effort. To reduce its disappointment, the rat says to itself, "Well, it's a pity there is no food here, but it's a very nice-looking goal box so it was really well worth running to after all." This process is referred to as "attributing cognitive attractions" to the goal box. There is of course evidence for the operation of dissonance-reducing processes in man (Festinger, 1957). Leaving out the anthropomorphic overtones, the theory states that when an animal expects reward for doing something and does not receive it, the situation in which it finds itself acquires reinforcing properties; the additional reinforcing properties will be stronger the more time and effort has been expended in reaching the situation. The cognitive attractions acquired by the goal box when animals are trained under PR continue to reinforce running behavior during extinction. Since CR animals will not have attributed extra attractions to the goal box during training, they extinguish more quickly.

The theory has much in common with the frustration hypothesis in that they both stress the consequences of a disappointed expectation. However, frustration theory postulates changes in the conditioning process in the runway as a consequence of such disappointed expectations, whereas cognitive dissonance theory stresses the additional reinforcing power acquired by the goal box. As the authors are aware, one of the weaknesses of cognitive dissonance theory is that it is not clear under what conditions cognitive dissonance will be reduced by attributing extra attractions to the goal box, and under what conditions it will be reduced by the organism saying to itself, "Well at least I can learn by my mistakes; I'll never go near that goal box again." Festinger points out that human beings take the latter way out when they have little to lose by it. A man who discovers he has paid more for a house than it is worth cannot simply sell the house, since he would incur further financial loss; therefore, he pretends to himself that the house has all sorts of attractions that he did not notice at the time he purchased it. Similarly, the PR rat cannot simply stop running to the goal box since it has too much to lose; if it ceases to run it will fail to obtain food even on trials on which the food is present.

Despite its weaknesses, the theory is highly original and ingenious. Moreover, it has led to many new and interesting experiments, the more important of which will be reviewed here. The experiments performed by Lawrence and Festinger (1962) themselves will be referred to by the number the authors allot to each in their book (e.g., L & F 1).

### B. (E-16) Delay of Reinforcement

One area in which cognitive dissonance theory makes interesting predictions is that of the effects of delay of reward on extinction. Lawrence and Festinger argue that delay of reward is likely to set up cognitive dissonance and this results in extra attractions being attributed to the situation in which delay occurs. Fehrer (1956), Holder, Marx, Holder, and Collier (1957), Marx, McCoy, and Tombaugh (1965), Peterson (1956), and Sgro and Weinstock (1963) all found that rats under CR extinguish more slowly if they are delayed in the goal box before receiving food than if food is given immediately. It should be noted that Sgro, Dyal, and Anastasio (1967) failed to find any effect of constant delay of reward on extinction after extended acquisition training (cf. also Renner, 1964). Wike and McNamara (1957) found that if trials on which a delay was introduced were interspersed with trials on which no delay occurred, resistance to extinction was greater than if reinforcement was always given without delay (cf. also Wike, Kintsch, & Gutekinst, 1959).

In a direct test of their own explanation of the effects of delay, Lawrence and Festinger (L & F 7) ran animals in a runway that contained a delay box half way up; all animals received 100% reinforcement in the goal box, and a delay of 20 sec in the delay box. The animals were split into three groups that received respectively 0, 50, and 100% reward in the delay box. Subsequently, animals were run from the start to the delay box with no reward present in that box. The group that had never received reward in the delay box was the most resistant to extinction, while the group that had always been rewarded in the delay box extinguished fastest. They interpreted this result to mean that animals not rewarded in the goal box had experienced more cognitive dissonance in it than animals who did receive reward there; hence the former group attributed more cognitive attractions to the delay box than the latter. Lawrence and Festinger (L & F 11) also found that if two delay boxes were placed in series in a runway, and half the animals were delayed only in the first box, and the remainder only in the second, the latter group ran faster during extinction from the first to the second box. They claim that the second goal box had acquired cognitive attractions for the group delayed in it.

The predictions the theory makes about delay are, in fact, not specific to the theory. The discrimination hypothesis can explain these results equally well: obtaining no reward after running could be regarded as equivalent to an infinite delay of reward so that extinction conditions are more similar to conditions during training for a group that has had delayed reward than for animals given immediate reward. (Lawrence and Festinger put up a curious argument to the opposite effect, claiming that no delay is involved in extinction.) Again, if an animal has been delayed in a goal box halfway up the runway and is then extinguished just on the section of the maze leading to that goal box, extinction conditions will be more similar to training conditions than for an animal that has never been blocked in the delay box; the extinction procedure itself involves blocking the animal in the delay box. On the assumption that delay involves frustration, frustration theory is also able to explain these results. It must be admitted that our own theory has difficulty in accounting for the effects of delay. It could be that delaying reward causes uncertainty and, hence, leads to more analyzers being switched-in, but this is special pleading.

Lawrence and Festinger (L & F 13) were unable to demonstrate directly that a delay box acquires extra attractions. They again trained animals with two delay boxes in series, delaying one group in the first, the other in the second. The two delay boxes were then placed at opposite ends of a T maze and the animals were run with no primary reward. The prediction was that each group of animals should tend to go to the delay box in which it had been delayed, since only that delay box would acquire extra attractions. In fact, animals of both groups showed equally strong tendencies to go to the second delay box; we would expect this on the basis of secondary reinforcement, since the second delay box was, of course, nearer to the food during training, and should, therefore, have acquired greater secondary reinforcing value than the first. This experiment was the most direct test made of predictions from cognitive dissonance theory concerning the effects of delay and since the prediction was not verified, it would appear that both the discrimination hypothesis and frustration theory actually account better for the effects of delay than does cognitive dissonance.

Moreover, there are three experiments performed by Capaldi (1967) and his associates designed to test specific predictions from his own theory about the effects of delay of reinforcement. They trained animals under partial delay and used two different lengths of delay. They varied the sequence of delayed and immediately reinforced trials. All animals had the same number of trials with each delay length, but for some animals within a daily sequence of trials, the trial with the long delay was followed by an immediate-reinforcement trial, for others, only the trial with the short delay was followed by an immediately reinforced trial. The former group were more resistant to extinction than the latter. Capaldi, of course, predicts this result since the memory stimulus for a long delay trial is more similar to the memory stimulus operating in extinction than is that for a short delay. Animals that receive reinforcement in the presence of the memory stimulus for long delay should therefore be more resistant to extinction than animals receiving reinforcement only in the presence of the memory stimulus for a short delay trial. Using similar reasoning, Capaldi (1967) argued that since the memory stimulus for non-reinforcement is nearer to the memory stimulus present in extinction than is the memory stimulus for delayed reinforcement, animals experiencing the sequence DNR (where D stands for delay) should be more resistant to extinction than animals experiencing the sequence NDR. This prediction was confirmed by Capaldi and Olivier in an experiment cited in Capaldi (1967). Finally, Capaldi and Spivey (1965) showed that, with a small number of training trials, alternating delayed trials with immediately reinforced trials led to greater resistance to extinction than giving a random sequence of immediate and delayed reinforcement. This follows from Capaldi's theory, since, when only a small number of training trials is given, resistance to extinction should increase as the number of reinforced N lengths is increased (cf. Capaldi & Hart, 1962, described on page 358). It is impossible for either cognitive dissonance theory or frustration theory to explain any of these three results, and it would appear therefore that Capaldi's theory gives the most adequate account of the effects of delay of reinforcement on extinction.

### C. (E-17) Effects of Effort

A second area in which the theory makes new predictions is the effect of varying the effortfulness of an instrumental response. Lawrence and Festinger argue that even when reward is present there will be some cognitive dissonance if a response is effortful and hence cognitive attractions are likely to be attributed to the goal box. Several investigators have found that animals trained under conditions of high effort are slower to extinguish than animals trained under conditions of low effort, both when the two groups are extinguished on the high effort condition and when they are extinguished under low effort (Aiken, 1957; Stanley & Aamodt, 1954). As Trotter (1956) has pointed out, however, a different response is learned under high and low effort conditions and whereas a response involving much effort will normally ensure that the response continues to be made when little effort is required, the reverse is not true. To overcome this problem Lawrence and Festinger (L & F 14) trained two groups of rats on runways inclined at 50 and at 25° to horizontal; the first condition is more effortful than the second. They found that when their animals were extinguished with the same runway inclination that had been used for each in training, the high-effort group extinguished more slowly than the low. Unfortunately no other investigator has obtained this result: Aiken (1957), Appelzweig (1951), Maatsch *et al.* (1954), and Mowrer and Jones (1943) all failed to find any difference in trials to extinction between rats trained and extinguished under high effort and rats trained and extinguished under low effort, while Stanley and Aamodt (1954) found that the higher the effort the lower was the resistance to extinction. A further experiment by Lawrence and Festinger (L & F 15) on the effects of effort on extinction under PR and CR is rather unconvincing, since an effect of effort could only be demonstrated by taking a much less stringent criterion of extinction for the effortful groups than for the noneffortful.

There is, however, one experiment that does provide direct confirmation for Lawrence and Festinger's position on effort. Lewis (1964) trained rats to run to a food substance with a distinctive taste under different conditions of effort; the effort was varied by making different groups of rats pull different weights in order to obtain the food. After rats were trained in this way, they were tested in a new situation to discover whether the reinforcing value of the food itself had been influenced by the amount of effort the animals had had to expend in order to obtain it at the first stage. At the second stage of the experiment, the animals were trained to run a runway for the food reward. Lewis discovered that the larger the effort required in initial training, the faster were animals' running speeds at the second stage; moreover, animals trained with large weights ate faster and ate more food than animals trained with small weights. This appears to be a rather convincing demonstration that the more effort an animal makes to obtain a reward, the greater the reinforcing value of that reward becomes for the animal. The experiment is, however, open to a different interpretation. It may be that effort expended can itself be directly conditioned to a situation. If this is true, we might expect the extra effort exerted by animals trained with large weights to generalize to further situations involving the same reward, and this would explain why animals trained under high effort make more vigorous responses in the second situation involving the same food reward.

Since it is not possible to resolve the conflicts in the literature upon the effects of effort it is not worth prolonging the discussion of this problem here.

It should be added, however, that if effort is to some extent frustrating, frustration theory makes much the same predictions about the effects of effort as does cognitive dissonance theory. Our own theory could also account for greater resistance to extinction with additional effort. The slowing down of performance under high effort may allow the response to be attached to more different cues or alternatively animals may actively seek to minimize effort by trying alternative solutions to an effortful problem, and, hence, learn about more cues.

### D. (E-18) Preference for PR Goal Box

Perhaps the most interesting predictions made by Lawrence and Festinger hinge on their assumption that the PRE is due to extra attractions in the goal box rather than to anything that happens in the runway. The most direct way of testing this is to see whether a goal box associated with PR is preferred to one associated with CR. Lawrence and Festinger undertook several experiments along these lines. They showed (L & F 1) that after initial training under CR or PR, PR animals learned a new task more readily in order to reach their original goal box than did CR animals; to produce any learning of the new task they had to give a few reinforcements at this stage. Both frustration theory and the discrimination hypothesis are also able to account for this result. According to frustration theory, more frustration should be evoked by an empty box previously associated with CR than one associated with PR and, hence, PR animals will learn a new task more readily in order to reach the empty box. On the discrimination hypothesis, the sudden change in schedule when CR animals are run with occasional rewards to the goal box should reduce its reinforcing properties more rapidly than the lesser change that occurs when animals trained under PR are run on the new task.

In two further experiments (L & F 12 and 13), Lawrence and Festinger were unable to demonstrate that animals will learn to run more readily to a box in which they have not been rewarded than to one in which they have been previously rewarded. The box had been placed halfway down the runway during initial training with CR given in a second box at the end of the alley. Animals learned a new instrumental response equally well to get to the mid-box whether or not they had been reinforced in it. Saltzman (1949) and Klein (1959) have used a choice situation to test whether animals prefer a goal box associated with reward to one not associated with reward. During training, they ran animals under PR but one goal box was always present on reinforced trials, the other on nonreinforced trials. When the goal boxes were placed on either arm of a T maze, animals showed a preference for the box associated with reinforcement. The first stage of these experiments is similar to that of Bitterman *et al.* (1953) (E-8) but these investigators used only a single goal box in extinction and ran one group with the box associated with reward, the other with the nonreinforcement box; under these conditions, the latter group extinguished more slowly (but this finding has been difficult to repeat, see page 363). It is difficult for Lawrence and Festinger to explain why it is that a preference for the CR box should emerge when animals are offered a choice between the two goal boxes, while when different groups are extinguished on each box, the group run to the nonreinforced box may sometimes extinguish more slowly. Both frustration theory and the discrimination hypothesis can, however, handle these findings: when offered a choice between the two goal boxes animals should, at least initially, prefer the one associated with reinforcement, because it will have acquired secondary reinforcement properties. If only one box is used with each animal, however, running should be maintained longer on the goal box not associated with food, either because it is more difficult for the animal to detect a change in schedule (discrimination hypothesis) or because less frustration is evoked by discovering no food in the nonreward goal box than by discovering no food in the reward goal box.

The experiments just discussed investigated preferences for a goal box not associated with reward and one associated with reward. Similar experiments have been performed using a goal box always associated with food and one associated with food on a PR schedule. Once again, Lawrence and Festinger should predict that when offered a choice between the two boxes animals will prefer the PR box. D'Amato, Lachman, and Kivy (1958) ran this experiment and found that when animals were offered a choice between the two boxes they showed a preference for the CR box over the first 15 trials of testing, and thereafter, performance reverted to chance. When testing was conducted with separate groups, one extinguished on the CR box paired with a neutral box, the other extinguished with the PR box paired with a neutral box, animals tested with the CR box showed a larger preference over the first 15 training trials than did animals tested with the PR box; but over the second 15 trails, animals tested with the PR box maintained their preference at a higher level. Mason (1957) also found that in a choice situation animals initially prefer a box associated with CR to one associated with PR.

At first sight, these results are against cognitive dissonance theory since it could be argued that the original goal box associated with PR would gain extra attractions and would therefore be preferred when animals had a choice of running to it or to a goal box associated with CR. Lawrence and Festinger, however, argue that early in extinction the additional expectation of reward associated with a CR goal box may more than counteract the extra attractions attributed to the PR goal box; they would therefore only expect the extra attractions due to PR training to show if sufficient preference trials are given for the expectation of consistent reward to weaken. Although

D'Amato *et al.* found that the preference for the CR goal box wore off after about 15 test trials in the choice situation, there is no evidence that eventually the PR goal box comes to be preferred to the CR goal box. These results are therefore in conflict with predictions made by dissonance theory, and apart from L & F 1 the results fit well with expectations based on the concept of secondary reinforcement.

## E. (E-19) PRE Determined by Alley or Goal Box

Perhaps the most ingenious of Lawrence and Festinger's experiments are L & F 3 and 4. They trained rats on the combinations of goal boxes and alleys shown in Figure 10.6. Animals were rewarded 100% of the time on the

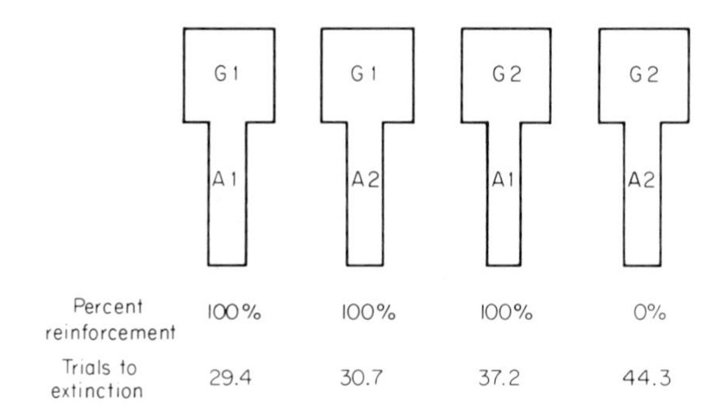


**Fig. 10.6.** Lawrence and Festinger, experiments L & F 3 and 4.

combinations $A_1$–$G_1$, $A_2$–$G_1$, $A_1$–$G_2$, but were never rewarded when the combination $A_2$–$G_2$ was present. The animals were then split into four groups and each group was extinguished on a separate combination of alley and goal box. Trials taken to a criterion of extinction are shown in Figure 10.6. The training procedure ensures that animals will experience PR on Alley 2 and Goal Box 2, but CR on Alley 1 and Goal Box 1. It is clear that animals are more resistant to extinction when they are extinguished with either the partial alley or the partial goal box present than when they are extinguished with the alley or goal box associated with CR during training. Moreover, the difference in extinction scores between Goal Box 1 and Goal Box 2 is slightly larger than the difference between Alley 1 and Alley 2. The difference in extinction times when the two different goal boxes are used is predicted by Lawrence and Festinger, and it is important for them to establish that such a difference will occur when the alley is held constant but different goal boxes associated with CR and PR are used. It seems to us, however, that it is difficult for Lawrence and Festinger to explain why the PR alley should result in greater resistance to extinction than the CR. If the PR alley does acquire cognitive attractions (and it is hard to see how it can since animals do not have expectations disappointed in the alley itself) then animals should actually run more slowly in Alley 2 than in Alley 1, since the extra cognitive attractions should make them loath to leave the alley associated with PR. On the other hand, if extra attractions are not attributed to the PR alley, we would expect there to be no difference in running times in the PR and CR alley.

The discrimination hypothesis seems able to account for why both the PR alley and the PR goal box should produce a PRE: whether the CR alley or the CR goal box is used in extinction, there will clearly be a bigger change in schedule than when the PR alley or goal box is used. The frustration hypothesis can account well for why animals should be slower to extinguish in the PR alley, since running will have been conditioned to anticipatory frustration in that alley. It can account for the difference in the PRE produced by the two goal boxes by supposing that greater frustration is experienced in the CR goal box than in the PR one and hence aversion to it builds up faster in extinction. Our own theory also accounts well for why the PR alley should be effective in producing a PRE since animals should have learned to run to many different cues in this ally. It is, however, impossible for us to explain why the PR goal box should produce the effect.

### F. (E-20) Nonrewarded Intertrial Placements

In Section (E-7), we discussed the effects of rewarded intertrial placements. Lawrence and Festinger's theoretical ideas led them to undertake experiments on the effects of unrewarded placing. They ran (L & F 6) three groups of animals: one under CR, one under PR, and the third with CR for running, but with an equal number of interpolated trials on which rats were manually placed in the goal box without reinforcement. Since the third group extinguished slightly more quickly than the CR group with no placements, there was no PRE due to placing without reward. On the assumption that on placement trials, no effort was expended, Lawrence and Festinger can predict this result.

When this procedure is used, animals can always predict reward conditions: they are always rewarded when run and never when placed. The question arises of what happens when the placements themselves are given on a PR schedule. In a further experiment (L & F 16, see also Theios and Polson, 1962, where apparently the same experiment is reported), Lawrence and Festinger attempted to resolve this issue. They repeated the previous experiment but added a fourth group that was given CR for running but received placed trials on which reward was present on one placement in four. This group showed a marked PRE: they were much slower to extinguish than an ordinary CR group, and took almost as long as a normal PR group. Lawrence and Festinger interpret this result as being favorable to their hypothesis, but since no obvious effort was involved on placed trials, it is hard to see why any cognitive dissonance should have been produced.

Several further experiments have been performed on the effects of placing animals on a PR schedule. D'Amato and D'Amato (1962) found that if rats were first given a few CR trials with reinforcement for running, and subsequently given 64 placements in the goal box either under a PR or CR schedule, there was no PRE. Essentially the same result has been obtained by Lewis and Cotton (1958). Jones (1966), on the other hand, gave PR or CR placements without ever allowing animals to run to the goal box during training and obtained a PRE. Trapold and Doren (1966) and Trapold and Holden (1966) argued that the PRE obtained with PR placements by Theios and Polson may have been due to the animals having to run a short distance when placed in the goal box in order to discover whether there was food or not. They obtained a PRE when rats had to run 8 in. after PR placements, but found no effect when animals were placed in such a position that they could perceive the food without locomotion. The results were the same whether the placements were interspersed with normal CR training or not. It seems clear that placement under PR can sometimes produce a PRE, but more experiments are needed to determine the exact conditions under which the effect is obtained.

If Trapold is right in his claim, the results fit with cognitive dissonance theory, since a PRE is obtained only when the animal has to make some effort before discovering that there is no food present. Frustration theory could also explain why PR placements should increase resistance to extinction since presumably less frustration would be evoked during extinction if animals had already had some experience of a goal box not containing food. The frustration hypothesis might also explain why a bigger effect is obtained if the animals have to run to discover whether food is present when placed: the running response in the goal box will be conditioned to frustration and might generalize to anticipatory frustration in the runway under extinction. Frustration theory should predict that resistance to extinction will be greater if PR placements are interspersed with or followed by normal CR runway training since with this procedure the running response in the alley should become conditioned to anticipatory frustration, but Trapold's experiments disconfirm this prediction. Capaldi's version of generalization decrement could explain why PR placing results in greater resistance to extinction, if it is assumed that there is a memory stimulus for such nonrewarded placements. However, if the theory is used in this way to explain the effect of partial placement, it must also predict that when all placements are unrewarded, resistance to extinction will increase and this prediction has been falsified.

Our own theory cannot adequately explain why resistance to extinction should be greater after PR placements, unless we assume that this increases uncertainty in the runway and hence increases the breadth of learning.

### G. (E-21) Classical Conditioning

It is clear that neither dissonance theory nor frustration theory can account for the occurrence of a PRE in an aversive classical conditioning paradigm: it could hardly be argued that the omission of an air puff or electric shock was in any way disappointing or frustrating for the subject. A PRE can certainly be obtained with eyelid conditioning in man (Humphreys, 1939; Spence, Rutledge, & Talbott, 1963; Moore & Gormezano, 1963; Perry & Moore, 1965). However, it is probable that the effect in man is determined by verbally mediated expectancies; the effect disappears when the experiment is conducted in such a way that the subject does not know he is being conditioned (Spence, 1966). Moreover, it has often been found impossible to obtain a clear-cut effect of PR on extinction in animal studies.

Fitzgerald (1966) training dogs with shock as US and change in heart rate as conditioned response obtained no effect when the same number of trials was given to CR and PR animals, but did find a PRE when numbers of reinforcements were matched. However, he gave very few reinforcements (only 12). Fitzgerald, Vardaris, and Teyler (1966) also found a PRE with matched number of reinforcements; they gave an increased number of training trials (30 reinforcements) but did not include a PR group matched for number of trials. In both experiments Fitzgerald controlled for the possibility that the change in heart rate was produced as a consequence of the occurrence of an instrumental struggling response.

Fitzgerald (1966) found a small PRE in salivary conditioning of dogs. Wagner, Seigel, Thomas, and Ellison (1964), also studying conditioned salivation in dogs, found that within each day's extinction session the strength of the conditioned response tended to weaken more rapidly after CR training than after PR but on early trials of each day the response was stronger in CR animals than in PR. A similar finding has been obtained by Berger, Yarczower, and Bitterman (1965) working with classically conditioned activity in goldfish and equating reinforcements not trials. The reason for obtaining a within-days effect may be that CR animals were able to discriminate the within-days extinction schedule more readily than PR animals.

Several other experiments have failed to find any evidence of a PRE using classical conditioning (e.g., Thomas and Wagner, 1964, eyelid conditioning in the rabbit; Vardaris and Fitzgerald, 1969, eyelid conditioning in dogs; Longo, Milstein, and Bitterman, 1962, activity conditioning in pigeons).

It seems possible that a PRE can sometimes be obtained under classical conditioning, particularly if reinforcements rather than trials are equated. It is, however, certain that the effect is much weaker and more elusive than with instrumental training.

Neither cognitive dissonance theory nor frustration theory can explain why there should ever be any PRE under aversive classical conditioning. The discrimination hypothesis can explain the occurrence of the effect, but fails to explain why it should be weaker with classical than with instrumental conditioning. At first sight this appears to be a major weakness in Capaldi's theory. It may, however, be a source of strength: it is not implausible to suggest that classical conditioning may be much more "stimulus bound" than instrumental conditioning, i.e., that classical conditioning occurs only to the immediate physical stimulus and cannot be brought under the control of a memory stimulus. This would go some way to explaining the "involuntary" nature of the classically conditioned response. Some direct support for this suggestion comes from Leonard and Theios's (1967) finding that no patterning effect occurs when rabbits receive eyelid conditioning on an alternating schedule; this strongly suggests that classically conditioned responses cannot be brought under the control of a memory stimulus. Our own theory goes some way to explaining both why the effect should occur and why it should be weaker. It should occur only if the CS provides at least two different cues (i.e., can be detected by more than one analyzer); it should be weaker under classical conditioning than under instrumental since all classical conditioning experiments on the effects of PR have employed very simple stimuli that could only be detected by a few analyzers.

## F. Conclusions

Dissonance theory accounts for some but not all of the results given in earlier sections. Lawrence and Festinger claim that the effects of patterning (E-6) will be accounted for if no cognitive dissonance is aroused on trials where animals do not expect reward; since patterning allows the animal to predict when no reinforcement will occur, resistance to extinction should be less with patterned PR than with random. The rationale of this explanation is not altogether clear since the animal has made the effort of running on non-reinforced trials and on a cognitive dissonance theory one might have expected dissonance to be set up due to the expenditure of effort for no direct reward. However, the result is predicted if less dissonance occurs when no reward is expected than when reward is expected.

The theory can also explain why increasing the size of the reward decreases resistance to extinction after CR and increases it after PR (E-10). In the latter case dissonance during training will be stronger with large rewards than with small, and hence animals run with large rewards will attribute more attractions to the goal box than animals run with small rewards. When CR is given, the deterrent effect of finding no reward in the goal box on extinction will presumably be greater if large rewards have been given during training. The same sort of explanation can be given of why increasing the number of trials during CR training reduces resistance to extinction, but it cannot explain why increasing the number of trials during PR training should sometimes reduce resistance to extinction (E-11). The theory could account for the faster running times after prolonged training under PR than under CR (E-9), if the extra cognitive attractions acquired under PR outweigh the extra incentive value under CR. However, if this is the correct explanation of faster latencies under PR, we would expect the PR goal box to be preferred to the CR box under choice extinction and we have seen above that this is not the case. The theory can also explain some of the effects of drugs on extinction if we assume that depressant drugs decrease dissonance (E-15).

However, we have already seen that in many cases where predictions have been drawn directly from the theory, the predictions have been falsified. Moreover, the theory fails to account for the results of experiments on the generality of the PRE (E-12), for some of our own findings, and for many of Capaldi's results. Cognitive dissonance theory should predict that giving CR before PR would lead to greater resistance to extinction than giving CR after PR since cognitive dissonance at the start of PR will be greater if it has been preceded by CR; we obtained the opposite result (E-2). Moreover, there is no reason why the breadth of learning about runway cues should be greater with PR training than with CR (E-1). It should, however, be noted that dissonance theory might well predict that breadth of learning about goal-box cues would be greater after PR than after CR: animals that experience cognitive dissonance in the goal box might well examine it more thoroughly in order to discover extra attractions. This prediction has not been tested. Finally, the theory cannot explain the results of Capaldi and his associates on sequential effects, e.g., the effects of varying the position at which an intertrial reinforcement is given (E-8).

It appears then that there is little direct evidence to show that cognitive dissonance affects the behavior of rats. Many predictions made by the theory have been falsified, and there are no experimental findings that can be accounted for by dissonance theory and cannot be explained in terms of the discrimination hypothesis, frustration theory, or our own theory. Moreover, the operation of frustration and selective attention receive support from many experiments not directly concerned with the PRE. In these circumstances, it seems safe to reject the cognitive dissonance account of the PRE.

## VII. Conclusions

The experiments reviewed in this chapter are extremely complicated. It may be as well therefore to summarize the main results.

(1) Animals learn to attach the instrumental response to more cues under PR than under CR.

(2) Giving CR training after PR leads to only a small reduction in the PRE, whereas giving CR training before PR leads to a large reduction in trials to extinction.

(3) There is some evidence to suggest that resistance to extinction after PR training is greater with increases in the number of cues to which the instrumental response can be attached.

(4) In a free-operant bar-pressing situation animals may learn a response unit of several bar-presses; thus, if food is given on average every five presses, in extinction animals run to the food hopper only after making several presses.

(5) A large PRE can be obtained when the intertrial interval during training is 24 hr.

(6) With a moderately large number of training trials, the PRE increases as the length of the sequences of nonreinforced trials increases. This variable is more important in determining the effect than is the percentage of nonreinforced trials. With a small number of training trials, the size of the PRE is positively correlated with the number of NR transitions rather than with N length. If reinforced and nonreinforced trials are alternated, animals come to run fast on R trials and slowly on N trials. This effect is not due to external situational cues with short intertrial intervals, but it is not yet known whether it occurs with long intertrial intervals when situational cues are eliminated.

(7) When animals are placed in the goal box and fed in the intertrial interval succeeding each N length, the PRE is weakened or abolished. The effects of such rewarded placements are reduced if a large number of trials is given or if placements are made in a goal box differing from that to which the animals are run.

(8) There is some evidence that when animals are run under PR, and different goal boxes are used on reinforced and nonreinforced trials, animals are more resistant to extinction if extinguished with the goal box associated with nonreinforcement than when the goal box associated with reinforcement is present in extinction. Results on this issue are, however, conflicting. There is evidence that, other things being equal, animals extinguish more slowly when the same intertrial interval is used in extinction as that used between an N and an R trial in acquisition. Similarly extinction is slower (at least initially) when the same goal-box confinement time is used in extinction as was used in acquisition on nonreinforced trials. Where two different confinement times are used for all rats during acquisition, but one is followed by a trial with large reward, the other by a trial with small reward, animals are more resistant to extinction if the confinement time in extinction is the same as that followed by large reward in acquisition.

(9) Asymptotic running times are faster under PR training than under CR, particularly towards the start of the runway.

(10) Increasing reward size reduces resistance to extinction with CR training but increases it with PR training. Resistance to extinction is greater after CR training with varied reward than after CR training with constant reward. When PR training is given with two sizes of rewards on reinforced trials, resistance to extinction is greater when a large reward follows an N trial, and less when a large reward precedes an N trial.

(11) Increasing the number of training trials usually reduces resistance to extinction after CR training; although the same effect is sometimes produced after PR training, results on this question are rather conflicting. A PRE can be obtained with as little as two training trials.

(12) The PRE transfers readily from one situation to another when the response learned in the first helps to maintain behavior in the second.

(13) If rats are given PR training to one stimulus and are then taught to discriminate between that stimulus and another stimulus, the effects of the PR training on the rate of learning the discrimination depend both on whether the PR training is on the positive or negative discriminatory stimulus and upon how many PR trials are given.

(14) When CR training is given, punishing the animals in the goal box during acquisition increases resistance to extinction. The higher the frequency of punishment, the greater the resistance to extinction.

(15) Barbiturate drugs increase resistance to extinction after CR training. When animals are trained under barbiturates, PR training does not lead to higher asymptotic running times than does CR training. However when animals are both trained and extinguished under barbiturates, the drugs do not abolish or even weaken the size of the PRE nor do they reduce the energizing effects of frustration as measured in the Amsel and Roussel double runway.

(16) Delay of reward increases resistance to extinction. There is no evidence that animals come to prefer a box in which they have been delayed to one not associated with delay. When rats are given a number of trials involving immediate reinforcement, short delay of reinforcement and long delay of reinforcement, they are more resistant to extinction when trials with a long delay are succeeded by immediately reinforced trials than when trials with a short delay are succeeded by immediately reinforced trials.

(17) Increasing the effortfulness of the instrumental response does not increase resistance to extinction when animals are trained and extinguished under the same conditions of effort. If trained under high effort, however, animals may take longer to extinguish under low effort than animals trained and extinguished under low effort; the amount of effort exerted by the animal may itself be conditioned to a situation.

(18) Although animals may take longer to extinguish when run to a PR goal box than to a CR goal box, when given preference tests between the two after training they initially tend to have a preference for the CR goal box.

(19) When the same animals receive training involving PR in one alley and CR in another, and PR in one goal box CR in another, resistance to extinction is greater both when the PR alley is used in extinction and when the PR goal box is used.

(20) Giving goal-box placement trials with no reward does not increase resistance to extinction of CR animals. Giving placement trials on a PR schedule increases resistance to extinction under some circumstances, possibly when animals have to locomote on placement trials in order to discover whether or not food is present.

(21) A PRE can be obtained with a classically conditioned eye-blink in man only when the subject is aware that such conditioning is taking place. Although there are some reports of a PRE with classical conditioning in animals, there are several negative findings and if the effect exists it is weaker and more elusive than the PRE obtained after instrumental learning.

Of the theories discussed, two must be rejected because they cannot account for PRE when there are long intertrial intervals: these theories are the response units hypothesis and the Sheffield version of the stimulus after-effects hypothesis. Of the remaining theories, cognitive dissonance appears to make many wrong predictions; it is particularly difficult for it to explain why the presence of the alley associated with PR should be important in determining the effect. Moreover, efforts to demonstrate the predicted preferences for a PR goal box and for goal boxes associated with delay have failed. Frustration theory has received considerable confirmation from experiments in other areas and handles well many of the findings on PR. It is the only theory that appears to account for the generality of the PRE and for faster running times during PR acquisition. Capaldi's version of the discrimination hypothesis also appears to stand up very well. It has led to a great many predictions relating to the sequence in which trials of various types are given and no other theory comes near to explaining these results in detail.

Our own theory has led to predictions on breadth of learning and the effects of giving CR and PR in different orders; no other theory appears to be capable of explaining these results. On the other hand, there are many findings that our own theory has difficulty in predicting, but that can be accounted for by either frustration theory or the discrimination hypothesis. There seem to be no findings that cannot be explained by one or other of these three theories.

It is, of course, likely that several factors contribute to the PRE, and to this extent, we believe that all three hypotheses are correct. This makes it very difficult to make exact predictions. There are three different cases. First, all three theories may predict a given phenomenon; e.g., they all make much the same predictions about the effect of alternating reinforced and nonreinforced trials. Where all three theories make the same prediction, we are unable to assess the relative contribution of each factor to the results. Second, one theory may predict a given result and the others may make no prediction. Our own theory predicts the additional breadth of learning under PR training, the other two do not specifically predict this effect, but are not incompatible with it. Results of this nature offer direct confirmation for the theory making the prediction. Third, the theories may actually make opposing predictions. Both the discrimination hypothesis and frustration theory make opposite predictions to our own theory about the effects of giving PR and CR training in different orders. In such cases, obtaining a given result supports the theory predicting it, but does not necessarily disconfirm those making the opposite prediction. In this situation, it may be that one result is obtained under one set of circumstances, the opposite under another and this opens up the possibility of testing two or more theories simultaneously. Our own theory predicts that when both CR and PR training are given, giving CR before PR reduces resistance to reinforcement more than giving it after PR. Frustration theory makes the opposite prediction. Now, the larger the reward given, the bigger the role that frustration should play in determining extinction effects. We might find therefore that whereas our own theory made correct predictions with low levels of reward, frustration theory made the correct prediction with high levels of reward and such a finding would simultaneously corroborate both theories.

In this chapter, we have been concerned mainly with the effects of giving PR on extinction. There is, however, another situation in which giving PR affects behavior, namely, in probabilistic discrimination learning where animals are placed in a choice situation and given inconsistent reward on two different schedules for each choice; e.g., they may be reinforced 70% of the time for choosing one alternative, and on 30% of trials on which they chose the other alternative. Two main effects of such schedules have been examined: the asymptote of choice responses during acquisition and the effects of such schedules on choice extinction and on choices during reversal learning. Our own model has led to some interesting predictions about these phenomena; the predictions and the data will be reviewed in Chapter 11, and the success of selective attention theory in handling the results of experiments on probabilistic learning is further evidence for its role in determining the PRE.

CHAPTER 11

# Partial Reinforcement and Choice Behavior

## I. Introduction

In Chapter 10, we argued that one reason an instrumental response acquired under PR takes longer to extinguish than one acquired under CR is that PR animals switch-in more analyzers than do CR animals. According to Rule 2 of our model, an analyzer is strengthened to the extent that outputs are consistently correlated with the presence and absence of reinforcing events. If an animal is rewarded on a random 50% of trials for running down an alleyway, then no analyzer will consistently predict reinforcement, and the animal will try out a number of different analyzers, attaching the running response to the outputs of each of the analyzers used. In extinction, therefore, the running response must be extinguished to each of the analyzers before the animal stops running. A CR animal, on the other hand, will have used only one or two analyzers already dominant in the situation and will stop running as soon as the response is extinguished to the output of these few analyzers.

The most direct support for this interpretation of the effects of PR is provided by experiments showing that, when trained on a multiple-cue discrimination problem, PR animals learn about more cues (switch-in more analyzers) than CR animals (Sutherland, 1966b). In this chapter, we consider a second implication of this account. Inconsistently reinforced animals switch-in a number of different analyzers during the course of training; this is because no one analyzer is consistently strengthened. In a single-cue discrimination problem, therefore, inconsistently reinforced animals should never learn to switch-in the single relevant analyzer as strongly as CR animals. According to Rule 4, performance is a joint function of analyzers having high strengths; in the extreme case, therefore, where the relevant analyzer is only switched-in very weakly, accuracy of performance should suffer from inconsistent reinforcement.

When rats are trained on a discrimination problem with 50% reinforcement on one stimulus and zero reinforcement on the other, they take longer to reach criterion than CR animals, but there is not usually any significant difference in the asymptotes reached. There is, however, a second method of providing inconsistent reinforcement in a discrimination problem: this is to reinforce one alternative on, say, a random 75% of trials and reinforce the other alternative on the remaining 25% of trials. Studies of "probability learning," as experiments utilizing such a procedure have been called, have generated a startling amount of controversy, but it is worth stating at the outset that when trained by a correction or guidance procedure on a 75 : 25 problem, rats attain a lower asymptote than when trained on a conventional 100 : 0 problem. Performance under these conditions of inconsistent reward is not perfectly accurate. According to our model, this should be (at least in part) because inconsistently reinforced animals do not learn to switch-in the relevant analyzer as strongly as CR animals, and the first half of this chapter will be devoted to a discussion of this theoretical position. First, however, it will be necessary to retrace our steps by justifying the bold assertion we have just made about asymptotic performance.

## II. Probability Learning

### A. Asymptotic Performance in Probability Learning Experiments

First, it should be made clear that we are only interested in experiments on probability learning in which animals are trained by a correction or guidance procedure, such that each trial terminates with a rewarded response. When trained by noncorrection, rats indisputably "maximize," i.e., select one alternative on approximately 100% of trials (Bitterman *et al.*, 1958; Parducci & Polt, 1958; North & McDonald, 1959). Virtually all animals absorb on the "majority stimulus,"—i.e., the alternative with the higher probability of reinforcement, although some animals are occasionally caught on the minority stimulus (see Weinstock, North, Brody, & LoGuidice, 1965). This maximizing behavior in a noncorrection procedure is not particularly surprising: as soon as the subject starts selecting the majority stimulus on more than 50% of trials, the probability of its actually receiving a reinforcement on the minority stimulus decreases rapidly, and the *effective* schedule of reinforcement for the animal changes from 75 : 25 into 75 : 0. In order to ensure that the subject experiences a 75 : 25 reinforcement schedule, it is necessary to use a correction or guidance procedure, such that if the subject's first choice on any trial is of the alternative not reinforced on that trial, it will then receive reinforcement on the other alternative. It is experiments utilizing this procedure that we shall be talking about, and it is only in such experiments that rats generally fail to maximize.

Paradoxically, this statement may encounter resistance largely because of a previous and entirely different statement made (notably by Estes) about asymptotic performance in probability learning situations. Estes (1959, 1962) suggested that rats never attain an asymptote in probability learning experiments greater than the probability of reward on the majority stimulus. In other words, rats trained on a 75 : 25 problem come to select the majority stimulus on approximately 75% of trials; they "match" the probability of reinforcement. At no time has the confidence with which the statement has been made borne more than a tenuous relationship to the evidence. Estes (1962) cited the following studies as showing matching behavior by the rat in correction experiments in a T maze: Brunswik (1939); Parducci and Polt (1958); Witte (1959); and Hickson (1961). We shall discuss these experiments in chronological order.

(1) Brunswik (1939) was not interested in the question of whether rats would match the probability of reinforcement; he was interested in the prior question of whether they were even capable of learning to select the majority stimulus above chance, i.e., whether they could learn to "distinguish a better chance of reward from a less good one [p. 176]." After 24 trials, a group trained on a 75 : 25 schedule were selecting the majority stimulus on 79.8% of trials (presumably not significantly better than matching), and at this point, the animals having clearly answered Brunswik's question in the affirmative, they were trained on the reversal of their original problem. It can hardly be supposed that 24 trials is sufficient for asymptote to have been reached, and that it is not sufficient in all cases is shown by the results of a second group trained by Brunswik on a 67 : 33 schedule: by the end of this stage of the experiment, this group was only selecting the majority stimulus on 53.1% of trials.

(2) Parducci and Polt (1958) trained four groups of rats in a T maze. Two groups were trained on an 85 : 15 schedule, two on a 50 : 50 schedule, and one of each of these groups was trained by a noncorrection technique, the other by correction. After 60 trials, all groups were reversed to a 30 : 70 problem either by correction or noncorrection. In the first 60 trials of the experiment, the 85 : 15 group reached an approximately matching asymptote, and it is, presumably, this result that Estes is quoting. In the second stage of the experiment, when the new majority stimulus was reinforced on 70% of trials, correction trained animals learned to select this stimulus on approximately 85–90% of trials. The group initially trained on 50 : 50 reached this

asymptote within 100 trials; the group initially trained on 85 : 15 took nearly 200 trials to reach it. This study therefore, so far from supporting Estes, provides the first clearcut evidence that rats do not necessarily match the probability of reinforcement.

(3) Witte (1959) trained rats on a 50 : 50 schedule. The most striking feature of his results is that 58% of his subjects became virtually absorbed on one or other alternative, i.e., selected that alternative on nearly 100% of trials. Again the results do not provide very good evidence of matching.

(4) Hickson (1961) trained a group of rats by a correction technique on a problem where the probability of reward was initially 50 : 50 on each alternative, but was gradually increased to 67 : 33. Training continued for 90 trials, at the end of which time the rats were matching quite closely. Again it is far from clear that the rats were given long enough to adjust to such a gradually changing reward schedule.

The results even of those experiments quoted by Estes, therefore, provide little evidence to suggest that the asymptote of rats' performance matches the probability of reward. Other experiments yielding a matching asymptote are equally unconvincing. Weitzman (1967) reported that rats matched under one set of conditions, but the group average was depressed by the inclusion of one subject (out of a total of five) that selected the majority stimulus on less than 30% of the final 200 training trials. A similar group matching average, artificially depressed by the inclusion of subjects scoring significantly below chance, was reported earlier by Estes (1957).

With spatial discriminations, rats have attained asymptotes significantly above matching in the following experiments: Bitterman *et al.* (1958); Solomon (1962); Uhl (1963); Gonzalez *et al.* (1964); Cole, Belenky, Boucher, Fernandez, and Myers (1965); Roberts (1966); Wright (1967); Calfee (1968). With visual discriminations, the following studies have obtained significantly above matching asymptotes: Bitterman *et al.* (1958); Gonzalez *et al.* (1964); Roberts (1966). To these two lists may be added the results of further studies described in this chapter.

It is certain that rats do not typically match in probability learning experiments. It is equally certain, however, that they do not typically maximize. The term "maximize" may, of course, be given a number of different interpretations, but in the two most obvious senses of the term, maximizing has rarely been observed. The first sense is that rats learn to select the majority stimulus on 100% of trials; or, at any rate, their asymptotic performance is not significantly different from 100%. A casual inspection of any of the published learning curves (including our own shown on pages 413–414) proves that this is false: typically, even after several hundred training trials on a 70 : 30 or 75 : 25 problem, rats choose the minority stimulus on 5–25%

of trials. The only exception to this rule that we are aware of comes from a study by Roberts (1966) on spatial probability learning in immature rats. He found that the average asymptote attained by a group of 10 subjects was 99.4% choice of the majority stimulus. By any reasonable definition, this constitutes maximizing behavior. The second interpretation of the term would be to say that the asymptotic performance of rats trained by a correction technique on a 75 : 25 problem is as accurate either as performance on the comparable 100 : 0 problem, or as performance on a 75 : 25 problem when noncorrection training is given. Even this weaker interpretation, however, is not borne out by the evidence. In the second stage of their experiment, Parducci and Polt (1958) found that after 200 trials, groups trained by correction selected the majority stimulus on significantly fewer trials than did groups trained by noncorrection. Bitterman *et al.* (1958) found a relatively sharp increase in choice of the majority stimulus on a 70 : 30 problem when rats were shifted after 600 trials from a correction to a noncorrection training procedure. Uhl (1963) found differences of up to 20% in percentage choice of the majority stimulus between correction and noncorrection groups after 800 trials. In neither of these senses, therefore, can rats generally be said to maximize; only if the term maximize is given the weak sense of "not match" is it true to say that rats maximize in probability learning experiments.

It is hard to see how anyone could dispute the conclusions reached in the preceding paragraphs. Certainly, we suspect that the reader unacquainted with the research and theorizing on probability learning of the past 15 years will wonder that we should devote so much space to establishing the point that rats make some (but not very many) errors when trained on probability problems. We should emphasize that this is all that, for the moment, we are asserting. We are not, in fact, interested in using the terms "matching" and "maximizing"; so far as we are concerned, the first has appeared in the psychological literature solely because of the predictions which, *with certain assumptions*, can be shown to follow from linear stochastic learning models; while the second has arisen from a reaction against the claims made by proponents of these models.

These predictions derivable from the linear models have focused attention on what is essentially a false issue, for there is nothing intrinsically special about a matching asymptote that makes the behavior of a group of rats that selects the majority stimulus on a 75 : 25 problem on 75% of trials, *qualitatively* different from the behavior of another group that selects the majority stimulus on 80% of trials. The first group is performing somewhat less accurately than the second—that is all. Nor is there anything to prevent the experimenter from arranging conditions so as to depress the performance of the second group until it is exactly matching. Hara and Warren (1961), for example, were able to obtain any level of accuracy they chose when training cats on 100 : 0 visual discrimination problems, simply by varying the discriminability of the stimuli (the results have been replicated with rats by McGonigle, 1967). Analogous procedures will surely have analogous effects in probability learning experiments; if rats reach an asymptote of 90% selection of the majority stimulus in a 75 : 25 black–white discrimination, it will be a simple matter to find two shades of gray on which their performance drops to a matching level. Indeed, it is highly probable that performance on a 75 : 25 schedule is much more sensitive to variations in task difficulty than is performance on a 100 : 0 schedule. A similar argument will apply in the reverse direction: if a group of rats reaches an asymptote of matching the probability of reward in a given situation, this may imply little about the limits of rat performance, and a great deal about the inadequacy of the experimental situation. The stimuli may seem fully discriminable in the sense that rats trained on a 100 : 0 schedule may reach a 100% asymptote, but, from what we have just said about the relative sensitivity of performance on probabilistic and consistent schedules to task difficulty, it would follow that increasing the discriminability of the relevant stimuli could still be expected to improve performance.

### B. Analysis of Asymptotic Errors

Having cleared the ground, we can return to the questions posed earlier in this chapter. Why do rats not learn to select the majority stimulus of a probabilistic discrimination on 100% of trials? What procedures will affect the accuracy of performance in a probability learning situation?

Although we do not wish to claim that other factors are of no importance, our answer to the first question is: because they do not learn to switch-in the relevant analyzer strongly enough to ensure that their behavior is controlled exclusively by the relevant cue. The answer to the second question is that accuracy of performance can be affected by any procedure that affects the strength of the relevant analyzer.

We are suggesting that animals make errors, or fail to select the majority stimulus, not because of a momentary preference for the minority stimulus, but because of a momentary failure of attention to the relevant cue. The relevant analyzer is too weak to exert exclusive control over behavior on all trials. This is equivalent to stating that animals make errors on certain trials because their behavior is controlled by an irrelevant analyzer. If a rat is being trained on a simultaneous brightness discrimination, then it is safe to assume that the irrelevant cue controlling behavior will be position. Therefore, rats make errors in a 75 : 25 brightness discrimination because, on certain trials, their performance is controlled by a position analyzer. It should be possible, from an analysis of trials on which errors are made, to see whether they occur under control of the relevant visual cue or under control of irrelevant position cues.

If errors occur because of a momentary preference for the minority stimulus, they should occur when that preference is presumably greatest, i.e., immediately after minority rewards. Subjects would be expected to "reward follow" on the relevant cue. If, on the other hand, errors occur under control of position cues, then we could predict either of two possible patterns: subjects might have a permanent bias in favor of one side, and the majority of errors should occur to that side; or they might show a momentary preference for the most recently rewarded position, i.e., reward follow on the irrelevant cue.

TABLE 11.I

**Analysis of Errors over Final 100 Trials in Three Experiments on Brightness Probability Learning with a 75 : 25 Schedule**[a]

| | | | Percentage of errors occurring on trials | | |
|---|---|---|---|---|---|
| Experiment | $N$ | Overall probability of error | Following minority stimulus rewards (chance = 25%) | When minority stimulus is in last rewarded position (chance = 55%) | When minority stimulus is in preferred position[b] (chance = 50%) |
| I | 28 | .198 | 26.2 | 66.2[c] | 66.8[c] |
| II | 20 | .155 | 32.1[c] | 79.4[c] | 74.9[c] |
| III | 10 | .112 | 25.7 | 70.1[c] | 84.5[c] |

[a] Data obtained from Mackintosh (1970).

[b] For each subject, preferred position was the side chosen more frequently over the first 20 trials on which the present analysis was performed.

[c] Different from chance at .01 level.

Table 11.I presents an analysis of the errors made over the final 100 trials of training in three separate experiments on brightness probability learning on a 75 : 25 schedule in rats (Mackintosh, 1970). The table shows the overall probability of an error, together with the percentage of errors occurring on trials following minority rewards, on trials on which the minority stimulus appears in the position rewarded on the preceding trial, and on trials when the minority stimulus is in each subject's preferred position. All three experiments show the same general trend. The proportion of errors occurring on trials following minority stimulus rewards is not greatly above chance (in only one of the three experiments was the difference significant). Choice behavior is not, therefore, particularly well predicted by a knowledge of which value of the relevant cue was last rewarded. It is, on the other hand, well predicted by a knowledge of which value of the irrelevant cue was last rewarded: in all three experiments, animals showed a significant tendency to make errors by responding to the position rewarded on the preceding trial. In addition to this, animals continued to show systematic preferences for one position rather than the other, and were more likely to select the minority stimulus when it was on their preferred side.

These results suggest that errors occur under the control of irrelevant rather than of relevant cues. The conclusion is strengthened by an analysis of the performance of rats trained on a 75 : 25 spatial discrimination (Mackintosh, 1970). Two groups of rats were run on this problem, one with an irrelevant brightness cue present, the other with no irrelevant cues. The error analysis for the two groups is shown in Table 11.II. Subjects trained with

TABLE 11.II

**Analysis of Errors over Final 100 Trials of Spatial Probability Learning on a 75: 25 Schedule, with or without Brightness Irrelevant[a]**

| | | | Percentage of errors occurring on trials | | |
|---|---|---|---|---|---|
| Group | $N$ | Overall probability of error | Following minority side reward (chance = 25%) | When minority side is last rewarded color (chance = 55%) | When minority side is preferred color[b] (chance = 50%) |
| Control | 10 | .050 | 50.9[c] | — | — |
| Brightness Irrelevant | 10 | .102 | 31.5 | 56.8 | 87.9[c] |

[a] Data obtained from Mackintosh (1970).
[b] This turned out to be black for all subjects.
[c] Different from chance at .01 level.

brightness irrelevant again showed no significant tendency to reward follow on the relevant (position) cue, nor, in this instance, did they reward follow on the irrelevant cue. However, all subjects showed a significant preference for one value of the irrelevant cue (in all cases black), and the large majority of errors occurred when the minority position was black. The overall probability of an error was lower in the group trained without the irrelevant brightness cue (a result to be discussed later), and it was not, of course, possible to analyze any responses to irrelevant cues. These subjects did, however, show a highly significant tendency to reward follow on the relevant cue: just over half of their errors occurred following minority rewards. This result not only implies that errors *may* occur because of a momentary preference for the minority stimulus, provided there are no irrelevant cues competing for control of behavior; it also shows that the measure we have employed is sufficiently sensitive to detect relevant-cue reward following, provided a substantial tendency exists.

When irrelevant cues are present, however, errors occur because of a momentary or permanent preference for one value of the irrelevant cue, rather than because of a momentary preference for the minority stimulus. This finding may not force acceptance of an attentional analysis of discrimination learning (any more than the hypothesis behavior described by Krechevsky); it is, however, consistent with such an analysis, and, if interpreted within the present framework, proves that the major cause of errors in probability learning is a failure to attend correctly rather than a failure to learn the correct response.

## C. Factors Affecting Asymptotic Performance

If inaccurate performance in probability learning experiments is mainly due to a failure to switch-in the relevant analyzer strongly, it should be possible to influence accuracy by procedures affecting the strength of the relevant analyzer.

### 1. *Pretraining*

Chapter 6 reviewed a number of experiments (mostly stemming from the work of Lawrence) in which subjects pretrained to switch-in or switch-out various analyzers were tested on discrimination problems involving cues detected by those analyzers. Lawrence (1950), for example, demonstrated that if rats were trained to utilize Cue A and ignore Cue B in solving a successive conditional discrimination, then Cue A was more likely than Cue B to control behavior when the subjects were subsequently trained on a simultaneous discrimination. In order to see whether similar pretraining procedures would have comparable effects on *accuracy* of performance in a probability learning situation, Mackintosh and Holgate (1967) trained three groups of rats on a 75 : 25 brightness discrimination. Two groups had received pretraining on 100 : 0 successive conditional discrimination problems; the third (control) group had received no pretraining. Of the two pretrained groups, one (Group B–W) had been trained on a successive brightness discrimination, i.e., to switch-in a brightness analyzer; the second (Group H–V) had been trained on a successive orientation discrimination with brightness irrelevant, i.e., to switch-in an orientation analyzer and to switch-out brightness. The performance of the three groups over 500 trials of brightness training is shown in Figure 11.1. Two predictions follow from the theory. First, Group B–W should perform more accurately than the control group. Except possibly

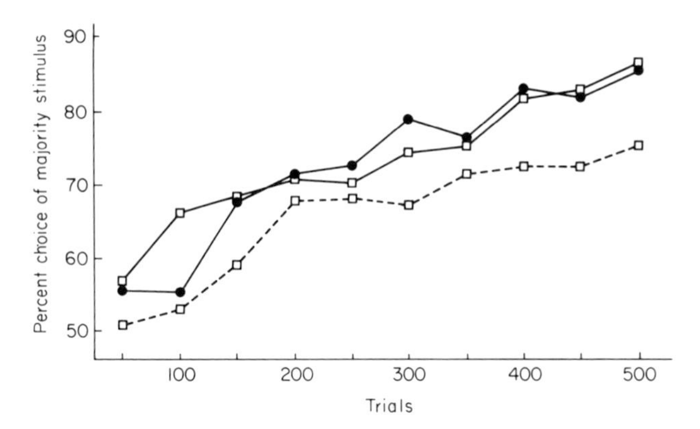


**Fig. 11.1.** Performance of three groups of rats trained on a 75: 25 brightness discrimination. Group B-Rel was pretrained on a conditional successive brightness problem; Group B-Irrel on an orientation problem with brightness irrelevant; the control group received no pretraining. □—□: brightness relevant; □- - -□: brightness irrelevant; ●—●: control. [*After* Mackintosh and Holgate (1967). Reprinted with permission of author and publisher: Mackintosh, N. J., & Holgate, V. Effects of several pre-training procedures on brightness and probability learning. *Perceptual and Motor Skills*, 1967, **25**, 629–637, Figure 1.]

over the first 100 trials—this prediction is clearly not confirmed. Second, Group H–V should perform less accurately than the control group—this prediction is abundantly confirmed. Even after 500 trials, this group was performing at a level of accuracy 10–15% lower than that of the other two groups. These results do not, of course, prove that Group H–V would *never* have learned to perform as accurately as the controls (nor does the theory require any such extreme claim). The point is, rather, that over a large number of trials, the performance of Group H–V was adversely affected by the pretraining they had received.

The pretraining designed to strengthen the relevant analyzer, however, did not lead to any increase in accuracy. One possible explanation of this failure can be largely discounted. It seemed reasonable to suppose that even if the brightness analyzer were strengthened before probability training began, it would be gradually weakened during probability learning, and therefore have no beneficial effect on performance after the first few trials. To test this, another group was trained on a successive brightness discrimination, and then on a brightness probability problem, with retraining on the successive problem at the outset of each day. Again there was some evidence of superior performance at the beginning of probability learning, but again this superiority rapidly disappeared.

A second relatively plausible cause for the failure of the prediction can be suggested. A successive brightness discrimination is more properly called a conditional brightness and position discrimination (see Chapter 3). Solution demands the joint utilization of brightness and position analyzers. Learning to use a brightness analyzer may benefit brightness probability learning; but learning to respond in terms of position could only depress performance. To the extent that one of the main causes of transfer effects is the suppression of irrelevant cues (see Chapter 6), it is not surprising that successive pretraining may not always enhance simultaneous discrimination performance.

Mackintosh and Holgate (1967) also investigated the effect of absolute brightness training on brightness probability learning. In two studies, experimental subjects were pretrained on an absolute brightness discrimination (black or white versus gray) and their performance on a 75 : 25 black–white discrimination compared with that of a control group. The results are shown in Figure 11.2. In both cases, experimental subjects appear to have learned

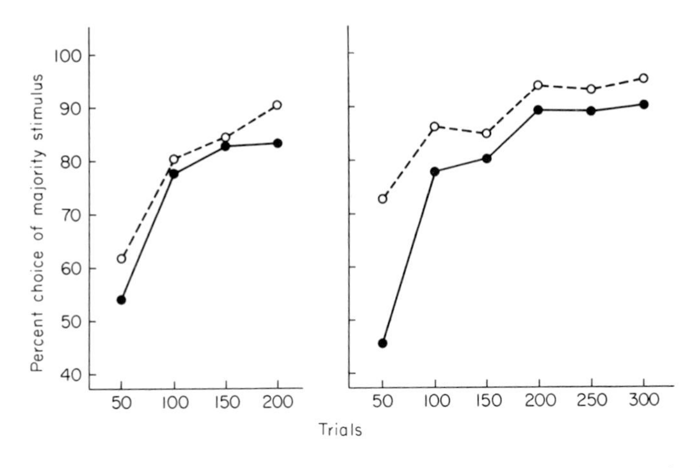


**Fig. 11.2.** Brightness probability learning by controls and by rats pretrained on an absolute brightness discrimination. The first replication is shown on the left; the second on the right. ○: absolute brightness; ●: control. [*After* Mackintosh and Holgate (1967). Reprinted with permission of author and publisher: Mackintosh, N. J., & Holgate, V. Effects of several pre-training procedures on brightness probability learning. *Perceptual and Motor Skills*, 1967, **25**, 629–637, Figure 2.]

more rapidly, and to have reached a higher asymptote. The results are, however, far from impressive; in the second experiment the difference between the two groups over the final 150 trials was not quite significant and two experimental subjects (whose results are not included in the figure), having learned the problem quite rapidly, showed a progressive decline in performance over the final 200 trials, and ended the experiment selecting the minority stimulus above chance.

Despite this aberrant behavior on the part of some experimental subjects, one feature of the results strongly suggests that performance is influenced by the strength of the relevant analyzer. As was to be expected, the control animals in this experiment showed no significant tendency to reward follow on the relevant cue, but did reward follow on position. The experimental animals, however, made errors for exactly the opposite reason, i.e., following minority stimulus rewards, but not on trials when the minority stimulus occupied the last rewarded position. Whereas, therefore, animals normally make errors under the control of irrelevant cues, animals pretrained to attend to the relevant cue make errors under relevant-cue control.

### 2. *Salience of Relevant and Irrelevant Cues*

If probability learning can be affected by pretraining procedures designed to alter the relative dominance of the relevant cue, it should equally be affected by the natural salience of relevant and irrelevant cues. In the absence of strong irrelevant cues, the relevant analyzer will still control behavior, even if its absolute strength is not high. As mentioned above, two groups of rats trained on a 75 : 25 position problem, one with and the other without an irrelevant brightness cue, attained significantly different asymptotes (Mackintosh, 1970). Without the irrelevant brightness cue, animals came to select the majority side on 95% of trials; with it they reached an asymptote of just less than 90% choice of the majority side.

Animals trained on the simple position problem not only performed more accurately than those trained with brightness irrelevant, they also learned much more rapidly, and to a higher asymptote, than animals trained in other experiments on visual problems. The typical asymptote attained under the conditions of our experiments has been about 90% on visual problems, and in some cases, animals have required up to 500 trials to reach this level of performance. Although it is possible that further training would have led to better performance, over a relatively large number of trials performance is better with a more salient relevant cue and in the absence of irrelevant cues.

Two other studies have found that rats attain a higher asymptote more rapidly when trained on spatial probability problems than when trained on the same schedule in a visual problem. Bitterman *et al.* (1958) found that over trials 101–200 of a 70 : 30 spatial problem, subjects selected the majority stimulus on 95.33% of trials; while on a 70 : 30 visual problem they only reached an asymptote of 91.0% choice of the majority stimulus over trials 501–600. Similarly, Roberts (1966) trained immature rats on 70 : 30 visual and spatial problems, and found an asymptote of 99.4% choice of the majority stimulus over trials 141–200 of the spatial problem, but one of only 93.3% over trials 301–400 in the visual problem. The differences are small, but in both cases significant.

### D. Conclusions

Although these studies may have left a number of ambiguities unresolved, they have provided good evidence for the applicability of our model to performance in probability learning situations. First, errors are neither made at random, nor are they, in general, due to subjects selecting the last rewarded relevant stimulus. The most comprehensive account of errors is to say that they occur when subjects respond in terms of irrelevant cues—whether by displaying an absolute preference for one value of the irrelevant cue over the other, or by selecting the last rewarded irrelevant stimulus. Second, accuracy of performance in probability tasks is a function of a number of variables most reasonably interpreted as affecting the strength of the relevant analyzer —whether by specific training procedures or by variations in the nature of the relevant and irrelevant cues.

## III. Inconsistent Reinforcement and Extinction of Choice Behavior

We turn now to the more general question of the effects of inconsistent reinforcement on the extinction of choice behavior. What is the effect of a 50 : 0 or 75 : 25 training schedule on the extinction of a discrimination habit? By comparison with the amount of data collected on the effects of inconsistent reinforcement on the extinction of running or bar-pressing behavior, remarkably few experiments have been undertaken on choice extinction. In his 1960 review of PR, Lewis cited only five studies under this heading. Lack of interest in this question is presumably justified by the assumption that PR has identical effects on running time and choice behavior: in both cases, PR will lead to slower extinction. If this were indisputably true, there would be reasonable cause to concentrate effort on an investigation of one rather than the other; and the tendency to investigate supposedly simpler situations would undoubtedly dictate a preference for the runway over the T maze. Unfortunately, the few early studies of choice extinction were misleadingly interpreted as proving just this similarity in the effects of PR, and despite some subsequent contradictory evidence, there the matter has rested.

### A. Predictions

Our own model predicts that PR, while increasing resistance to extinction on a latency measure, will (under certain circumstances), actually decrease resistance to extinction on a choice measure. The first prediction derives from the assumption that PR increases the number of analyzers used by the subject, and, therefore, increases the number of connections that must be broken in extinction. The second prediction follows from the converse of this: if an animal is trained under PR on a problem with a single relevant cue, it will try out a number of different analyzers, and, therefore, not fully strengthen the relevant analyzer. Hence in a probability learning experiment the animal will fail to reach a 100% asymptote. Hence, in extinction, the relevant analyzer will soon be weakened, and the subject's behavior will be controlled by other, irrelevant analyzers. Continued selection of S+ over S−, however (i.e., resistance to choice extinction), depends upon continued use of the relevant analyzer; as soon as this is switched-out, the subject will respond at chance with respect to the relevant cue and will have reached a criterion of choice extinction.

This prediction about choice extinction depends upon exactly the same argument as that developed in Chapter 10, where we suggested that the magnitude of the PRE in a runway situation would depend upon the number of available cues. In an impoverished environment, the PRE was found to be smaller than that obtained in a more complex runway. If PR simply increases the range of analyzers used by preventing any single analyzer from becoming very strong, its effects on extinction will depend upon the number of cues that can control the response being measured in extinction. Thus, while we expect PR to decrease the resistance to extinction of choice behavior in a situation with only a single relevant cue, it could well increase resistance to extinction if a large number of relevant cues were available. This argument will not only apply to situations where the experimenter deliberately manipulates the number of relevant cues. A spatial discrimination in a T maze is soluble on the basis of numerous cues: visual, tactile, olfactory, kinesthetic, as well as (in an elevated maze) an indefinite number of extramaze cues. It is only where there is good reason to believe that the discrimination problem involves a single relevant cue, that our model unequivocally predicts that PR will decrease resistance to choice extinction.

The suggestion that PR might have opposite effects on the extinction of choice behavior and of a simple running response has perhaps an air of paradox. It should not seem too surprising, however, since there are other cases where choice and latency measures fail to agree. Several investigators have reported that asymptotic running speed in a runway may be enhanced by PR (e.g., Weinstock, 1958; Goodrich, 1959); yet the experiments reviewed in the early part of this chapter have all been concerned with the problem of why inconsistent reinforcement tends to depress asymptotic performance in a discrimination problem. Despite this, most theoretical accounts of the PRE obtained in runway situations provide no grounds for predicting a different outcome in a choice situation. All the theories considered in Chapter 10 appear to predict that PR will retard choice extinction in just the same way that it retards extinction of a running response. Our own account seems to be the only one that predicts opposite results in the two situations.

Unfortunately, the fact that unambiguous predictions can be derived from different theoretical positions does not imply that the results of experimental tests of these positions will be equally unambiguous. For, to the extent that inconsistent reinforcement has effects on behavior inferable both from our own theory and from, say, some version of the discrimination hypothesis, we are in the familiar position of being unable to make exact predictions until we know the relative weights of each factor. Discouraging as this may seem, it is still possible to make a weak prediction: to the extent that our own account of PR is correct, PR should at least not retard extinction of choice behavior as much as it retards extinction of a running response. As we shall see, there is some evidence supporting this weak prediction.

## B. Evidence

Of five published experiments on the effect of PR on extinction of choices in a T maze, two found that PR increased resistance to extinction, two found no effect, and in one a reversed PRE occurred. Lewis and Cotton (1958), and Cotton, Lewis, and Jensen (1959) both found that PR significantly increased resistance to extinction of choice behavior. In the latter experiment, it should be noted, the apparatus was an elevated T maze with one goal arm black and the other white, so that there were many relevant cues. In the remaining studies, however, no such PRE was observed. Denny (1946) and Pavlik and Lehr (1967) found no difference between partially and consistently rewarded subjects over a series of free extinction trials. Pavlik and Lehr also recorded speed of responding during extinction, and on this measure did observe a significant PRE. Finally, Spear and Pavlik (1966) found that PR subjects extinguished significantly more rapidly than CR subjects.

The impression derived from these studies, that the PRE standardly obtained in runway experiments is much more elusive in a choice situation, is abundantly confirmed by the results of visual studies. Of those experiments in which subjects have been trained on visual discrimination problems (with position irrelevant), only one has found that PR increases resistance to extinction. Pavlik and Born (1962) trained four groups of rats for 180 trials on a brightness discrimination, giving blocks of one free and three forced trials that ensured equal experience with both alternatives. The schedules of reinforcement for the four groups were as follows: 100 : 0; 50 : 0; 100 : 50; and 67 : 33. In 60 extinction trials, Group 100 : 0 made more choices of their original positive stimulus than any other group. This was largely because they had reached a higher asymptote in original learning. Even a covariance analysis, however, failed to reveal a normal PRE. There is some difficulty in interpreting these results, since none of the groups showed any marked signs of extinguishing at all, presumably because of the extensive forcing given to the less favorable alternative during original learning. Nevertheless, it is clear that PR did not increase resistance to extinction.

In two separate experiments, we have found PR to decrease resistance to extinction. Sutherland (1966b) trained rats under either 100 or 50% reinforcement for 200 trials on a combined brightness and orientation discrimination. Over 60 extinction trials, the CR group failed to respond within 2 min on 35% of the trials, the PR group only failed to respond on 25% of trials. However, on trials on which subjects did respond, the CR group selected S+ 98% of the time, while the PR group selected S+ on only 79% of trials.

Mackintosh and Holgate (1968) trained two groups of rats for 200 trials on a brightness discrimination, one on a 100 : 0 schedule, the other on a 75 : 25 schedule. Both groups were then extinguished to a criterion of equal choice of black and white over 10 trials. Group 75 : 25 reached this criterion in fewer (but not significantly fewer) trials. Latencies were recorded during extinction, and revealed a large and highly significant PRE. The two sets of data (choice and latency measures) are shown in Figure 11.3. The fact that from the same subjects a PRE may be obtained on one measure and not on another provides important evidence in our favor.

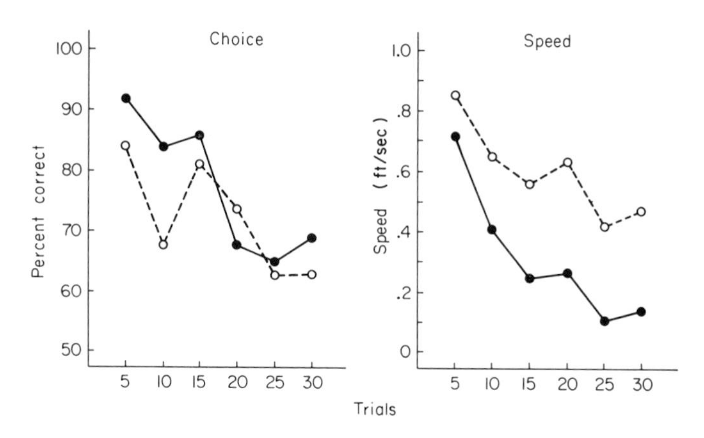


**Fig. 11.3.** Choice behavior (percentage choice of the former S+) and speed of responding during extinction following consistent or inconsistent reinforcement. ●: Group 100:0; ○: Group 75:25. [*After* Mackintosh and Holgate (1968). Copyright by the American Psychological Association. Reproduced by permission.]

Finally, in two experiments, an attempt has been made to examine the effect of PR on choice extinction as a function of the number of relevant cues. As mentioned in Chapter 10, McFarland and McGonigle (1967) trained rats on two visual discriminations and then extinguished them with either one or both relevant cues present. They found that PR during acquisition increased the number of correct responses in extinction, irrespective of whether one or two cues were present. Sutherland and Andelman (unpublished) trained PR and CR groups of rats either on a single-cue discrimination, or on one with six relevant cues. They found no effect of PR on choice extinction, but again, the result was not affected by the number of relevant cues. These results, in particular McFarland and McGonigle's demonstration of a PR choice extinction effect, do not accord easily with our position. It is possible, however, as was also noted in Chapter 10, that the main determinant of choice extinction is the relative strength of relevant and irrelevant analyzers: in rats trained on a visual problem, chance performance (complete choice extinction) is correlated with the onset of position responding. If PR fails to suppress irrelevant position analyzers as effectively as CR, then no number of relevant cues will produce a PR choice extinction effect when position is irrelevant.

### C. Conclusions

Our own account of PR leads to the prediction that inconsistent reinforcement may result in more rapid extinction of choice behavior than CR. Only one study (Spear & Pavlik, 1966) has obtained a significant difference in this direction. Other accouts of PR, however, predict exactly the opposite result; i.e., that inconsistent reinforcement should result in slower extinction of choice behavior. Only two studies of spatial discrimination learning (Lewis & Cotton, 1958; Cotton *et al.*, 1959), and one of visual discrimination learning (McFarland & McGonigle, 1967) have obtained a significant difference in this direction. In the remaining studies (a total of six), no significant differences were observed.

The prediction derivable from our account may depend upon the number and relative salience of relevant and irrelevant cues. Where there are several relevant cues and no strong irrelevant cues, we predict that PR may increase resistance to extinction; one of the three studies obtaining this result (Cotton *et al.*) must have satisfied these conditions, since rats were trained on an elevated T maze with black and white goal arms. While we certainly cannot account for the results of McFarland and McGonigle,[1] we can reasonably suggest that the overall trend of the evidence on choice extinction is so different from the evidence derived from runway experiments as to offer sub-

[1] We cannot account for these results unless we adopt the somewhat speculative suggestion, noted in Chapter 9, that just as responses become conditioned to the memory of nonreinforcement, so also may analyzers. If this were so, then although a PR analyzer would never attain the same strength as one consistently reinforced, it would be more resistant to extinction. This, of course, would lead us to predict a PR choice extinction effect, a prediction not supported by the balance of the evidence. It is possible that an important variable here is the amount of acquisition training (McFarland and McGonigle gave more than twice as many acquisition trials as have other investigators); conceivably the conditioning of an analyzer to the memory of nonreinforcement only occurs after extended training.

stantial support for any theoretical position that makes different predictions for the two situations. So far as we are aware, with the exception of our own, the theories described in the previous chapter predict that PR will have the same effect on extinction whether a choice or a latency measure is used. Whatever else may be said about the results reviewed in this section, it is at least clear that this is not true, and to this extent, it is likely that our own analysis provides part of the explanation of the effects of inconsistent reinforcement.

## IV. Inconsistent Reinforcement and Reversal Learning

Of the five studies mentioned by Lewis (1960) on the effects of PR on choice extinction, two are studies of reversal learning following consistent and partial acquisition. As Lewis recognizes, the fact that reversal learning was, in both cases, found to be impeded by PR acquisition, provides, at best, ambiguous information about the effects of PR on extinction. The reversal of a 100 : 0 discrimination problem is frequently learned more slowly by animals trained only to criterion than by animals overtrained on the initial problem; and although it has been suggested that this is because overtraining decreases choice resistance to extinction, it is certain that this explanation is incorrect (see Chapter 8, pages 287–293). Overtraining, in fact, facilitates reversal learning despite retarding extinction. Equally, therefore, CR animals may reverse more rapidly than PR animals, but still extinguish more slowly than PR animals.

If inconsistent reinforcement leads to incomplete strengthening of the relevant analyzer in a discrimination problem, then it follows from our explanation of the ORE that inconsistent reinforcement will also retard reversal learning. Animals trained only to criterion are likely to reverse more slowly than overtrained animals because analyzer learning proceeds more slowly than response learning, and training only to criterion will mean that the relevant analyzer is substantially below asymptote. If the relevant analyzer is only weakly switched-in, animals will reverse slowly because the analyzer is soon extinguished during early reversal trials, and the less the analyzer is switched-in, the less subjects can learn about the new correlation of the relevant cue with reward. Since the relevant analyzer is never switched-in strongly if inconsistent reinforcement is given, it follows that such a training procedure is bound to produce slow reversal.

### A. Speed of Reversal

That reversal learning is, in fact, retarded by inconsistent reinforcement during acquisition is indisputable. There appears to have been only one T maze study of PR and reversal of a spatial discrimination (Hill, Cotton, &

Clayton, 1962), and in this case, PR led (in general) to slower reversal learning: PR subjects made fewer correct choices in reversal training.

There has, however, been a relatively large number of studies of visual discrimination reversal following consistent and inconsistent reinforcement. Grosslight *et al.* (1954), Grosslight and Radlow (1956, 1957), Kendler and Lachman (1958), Wise (1962), and Erlebacher (1963) have all shown that inconsistent reinforcement during acquisition retards the reversal of a simultaneous brightness discrimination in rats. Elam and Tyler (1958) showed that reversal of a shape discrimination in monkeys was slower following a 60 : 40 reinforcement schedule than a 60 : 0 schedule. The only exception to this general rule comes from an experiment with cats (Warren & McGonigle, 1969): three groups of cats trained with either a 100 : 0, 80 : 20, or 60 : 40 reinforcement schedule on a brightness discrimination, did not differ in the number of trials required to reach criterion on the 0 : 100 reversal of the original problem.

### B. Pattern of Reversal Learning

Although, as we have said, we predict that inconsistent reinforcement will retard reversal, our interpretation differs sharply from that normally put upon these results. According to our model, PR subjects should reverse more slowly than CR subjects because, although they extinguish more rapidly, they switch-in other analyzers during reversal, and are more likely to respond to irrelevant cues. The implicit assumption underlying most of these experiments has been that PR subjects reverse slowly simply because they extinguish slowly, i.e., because they continue to select the formerly more favorable alternative for a longer time. In principle, there is no difficulty in deciding between these two interpretations by a detailed examination of the course of reversal learning. Unfortunately, this has not often been undertaken, and the results of the relevant experiments have frequently been presented simply in terms of the total number of correct responses during a fixed number of reversal trials.

In four studies, however, reversal learning curves have been plotted, and three of these sets of curves (those obtained in the experiments of Grosslight *et al.*, 1954; Grosslight & Radlow, 1956; Wise, 1962), appear to favor the orthodox interpretation rather than our own. As an example, Figure 11.4 (Panel A) shows the reversal learning curves of the two groups of rats run in the study by Grosslight *et al.* It is clear that the PR group is inferior in performance to the CR group at all stages of reversal. This implies that so far from PR leading to more rapid initial extinction, as our model demands, it significantly retarded extinction. In the fourth study, however, the shape of the curves equally clearly supports our account rather than the usual interpretation. Elam and Tyler (1958) compared reversal learning after 60 : 0

and 60 : 40 schedules of reinforcement in acquisition, and found that Group 60 : 0 reversed more rapidly. They suggested that this was because, "subjects in this group were able to discriminate the transition from initial training to reversal training more rapidly than subjects in Group 60 : 40 [p. 586]." Inspection of their reversal learning curves shown in Figure 11.4 (Panel B)

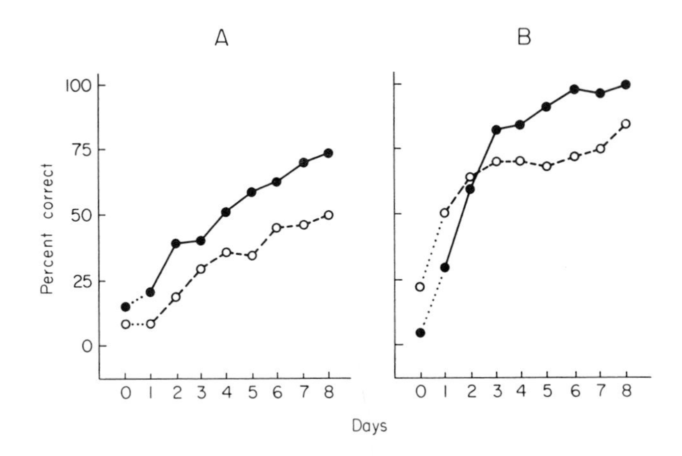


**Fig. 11.4.** Reversal learning curves after different reinforcement schedules in acquisition. Day 0 = last day of acquisition. Panel A: ●, Group 100:0; ○, Group 66:0. Panel B: ●, Group 60:0; ○, Group 60:40: [Panel A *after* Grosslight, Hall, and Scott (1954). Copyright by the American Psychological Association. Reproduced by permission. Panel B *after* Elam and Tyler, (1958).]

casts considerable doubt on this interpretation: Group 60 : 40 are not reversing more slowly because they continue predominantly selecting the former majority stimulus for a longer time than Group 60 : 0. On the contrary, Group 60 : 40 have completely extinguished the original discrimination habit by the very first day of reversal. In fact, on the first day of reversal, Group 60 : 40 selected the former majority stimulus on only 50% of trials; while Group 60 : 0 selected it on 70% of trials. The difference was significant at the .05 level.

Three further studies have reported detailed evidence concerning the course of reversal learning. Kendler and Lachman (1958) trained rats on a brightness discrimination on either a 100 : 0 or 50 : 0 reinforcement schedule, and then reversed them. For reversal, each group was divided into two subgroups, one run under high drive, the second under low drive. In addition to showing that inconsistent reinforcement during acquisition led to fewer correct responses during the course of reversal, they provide two further important details. The first is the number of perseverative errors at the outset of reversal; this, of course, is one measure of resistance to extinction of choice behavior. Consistently reinforced animals made an average of 7.2 perseverative errors; partially reinforced animals made an average of 11.8 errors. The difference is in the opposite direction to that predicted by us, but was not, in fact, significant. The authors also present an analysis of systematic responses to position during the course of reversal, with results more favorable to our account. As we should expect, the PR animals were more likely to develop position habits during reversal than were CR animals. Furthermore, among animals reversed under high drive, at any rate (for the opposite was true of animals reversed under low drive), PR animals reached a 50% criterion of extinction sooner than CR animals.

Warren and McGonigle (1969) analyzed the data of their cat experiment in terms both of trials and errors to the criterion of reversal learning, initial errors, and trials to 50% choice of the former positive stimulus (this last as a measure of extinction of the original discrimination). Although, as mentioned above, they found no effect of reinforcement schedule on total trials to reversal criterion, on all other measures CR interfered with reversal. Consistently reinforced animals, for example, took significantly longer to reach a chance level of performance during reversal; i.e., as we should predict, CR increased choice resistance to extinction. Although they do not report the exact figures, it is possible to estimate the number of trials required to attain criterion after the 50% level had been reached: CR animals required approximately 150 trials; both inconsistently reinforced groups required over 200 trials. In general, therefore, in line with our predictions, CR retarded extinction of the original discrimination, but facilitated the acquisition of the reversed discrimination.

In an attempt to provide further detailed information on the effect of reinforcement schedule on the course of reversal learning, Mackintosh and Holgate (1968) trained two groups of rats on a brightness discrimination with either a 100 : 0 or a 75 : 25 reinforcement schedule. After 200 trials (sufficient to have given Group 100 : 0 extensive overtraining) both groups were extinguished to a criterion of equal choice of black and white over 10 trials. As mentioned previously, Group 75 : 25 reached this criterion somewhat (but not significantly) faster than Group 100 : 0. After meeting this extincton criterion, animals were trained by a noncorrection procedure on a 0 : 100 schedule, i.e., on the reversal of the original discrimination. The results are shown in Figure 11.5. There are three important features of these results: First, Group 100 : 0 reached criterion on the reversal significantly faster than Group 75 : 25; thus, inconsistent reinforcement retarded the relearning stage of reversal (but not the extinction stage). Second, Group 75 : 25 made more responses to position during reversal than Group 100 : 0; thus, inconsistent reinforcement retarded reversal by increasing the probability of responses to irrelevant cues. Third, Group 75 : 25 showed the initial dip upon reversal

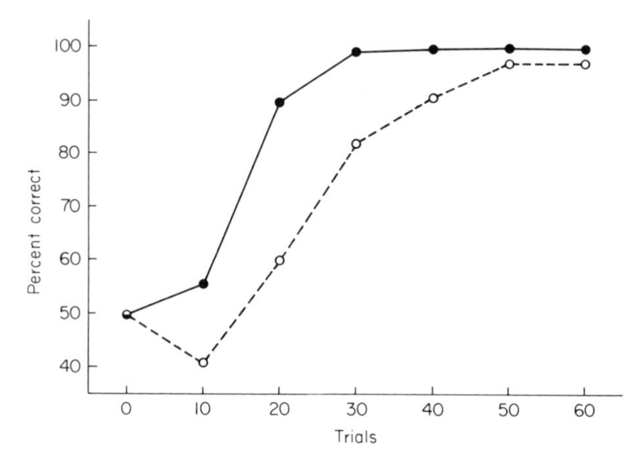


**Fig. 11.5.** Reversal learning curves following consistent or inconsistent reinforcement and extinction. ●: Group 100:0; ○: Group 75:25. [*After* Mackintosh and Holgate (1968). Copyright by the American Psychological Association. Reproduced by permission.]

after extinction that is characteristic of rats trained only to criterion on a 100 : 0 brightness discrimination. We interpret this dip (see page 268) as showing that, during extinction, the relevant analyzer extinguishes significantly sooner than its response attachments. If this happens, animals will reach a criterion of extinction by switching-out the relevant analyzer with its responses still attached; if the reintroduction of reward causes the analyzer to come back in, such animals will immediately respond in the direction in which they were initially trained. The occurrence of this dip in Group 75 : 25, and its absence in Group 100 : 0, indicates, therefore, that by the end of initial training the relevant brightness analyzer was only weakly switched-in by subjects receiving inconsistent reinforcement.

Mackintosh and Holgate's results show that the effects of PR on reversal learning are at least in part due to its effects on the strength of the relevant analyzer. Just as animals trained only to criterion on a 100 : 0 brightness discrimination reverse more slowly than animals extensively overtrained because they extinguish the relevant analyzer more rapidly at the outset of reversal, so animals that have received inconsistent reinforcement during acquisition reverse slowly for exactly the same reason.

## C. Inconsistent Reinforcement and Nonreversal Shifts

One result that supports our interpretation of the ORE is the finding that in certain situations overtraining retards nonreversal (extradimensional) shift learning. If subjects are trained in Stage 1 with Cue A relevant and Cue B irrelevant, and in Stage 2 with Cue B relevant and Cue A irrelevant, then overtraining at Stage 1 often retards (and never facilitates) Stage 2 learning (see Chapter 9). This finding provides conclusive evidence against any attempt to explain the ORE in terms of the effects of overtraining on the extinction of the original discrimination habit, and supports our contention that a major effect of overtraining is to strengthen the relevant analyzer. Similarly, our account of the effects of inconsistent reinforcement on reversal learning predicts that inconsistent reinforcement will facilitate such shift learning; if such a result were obtained, it would provide similarly conclusive evidence against any attempt to ascribe the reversal results to the effects of inconsistent reinforcement on extinction. Accordingly, Mackintosh and Holgate (1968) trained two groups of rats for 200 trials on a simultaneous brightness discrimination, one on a 100 : 0, the other on a 75 : 25 reinforcement schedule. In Stage 2, all subjects were trained on a 100 : 0 spatial discrimination with brightness irrelevant. The results are shown in Table 11.III. It is clear that

TABLE 11.III

**Nonreversal Shift Learning to Position following 100 : 0 or 75 : 25 Brightness Discrimination Learning**

| Groups | Nonreversal shift trials to criterion | $p$ |
|---|---|---|
| 100 : 0 | 52.9 | $<.01$ |
| 75 : 25 | 38.5 | |

inconsistent reinforcement facilitated shift learning in this experiment and we can conclude that the major determinant of the effects of an inconsistent reinforcement schedule on reversal and shift learning is that such a schedule fails to strengthen the relevant analyzer.

### D. Conclusions

Although there is general agreement (only one dissenting result) that inconsistent reinforcement in acquisition retards subsequent reversal learning, there is less agreement both at an empirical and at a theoretical level as to the causes of this effect. The critical question is whether inconsistent reinforcement affects speed of reversal by slowing down extinction of the original discrimination or by interfering with the acquisition of the new problem. Some of the published reversal learning curves suggest that the effect is largely on extinction, but the two studies, which have explicitly distinguished between the two phases of reversal learning, have both shown that inconsistent reinforcement tends to hasten extinction, but to retard subsequent acquisition. Once again, our own analysis receives support, although again it is probable that other factors are involved.

## V. Summary

The acquisition of a visual or spatial discrimination is usually retarded if responses to the positive stimulus are inconsistently rather than consistently reinforced. If a reinforcement schedule involving not only less than 100% reinforcement of S+, but also greater than 0% reinforcement of S− is employed, both rate of learning and asymptotic performance are adversely affected. The accepted explanation of these results has been to appeal to the effect of the reinforcement schedule on differential response strength: it has been assumed that a probability learning schedule decreases response strength to S+ (the majority stimulus), and increases response strength to S− (the minority stimulus), with the result that differential response strength (the determinant of choice behavior) is too small to dictate consistent choice of the majority stimulus.

The evidence presented in this chapter suggests that this explanation is incomplete and ignores a probably more important source of errors in most probability learning experiments. Animals do not appear to select the minority stimulus solely because of an occasional, momentary preference for that stimulus, but because of an occasional failure of control by the relevant dimension. Errors occur because the relevant analyzer is not sufficiently strong, and irrelevant analyzers are not sufficiently weak, to ensure exclusive control by the relevant analyzer on all trials. Evidence for this statement is provided by an analysis of the trials on which errors are made, by an examination of the probability of errors on different types of tasks and following different pretraining experiences, and (less directly) by a study of the effects of inconsistent reinforcement on choice extinction and reversal.

This conclusion, if accepted, has one interesting implication. The evidence is clear-cut that the same reinforcement schedule (say, 70 : 30) results in a much higher level of performance if a noncorrection, rather than a correction or guidance, procedure is used. Now, once an animal is selecting the majority stimulus on the majority of trials, the only substantial difference between these procedures is that a correction or guidance procedure ensures that the animal continues to receive reinforcement in the presence of the minority stimulus, whereas, eventually, the animal may receive no such reinforcements when trained by noncorrection. The difference between the effects of the two procedures would be quite intelligible if we could accept that continued choice of the minority stimulus under correction training reflected an occasional preference for that stimulus. However, if it reflects a failure of attention to the relevant dimension, we appear to be forced to the conclusion that the occurrence of extra reinforcements (in the presence of the minority stimulus) weakens the relevant analyzer (Mackintosh, 1970). This conclusion is not easy to accept unless we take seriously the idea that *consistency* of reinforcement is a major determinant of the strength of analyzers. In the form in which it is outlined in Chapter 2, Rule 2 of our theory enables us to account for these data: according to Rule 2, in a two-choice situation an analyzer is strengthened both when the stronger of its two response attachments is reinforced, and when the weaker is not reinforced. Reinforcement of the weaker response attachment (in the present context, the response to the minority stimulus) is assumed to weaken the analyzer. As we saw in Chapter 2, there are several reasons for preferring our formulation to the simpler one (that analyzers are directly strengthened and weakened by reinforcement and nonreinforcement) adopted by Zeaman and House (1963) and Lovejoy (1966). If the present analysis is accepted, the data from probability learning experiments provide further empirical support for our formulation.

Indirect evidence that inconsistent reinforcement in a discrimination situation prevents the strong establishment of the relevant analyzer is provided by the results of experiments on choice extinction and reversal. Although the evidence is conflicting, it is at least clear that the PRE observed on running speed measures in an alleyway is much less reliably observed when a choice measure of extinction is used following discrimination training. The strong prediction derivable from our position, namely, that inconsistent reinforcement decreases the resistance to extinction of choice behavior, has only rarely been supported at a statistically significant level. This is not entirely surprising. In the previous chapter we saw that in order to explain the effects of PR on the extinction of running behavior, it was necessary to supplement our account with some further principles. We suggested in fact that responses must be assumed to be conditionable to the memory of nonreinforcement. If this occurs in a choice situation also, it will obviously serve to retard choice extinction. If this is so, then PR has two effects: it increases the similarity of acquisition and extinction stimuli, and also decreases the strength of any one analyzer while increasing the total number of analyzers used. In runway studies, where there is a multitude of relevant cues, these effects act together to produce greatly increased resistance to extinction. In choice studies, however, the two effects act in opposition, thus producing small and variable effects of PR on extinction.

CHAPTER 12

# Some Comparative Psychology

## I. Aims and Problems of Comparative Psychology

Psychologists have only rarely displayed much interest in pursuing a comparative approach to their subject matter. The early animal psychologists, it is true, investigated similarities and differences in the behavior of a relatively wide variety of animal groups, and in recent years several voices have been raised in complaint against the long domination of the laboratory rat (e.g., Beach, 1950; Bitterman, 1960). Yet throughout most of the history of psychology, the large majority of psychologists has had not much understanding of, and even less sympathy with, the aims of comparative research. Nor are the reasons for this hard to find. It is obvious, of course, that research on the rat is not going to answer questions about the mechanisms controlling behavior in arthropods, but although use of the rat is not justified by calling it a "representative mammal" (an extraordinary concept), there are many good reasons for using the rat in psychological research.

Rats are convenient animals with which to undertake behavioral experiments. They are cheap, of a suitable size, and easy to maintain. They are easily tamed and motivated, and this means that they are easy to train. These are not idle reasons; in many ways the squirrel is obviously better suited than the rat to studies of visual discrimination, but anyone who has tried to tame and handle squirrels knows why they have not been used more frequently. Similarly, the problems involved in motivating amphibia make their use in learning experiments a time-consuming affair. Psychologists need hardly be ashamed of admitting such considerations of convenience. Convenience has influenced the choice of subject in many areas of biology; early geneticists studied *Drosophila* not because they knew that the laws of genetics were to be found in their purest form in such insects, but because

*Drosophila* breeds rapidly and has large chromosomes. We may study learning in the rat, not because the rat's brain is typical of anything else, but because rats are reasonably intelligent small mammals.

Furthermore, the use of a particular subject for research is, in a sense, self-perpetuating. On the one hand, discovery of certain facts about the behavior of an animal is a prerequisite for the study of other problems, for until one has discovered how to maintain and motivate, reward and punish an animal, it is impossible even to begin interesting studies of learning. The acquisition of such knowledge is a tedious business, but the rat psychologist is in the fortunate position of being able to rely on the labors of previous generations for the information he needs. On the other hand, the discovery of a particular fact about a particular animal often poses further questions that can only be answered by further work with that animal; e.g., the question why the PRE in rats is a function of reward size is unlikely to receive a definitive answer from experiments with pigeons.

Reluctance to deprecate the use of the rat in psychological research, however, need not imply a failure to appreciate the role and importance of comparative research. We are fully convinced both that it is the job of the comparative psychologist to understand and explain differences and similarities in the behavior of different animal groups, and that the investigation of such similarities and differences provides an important and distinctive means of gaining insight into the mechanisms controlling behavior. We shall elaborate this claim later (although our major justification lies in our experiments), but we should first consider some of the problems that arise in comparative research.

It is impossible to draw unequivocal conclusions from an investigation of a given problem in different animals, unless there is good reason to suppose that such investigations are taking place in equivalent conditions for the animals concerned. This demand for equivalent conditions can cover an unlimited range of variables. Even Skinner (1959) accepts that equivalence of demands on sensory and motor apparatus is a necessary prerequisite for comparative studies. He seems to imply however that once this has been satisfied, there is nothing more to be done, as if there were no further relevant differences between animals. There may not be in their adjustment to some of the simpler schedules of reinforcement, but there is no reason to believe that there are none in other areas. For instance, although one reason why rats, by comparison with rhesus monkeys, show an extremely rudimentary capacity to form learning sets, may well be that their sensory capacities are inferior, it is likely that learning set formation is one quite good measure of phylogenetic advance. Pigeons, with visual ability indisputably superior to that of the rat, are even worse than rats at learning set formation (Zeigler, 1961).

We know perfectly well that there are variables other than sensory and motor capacity determining performance in learning situations. In the case of many of these variables, moreover, it is no easy matter to ensure equivalence of conditions for different species. One can perform equivalent operations, but there is no reason to suppose that similar operations will have similar effects on different animals. For example, a magnitude of food reward that would act as a suitable incentive to a rat would hardly be noticed, let alone worked for by an elephant. Conversely, a feeding schedule just severe enough to keep a rat well motivated, would kill a shrew within hours. The absurdity of such instances should not blind us to the fact that when we reduce rats and pigeons to an equal percentage of *ad lib*. weight, we have no idea whether we are producing equivalent levels of motivation.

It must be borne in mind that the effects of some of these uncontrollable variables may well be complex. In simple learning situations, perhaps, changes in magnitude of reward may have simple, unsurprising effects; both rats and chicks are liable to run faster for large rewards than for small. In less simple situations, however, the effects of changes in these variables may also be less simple. Magnitude of reward has been shown to affect the resistance to extinction of a CR response in one direction (Armus, 1959) and of a PR response in the opposite direction (Hulse, 1958; Wagner, 1961). It is certainly one determinant of the effect of overtraining on reversal learning (Hooper, 1967; Mackintosh, 1969a), and may influence the effect of over-training on runway extinction (Ison & Cook, 1964). To take one example, therefore, a comparison of the effects of PR on resistance to extinction in rats and fish will present considerable problems of interpretation. If the fish fail to show a PRE, is this due to differences in the mechanisms underlying learning in rats and fish, or is it simply because the reward given to the fish is functionally equivalent to the 45-mg reward that sometimes fails to produce the effect in rats?

We do not claim that these problems are insuperable. Bitterman (1960, 1965) has suggested one technique, which he calls "systematic variation," that can be used to answer this type of question. If it is not known how to give equivalent rewards to rats and fish, and it is known that the rat's behavior in a given situation depends on the magnitude of reward, then several groups of fish can be run, each receiving a different reward size; if the difference between rat and fish holds up for all sizes of reward used, then it can hardly be due to differences in the rewards used for the two classes of animal. This is obviously one, albeit time-consuming, way of providing a solution to the general problem. In the case of the PRE, e.g., it is now certain that at least some of the differences that appeared to exist between rats and fish (e.g., Gonzalez, Eskin, & Bitterman, 1963) disappear when fish are trained with a larger reward (Gonzalez & Bitterman, 1967).

Systematic experiments of this sort will always be of the utmost importance, and it is unfortunate that we do not have more experimental evidence on the effects of variations in reward size, drive level, and other such factors on the behavior of birds and fish. Nevertheless, important as such evidence will be, we are not convinced that it will always provide the most satisfactory means of answering criticism. The criticism, it should be remembered, is an unlimited one, for there is an indefinite number of conditions that might not have been equated in any single comparison of rat and fish. To have shown that variations in reward size cannot account for differences, say, in serial reversal learning between rat and fish, is to invite the retort that the differences could be due to variations in drive level, or to an interaction of reward size and drive level, or to variations in stimulus–response–reinforcement contiguity.

An alternative approach is to utilize theoretical understanding of the situations investigated in the comparative research program, to make predictions as to what further similarities should be obtained if some initial similarities in behavior of the two groups of animals have been found, and also what further differences should follow once certain initial differences have been found. The justification for any single claim to have shown a certain difference between, say, the rat and the domestic chick then rests upon the internal consistency of the total series of claims made by the experimental program. It seems to us unrealistic to undertake comparative research without some prior theoretical understanding of the processes one plans to investigate; with such understanding, it becomes possible to apply this strategy. The theory advanced here provides a relatively detailed explanation of such experimental phenomena as the ORE; therefore, if we obtain the ORE in a given animal group, we can predict what further results should be obtained with this group; conversely, and more interestingly, if we fail to find the ORE, we can predict what further differences we might expect to find between the behavior of this species and that of the rat. Such a research program in depth seems to us to promise greater reliability of results, as well as greater theoretical importance.

We shall describe experiments using octopuses, goldfish, and domestic chicks, in which we have attempted to implement these ideas; we shall also offer an explanation of some of the results obtained by other workers in the field of comparative psychology. Our general claim is that the model of learning we are putting forward to account for the behavior of the rat in a variety of experimental situations will also, once certain parameters have been changed, account for the different behavior shown by some of these other animals in these situations.

## II. Overtraining and Reversal Learning

The conditions under which overtraining facilitates reversal learning in rats are, as we have seen in Chapter 8, relatively restricted. At the least it appears to be necessary to use a large reward and to train on a moderately difficult discrimination. Since the effect is not universal in rats, a failure to obtain it in some other animal group need not be taken as evidence of any important differences between them and rats.

A large number of studies with primates has failed to produce any evidence of an ORE; the primates studied include rhesus monkeys (Tighe, 1965; Beck, Warren, & Sterner, 1966; Cross & Boyer, 1966); stump-tailed monkeys (Boyer & Cross, 1965); Capuchin monkeys (D'Amato, 1965); and squirrel monkeys (Cross & Brown, 1965; Cross, Fickling, Carpenter, & Brown, 1964). In many of these experiments, however, the problem has been a relatively easy object discrimination, while in others the subjects have been experimentally sophisticated. In either case, the initial strength of the relevant analyzers may have been too high for an attentional analysis to predict an ORE. Although this argument cannot reasonably be applied to all of these studies (for in D'Amato's experiment the subjects had been pretrained to attend to a cue that was made irrelevant for the test problem), it may well account for many of the failures. To the best of our knowledge, moreover, the reward used in all of these experiments was a relatively small one. It is, therefore, too early to assert with confidence that the effect of overtraining on reversal is different in rats and monkeys, and unnecessary to speculate on the nature of any underlying differences.

The ORE has also been sought unsuccessfully in cats. In the first such experiment, however, the relevant stimuli were stereometric objects providing a number of relevant cues, and the finding that overtraining retarded reversal is not entirely surprising (Beck *et al.*, 1966). In a subsequent experiment, Hirayoshi and Warren (1967) trained cats both on a black–white discrimination and reversal with three irrelevant stimuli, and then on a shape discrimination (with brightness irrelevant), and found no ORE under any conditions. Since the reward used was large (3-gm pieces of kidney), and since there was no tendency for overtraining even to decrease responses to irrelevant cues during reversal, it seems reasonable to suggest that there are, indeed, important differences between rats and cats. We do not know what these differences are, and, in the absence of further evidence, we do not intend to speculate. We shall show, however, that it is possible to manipulate the parameters of a two-process theory in such a way that no ORE is predicted, and then to go on to test further predictions from the revised model.

### A. Octopuses

Although a reasonable range of vertebrate species has been studied by the comparative psychologist, the study of learning in invertebrates has so far touched upon only a few groups. Leaving aside experiments whose major purpose has been to prove that learning does occur (e.g., the work of Gelber,

1965, on paramecia, and of McConnell, 1965, and his associates on flatworms), the most extensively studied invertebrates have been earthworms, various insects, and octopuses. Not surprisingly, many of these studies have uncovered interesting differences between the behavior of invertebrates and mammals (e.g., Maier & Schneirla, 1935; Datta, 1962); and it was with the hope of uncovering further differences that we undertook research on the octopus. Early results, indeed, appeared to suggest that the octopus behaves very differently from the rat. J. Mackintosh (1962) investigated the effect of overtraining on reversal learning, and found that speed of reversal was unaffected by degree of initial training. Mackintosh also found no evidence of improvement over a series of reversals in her octopuses, and this result has been confirmed by Young (1962).

Once again, however, the difference of experimental outcome may reflect a difference in procedure rather than a difference in learning mechanisms. Mackintosh's octopuses were trained on a successive, go–no-go discrimination. While overtraining does facilitate reversal learning by rats in this situation (Birch *et al.*, 1960; Sperling, 1970), we argued in Chapter 8 that this may represent a somewhat different effect from the ORE obtained in simultaneous visual discriminations. In terms of our theory, of course, no ORE might be expected in the successive situation, since there are no deliberately introduced irrelevant cues.

In a second experiment, designed to test for similarities or differences in the behavior of octopuses and rats, Mackintosh and Mackintosh (1963) ran three major groups of subjects. Group Successive was essentially a replication of the original experiment; subjects were trained on a successive discrimination between black and white vertical rectangles. Group Simultaneous was trained on a simultaneous discrimination between the same black and white rectangles. In order to discover whether any differences obtained between these two groups were due to the presence or absence of irrelevant cues, a third group was run. Group Successive: H–V was trained on a successive discrimination between black and white rectangles, but an irrelevant cue of orientation was introduced by showing the rectangles randomly either horizontally or vertically. Half of each group was trained to criterion only, the other half was overtrained on the original problem, before being trained on the reversal.

The results of this experiment are shown in Figure 12.1; it is apparent that overtraining had little effect on reversal in the simple successive group, but marked effects in the two groups with irrelevant cues. It is the nonovertrained subjects that showed the major effect of the introduction of irrelevant cues, and analysis of systematic response tendencies showed that nonovertrained subjects in groups Simultaneous and Successive: H–V showed a significant tendency to respond to irrelevant cues during reversal. The

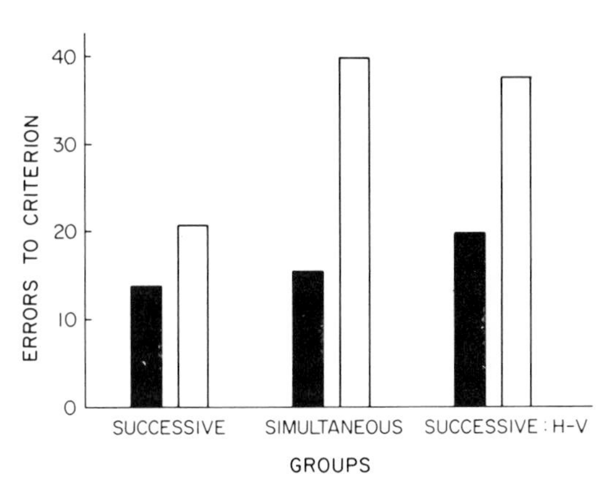


**Fig. 12.1.** Errors to reversal criterion in *Octopus*, as a function of overtraining and the presence of irrelevant cues. For further details see text. (■): overtrained; (□): nonovertrained. [*After* Mackintosh and Mackintosh (1963).]

results, therefore, are strikingly similar to those obtained with rats (e.g., Mackintosh, 1963a): an ORE occurred when irrelevant cues were present; irrelevant cues retarded reversal for nonovertrained subjects but not for overtrained subjects.

This result suggests a striking similarity in the behavior of rats and octopuses despite the wide differences between the experimental situations used for each animal. The obvious interpretation is to assume that a two-process theory can be applied to each species in a similar fashion and that there is no need to make changes in any of the assumed underlying mechanisms. Such a conclusion is not, of course, proved by the above results; in principle, it is possible that quite different sets of rules could generate the same results, but as a working hypothesis this assumption was perfectly reasonable and encouraged further investigation.

If overtraining facilitates reversal by selectively strengthening the relevant analyzer, it should not facilitate the learning of nonreversal shift problems. The results of experiments with rats were conflicting, but provided some support for this proposition. Mackintosh and Holgate (1965) tested the prediction with octopuses. The design used was that in which the Stage-1 relevant cue (in this instance, brightness) becomes irrelevant in Stage 2, while the Stage-1 irrelevant cue (orientation) becomes relevant. Mackintosh and Holgate found that animals receiving 100 overtraining trials on the Stage-1 problem took longer than nonovertrained controls to learn the Stage-2 orientation discrimination.

In a third stage of their experiment, Mackintosh and Holgate investigated the effects of overtraining on the relative extinction of analyzers and responses. If overtraining strengthens the relevant analyzer more than its response attachments, it should decrease the extent to which the analyzer undergoes extinction during the course of reversal. In Chapter 8, we discussed evidence which suggested that when rats were trained to criterion on a discrimination problem, the effect of a series of extinction trials was to weaken the relevant analyzer much more than its response attachments, and that this difference was reduced if the animals were overtrained on the original discrimination. Mackintosh (1963b) trained or overtrained rats on a brightness discrimination, extinguished them until they reached a criterion of equal choice of S+ and S− over 10 trials, and finally reversed them. Upon reversal, overtrained rats rapidly increased their choice of the new S+, but nonovertrained rats immediately increased their choice of the former S+. This latter result (initially observed by D'Amato and Jagoda, 1960) was interpreted to mean that nonovertrained subjects reached the criterion of extinction by switching-out the relevant analyzer with its responses attached, and switched it back in upon the reintroduction of reward, thus responding in the direction in which they were intially trained. Since overtraining strengthened the relevant analyzer more than its responses, it ensured that these responses were more nearly equalized during the course of extinction trials.

This experimental design was suitable for testing our prediction with rats, but would have been less suitable with octopuses as subjects, since it demands that subjects should reach a choice criterion of extinction before reaching a latency criterion that is within the limits of the experimenter's patience. Rats in a jumping stand will continue to jump in extinction with reasonably short latencies, possibly because merely getting off the jumping platform is mildly rewarding. In the situation used for training octopuses, however, the animals are required to make a response by leaving their "home." With few exceptions, octopuses quickly stop attacking Perspex stimuli when no longer rewarded for doing so; it is likely, therefore, that they would stop attacking altogether before reaching a choice criterion of extinction.

In Mackintosh and Holgate's study, octopuses were initially trained on a brightness discrimination, and in the second stage of the experiment this brightness discrimination was extinguished by training subjects on an orientation discrimination with the brightness cue present but irrelevant. In order to solve the orientation problem, subjects had to extinguish the brightness analyzer, and learn to switch-in an orientation analyzer. Subjects trained only to criterion on the brightness problem should have switched-out the brightness analyzer with its responses still largely intact; but this effect should have been reduced by overtraining. In order to test this, subjects were given preference tests in the third stage of the experiment between black and white stimuli. The prediction was confirmed: overtrained subjects showed no preference for their originally positive brightness, selecting it on only 45.7% of trials; but nonovertrained subjects selected their original positive brightness on 72.5% of test trials.

The design of this experiment differs considerably from that of the rat experiments that provided the initial evidence on the effects of overtraining on analyzer and response extinction. The fact that the predictions made by our model can be confirmed with different animals in different situations provides support for the model not easily obtainable from further experimentation with rats. We were, indeed, surprised that it should be possible to obtain similar results with rats and octopuses, not least because of the failure to do so with birds and fish; but once some initial similarities had been uncovered, the discovery of further similarities provides strong support for the theory.

### B. Birds

Brookshire *et al.* (1961) compared the effects of overtraining on brightness discrimination reversal in rats and chicks. In rats, overtraining facilitated reversal; in chicks it retarded reversal. This would certainly appear to provide good evidence for a phylogenetic difference—an impression strengthened when we find the result to be readily reproducible. Mackintosh (1965b) repeated the same experiment (in a somewhat different apparatus) and obtained essentially similar results. Of further interest was the reason for the effects of overtraining on reversal in chicks: in this experiment as in other ORE studies with rats, overtraining facilitated reversal by reducing responses to position; in chicks overtraining retarded reversal partly because it *increased* responses to position. A basically similar result has been obtained by Schade and Bitterman (1965) with pigeons; they found that overtraining significantly retarded the reversal of a red–green discrimination.

Once again, it is important to consider to what extent these different results can be explained by differences in the experimental conditions. Although no studies of overtraining and reversal in lower birds have deliberately varied reward size, it does not, in fact, seem likely that overtraining failed to facilitate reversal simply because a very small reward was used. Schade and Bitterman's pigeons, for example, received 2.5-sec access to food. It is our experience that about 75 such reinforcements are more than sufficient to maintain a pigeon's weight. Taking a rat's daily ration as about 12–15 gm, an equivalent proportion equals about 160—200 mg, which is quite large enough a reward to produce an ORE. Mackintosh's chicks received 5-sec access to food, itself a large enough reward to produce an ORE in rats.

The second condition under which no ORE occurs in rats is if the problem is a very easy one. This, indeed, seems a much more likely explanation of the adverse effects of overtraining in the bird experiments. In both Brookshire *et al.*'s and Mackintosh's experiments, the chicks learned the brightness discrimination much more rapidly than the rats. In Schade and Bitterman's experiments, pigeons learned the color discrimination with less than three errors. This is hardly surprising. There is no need to document the observation that most birds are markedly more visual animals than are rats. If this is so, then just as rats do not, for example, have to learn to attend to spatial cues, so birds may not have to learn to attend to simple visual cues. (Indeed, evidence reviewed in Chapter 8 indicated that even rats would show no ORE when trained on very obvious black–white discriminations.)

That differences in problem difficulty (existing even when both classes of animal are trained on exactly the same problem) are one cause of the different outcomes of rat and bird experiments has been shown by the results of further experiments reported in the above studies. In all three studies, birds were also trained on more difficult discriminations, and in all cases the adverse effect of overtraining on reversal vanished (and in one case turned into a significant ORE). Brookshire *et al.* trained their chicks on what was for them a much harder spatial (response-learning) discrimination, and found no effect of overtraining on reversal. Mackintosh trained his chicks on an extremely difficult orientation discrimination (with both brightness and position irrelevant), and found that overtraining significantly facilitated reversal. Schade and Bitterman trained their pigeons on a shape discrimination and, except after several reversals, found that speed of reversal was unaffected by overtraining. A further experiment with pigeons (Marsh & Johnson, 1968) confirmed that if they are trained on a moderately difficult visual discrimination, overtraining does not retard reversal.[1]

The suggestion, therefore, that part of the difference between rats and birds in the original experiments was due to a difference in the initial strength of the relevant analyzer, receives nice support. It seems likely, however, that more must be involved than this. It is relatively difficult to obtain a positive ORE in birds, and when it occurs, it is relatively small; conversely, overtraining more often than not significantly retards reversal of easy problems. On the other hand, overtraining has never been shown to retard significantly the reversal of a visual discrimination in rats, and only occasionally to retard reversal of a spatial discrimination.

These considerations suggest that there is some further difference between rats and birds; in addition to postulating differences in the starting values of the relevant analyzer in, say, a brightness discrimination, it is necessary to make some further alterations in the parameters of the model that will have

[1] Williams (1967b) reported a significant ORE with pigeons trained on an extremely difficult color discrimination (requiring over 200 trials to learn). As he notes, the effect probably depended upon allowing an uncontrolled number of intertrial responses to occur, which tended to eliminate position habits. When they were not permitted, overtraining had no effect on reversal. In no case, at any rate, did overtraining retard reversal of this difficult problem.

the effect of reducing the beneficial effects of overtraining on reversal.

It may be recalled from the discussion of reward size effects in Chapter 8 that small quantitative changes in the parameters of a two-process model may have marked effects on the type of prediction generated by the model. The prediction of an ORE depends upon overtraining strengthening the relevant analyzer more than its response attachments; this is most easily represented by assuming that analyzers take longer than responses to reach asymptote. It is clear that appropriate changes in the parameters controlling the relative rates of increase of analyzers and responses will abolish the prediction of an ORE, even when the initial strength of the relevant analyzer is low.

While this is quite an easy matter, it is also too arbitrary to carry much conviction by itself. What is needed to justify such parameter changes is the demonstration that they lead to further testable predictions. The evidence that we have taken as most directly relevant to our analysis of the ORE is that provided by experiments on the extinction of discrimination habits. If overtraining strengthens the correct analyzer relative to its responses, then it should also decrease the extent to which the analyzer is extinguished over a series of extinction trials and increase the extent to which the formerly correct responses are extinguished. The "dip effect" (see page 268) was taken as evidence supporting this proposition. Conversely, therefore, if the effect of overtraining even on the reversal of difficult problems is small and not entirely reliable, then the dip effect too should be small (or more accurately, the *difference* between overtrained and nonovertrained subjects should be small), while if overtraining significantly retards reversal, then we might expect a reverse dip effect also.

Mackintosh's experiment with chicks involved tests of these predictions. In the first test, chicks were trained to criterion or overtrained, then extinguished, and finally reversed, all on the easy black–white discrimination that yielded a reverse ORE. In the second test, the entire procedure was repeated using the difficult orientation discrimination which yielded a small but significant ORE. The results of these two experiments are shown in Figure 12.2. The results for the brightness discrimination are the exact opposite of those obtained with rats using the same procedure, apparatus, and problem: only overtrained chicks reverted to selecting their former S+ upon reversal after extinction. The results for the orientation discrimination are more similar to those obtained with rats (see Figure 8.4). Nonovertrained animals showed a substantial increase in choice of their former S+ (greater than that shown by overtrained subjects). However, overtrained chicks, unlike overtrained rats, also showed a significant dip, and, as a consequence, the difference between overtrained and nonovertrained subjects was smaller in the case of chicks.

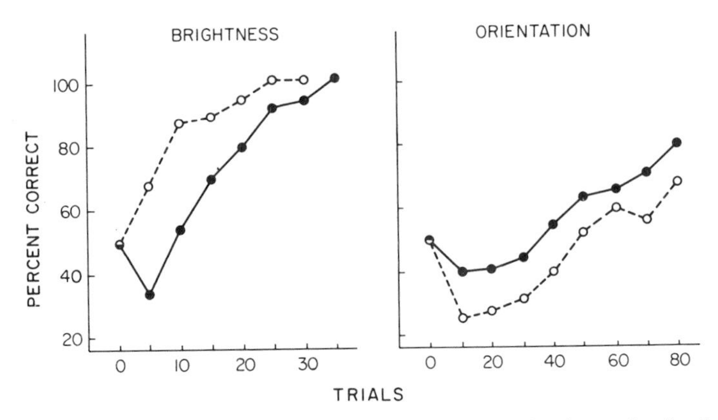


**Fig. 12.2.** Brightness and orientation reversal learning following extinction by overtrained (●) and nonovertrained (○) chicks. [*After* Mackintosh (1965b). Copyright by the American Psychological Association. Reproduced by permission.]

## C. Fish

Warren (1960a) trained paradise fish on a combined brightness and spatial discrimination, and found that overtraining retarded reversal. Since this is a situation in which it is certain that overtraining will not benefit reversal learning in rats (D'Amato & Schiff, 1964), this experiment does not provide evidence of any crucial difference between rats and fish. At least one more recent experiment, however, has shown that overtraining will not benefit reversal learning by goldfish when they are trained on a simultaneous visual discrimination (Mackintosh, Mackintosh, Safriel-Jorne, & Sutherland, 1966). The problem was a shape discrimination, of sufficient difficulty to maximize the chances of finding an ORE. Despite this, 100 overtraining trials slightly (but not significantly) retarded reversal.

This failure to obtain an ORE cannot reasonably be ascribed to a failure to use a difficult enough problem, nor is it likely to have been due to the use of too small a reward. In terms of their satiating effects, the rewards given to the fish were very much larger than those usually given to rats. We ran 10 spaced trials a day with fish, because although it would have been desirable to run more, with the size of reward used it was impossible to do so and maintain adequate motivation. In previous studies, we had found that the fish refused to respond after receiving about 15 pieces of reward and appeared to be fully satiated.

Overtraining, therefore, does not facilitate (indeed is more likely to retard) reversal learning in fish, even when the conditions used are those most favorable for producing an ORE in rats. If the results are to be explained satisfactorily, it will only be by postulating differences in the parameters of a theory of the ORE. We have already shown that it is a relatively simple matter to produce appropriate parameter changes in a two-process model, and it is not entirely surprising that if some changes must be made in the model to account for differences in the behavior of rats and lower birds, yet greater changes must be postulated to account for the differences between rats and fish. Let us assume then, that in fish, analyzers increase in strength at least as rapidly as responses; overtraining, therefore, will strengthen analyzers and responses equally, and have little or no effect on speed of reversal.

Once again, it is necessary to provide some independent evidence for such an arbitrary assumption, and once again it can be provided by examining the effects of overtraining on extinction. In fish, we predict that overtraining will *not* affect the extent to which analyzers and responses are weakened by a series of extinction trials, but that both overtrained and nonovertrained subjects will show an equal preference for their former S+ after extinction. To test this prediction, Mackintosh *et al.* used essentially the same procedure as that used by Mackintosh and Holgate (1965) in their experiment with octopuses. Two groups of fish were trained on an orientation discrimination with both brightness and position irrelevant. One group was trained to criterion, the other received 100 overtraining trials. Both then learned a brightness discrimination with the original orientation stimuli present but irrelevant, and were finally given preference tests with stimuli differing only in orientation. In the octopus experiment, nonovertrained subjects showed a marked preference for their original S+, but overtraining abolished the preference; in the fish experiment, *both* groups showed a significant preference for the original S+, and the scores (75.6% choice of S+ for overtrained animals, 71.9% for nonovertrained) did not differ. These results are consistent with the suggestion that in fish overtraining does not differentially affect analyzer and response strengths, and that it is for this reason that it fails to facilitate reversal.

### D. Conclusions

Experiments on overtraining and reversal learning in animals other than the rat have rarely produced results similar to those obtained with rats. It is true, of course, that the results of experiments with rats are themselves confused, and to the extent that we do not know the exact conditions under which an ORE occurs in rats, it is idle to speculate whether other animals have failed to show an ORE simply because they have not been trained under appropriate conditions. This does, however, seem an unduly pessimistic evaluation; there is reasonable evidence that the ORE is more likely to occur in rats under some conditions rather than others, and there is also reasonable evidence that overtraining is less likely to facilitate reversal in either lower birds or fish even when these conditions are satisfied. If this is accepted, then it is the clear task of the comparative psychologist to offer some explanation of these differences, either by proposing an entirely different theory to account for the behavior of inframammalian vertebrates (and, perhaps, cats), or to suggest which new processes come into effect to determine the behavior of the rat, or finally, to propose modifications to an existing theory which will encompass both kinds of data. It is this last possibility (proposed, e.g., by Hull, 1945) that we have followed here. The specific suggestion has been that, in birds and fish, the relative rates of response and analyzer acquisition are more similar than they are in rats, and therefore, that overtraining does less to increase the correct analyzer relative to its response attachments.

Evidence to support this suggestion was provided by the demonstration that overtraining has a smaller effect on the relative extinction of analyzers and responses in chicks than in rats, and has no effect at all in goldfish. These experiments, in fact, revealed a further difference between these three classes of animal. The only prediction that follows from the failure of overtraining to facilitate reversal in fish is that overtraining should also fail to affect the extent to which preference for S+ is abolished by a series of extinction trials. Nothing is said about whether both overtrained and nonovertrained groups should extinguish or fail to extinguish this preference, only that they should both behave alike. In fact, the preference remained significant in both groups. In terms of the theory, this implies that in fish (unlike rats), even after overtraining has increased both the relevant analyzer and its responses to asymptote, the analyzer extinguishes more rapidly than its responses. To the extent to which overtrained chicks also failed fully to extinguish a preference for their former S+, we can infer that in chicks also, analyzers extinguish substantially more rapidly than responses.

In rats, therefore, analyzers are strengthened and weakened slowly relative to responses, in lower birds, rather more rapidly, in fish, much more rapidly. These parameter changes are dictated by the results discussed in this section, and are sufficient to explain such differences between these animal groups as have so far been described. We shall now show that exactly the same set of parameter changes will explain at least two further sets of differences between these animals.

## III. Serial Reversal Learning

The above suggestion may be regarded as paradoxical, if taken to imply that analyzers change faster in a lower vertebrate than in a mammal. Rapid learning to use a new analyzer, or rapid extinction of an old analyzer might be regarded as an advanced capacity, of advantage to the animal possessing it. We should stress that we are not stating anything about the *absolute* rate of change of analyzers, only about the *relative* rates of change of analyzers and responses; we could as well assume that responses change more slowly as that analyzers change more rapidly. It is a *relative* difference in analyzer and response changes that can be shown to predict what is usually regarded as "less efficient" performance in a number of experimental situations.

## A. Causes of Serial Reversal Improvement

If mammals are trained on a series of reversals of a discrimination problem, they show a significant improvement over the series. In Chapter 9, we considered three possible explanations of this serial reversal improvement. The simplest account is that suggested by Gonzalez *et al.* (1966): with repeated reversals, errors per reversal decline because of a decrease in negative transfer from one problem to the next, and negative transfer decreases because proactive interference develops, and causes a progressive failure to remember from one session to the next which alternative was last correct. There is excellent evidence to support this analysis, although it is certain that other factors must also be brought in to account for the extent of serial reversal improvement in rats. One possible factor is that animals develop an appropriate strategy, based on the retention of their previous choice and its outcome, of repeating rewarded choices and shifting choices after nonreinforcement. We argued that there is no direct evidence that rats are either capable of forming such a rule, or do, in fact, do so when trained on a series of reversals. At the level of the rat, the effects of serial reversal training appear to be specific to the dimension on which training occurs, and can largely be accounted for by saying that such training causes a progressive strengthening of the relevant analyzer. Early reversals are learned slowly because the relevant analyzer is extinguished, while later reversals are learned more rapidly as the subject learns to maintain the relevant analyzer, even when its use does not lead to 100% reinforcement.

## B. Rule Learning by Primates

In an experiment designed to investigate the basis of learning set performance in chimpanzees, Schusterman (1962) trained one group of subjects on a series of object discrimination reversals and another group on an object alternation problem (in which the solution was to alternate choices of the two stimuli). Both groups were then trained on two more discrimination reversals, and finally on a series of six-trial object discriminations. The reversal learning group performed very much better than the alternation learning group on the two new reversal tasks, and also showed near perfect transfer to the object learning set task. Schusterman concluded that serial reversal training established a win–stay, lose–shift strategy, which was transferred to new discrimination problems regardless of the differences between the stimuli in the various problems.

This conclusion is reinforced by the results of a study by Warren (1966). Warren trained two groups of rhesus monkeys on a series of reversals of a single discrimination problem. For one group, the problem was an object discrimination; for the other group the problem was a spatial discrimination with object cues irrelevant. Both groups were then shifted to a series of six-trial discrimination problems, and their performance was compared with that of an untreated control group. Both groups showed significant positive transfer, performing with virtually complete accuracy by the end of the first day of testing. That transfer should have occurred between position reversal training and object discrimination learning is striking, and entirely inconsistent with an attentional analysis. Although it is only indirect evidence for the reality of rule learning by primates, it is not easy to think of any other explanation of such a general transfer effect.

In the same study Warren also examined the effect of reversal training on learning set formation in cats. He found no evidence of any positive transfer from serial reversal training to discrimination learning, and concluded that cats do not acquire any response rule during the course of serial reversal learning. The conclusion may be too strong, since it is possible that such learning occurred, but failed to transfer to new stimuli; nevertheless the difference in the outcome of the experiment between monkeys and cats is an important one, and is most naturally interpreted as evidence of the importance of rule learning in primates, and its rarity (or at least specificity) in lower animals.

## C. Serial Reversal in Vertebrates

### 1. *Fish*

Although most mammals learn a series of reversals at comparable rates, attaining similar, extremely efficient levels of performance, Warren's evidence suggests that this similarity of behavior conceals important differences in underlying mechanisms. Below the level of mammals, it is even more certain that theoretical differences must be postulated, since the behavior itself changes markedly. Although octopuses show significant improvement if trained on a series of simultaneous discrimination reversals (Mackintosh & Mackintosh, 1964), an extensive series of studies has failed to find reliable evidence of improvement in fish. Bitterman *et al.* (1958) Behrend, Domesick, and Bitterman (1965), Behrend and Bitterman (1967), and Gonzalez, Behrend, and Bitterman (1967) have conducted a total of 10 experiments with goldfish and African mouthbreeders, none of which yielded significant evidence of improvement. Warren (1960a) trained paradise fish on a series of combined visual and spatial reversals, and also found no sign of improvement. If it is borne in mind that serial reversal improvement in rats is a highly reliable phenomenon, occurring under a wide range of conditions, it is implausible to suggest that the failure of fish to show such improvement is simply a consequence of the experimental conditions under which they have been studied (which include visual or spatial problems, large or small reward, massed or spaced trials).

This should not be taken as proving that fish are never capable of showing any significant improvement. In a recent study by Setterington and Bishop (1967), mouthbreeders were given daily reversals of a position discrimination for 80 days and showed a significant reduction in errors per problem over the course of the experiment. It is probable that one cause of this different outcome is that Setterington and Bishop used a very short (2 sec) intertrial interval (cf. Behrend & Bitterman, 1967). Nevertheless, the results do not demonstrate that rats and fish behave identically in serial reversal situations. Even though they showed significant improvement, Setterington and Bishop's subjects were still averaging about five errors per (20 trial) reversal at the end of the experiment. Rats trained on a comparable series of problems perform more accurately than this after 10 or 20 reversals; e.g., Stretch, McGonigle, and Rodger (1963), and Stretch *et al.* (1964) trained rats on daily, 10 trial reversals, and found that, after 10 days, their subjects averaged only one or two errors per reversal.

Although many psychologists are extraordinarily reluctant to accept such a conclusion, it seems to us clear that these data show differences in serial reversal performance between rat and fish that cannot plausibly be ascribed solely to situational differences. The implication, therefore, is that some or all of the factors responsible for serial reversal improvement in rats operate in an attenuated manner or not at all at the level of the fish. It is, indeed, possible to argue that differences in each of the factors we have considered may be responsible for the behavioral differences.

Gonzalez *et al.* (1967) have suggested that there may be differences in retention between fish and other animals that show serial reversal improvement. They trained a group of pigeons and a group of goldfish on a series of red–green reversals. Each problem was presented for two days (40 trials per day), and then subjects were reversed regardless of the criterion reached. Pigeons showed a significant reduction in errors over the course of the experiment; the fish showed no such trend. To measure retention, Gonzalez *et al.* compared the proportion of choices of one stimulus that remained the same from the last five trials of one day to the first five trials of the next day. On the first day of each reversal, subjects were, of course, being reinforced for responding to the stimulus not reinforced on the preceding day; any decline in this measure of retention, therefore, may be as well ascribed to an increase in reversal proficiency. However, during the course of the experiment, pigeons also showed a decline in retention from the first to the second day of each reversal, while fish showed no such decline. The authors suggest that this difference in forgetting may have caused the difference in reversal performance. The issue is complicated by the use of a relative retention measure. The maximum possible retention loss was 2.5; this assumes that the subject responded perfectly at the end of one day, (5 choices of, say, red) and at chance at the beginning of the next day. This score, however, depends upon the level of performance attained by the end of the first day of each problem; in the extreme case, a subject responding at chance could only show zero forgetting. Since subjects were given a fixed number of trials per day, it is just as possible to ascribe the increase in forgetting shown by pigeons during the experiment to an improvement in reversal learning, as to ascribe the improvement in reversal to an increase in forgetting. Equally, the fish (which continued to average about 20 errors over the first day of each reversal) may have shown no increase in forgetting, because they failed to improve reversal performance.

Since a progressive reduction in negative transfer is not a sufficient explanation of serial reversal improvement in rats, it follows that whether or not proactive interference develops in fish, other differences between them and rats must be postulated. If one cause of improved reversal by the rat is the attainment of an appropriate response rule, then it is not unreasonable to suggest that such an intellectually demanding ability is better developed in mammals than in lower vertebrates (just as it is better developed in primates than in other mammals), and that this difference is a further cause of reversal differences. The suggestion is supported by two lines of evidence. First, the only experiment in which unequivocal evidence of serial reversal improvement in fish was found (Setterington & Bishop), used an exceptionally short intertrial interval. As Bitterman (1969) has argued, this could mean that fish are capable of using the memory of their last choice and its outcome to guide their current choice, but that such short-term memory decays very much more rapidly in fish than in higher animals. While this is a possible interpretation, however, it is not a necessary one; as discussed in Chapter 9, Clayton (1966) has argued that reversal learning may improve with shorter intertrial intervals, not because short-term memory traces are used as stimuli to control current choice behavior, but because the incorrect response, partially extinguished by the outcome of one trial, may spontaneously recover during a long intertrial interval to interefere with performance on the next trial.

A second source of support for this possibility was briefly mentioned in Chapter 10. Partial reinforcement effects in fish are somewhat more elusive than they are in rats. To the extent that this cannot be entirely ascribed to differences in reward size (Gonzalez & Bitterman, 1967), and if it is accepted that one cause of the PRE in rats is that subjects trained under PR learn to use the memory of nonreinforcement as a stimulus to control current performance, we could argue that fish show a weaker PRE precisely because of differences in the stability of their short-term memory, or because of differences in their ability to use such memory stimuli to control behavior. While this conclusion rests upon rather insecure foundations, it represents an important type of argument: theoretically, the most interesting conclusions that can emerge from comparative studies, is when one attempts to explain an apparently heterogeneous set of behavioral differences by appealing to a single underlying difference in learning mechanism. Our own explanation of the fish's poor serial reversal performance, although it successfully accounts for other behavioral differences between rat and fish, is unable to account for possible differences in the effect of PR on extinction.

If serial reversal improvement depends upon the development of stable attention in the face of changing reinforcement conditions, then our previous assumption, namely, that analyzers are less stable in fish than in rats, leads to the prediction that fish should be poorer than rats at serial reversal learning. There are, therefore, independent grounds for supposing that differences in this third factor responsible for serial reversal improvement exist between rats and fish. Moreover, the assumption leads to several further predictions; we assumed not only that there were attentional differences between rats and fish, but that there was an orderly decrease in the stability of attention from rats to lower birds to fish. We can predict, therefore, that there will also be orderly differences in serial reversal efficiency between these three animal groups.[2]

### 2. *Birds*

Bitterman has used the serial reversal task as a major test of learning capacities in his comparative research program. We have already mentioned the results of his experiments with fish: with the exception of a single, early experiment (Wodinsky & Bitterman, 1957), whose results he has since discounted (Bitterman, 1969), none of Bitterman's studies has revealed reliable evidence for improvement. Along with other investigations, his studies of reversal in rats (e.g., Bitterman *et al.*, 1958; Gonzalez *et al.*, 1964) have obtained substantial improvement. However, the results of experiments with

[2] Much of the remaining discussion in this section, as well as in the following section on probability learning, has been presented elsewhere (Mackintosh, 1969b). For a discordant note, see Bitterman (1969).

pigeons (e.g., Bullock & Bitterman, 1962a; Stearns & Bitterman, 1965; Gonzalez *et al.*, 1966), in which significant reversal improvement occurred have led Bitterman to classify pigeons with rats as animals capable of showing improvement over a series of reversals, and to contrast them with fish (and cockroaches) as animals incapable of showing improvement. This reliance on a qualitative distinction—either animals show significant improvement, or they show none at all—seems to us mistaken, since one of the major arguments of this chapter is that it is possible to explain differences in behavior between different animal groups by making quantitative changes in the parameters of a model. The fact that rats improve over a series of reversals and fish fail to improve implies to us only a quantitative difference in the operations of similar mechanisms, not a qualitative difference of mechanism. Our claim, therefore, that birds occupy an intermediate position between rats and fish with respect to the stability of analyzers, implies that their performance on a serial reversal task should also fall between that of the fish and the rat. In other words, they should show some improvement (more than the fish), but not as much as the rat. This means that we need to draw finer distinctions than has Bitterman, i.e., to make quantitative comparisons between the amount and rate of improvement shown by rats and birds.

There is, in fact, substantial evidence to suggest that lower birds are less efficient reversal learners than are rats. In the first place, they are less efficient than higher orders of birds such as passerines (Gossette, 1967; Gossette, Gossette, & Inman, 1966; Gossette, Gossette, & Riddell, 1966), and even passerines do not so rapidly reach the level of performance usually attained by rats. Second, in a number of experiments the improvement shown by pigeons and quail has been marginal or nonexistent. Gonzalez *et al.* (1966) only found improvement in quail given 40 trials per problem, none in birds given 20 trials per problem; Stettner, Schulz, and Levy (1967) obtained only "limited or 'marginal' improvement in quail [p. 4]." Reid (1958) found no improvement in pigeons, while Bullock and Bitterman (1962a) found that pigeons improved only "to a limited extent in the earlier sessions [p. 960]" of a series of simultaneous visual reversals, and hardly at all over 55 reversals of a successive, conditional discrimination.

Some more or less direct comparisons of the reversal performance of birds and rats serve to strengthen our conclusion. The results of three experiments performed by Bitterman on visual reversal learning in rats, pigeons, and mouthbreeders, are shown in Figure 12.3. The choice of these particular studies was guided by the fact that the initial discrimination was learned at a roughly comparable rate by each class of animal. The first impression from this figure is that the major difference in performance lies not between rats and pigeons on the one hand and fish on the other, but between pigeons and fish on the one hand, and rats on the other. This is not, of course, a reliable impression,

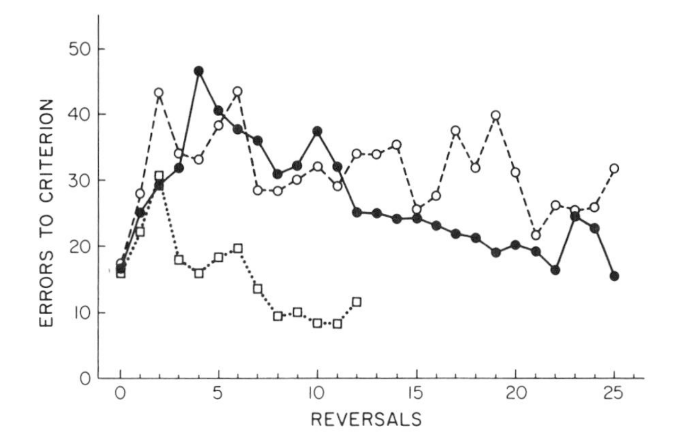


**Fig. 12.3.** Visual serial reversal in rats (□), pigeons (●), and fish (○). [*After* Gonzalez, Roberts, and Bitterman (1964); Stearns and Bitterman (1965); and Behrend, Domesick, and Bitterman (1965).]

since the pigeons showed a significant improvement, while the fish did not. Nevertheless our major claim is surely substantiated by these results: rats show very much greater improvement over a series of reversals than do pigeons, while pigeons show greater improvement than do mouthbreeders. The three groups of animal show precisely the ordering of performance predicted by the assumption of differences between them in the stability of analyzers.

Finally, Mackintosh (1969b) trained rats and doves on a series of either visual or spatial reversals, and also observed significant differences in reversal performance. All animals were trained in the same apparatus, a Grice box with dark-gray and light-gray goal boxes; half of each group learned a series of brightness reversals, while half learned a series of position reversals. The results are shown in Figure 12.4, averaged over brightness and position problems. The difference between the two classes of animal over the 10 reversals was highly significant.

This study also provided further tests of the present analysis of reversal differences. If the reason that rats reverse more rapidly than doves is that they have better learned to maintain the relevant analyzer, then they should find it harder than doves to shift analyzers and learn a nonreversal shift problem. After their tenth reversal, all subjects were, therefore, trained on such a shift (from brightness to position, or vice-versa), and the results are also shown in Figure 12.4. The superiority of rats to doves in the tenth reversal was itself reversed for the nonreversal shift problem. Further groups of rats and doves were also trained on a series of nonreversal shift problems (for a typical subject, black might be positive in the first problem, left in the second, then

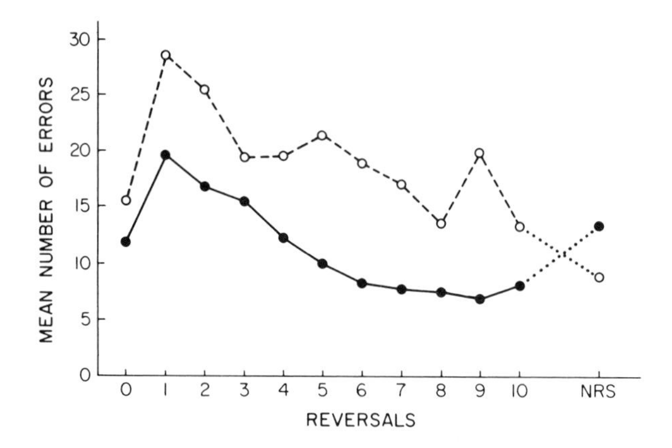


**Fig. 12.4.** Visual and spatial serial reversal learning followed by nonreversal shift learning in rats (●) and doves (○). [*After* Mackintosh (1969b).]

black again, then left again, etc.). In this series of problems, formally very similar to a series of reversals, except that the subject is not required to maintain attention to the same stimulus dimension, doves performed just as efficiently as rats. That the superiority of rats to doves was confined to situations requiring just such a maintenance of attention, strongly suggests that this is at least one underlying difference between these animals.

## D. Conclusions

By comparison with the effects of overtraining, the effects of prior reversal training on subsequent reversal learning are reliable and consistent. Although not all experiments with rats have yielded significant reversal improvement (e.g., if very long intertrial intervals are used), it is very unlikely that the repeated failures to obtain such an effect in fish are a consequence simply of inappropriate experimental conditions. To dismiss the evidence cited in this section as an artifact in this sense is to imply, first, that nearly a dozen studies with fish have coincidentally used just that set of conditions (so far unspecified) that yields negligible reversal improvement in rats; second, that approximately twice that number of experiments with birds has used conditions intermediate between those used in all the rat and all the fish experiments; third, that experiments with higher avian orders such as passerines have used a more rat-like set of conditions and experiments with lower avian orders such as columbiformes have used a more fish-like set of conditions. The act of faith demanded to accept these implications is one not easily overthrown by argument.

If the differences are real ones, then they must be explained by postulating differences in the factors underlying serial reversal improvement in rats. Such an implication should be less unwelcome in view of Warren's evidence of a difference in rule-learning abilities between primates and cats. If the development of a win–stay, lose–shift strategy were one cause of reversal improvements in rats, it would be natural to extrapolate Warren's conclusions and suggest that differences in the extent to which such strategies or rules are within the subject's capacity represent one cause of differences in reversal performance between rats and fish. While we do not believe that it is possible to rule out such a suggestion, there is little evidence that such strategies do play an important role in reversal learning by rats, and the evidence that might imply differences in such capacities between rat and fish is susceptible of alternative explanations.

Although it is also possible that there are differences between rats, birds, and fish in the way in which proactive interference develops during the course of serial reversal to reduce negative transfer from one problem to the next, the evidence for this (rather surprising) suggestion is not entirely satisfactory. There is, however, relatively good evidence that one cause of reversal differences between rats and birds is that the latter do not learn so effectively to maintain attention to the relevant cue. This difference was, of course, implied by the results discussed in the preceding section on overtraining, extinction, and reversal; it also serves to explain a further set of differences between these classes of animal, namely, differences in performance in probability learning experiments.

## IV. Probability Learning

Experiments on probability learning have formed the second major class of study undertaken by Bitterman and his associates in comparative psychology. On the basis of a classification of performance into two categories —matching versus maximizing—he has shown a number of differences between mammals, birds, and fish. In Chapter 11, we argued that this dichotomy is misleading, and that rats trained on probabilistic discriminations by a correction procedure normally perform somewhere between these two extremes, their exact score being a function of a number of factors influencing the strength of the relevant analyzer. If only for this reason, therefore, we do not accept Bitterman's argument that the performance of rats (said to maximize) differs *qualitatively* from that of fish (said to match). It seems more appropriate to say that rats simply perform more efficiently than fish. A second reason for this conclusion will appear as we discuss the performance of birds in these situations.

## A. Differences in the Performance of Rats, Birds, and Fish

The evidence for some behavioral difference between rats and fish is strong. We have already (on page 407) reviewed results showing that rats usually attain an asymptotic level of performance slightly below 100% choice of the majority stimulus. Although this is, in a sense, less than perfect, it is very much better than the performance of fish; in neither visual nor spatial experiments have fish ever performed at this level of accuracy. In visual problems, indeed, they usually perform below rather than above matching. Bitterman *et al.* (1958) found that mouthbreeders trained on a 70 : 30 visual discrimination reached an asymptote of 69% choice of the majority stimulus; Behrend and Bitterman (1961, 1966) found asymptotes of 68% and 66% choice of the majority stimulus in mouthbreeders and goldfish, respectively. Similar performance also occurs at other reinforcement ratios (Behrend & Bitterman, 1961), and matching is an individual, as well as a group phenomenon (Behrend & Bitterman, 1966). Although performance on spatial problems has often been slightly better than this (75% choice of the 70% majority stimulus both by Bitterman *et al.*'s mouthbreeders and by Behrend and Bitterman's goldfish), it has usually been about matching at all reinforcement ratios (Marrone & Evans, 1966; Weitzman, 1967), and has never attained the 90–95% accuracy typically shown by rats. In an unpublished study, we found that goldfish performed below matching on a 70 : 30 combined visual and spatial discrimination.

As with the other experimental situations discussed in this chapter, it is possible that these differences in performance reflect differences in the conditions under which rats and fish have been studied rather than inherent differences in efficiency. It is certainly likely that both drive level and reward size affect probability learning in rats (Solomon, 1962; Uhl, 1963; Wright, 1967), although since these experiments have yielded inconsistent results, we do not know what these effects are. In Wright's experiment, rats trained with a small reward (37 mg) and under a high drive (23-hr deprivation) performed at a level typically attained by fish. However, these conditions were nearly the same as those (45-mg reward; 22-hr deprivation) under which Calfee (1968) observed performance rather nearer maximizing than matching in rats trained on a difficult visual discrimination; and it is likely, therefore, that Wright ran too few trials (110) to obtain asymptotic performance. In another experiment, Weitzman (1967) trained rats under conditions designed to be similar to those used in experiments with fish, and found performance no better than matching. However, the group average in this experiment was depressed by inclusion of one subject (out of a total of five) that continued to select the minority stimulus on over 70% of trials; and Wietzman used a zero-delay, self-correction procedure which must have minimized the consequences of an error. Fish continue to match even if delayed before being permitted correction (e.g., Bitterman *et al.*, 1958).

We agree that the determinants of asymptotic performance in probability learning experiments should be studied in greater detail, and that a greater variety of conditions should be run with fish. We also agree that some of the differences in performance between rat and fish may be attributable to situational differences. Nevertheless, until the conditions are discovered under which goldfish will select the majority stimulus of a 70 : 30 spatial problem on about 95% of trials, there are good reasons to doubt all attempts to explain away in their entirety the behavioral results we have been discussing.

If we accept that fish perform less efficiently than rats in probability learning experiments, we can go on to ask about the performance of birds. Bitterman's insistence on a dichotomous classification of behavior as either matching or maximizing does not permit many options. The one selected (Bitterman, 1965) is that "the pigeon shows random matching in visual problems; in spatial problems it tends to maximize [p. 405]."

It is only if all above-matching performance is counted as maximizing (and, therefore, essentially similar), that lower birds can be said to maximize (and, therefore, behave in a rat-like manner) when trained on spatial probability learning tasks. The one published study of spatial probability learning in pigeons found that six subjects attained an asymptote of 82% choice of the majority stimulus in a 70 : 30 problem (Graf, Bullock, & Bitterman, 1964). If this is compared with the 95% asymptote attained by rats, and the 75% asymptote attained by fish in one 70 : 30 spatial experiment (Bitterman *et al.*, 1958), it is hard to resist the conclusion that pigeons are more efficient than fish, but less efficient than rats. Mackintosh (1969b) found significant differences in spatial probability learning between rats and chicks (although the latter performed above matching), a result which supports the general contention that birds show performance intermediate between rats and fish.

The evidence for the suggestion that birds match on visual probabilistic discriminations is even more tenuous. In two papers, Bitterman has reported the results of seven experiments with pigeons (Bullock & Bitterman, 1962b; Graf *et al.*, 1964). While two of these seven experiments produced relatively close approximations to matching, in the remaining five, performance was significantly better than matching (details of this analysis have been presented by Mackintosh, 1969b). Shimp (1966), training pigeons for 20,000 trials on a 75 : 25 successive, red–green discrimination found that subjects eventually came to select the majority stimulus on 86% of trials. Finally, in an unpublished study, Mackintosh and Little obtained an asymptote of just over 90% choice of the majority stimulus in a group of eight pigeons trained (by guidance) on a 70 : 30, red–green discrimination. Chicks also perform well above matching on visual probability problems. Figure 12.5 shows the results of an experiment by Mackintosh (1969b) on visual and spatial probability learning in rats and chicks. In this experiment, chicks performed virtually as accurately as rats on the visual problem, less efficiently only on the spatial problem. In general, however, the conclusion that pigeons do not usually match on visual problems should not be taken to imply that they typically perform as efficiently as rats. In only one of Bitterman's experiments has performance been substantially above matching; and Shimp's subjects responded only slightly above matching for several thousand trials.

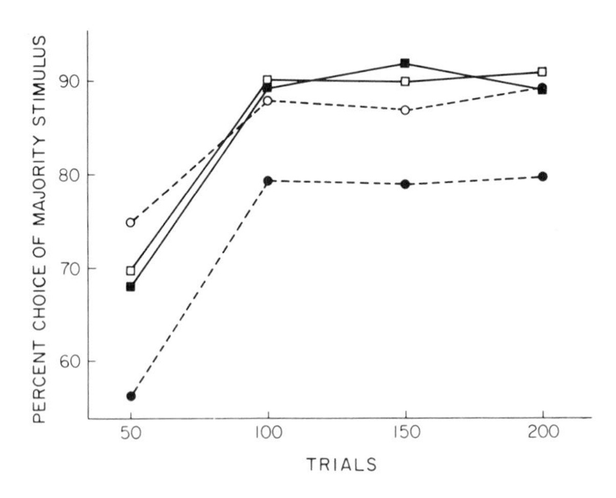


**Fig. 12.5.** Visual and spatial probability learning in rats and chicks. (■): position in rats; (□): brightness in rats. (●): position in chicks; (○): brightness in chicks. [*After* Mackintosh (1969b).]

We conclude, therefore, that rats, birds, and fish do not behave alike when trained on probabilistic discriminations. The differences in performance consist in differences in the asymptotic choice of the majority stimulus, with rats attaining higher asymptotes than birds, and birds attaining higher asymptotes than fish.

## B. Theoretical Interpretation

The ordering of performance in probability learning experiments is exactly the same as that which appears in comparisons of serial reversal performance: rats are more efficient than birds, and birds are more efficient than fish. Although Bitterman (1965) has argued, "experiments on habit reversal and experiments on probability learning tap somewhat different processes [p. 405]," we shall argue that precisely those changes already made in the parameters of our model will predict these differences in probability learning performance.[3]

While both probability learning and serial reversal learning can be said to demand an adjustment to inconsistent reward, it is worth noting that one of the possible strategies for serial reversal learning would actually prevent subjects from performing accurately in a probability learning situation. The utilization of a win–stay, lose–shift strategy, while appropriate to a reversal task, would tend to produce matching behavior in probability learning: here, in order to select the majority stimulus on a high proportion of trials, the subject must precisely *not* select the last rewarded stimulus. To the extent therefore, that we wish to postulate a single theoretical difference between rats, birds, and fish, which will encompass the largest possible proportion of the data, the fact that rats are also more efficient than fish at probability learning discounts the suggestion that differences in reversal learning are attributable to differences in the availability of appropriate rules.

There is one possible way out of this dilemma: although no simple difference in the type of strategies employed by these animal groups will account for both sets of behavioral differences, it might be the case that rats are more flexible than birds or fish, and adopt a win–stay, lose–shift strategy in reversal situations, but a win–stay, lose–stay strategy in probability learning situations. If this were so, then rats that had acquired one type of strategy should continue for a short time at least to use it (inappropriately) when shifted to the other type of problem. Mackintosh *et al.* (1968) trained rats on a series of black–white reversals, and then compared their performance on a black–white probabilistic discrimination with that of a group of naive control animals. So far from interfering with probability learning, prior reversal training produced much more rapid learning to an equally high (above matching) asymptote. It does not look, therefore, as if it will prove possible to explain behavioral differences in reversal and probability learning by talking of differences in response strategies.

Although, however, the response strategies appropriate to the two tasks are different, successful performance on both does depend on maintaining attention to a given cue, even when that cue's correlation with reward is either imperfect or constantly changing. Indeed, we have shown in Chapter 11 that a major cause of errors made by rats in probability learning experiments

[3] Bitterman's argument that differences in reversal and probability learning cannot reflect a common underlying cause rests on the following interpretation of his results: in both visual and spatial reversal experiments, birds behave in a rat-like way, while in probability learning experiments they behave in a rat-like way in spatial problems, but in a fish-like way in visual problems. The preceding discussion makes it obvious why we reject this argument.

is precisely that they fail to switch-in the relevant analyzer strongly enough to control behavior on all trials.

Our theoretical analysis, and the experimental results cited in Chapters 9 and 11, agree in stressing that the maintenance of attention to the relevant cue in the face of inconsistent reward underlies successful performance in both reversal and probability learning. From this it follows that the differences in the stability of attention between rats, birds, and fish, which have already been used to account for the difference in serial reversal performance of each class, will equally account for the observed differences in performance in probability learning experiments. Indeed, the fact that the three groups of animals are ordered in the same way in terms of their performance on the two tasks, is itself striking support for our analysis of the common processes underlying performance on the two. This illustrates the theoretical value of comparative research. Instead of developing and testing theories by deriving new predictions about the behavior of old animals, we provide evidence of a different type by showing how, once certain initial alterations in the model have been made, a whole range of behavioral differences between various animals falls into place.

## V. The Theoretical Relevance of Comparative Psychology

In the previous sections, we have concentrated on showing how our model will explain the behavior of animals other than the rat. We showed that the behavior of octopuses in overtraining experiments seems essentially similar to that of the rat, and that if certain changes are made in the parameters of the model, it will successfully explain the rather different behavior of other animal groups. There has been astonishingly little attempt to do this in the past, not only because of the long dominant assumption in American psychology (see Bitterman, 1960) that there are no important differences between different groups, but also because even those who have recently been responsible for discovering a number of such differences (e.g., Bitterman, Harlow, & Warren) have often been content to describe the differences, and occasionally to offer rather vague explanations in terms of such global concepts as differences in adaptability and intelligence. Surprisingly little attempt has been made to progress beyond this informal stage.

### A. The Discovery of Similarities

There is more of value to the results of comparative research, however, than the discovery of further problems for explanation. Discoveries of both similarities and differences of behavior are of unique theoretical importance: they provide evidence relevant to theoretical development which could not be provided in any other way.

To consider first the discovery of similarities: we have shown in a series of experiments that octopuses and rats behave in apparently similar ways in experiments on the effects of overtraining on extinction, and reversal and nonreversal shift learning. This finding has several implications of interest.

Since the results obtained with rats in jumping stands can be replicated with octopuses in tanks of seawater, there can be no question of the results being an artifact of the precise experimental situation employed. Some experimenters, for example, have suggested that, in rats, the ORE may be produced by the further handling to which the overtrained rats are subject (Hill, Spear, & Clayton, 1962), but the discovery of the effect in octopuses renders such a suggestion distinctly less plausible.

Not only does the discovery of similar behavior in octopus and rat show that the rat results cannot be due to incidental features of the experimental situation, it also shows that they cannot be due to incidental features of the experimental subject. It implies, e.g., that we have not biased our experiments in favor of producing results implying the importance of attention, by using a subject that does not naturally rely on visual cues, and must, therefore, learn to attend to them. Octopuses are visual animals in a sense in which the nocturnal rat is presumably not, and yet we can show clearly that octopuses do not attend to all features of the visual environment (Sutherland, Mackintosh, & Mackintosh, 1965), and that they must learn to attend to the relevant one.

Experiments demonstrating selectivity of learning or control in rats have often been interpreted by appealing to the concept of the overt orienting response. While it is certain that many such results cannot reasonably be interpreted in this way, it remains true that rats may not automatically fixate the center of the door in a jumping stand, and must learn to do so. However, the suggestion that, in the experimental situation used, octopuses do not naturally see the relevant visual stimuli, is, to say the least, implausible. The stimuli are small relative to the background against which they are exposed, and animals unhesitatingly swim the length of their tank to attack these stimulus objects. They could only locate the stimuli by vision, and, indeed, the most casual observation of the effects of introducing visual stimuli into an octopus' tank will reveal the instant effect on fixation movements of the eye. Octopuses do not have to learn to look at the stimuli in our experiments, and our demonstrations of selective attention with octopuses, show that the concept of attention cannot be solely understood in such terms.

As has been stressed before in this chapter, one of the major problems of comparative research is to ensure equivalent conditions for experiments with different animals. We argued, in effect, that it was never possible to guarantee that such equivalence had been obtained, and this is always liable to render suspect the discovery of differences between different groups. In compensation for this, however, the discovery of similar behavior in widely separated groups is all the more significant. Unless one makes the somewhat implausible assumption that the several differences in experimental situations, although individually having an effect on the behavior in question, cumulatively cancel each other out to leave no final effect, it is reasonable to dismiss the respects in which the experimental situations differ as of no great importance. In other words, although one can never be certain that one has provided equivalent experimental situations for investigating, say, reversal learning in rat and octopus, any failure to ensure equivalence is actually *valuable* if the behavior observed turns out to be similar, since this implies that the uncontrolled variables exert a negligible effect on the performance in question.

Furthermore, of course, the fact that octopuses do behave in a manner similar to rats in our experiments, makes us more confident that the differences we have found between rats, chicks, and fish are not merely the result of uncontrolled variables. As a specific illustration, there are a number of respects in which our training situation for goldfish differs markedly from that used for rats, but in many of these respects it resembles that used for octopuses. The level of motivation and size of incentive used for fish and octopus is probably rather similar, in that both animals receive 10 pieces of food during the course of their 10 daily trials, and this appears to be sufficient virtually to satiate them for this particular incentive. (Neither animal can easily be trained for appreciably more than 10 trials per day, since neither can be induced to make choices within a reasonable period of time after more than 10–15 trials.) Thus, even if, in our experiments on overtraining in fish, the conditions of drive and reinforcement differ from those used with rats, they probably do not differ from those used with octopuses, and such differences cannot be used to explain the observed differences in behavior between rats and goldfish.

### B. The Discovery of Differences in Behavior

It has generally been reckoned that the objective of comparative psychology is to discover differences in behavior between different animal groups, and that to the extent that only similarities are discovered, nothing has been achieved. For the reasons just given, we disagree with this conclusion, but it is obvious that the discovery of differences both presents a greater challenge to the investigator, and is of value for rather different reasons.

Boycott and Young (1950) have argued that although biologists are not yet in a position where, like chemists, they can synthesize new materials on which to test their theories, they are, nevertheless, presented with a sufficient variety of living organisms to make such artificial synthesis unnecessary. This variety can be put to use in two ways. First, a judicious search will usually bring to light a living animal peculiarly suited to the biologist's needs. Biologists in general have made use of this diversity when it has been a matter of finding an animal suitable for the investigation of a particular experimental problem; but the same can hardly be said of all psychologists, for example, those that have studied hoarding behavior in the laboratory rat—an animal that does not naturally hoard. Boycott and Young also suggested a second way of taking advantage of the variety of living animals. Specifically, as neuroanatomists interested in learned modifications of behavior, they wanted to attempt to correlate differences in anatomical structure between various groups with differences in their behavior, thus, not only throwing light on the function of such structures, but also providing insight into the neural mechanisms controlling behavior.

This is undoubtedly a strategy with exciting potentialities, and one that will, one hopes, become increasingly popular. Several applications of the strategy have already proved its value. Sutherland (1963a), for example, has attempted to correlate differences in the ways in which cats and octopuses categorize visual patterns with differences in the structure and functioning of their visual systems. The work of Hubel and Wiesel (1962) on the physiological organization of the striate cortex of the cat suggests that differences in orientation of visual stimuli are initially categorized in terms of the type of cortical cell predominantly fired by each orientation. Different cells are maximally sensitive to lines in different orientations, and from their work, there does not seem to be any predominance of cells sensitive to one orientation over cells sensitive to other orientations. By contrast, Young's (1960) histological investigations of the organization of dendritic fields in the plexiform zone of the octopus visual system have suggested that, in these animals, there is a predominance of horizontal and vertical detectors over oblique detectors. In accordance with these anatomical differences, Sutherland has found that, whereas octopuses discriminate horizontal–vertical differences far more readily than the difference between opposite obliques, cats find both types of discrimination equally easy.

In a number of studies of amphibian phototactic behavior, Muntz (1963a,b) has attempted to correlate both ontogenetic and phylogenetic differences in behavior with differences in visual receptors and photochemistry. Many amphibia show positive phototaxis in certain situations, especially towards blue light. It has been suggested that the green rods of the amphibian eye, since they are (confusingly enough) maximally sensitive to blue light, underlie this phototactic behavior. Muntz has provided a simple and elegant test of this hypothesis by comparing the phototactic behavior of two species of urodeles, one (*Triturus cristatus*) possessing such rods, the other (*Salamandra salamandra*) lacking them. He found that *Salamandra* show both a weaker positive phototaxis than *Triturus* and also no preference for blue over other colors. The logic of this experiment is similar to that of numerous experiments involving brain lesions: if removal of a given region of the brain destroys a given pattern of behavior, then it is inferred that the region is a "centre" for the behavior in question. Muntz's work shows that the comparative psychologist may often be in a better position than the physiologist to undertake studies of this kind.

It will be noticed that in these examples the correlation sought has been between differences in the anatomy of the visual system with differences in visually guided behavior, rather than between differences in the structure of "association areas" with differences in learned modifications of behavior. This is hardly surprising; our present understanding of the physiological mechanisms of learning is so rudimentary that it is difficult to imagine just where one would start the search for such correlations. Nevertheless, if, at the moment, it is impossible to correlate differences in learning with neuroanatomical differences, it ought to be possible to correlate them with differences in hypothetical models of learning, and such a program, if successful, should prove an important guide to the anatomist and physiologist in knowing what sort of neurological differences to look for.

It is precisely such a correlation of differences in the parameters of a theoretical model with differences in learned behavior that we have been trying to establish. Our hope is that this theoretical program will provide the same sorts of benefit as does the neuroanatomical program. In the preceding sections, we have shown how our assumption of an increase in the instability of analyzers as one progresses from rat to birds to fish (an assumption initially based on comparisons of the effects of overtraining on extinction and reversal), will serve to explain further differences in the behavior of these animals in serial reversal and probability learning experiments. In a similar manner, knowledge of differences in the organization of the receptive fields of visual cells serves to explain differences in the classification of visual patterns by cats and octopuses. Furthermore, just as the discovery of these latter behavioral differences strongly confirms the importance of these anatomical structures in visual classification, so in our theoretical program, the fact that the behavior of rats, birds, and fish does differ in these further experimental situations, strongly confirms our theoretical analysis of the processes involved in these situations. Our analysis of the dip shown upon reversal after extinction by nonovertrained, but not by overtrained rats, is supported by the fact that in an animal such as the goldfish that does not show the ORE, no such difference between overtrained and nonovertrained subjects appears. Our hypothesis that success in probability learning situations depends upon maintaining attention to the relevant cue in the absence of a perfect correlation of that cue with reward, is supported precisely by the fact that performance in such situations varies in the way it does across animal groups. Conversely, our analysis of the ORE in mammals is supported by the difficulty of obtaining such an effect in birds and the impossibility of obtaining it in fish, when these results are taken in conjunction with the further findings of serial reversal and probability learning experiments with these animals.

### C. Limitations of Present Approach

In this chapter, we have concentrated on showing the theoretical importance of a program of comparative research. By correlating a wide variety of behavioral similarities and differences across different animal groups, we have attempted to show that such behavior can be understood in terms of a single theoretical model. This is not, of course, the only sense in which the term comparative psychology has been used, let alone the only benefit that may accrue from comparative studies. Before concluding, it is worth stating explicitly some of the things we have *not* been trying to do.

Many ethologists (as well as some psychologists) have argued that knowledge of the behavior of an animal in its natural environment is a prerequisite to the laboratory study of that animal's sensory or learning capacities. Jensen (1965), for example, has suggested that much of the effort expended by Gelber, McConnell, and others in investigating learning in paramecia and planaria would have been better directed towards achieving a thorough understanding of their natural behavior. Ignorance of the normal effects of certain operations on the behavior of these animals, Jensen argues, vitiates many of the interpretations put upon their results by these investigators. While we should agree that there is often much force in this line of argument, it is obvious that we have not been concerned to remedy such deficiencies. It seems worth suggesting, indeed, that the term comparative psychology be reserved not for an ethological approach to psychology, but to describe studies in which the major aim is to make comparisons across animal groups.

In many areas of biology, the goal of comparative studies, although clearly to make comparisons across animal groups, differs sharply from that we have set ourselves. Lorenz (1950) has stated,

> Since the days of Charles Darwin the term "comparative" has assumed a very definite meaning. It indicates a certain rather complicated method of procedure which, by studying the similarities and dissimilarities of homologous characters of allied forms, simultaneously obtains indications as to the phyletic relationships of these forms of life and as to the historical origin of the homologous characters in question [pp. 239–240].

With such a definition behind him, it is not surprising that Lorenz feels able to flay psychologists for presuming to use the term comparative to describe their studies. It is misleading, however, to imply, as does Lorenz, that comparative research is designed to throw light on evolutionary processes ("the phyletic relationships" of different groups, and the "historical origins" of various characters) *to the exclusion* of all else; and even were we to accept Lorentz's definition, there would be a clear need for a new term to describe the approach advocated in the same symposium by Boycott and Young (who, needless to say, describe their approach as "The Comparative Study of Learning"). However this may be, our own research will obviously not fit into Lorenz's definition. In the first place, we have not been studying closely allied forms; the choice of octopuses, goldfish, domestic chicks, and laboratory rats has been dictated by considerations of availability and convenience, not because we have revolutionary ideas about zoological classification. Second, we have not been concerned to establish a behaviorally defined phylogeny of the distantly related forms we have been studying. On the basis of our experimental results, it is possible to order our subjects from fish to bird to rat, with octopuses possibly occupying the same niche as the rat. From a phylogenetic viewpoint this does not perhaps form a very plausible series: birds may be distantly descended from fish, but not from teleost fish; mammals are not even distantly descended from birds; and cephalopods do not fit into any vertebrate series at all. The ordering of these forms that emerges from our studies is noteworthy, but an interest in phylogenetic development would obviously have dictated a different choice of subjects: apart from anything else (as has frequently been pointed out by zoologists), the use of most living and highly specialized representatives of different animal groups is not appropriate to the study of the course of evolution. Finally, we have not even been particularly concerned to trace developments in the adaptiveness of the behavior of the animals we have studied. It may seem plausible to claim that by improving over a series of reversals and reaching a higher asymptote in probability learning experiments, rats are behaving in a more adaptive manner than are goldfish, but there are always hazards in such interpretations, and in our present state of knowledge there seems little to be gained by stressing this approach to the comparative psychology of learning.

## VI. Summary

The results of experiments comparing the behavior of different animal groups are often difficult to interpret, because it is never easy to exclude the possibility that any differences observed are the outcome of differences in experimental conditions. Nevertheless, such experiments are worth undertaking, not only because it is possible that a psychology based on the study of the rat will raise local idiosyncrasies to the status of universal principles, but also because the discovery of systematic differences between different animals provides an important testing ground for theoretical systems. The present chapter has reviewed a number of studies of the behavior of animals other than the rat in experiments on discrimination learning. We should emphasize that once again we have been highly selective: it is obvious that we have not attempted to discuss more than a small fraction of those experiments that could be called comparative.

In the first place, we have not discussed learning sets and other complex discrimination procedures used with primates. A large part of these data has been recently reviewed by Warren (1965). Moreover, primates appear to develop rules for the solution of learning-set, oddity, and conditional discriminations, which are beyond the capacity of most lower animals. We have touched on this in our discussion of serial reversal learning because it was relevant to deciding the basis of serial reversal improvement in other animals, but we have not entered into a detailed discussion, since such capacities lie well outside the scope of the type of theory we are considering. We have also excluded a number of studies, even though they are of situations used by us, where the absence of further data allows little more than speculation as to the causes of differences. For example, there have been several unsuccessful attempts to obtain the ORE in cats; but without further evidence (of the sort we have obtained with goldfish), one can conclude that the theory proposed here needs modification in order to apply to cats, but can only speculate on the nature of those modifications. Similarly, a wide range of animals has been studied in serial reversal experiments, with the general finding that vertebrates above the level of fish show improvement while many invertebrates do not. Since these differences have not been correlated with further behavioral differences, little is gained by applying the sorts of changes proposed here for rats, birds, and fish to these other groups of animals.

Our main concern has been to show that a theory designed to account for the behavior of one particular animal in a series of experimental situations can be profitably tested by experiments with other classes of animal. The theory can be modified to explain differences in one situation, and the modified theory can be tested since it will now predict further differences in other experimental situations. (Equally, similar behavior in one situation leads to the prediction of further similarities.) In experiments on overtraining and reversal, certain differences have appeared between rats, pigeons and chicks, and goldfish. These differences led to predictions of further differences in experiments on the extinction of discrimination habits as a function of overtraining. Finally, some further differences discovered in these experiments led to predictions about the type of behavior expected in experiments on serial reversal and probability learning. Within the limits we have set, we can claim that many of the differences between these classes of animal are economically accounted for by postulating differences in the parameters of a two-process model of discrimination learning.

# CHAPTER 13

# Formal Models

## I. Phenomena

In the previous chapters, we have discussed a wide range of phenomena. By no means all of these are susceptible of explanation in terms of a two-process model of selective attention. Before proceeding to consider formal models of selective attention, it is, therefore, worth listing those effects that we believe can and should be explained in terms of selective attention. The evidence for the various phenomena will be found in earlier chapters and no attempt is made to recapitulate it here. We shall, however, indicate points at which the evidence is insufficient to establish a given conclusion with certainty. It should also be understood that in what follows we cannot list all of the conditions determining a given effect and the phrase, "other things being equal," is implied throughout. We have of necessity been rather vague about the phylogenetic groups that exhibit the various phenomena; except for the laboratory rat, there is often no adequate evidence.

Later in the chapter, we shall discuss how well different models simulate some of the facts presented below; the main phenomena to be discussed are indicated by an asterisk and to facilitate subsequent reference to them, they are labeled E-1, E-2, etc. Although we believe the other effects are ultimately explicable in terms of selective attention, they involve further processes (such as generalization) that have not yet been built into any formal model of two-process learning.

### A. Position Habits and the Course of Learning

(E-1)* Animals trained on a simultaneous discrimination usually perform at chance for many trials, and then shift rather abruptly to an asymptote of performance that is often virtually errorless.

(E-2)* In the initial stages of training, when animals are performing at chance on the relevant cue, they usually exhibit systematic responding to any dominant irrelevant cue. For example, rats develop position habits.

(E-3)* Position habits tend to *develop* in the course of training. Over the first 10 trials of training, rats make only 68% of their responses to the side to which they ultimately develop a position habit.

(E-4) While rats are consistently responding to one side, they develop differential latencies; latencies become faster when the positive stimulus appears on the preferred side than when the negative stimulus appears on that side.

(E-5)* Over 90% of rats break position habits on a trial on which the positive stimulus is on the nonpreferred side.

(E-6)* Once a position habit is broken, errors to the nonpreferred side are almost as common as to the preferred side. Moreover, the side to which individual rats develop position habits in initial training does not correlate with the side to which position habits develop during some subsequent phase of training, such as reversal.

(E-7)* The more conspicuous the relevant cue, the shorter is the position habit in initial learning and the fewer errors are made between breaking the position habit and reaching criterion.

(E-8)* Giving reversal pretraining while animals' responses are controlled by a position habit retards subsequent learning.

### B. Learning with Several Relevant Cues

(E-9)* If animals are first trained with one relevant cue and an additional cue is subsequently made relevant, they learn something about the incidental cue even when behavior is fully under the control of the first cue; in the extreme case, this can result in errorless learning. However, such pretraining usually reduces the amount learned about the incidental cue (blocking): animals pretrained on one component of a compound cue normally learn less about the other component than animals trained from the outset on the compound.

(E-10) Changing reinforcement conditions (e.g., by increasing the strength of shock in a CER paradigm) when the incidental cue is introduced facilitates learning about that cue.

(E-11)* The amount learned about an incidental cue may be reduced by pretraining on another cue even when the response requirements change when the incidental cue is introduced. For example, if rats are trained on a successive discrimination with brightness relevant and then switched to a simultaneous discrimination with both brightness and orientation relevant, the successive pretraining decreases the amount learned at the second stage about the orientation cue. Less is learned about the incidental cue when it appears as an irrelevant cue during pretraining than when it is not present at this stage.

(E-12)* In general, the more training given on the blocking cue, the less the learning about an incidental cue.

(E-13) There is some evidence that increasing drive level reduces incidental learning.

(E-14) Incidental cues are better learned when they are introduced abruptly than when they are introduced gradually ("faded in").

(E-15)* The more difficult the original discrimination, the more incidental learning occurs.

(E-16)* Animals trained with two (or more) cues relevant learn faster than animals trained with either cue on its own relevant, except when one of the cues is very dominant.

(E-17)* The amount learned about a weak cue in a given number of trials is reduced by the presence of a strong relevant cue (overshadowing).

(E-18) The more strongly correlated is the overshadowing cue with reinforcement conditions, the stronger is the overshadowing effect. This is true both when the correlation with reinforcement is varied by altering the probability of reinforcement in the presence of the cue and when it is varied by altering the probability of reinforcement in the absence of the cue.

(E-19)* When there are two cues of roughly equal strength, the amount learned about one may vary inversely in a population of animals with the amount learned about the other.

(E-20) When animals are trained with two cues, performance with both cues presented simultaneously is usually related to performance with each cue presented on its own by the following formula:

$$P = p_1 \cdot p_2/[p_1 \cdot p_2 + (1 - p_1)(1 - p_2)].$$

(E-21) There are, however, occasions on which animals may learn to attach responses to a compound cue that are not attached to the components (e.g., in learning conditional discriminations).

## C. Intra- and Extradimensional Shifts and Generalization

(E-22)* Animals that have learned a successive (or simultaneous) discrimination to one cue subsequently learn a simultaneous (or successive) discrimination to that cue faster than to other cues (cf. E-11 above).

(E-23)* The previous phenomenon is an example of intra- versus extradimensional shift learning. Instead of manipulating response requirements from one stage of learning to the next, we may alter the values of the stimuli along a given dimension. Once again, IDS learning is easier than EDS learning. The usual paradigm is that animals are trained at the first stage to discriminate between two stimuli differing along a certain dimension with another cue present but irrelevant. At the second stage they are shifted to a problem involving new stimuli: they must learn a discrimination either between stimuli differing along the same dimension as at the first stage or between stimuli differing along the dimension that was irrelevant at the first stage.

(E-24) Animals trained to discriminate stimuli exhibiting a large difference along a dimension subsequently perform better on stimuli exhibiting a small difference than do animals trained throughout with the difficult stimuli. This effect is known as "transfer along a continuum," and occurs only if the stimuli used in the easy discrimination differ along the same dimension as those used in the difficult discrimination.

(E-25) Nondifferential reinforcement on a given cue (i.e., reinforcing on the same schedule, whichever value of the cue is selected) appears to have conflicting effects on subsequent learning of that cue. It may be that nondifferential reinforcement of a salient cue retards subsequent learning of that cue, whereas nondifferential reinforcement of a nonsalient cue facilitates later learning of that cue.

(E-26)* When animals are pretrained with one cue relevant, subsequent learning of a second cue is slower if that cue is present but irrelevant during pretraining than if it is not present.

(E-27) After training on a single stimulus, generalization gradients are sometimes relatively flat if subjects have had no previous experience of stimuli differing along the dimensions tested.

(E-28) Even when animals have had previous experience of a dimension, single-stimulus training sometimes results in flat generalization gradients. Sloping generalization gradients are more likely to occur when explicit (or implicit) discrimination training is given, though this may be between presence and absence of a stimulus rather than between two stimuli with different values on the dimension to be tested.

(E-29) Giving discrimination training on stimuli having different values on a dimension steepens the gradient on both sides of the positive stimulus.

(E-30) Under some circumstances generalization gradients after single stimulus training are steeper after prior discrimination training on a different dimension, in other circumstances they are flatter.

(E-31) Discrimination training depresses subsequent control of behavior by irrelevant stimuli present during the training (cf. E-25).

(E-32) In generalization testing to extinction after single-stimulus pretraining, the number of responses given to stimuli other than the training stimulus is maximal after a moderate number of training trials and is less after very small or very large numbers of training trials.

(E-33) If responding in the presence of a given exteroceptive stimulus is largely under proprioceptive control, generalization gradients along the exteroceptive stimulus dimension are usually flat.

### D. Effects of Overtraining and of Serial Training

(E-34)* Animals overtrained on a simultaneous discrimination often learn to reverse the discrimination in fewer trials than animals not overtrained (ORE).

(E-35) Large rewards increase the probability that an ORE will occur.

(E-36)* The more irrelevant cues are present during reversal, the larger the ORE.

(E-37)* The more difficult the discrimination, the larger the ORE.

(E-38)* Overtraining increases the number of trials at the outset of reversal learning over which animals continue predominantly to select the stimulus that was positive during initial training.

(E-39)* Overtraining decreases the number of responses to irrelevant cues made in the course of reversal learning, and results in a shorter plateau (at 50% correct) in the course of reversal.

(E-40)* Rats trained to criterion on a simultaneous discrimination, then extinguished until they are responding equally often to each stimulus, and then reversed, often show a significant increase in the frequency of responses to the original positive stimulus at the start of reversal (the "dip effect"). Overtrained rats do not show this effect. The effect occurs only in situations in which an ORE is obtained; nonovertrained chickens do not show a dip effect in situations where no ORE is obtained, though overtrained animals sometimes show the effect in such situations.

(E-41) Animals usually learn nonreversal shifts more readily than reversal shifts; in optional shift experiments when two cues are relevant during shift training—one a reversal cue and the other an extradimensional nonreversal cue—animals tend to solve the shift problem on the basis of the nonreversal cue. This result appears to be unaffected by the amount of training.

(E-42) Overtraining tends to retard nonreversal shift learning when the original cue appears as an irrelevant cue during shift learning. When it is omitted or takes neutral values during shift learning, overtraining sometimes facilitates and sometimes retards nonreversal learning. The controlling variable is not known.

(E-43)* If animals are trained to criterion and then reversed, reversal usually requires more trials than initial learning.

(E-44)* Mammals learn to reverse progressively faster in the course of a series of reversals. Some of this improvement is due to their making fewer repetitive errors at the beginning of each reversal, but, in the course of successive reversals, they also come to progress faster from a chance level of responding to asymptote.

(E-45)* Serial reversal training tends to slow down the subsequent learning of an extradimensional shift (EDS) problem.

## E. Effects of Different Schedules

(E-46)* Animals trained with several relevant cues tend to learn about more cues when trained under partial reinforcement (PR) than when trained under consistent reinforcement (CR).

(E-47) When CR training precedes PR training, resistance to extinction is less than when CR training is given after PR training.

(E-48) There is some evidence to suggest that resistance to extinction after PR is greater with increases in the number of relevant cues.

(E-49) When rats are trained on a probabilistic schedule under a correction method, the asymptotic frequency of individual subjects' choices of the more and less frequently rewarded stimuli does not normally match the reward frequency of the schedule used. The asymptote is normally greater than 50% and less than 100% choices of the majority stimulus.

(E-50) At asymptote, most of the responses to the less rewarded stimulus are caused either by reward following on a strong irrelevant cue (such as position) or by a preference for one value of an irrelevant cue.

(E-51) Under some circumstances, pretraining on the relevant cue reduces the number of errors made by following irrelevant cues and increases the number of errors produced by reward following on the relevant cue.

(E-52) The more dominant the relevant cue and the fewer the number of irrelevant cues the higher the asymptote of performance reached on a probabilistic schedule.

(E-53) Reversal learning is usually slower after training on such a schedule than after training on a 100 : 0 schedule.

(E-54) As compared with CR, probabilistic training does not have much effect on choice extinction; some studies have found that probability learning facilitates extinction, others that it retards extinction.

(E-55) Extradimensional shift learning is faster after training on a probabilistic schedule than after training with consistent reinforcement.

## F. Interspecies Differences

(E-56) Provided the relevant cue is not too dominant, octopuses show a clear-cut overtraining reversal effect (ORE); in situations in which this effect occurs, nonovertrained animals show a dip at the start of reversal if reversal is preceded by extinction training; overtrained animals do not show such a dip.

(E-57) Chicks only show a small ORE, even under favorable circumstances. When chicks are overtrained on the initial problem and then extinguished before reversal, they show a dip at the start of reversal; nonovertrained groups only show a dip in a situation in which the ORE occurs.

(E-58) Fishes do not show an ORE even under favorable circumstances. The dip effect in reversal following extinction occurs in both overtrained and nonovertrained fish.

(E-59) Fishes show very little improvement in rate of learning over a series of reversals; when the initial difficulty of the problem is equated, pigeons show less improvement than rats but more than fish.

(E-60) Other things being equal, rats are more efficient at probability learning than are birds, and birds are more efficient than fishes.

## II. The Zeaman and House Model

We now proceed to describe three models of two-process discrimination learning. The first such model to be published was proposed by Zeaman and House (1963); although it was intended to apply to the discrimination behavior of retarded children, the ideas behind it are very similar to many of those we have been expounding throughout this book. We shall also consider a model put forward by Lovejoy (1968). Lovejoy has, in fact, proposed three separate theories. His first model (Lovejoy, 1966) has already been described in Chapter 2 and is, in fact, a special case of the Zeaman and House model: it is delightfully simple, but for the reasons given in Chapter 2 will not cope with a wide range of phenomena. Lovejoy's (1965) second model was intended to be analytically soluble, and the constraints imposed by this requirement again result in its being applicable only to a narrow range of phenomena. It is his third model (Lovejoy, 1968) that we describe here. We shall also describe a model of our own. We shall not discuss Trabasso and Bower's (1968) model since, although it has many points of similarity to the models presented below, it was intended to explain the results of experiments on the formation of simple concepts by adults and involves all-or-none learning. As was shown in Chapter 5, such learning is not characteristic of animals.

In describing the models, we shall often write equations in terms of specific analyzer and response strengths. We shall let $A_{BR}$, $A_{OR}$ and $A_{POS}$ respectively, stand for the strengths of the brightness, orientation, and position analyzers; $R_B$, $R_W$, $R_V$, $R_H$, $R_L$, and $R_R$ will, respectively, stand for the strengths of the approach response to black, white, vertical, horizontal, left, and right. It is hoped that this device will render the exposition easier to follow even if it involves the loss of some formal generality.

## A. Description of Model

Zeaman and House (1963) describe their model in terms of observing responses; for simplicity we have substituted the term "analyzer." This does not of course affect the formal working of their theory.

The structure of their model can be understood from Figure 13.1. This illustrates a situation in which three two-valued cues are present. The subject has a probability of switching-in each analyzer; these probabilities sum to 1.0. Each analyzer has two responses (with probabilities summing to 1.0) attached to it, and in the first (single-look) version of the model, behavior is controlled by only one analyzer at a time.

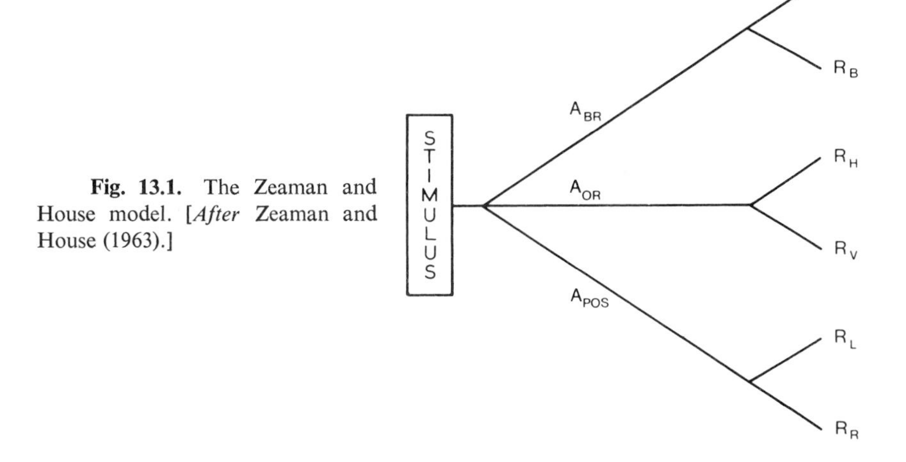


**Fig. 13.1.** The Zeaman and House model. [*After* Zeaman and House (1963).]

Response probabilities only change on trials on which an analyzer is switched-in. The probability of the response made is incremented after reinforced trials and decremented after nonreinforced trials applying the usual linear operators. The probability of switching-in the analyzer used on a given trial is changed in exactly the same way, although different operators were used for changing responses ($\theta_R$) and analyzers ($\theta_A$). Since analyzer strengths always sum to 1.0, any increment (or decrement) to the probability of the analyzer used on a given trial results in a decrement or increment to all other analyzers; these indirect increments and decrements are distributed to the analyzers not switched in on that trial in proportion to their current strengths.

Thus if the brightness analyzer is switched in on trial $n$ and no reward occurs, the position analyzer will be incremented according to the equation

$$A_{\mathrm{POS}(n+1)} = A_{\mathrm{POS}(n)} + \theta_{\mathrm{A}} \cdot A_{\mathrm{BR}(n)}[A_{\mathrm{POS}(n)}/(1 - A_{\mathrm{BR}(n)})].$$

Where reward occurs, the analyzers not used will be decremented using the equation,

$$A_{i(n+1)} = A_{i(n)} - \theta_{\mathrm{A}} \cdot A_{i(n)}.$$

## B. Discussion

Zeaman and House simulated their model and obtained the following results (among others).

(1) Where the initial probability of switching-in the relevant analyzer ($A_{\mathrm{rel}}$) was kept constant, the more unevenly the probabilities of the irrelevant analyzers were distributed, the faster was the rate of learning. Zeaman and House give no explanation of why this result was obtained, but the following account suggests itself. The amount a given analyzer increases when switched-in and rewarded becomes less the higher the analyzer probability, and the amount it decreases on nonrewarded trials becomes greater the higher the analyzer. It follows that the cumulative effects of trials on which irrelevant analyzers are switched-in will result in a lower $\sum A_{\mathrm{irr}}$ when one irrelevant analyzer has a high starting value, and is, hence, both often switched-in and also tends to increase relatively slowly but to decrease relatively fast. The limits of parameter settings under which this phenomenon is predicted have not been discovered, nor has this interesting prediction ever been tested behaviorally.

(2) Zeaman and House obtained reversal midplateaus when the correct response rate in reversal was about 60%. Plateaus became longer when $\theta_{\mathrm{A}}$ (the analyzer operator) was relatively high and $\theta_{\mathrm{R}}$ (the response operator) relatively low. In animals, plateaus are usually found when response rate is 50% correct, but Zeaman and House do not have any device for allowing for prolonged control by a single irrelevant analyzer, which is what appears to be behind this effect.

(3) Backward learning curves exhibited some initial stationarity followed by a steep rise.

(4) Nonreversal IDS learning was faster than reversal learning which in turn was faster than EDS learning (only a narrow range of parameter values was used in obtaining these results).

(5) In probability learning, the asymptote reached was usually below matching but above chance.

(6) Learning with two or more relevant cues was considerably faster than with one relevant cue. If a probabilistic cue (e.g., one rewarded on a 75 : 25 schedule) was added to a cue perfectly correlated with reward conditions, then only a very slight effect on speed of learning was found.

Zeaman and House indicate how it might be possible to extend their model to deal with compound learning as well as component learning (multiple-look model): they suggest that two or more analyzers can be used simultaneously and that the initial probability of this occurring may be simply the product of the probabilities of using the analyzers independently. This is an interesting idea, but it will not be further pursued here, since Zeaman and House have not attempted any simulation of a multiple-look model; they have, however, derived some predictions from it for the special case where the $\theta$ value for responses is 1 (all-or-none learning) and the $\theta$ value for change in analyzer strength is 0 (fixed values of analyzer strength). The single-look model was unable to account for the results of three-trial experiments with retardates in which stimuli varied along three dimensions and in which different combinations of the eight possible stimuli were presented on each trial. Zeaman and House were able to show that children can learn about more than one dimension on a single trial and the multiple-look model fitted their data reasonably well. Since animal learning is rarely all-or-none, and since we believe that the probability of using a given analyzer alters over trials, we shall not enter here into more detail about the Zeaman and House multiple-look model.

There is little doubt that further simulation would reveal that Zeaman and House's single-look model can account satisfactorily for many more phenomena than the six listed above. Indeed, Lovejoy (1965) has shown that his more limited version of the model can produce an ORE. On the other hand, there are many further phenomena for which it cannot account without considerable modification. Position habits cannot be predicted unless some device is built in to make a given response run away when it is rewarded on 50% of trials, and unless a dominant analyzer can control responding over long stretches of trials. Moreover, there is no device in Zeaman and House's model that "remembers" the initial starting value of an analyzer. This implies that with sufficient training on one cue, the strengths of all irrelevant analyzers will tend to equal one another. This is clearly unrealistic, since, when all else fails, a rat in a jumping stand will always fall back on a position habit. It must, of course, be remembered that the Zeaman and House model was intended primarily to account for data from experiments on children, not for data on animal discrimination learning. They were, moreover, the first workers to start serious exploration of the properties of a two-process model by simulation.

# III. Lovejoy's Model III

## A. Specification of Model

In describing Lovejoy's model we shall, in several instances, use a different notation from his own for the sake of uniformity. A key to the symbols used appears in Table 13.I.

TABLE 13.I

**Symbols Used in Lovejoy's Model**

| Symbol | Definition |
|---|---|
| [a] $B_i$ | Base strength of $i$th analyzer |
| $A_i$ | Additional strength of $i$th analyzer (=Lovejoy's $\Delta_i$, $0 < A_i < 1$) |
| $C_i$ | Control strength of $i$th analyzer |
| $pC_i$ | Probability of control by the ith analyzer |
| [a] $V_X$ | Response strength for approaching X ($0 < V < 1$) |
| $pR_X$ | Probability of choosing value X given that the analyzer to which response X is attached is controlling behavior |
| con | Index of analyzer controlling behavior on given trial |
| [a] $x$ | Probability of learning about analyzer controlling behavior on a given trial |
| [a] $y$ | Constant used to determine $pR_X$ from $V_X$ |
| [a] $\theta_1$ | Response operator for rewarded trials |
| [a] $\theta_2$ | Response operator for nonrewarded trials |
| [a] $\theta_3$ | Analyzer operator for rewarded trials |
| [a] $\theta_4$ | Analyzer operator for nonrewarded trials |

[a] Parameter of model.

### 1. *Changes in Analyzer Strength*

Lovejoy overcomes the problem of ensuring that an initially dominant analyzer secures control of performance when all else fails by preserving the base strength $B$ of each analyzer throughout. The value of $B$ can be any positive real number; in practice, Lovejoy has used values varying from 0.0 (dimension not present) to 2.0. The variable $A_i$ represents that proportion of the strength of the $i$th analyzer that can vary as a result of training. At any instant in time, $\sum A = 1.0$. The base values of each analyzer are a parameter of the model and the initial values of $A$ are distributed over the different analyzers in the same proportions as the base value. Hence, on Trial 1,

$$A_i = B_i / \sum B. \tag{13.1}$$

Only one analyzer controls behavior on a given trial (see page 476). With probability $x$, the value of $A$ for the controlling analyzer ($A_{con}$) will be changed as the result of the trial outcome. With probability $1 - x$, no change is made to the $A$ values (the animal has "forgotten" which analyzer it used to control behavior). On rewarded trials on which a change occurs, $A_{con}$ is increased according to the equation

$$A_{con(n+1)} = A_{con(n)} + \theta_3(1 - A_{con(n)}). \tag{13.2}$$

In order to keep $\sum A = 1.0$, the change in the strength of $A_{con}$ is deducted from all other $A$ values in proportion to their strength, using the equation

$$A_{i(n+1)} = A_{i(n)} - \theta_3 \cdot A_{i(n)}, \qquad \text{where} \quad A_i \not\equiv A_{con}. \tag{13.3}$$

On nonrewarded trials on which a change occurs, all $A$ values revert towards their value on Trial 1 according to the equation

$$A_{i(n+1)} = A_{i(n)} - \theta_4 \cdot A_{i(n)} + \theta_4 \cdot A_{i(1)}. \tag{13.4}$$

This constitutes a departure from Zeaman and House's model, since they assumed that when no reward occurred, the loss in strength to the control analyzer was distributed as an increment to all remaining analyzers in proportion to their existing strengths. Lovejoy's formulation involves the paradox that when an analyzer has an $A$ value below its starting value, it will actually increase in strength on nonrewarded trials. However, this assumption has the virtue that, during extinction, all analyzer values will drift back towards their starting values. It is not completely clear why, having invented this ingenious device for restoring the *status quo ante* to analyzer values in extinction, Lovejoy also needed to keep separate $B$ and $A$ values for analyzers.

## 2. *Changes in Response Strength*

On each trial learning takes place about the response attachments of one and only one analyzer. On trials on which $A_{con}$ is selected for learning (occurring with probability $x$), learning also occurs about the response strengths of the controlling analyzer. On the remaining trials (occurring with probability $1 - x$), analyzers are resampled at random with the probability of sampling a given analyzer proportional to its total strength,

$$P(\text{sampling } i\text{th analyzer}) = (A_i + B_i)/\sum (A + B). \tag{13.5}$$

On these trials, learning occurs about the response strengths of whichever analyzer is selected by this procedure.

Although $V$ represents a response strength rather than the probability of making the response, the two response strengths attached to a given analyzer sum to 1.0. The equations governing changes in response strengths are set out in Table 13.II using the case where the brightness analyzer with response attachments W and B has been selected for learning as an illustration. The equation used depends on whether response W or B occurred and on whether or not the reward was received. Since $V_W$ and $V_B$ sum to 1.0, the new value of $V_B$ is, of course, in each case $1 - V_{W(n+1)}$.

TABLE 13.II

**Equations Determining Changes in Response Strength**[a]

| Response made | Reward condition | Equations used to alter $V_W$ | |
|---|---|---|---|
| W | + | $V_{W(n+1)} = V_{W(n)} + (1 - V_{W(n)}) \cdot \theta_1$ | (13.6) |
| W | − | $= V_{W(n)} - V_{W(n)} \cdot \theta_2$ | (13.7) |
| B | + | $= V_{W(n)} - V_{W(n)} \cdot \theta_1$ | (13.8) |
| B | — | $= V_{W(n)} + (1 - V_{W(n)}) \cdot \theta_2$ | (13.9) |

[a] The equations are those used by Lovejoy. In Sutherland's model, the equations are modified by multiplying $\theta$ by the $C$ value of the analyzer to which the response is attached.

3. *Performance Rules*

Only one analyzer controls behavior on each trial. To determine which analyzer is used, control strength values for each analyzer are first calculated. The equation used is the same for all analyzers, but it is illustrated for the case of the brightness analyzer,

$$C_{BR} = (B_{BR} + A_{BR})/(V_W)(V_B). \qquad (13.10)$$

The probability of control by the $i$th analyzer is then given by the formula

$$pC_i = C_i/\sum C. \qquad (13.11)$$

The effect of these two equations is that (other things being equal) there will always be a strong tendency for analyzers with high $B$ values to regain control if there is no consistent reward schedule. As already pointed out, Lovejoy uses two separate means to achieve this effect. First, the base value of the analyzer appears in Equation (13.10) and this cannot be changed by learning. Second, under conditions of random reinforcement, all $A_i$ will tend to revert to their Trial 1 values, which are, in turn, determined by the base values. Since the denominator in Equation (13.10) is maximal when the response strengths are equal and minimal as one or other tends to zero, the learning of a response attachment to an analyzer greatly increases the chance that that analyzer will control behavior. This makes reasonable intuitive sense, although it is not completely clear whether or not this added complication is essential to the model. Lovejoy may have been worried lest, unless the values of $A_i$ were allowed after learning greatly to exceed those of $B_i$, there could be no guarantee that a given analyzer would ever come to control behavior with a very high probability, and, hence, retain control for long sequences of trials.

Once the controlling analyzer is selected, the response to be made is determined by one or other of its response strengths. For example if the brightness analyzer is controlling behavior, the probability of selecting white ($p$W) is determined by applying the appropriate equation of those listed:

$$pW = 0 \qquad \text{if} \quad V_W < y, \tag{13.12}$$

$$pW = (V_W - y)/(1 - 2y) \qquad \text{if} \quad y < V_W < (1 - y), \tag{13.13}$$

$$pW = 1.0 \qquad \text{if} \quad V_W > (1 - y). \tag{13.14}$$

The constant $y$ appearing in these equations must lie between 0 and .5. Figure 13.2 plots $p$W as a function of $V_W$ according to these equations. The effect

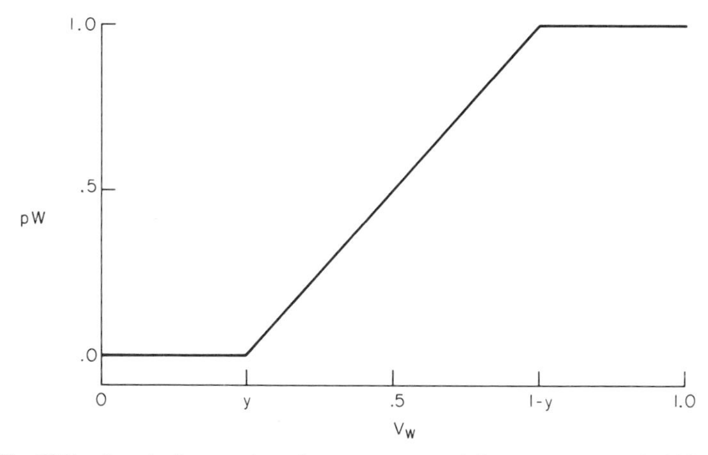


**Fig. 13.2.** Lovejoy's mapping of response strength into response probability. [*After* Lovejoy (1968).]

of the equation is to make $p$W remain at 1 or 0 over a range of changes in $V_W$. Hence, long runs of responses to a single position will occur if the position analyzer is controlling behavior and one or other of its response attachments is high; the same device will also generate a sequence of successive responses to the original positive stimulus at the beginning of reversal or extinction. The higher the value of $y$, the more readily will response probabilities tend to absorb on 1 or 0.

## B. Results Obtained from Lovejoy's Model III

Lovejoy simulated his model with three analyzers; under this condition, the model contains 12 parameter values (2 for each analyzer and the 6 last parameters listed in Table 13.I). Since he set all $V$ values at .5 for all analyzers at the start of each experiment, he explored the effects of changes in 9 parameters.

Lovejoy successfully simulated nine of the phenomena listed at the beginning of this chapter. In order to explore his model further, Sutherland used Lovejoy's equations in a program for running stat-rats. The results of both Lovejoy's and Sutherland's simulations are described below; numbers preceded by a capital E refer to the experimental phenomena listed at the beginning of the chapter. The parameter settings used during most of Sutherland's simulations are shown in Table 13.III. It is assumed that all analyzers have negligible strengths apart from those for brightness, position, and orientation. The parameter settings shown in the table represent the case where animals are trained on brightness with no orientation cue present. When any of these parameters were different, the changes are specified in the text. The settings shown represent an easy brightness discrimination since stat-rats learned to a criterion of 18 correct out of 20 successive trials in an average of 69 trials.

TABLE 13.III

**Standard Parameter Settings Used in Simulation of Lovejoy's Model**

| | | | |
|---|---|---|---|
| $B_{BR} = .5$ | $B_{POS} = 1.5$ | $B_{OR} = 0.0$ | |
| $V_W = .5$ | $V_L = .5$ | $V_H = .5$ | |
| $\theta_1 = .2$ | $\theta_2 = .02$ | $\theta_3 = .03$ | $\theta_4 = .03$ |
| $x = .7$ | $y = .2$ | | |

### 1. *Position Habits*

Stat-rats trained on a nondominant analyzer performed at chance for many trials (E-1); position habits developed in the course of learning, and were almost invariably broken when the positive stimulus was on the nonpreferred side (E-2–E-4). Moreover, Sutherland found that decreasing the base value of the relevant cue both prolonged the duration of position habits and resulted in more errors between breaking a position habit and reaching criterion (E-7). Although no simulation was attempted, it is clear that giving reversal pretraining while behavior is controlled by a position habit will retard subsequent learning since when stat-rats are under the control of one cue some learning about the response attachments to other cues does occur (E-8).

Although Lovejoy's model correctly reproduces most of the phenomena associated with initial learning and position habits, it does not reproduce the fine detail of such habits. Many stat-rats occasionally selected the nonpreferred side in the course of a long position habit; this behavior is rarely seen in real rats, at least in a jumping stand. Moreover, too many errors were made by stat-rats between the trial on which a position habit was first broken and the criterial run. If we deem that a position habit is finally broken when a rat goes to the nonpreferred side and thereafter does not run more than 10 trials in succession to the same side (thus discounting the sporadic breaks in the middle of position habits), then stat-rats make an average of 11 further mistakes before reaching criterion, whereas real rats make only 3 or 4.

Moreover, stat-rats never made errors after breaking a position habit by selecting the negative stimulus when it was on the nonpreferred side, whereas about one-third of errors made by real rats after breaking a position habit are of this sort (E-6). Lovejoy (private communication) reports that he found that 13% of errors made by stat-rats after breaking a position habit were to the nonpreferred side, but he was using parameter settings that resulted in rats taking about twice as many trials to learn, and 13% is still considerably too low.

Sutherland found that 83% of stat-rats trained to criterion and then reversed developed position habits during reversal to the same side as in initial training. Real rats exhibit no correlation between the side to which a position habit is developed during training and reversal. There was, however, no correlation between the side to which position habits were developed when stat-rats were overtrained before being reversed (E-6).

Errors to the nonpreferred side after breaking a position habit and a lack of correlation between position preference in training and reversal can only be achieved in a path-independent model if a massive response operator for position is used on rewarded trials. However, Sutherland found that the use of such a large operator sent stat-rats into position habits after only one or two training trials and this is not characteristic of the behavior of real rats.

### 2. *Learning with Several Cues*

Sutherland simulated learning about an added incidental cue (orientation with horizontal positive). In so doing, the parameter settings in Table 13.III were used for initial training on the brightness cue. When stat-rats met criterion on initial training, $B_{OR}$ was given a value of .5 and new $A$ values were calculated. $A_{OR}$ was set equal to $B_{OR}/(B_{BR} + B_{POS} + B_{OR})$ and $A_{BR}$ and $A_{POS}$ were set equal to their values at the end of initial training multiplied by $A_{OR}$.

This procedure involves the assumptions that an added cue will be as dominant when introduced after training as when introduced at the beginning of the experiment and that the ratios of other $A$ values will remain the same when the new cue is added though the absolute value of each is decreased. New response attachments to the incidental cue were learned; after 100 trials with that cue present, the value of $V_H$ was .76. However, the value of $A_{OR}$ was reduced during retraining to .02 after 100 trials; this occurred because at the beginning of retraining the brightness analyzer already had a very high control value and was, therefore, increased on all trials except some of those on which the animal forgot which cue controlled behavior. When in a third stage of the experiment $A_{BR}$ was removed, it required an average of only 21 trials for stat-rats to reach criterion on the orientation cue. It is clear that if sufficient training is given with an incidental cue, errorless learning could occur, since $V_H$ would tend to 1.0 and $V_V$ would, therefore, tend to zero, and as it does so, the analyzer for the incidental cue (orientation) will gain control as soon as the initial relevant analyzer (brightness) is removed [cf. Equation (13.10)].

It was found that pretraining on brightness retarded the learning of the additional cue: when both cues were present and relevant from the outset of training, the value of $V_H$ was .95 after 100 trials, as against .76 when stat-rats were pretrained with brightness relevant, and then given a further 100 trials with both cues relevant. The model, therefore, accounts for blocking (E-9).

Sutherland also found (as did Lovejoy) that the more difficult the original cue was made (setting $B_{BR} = .1$ instead of .5), the more rapidly learning about an added incidental cue occurred when it was made relevant at a second stage of the experiment (E-15). Stat-rats were trained to criterion on brightness with orientation absent, and then given 100 trials of retraining with both cues relevant. $B_{OR}$ was again set equal to .5; with $B_{BR} = .1$, the value of $V_H$ at the end of two-cue training was .85, whereas with $B_{BR} = .5$, the value of $V_H$, as stated above was .76.

Two additional phenomena related to incidental learning merit brief discussion. It is clear that, on Lovejoy's model, the more pretraining is given about a blocking cue, the less will be learned about an added incidental cue over a given number of trials (E-12). This happens because the additional training will both increase the $A$ value of the blocking cue and will also drive the negative response attachment ($V_B$) closer to zero. Hence, the blocking cue will both have a higher control value and will also be selected for learning more frequently on trials on which the controlling analyzer is forgotten. Learning about the added cue will, therefore, occur on a smaller percentage of trials.

It is less clear what the model predicts when animals are trained on one cue, and the response requirements are changed when a second cue is added

(E-11). If the incidental cue is not present during pretraining, then, on the assumptions we have made above, its $A$ value at the start of the second stage of training will be unaffected by the pretraining on the original cue. However, if despite the change in response requirements (usually accompanied by a change in the stimulus configuration, e.g., in changing from simultaneous to successive discrimination learning), the $V$ strength of the analyzer for the Stage-1 relevant cue is carried forward from the first to the second stage of training, some blocking of learning about the incidental cue will occur because of the high control value of the first analyzer. Such blocking is known to occur (E-11). Moreover, if the incidental cue is present but irrelevant during pretraining, the blocking will be greater than when it is not present since its $A$ value will be less at the start of the second stage of training. Lovejoy's model therefore copes well with the facts of incidental learning.

It is clear without simulation that the model will also produce overshadowing effects since adding a strong cue (with a high $B$ value) will reduce both the probability of a less strong cue being selected for control and also the probability of that cue being selected for response-attachment learning on trials on which the controlling cue is forgotten.

The model fares less well on additivity of cues (E-16) since stat-rats trained with two cues relevant ($B_{BR} = B_{OR} = .5$) required almost as many trials to reach criterion (60) as animals trained with only one cue relevant (69); there were 20 stat-rats in each group and the difference was not significant. However, it is extremely unlikely that real rats enter training with the same $B$ values as one another, and if in individual animals $B_{BR}$ was sometimes higher and sometimes lower than $B_{OR}$ this could result in a larger reduction in trials to criterion with two cues relevant since individual animals will now learn about whichever cue is more dominant (cf. the arguments on page 143). Lovejoy himself showed that when stat-rats were run with two cues relevant, there was a negative correlation between the amount learned by individual animals about each cue (E-19).

### 3. *Intra- and Extradimensional Shifts*

Lovejoy's model will also handle the phenomena associated with intra- and extradimensional shift learning (E-22). Positive IDS transfer from simultaneous to successive discriminations will be mediated by the increase in the $A$ value of the relevant cue produced by initial learning. For the same reason IDS should, in general, be easier than EDS (E-23 and E-26).

### 4. *Overtraining and Serial Training*

Lovejoy showed that his model does produce an ORE. He trained stat-rats to criterion and preserved all the values of all variables for each rat on meeting criterion. Using these values, he then computed trials to reversal for each rat both when reversal was given immediately after criterion was reached and when 100 trials of overtraining intervened before reversal. This procedure enabled him to test for the significance of the ORE by using a matched-pairs test: out of 40 rats, 29 reversed faster after overtraining, and 11 reversed faster when not overtrained. When experimenting on real rats, one obviously cannot compare trials taken to reverse, with and without overtraining, within a single subject. Sutherland simulated Lovejoy's model, but used different stat-rats in the criterion trained and overtrained groups. The results are shown in Table 13.IV. For comparison, some data from real rats (Mackintosh, 1962) are given in the first two rows of the table.

It is clear that the model does produce (as Lovejoy showed) an ORE (E-34). Moreover, the effect tends to increase as the discrimination is made more difficult by decreasing the $B$ value of the relevant cue (E-37; cf. experiments 2, 4, and 5 in Table 13.IV). Hence, the effect should also increase when irrelevant cues are added (E-36). Unfortunately, however, because of the size of the standard deviations the effect was not actually significant for any parameter setting used in the experiment shown in the table, even though 20 stat-rats were run in each condition (as against only 8 or 16 in most experiments with real rats).

The standard deviations of stat-rats are, in fact, considerably larger than those found with real rats and this despite the fact that all stat-rats (within a group) are run with the same starting parameters, whereas in real rats there is presumably considerable variance in the starting parameters. The large standard deviations in the model occur because of the use Lovejoy makes of probabilities and the fact that certain processes tend, once started, to "run away." For example, the rate at which a position habit develops is probabilistically determined (by an interaction between the responses the animal makes and the Gellermann sequence): once $V_L$ (or $V_R$) becomes high, it will tend to increase; as it increases the probability of learning occurring about the relevant cue decreases [cf. Equation (13.10)] and the higher the $V$ value on position, the more learning has to occur about the relevant cue before the relevant analyzer can gain control of behavior. Moreover, the more frequently (by chance) the relevant analyzer is selected for learning on trials on which the controlling analyzer is forgotten, the more probable it is that that analyzer will be selected on future trials. The very high standard deviations produced on Lovejoy's model must be considered a serious defect.

The model correctly reproduces the ratio between number of training trials and number of trials to reversal, both for overtrained and nonovertrained animals (E-43); compare the results of Experiment 0 with those of Experiment 4 in Table 13.IV. There is, however, a further point at which Lovejoy's model (at least with the parameter settings investigated) fails to

TABLE 13.IV

**Simulation of ORE**[a]

| | Experiment | Group | Trials to criterion in training | Trials to criterion in reversal | Successive errors in reversal | $B_{BR}$ | $B_{POS}$ | $B_{OR}$ | $\theta_1$ | $\theta_2$ | $\theta_2$ | $\theta_4$ |
|---|---|---|---|---|---|---|---|---|---|---|---|---|
| 0 | Real rats, | CRIT | 77 (22) | 125 (21) | 9 (5) | | | | | | | |
| | Mackintosh (1962) | OT | 76 (13) | 90 (10) | 19 (19) | | | | | | | |
| 1 | Lovejoy | CRIT | 121 | 229 (78) | | .25 | 1.5 | 0 | .2 | .02 | .03 | .03 |
| | | OT | 121 | 158 (72) | | | | | | | | |
| 2 | Lovejoy–NSS | CRIT | 131 (60) | 237 (70) | 6 (3) | .25 | 1.5 | 0 | .2 | .02 | .03 | .03 |
| | | OT | 123 (40) | 157 (60) | 7 (3) | | | | | | | |
| 3 | Lovejoy–NSS | CRIT | 92 (50) | 156 (60) | 6 (2) | .25 | 1.5 | 0 | .3 | .03 | .045 | .045 |
| | | OT | 105 (40) | 105 (60) | 7 (1) | | | | | | | |
| 4 | Lovejoy–NSS | CRIT | 72 (29) | 128 (53) | 7 (4) | .5 | 1.5 | 0 | .2 | .02 | .03 | .03 |
| | | OT | 74 (24) | 107 (35) | 8 (3) | | | | | | | |
| 5 | Lovejoy–NSS | CRIT | 48 (16) | 93 (27) | 8 (3) | .75 | 1.5 | 0 | .2 | .02 | .03 | .03 |
| | | OT | 48 (13) | 77 (25) | 7 (4) | | | | | | | |
| 6 | NSS | CRIT | 81 (3) | 131 (4) | 7 (5) | | | | | | | |
| | | OT | 80 (2) | 106 (3) | 17 (4) | | | | | | | |

[a] Standard deviations are given in parentheses.

reproduce the fine details of reversal learning: in real rats overtraining usually significantly increases the number of successive errors made at the start of reversal (E-38, cf. Experiment 0), whereas no such increase is detectable in the stat-rat data shown in Table 13.IV. The reasons for this appear to be as follows. At the start of reversal in Experiment 4, the mean strength of the positive response attachment to the relevant analyzer ($V_W$) was .993 for criterial trained animals and 1.0 for overtrained animals. This difference is too small to produce any appreciable difference in the probability of the relevant analyzer being selected for control, after the first trial or two of reversal training. However, at the start of reversal, the $A$ strength of the relevant analyzer averaged .52 for criterion trained and .94 for overtrained stat-rats. This difference is sufficient to produce a considerable difference in the probability of learning occurring about the response attachments to the relevant analyzer, particularly on trials on which rats forget which analyzer was used for control. As a result, the overtrained animals relearn the correct response at the beginning of reversal faster than the nonovertrained. On the trial on which the first correct response occurred in reversal the average value of $V_W$ (strength of choosing the stimulus that was positive during original training) was .92 for criterial trained stat-rats and .78 for overtrained animals. The corresponding $A$ strengths of the relevant analyzer were .46 and .78. It should be noted that in both criterial trained and overtrained stat-rats the first correct response tends to occur because the position analyzer takes control; it occurs before the $V$ values of the relevant analyzer have changed sufficiently to allow the new positive stimulus to be selected under the control of that analyzer. The reversal trial number on which the position analyzer first starts to gain control is about the same for overtrained and nonovertrained animals because the reduction in probability of control for overtrained animals brought about by more rapid changes in $V$ strengths just about compensates for the higher $A$ value on the relevant cue.

Because the $A$ value of the relevant analyzer remains higher during reversal in overtrained stat-rats than in nonovertrained animals, overtrained stat-rats showed fewer position habits in reversal than nonovertrained and also gave a shorter plateau at the 50% correct response level (E-39).

Connected with the ORE is the "dip effect" (E-40). If, after training, rats are given extinction to a criterion of 5 out of 10 responses to their original negative stimulus, and are then reversed, it is found that nonovertrained animals at the start of reversal select their originally positive shape on well over 50% of trials, whereas overtrained animals continue to perform at chance (or better); i.e., they do not revert to selecting the originally positive (now negative) stimulus. We have interpreted this to mean that at the point when animals meet the criterion of choice extinction, nonovertrained subjects have switched-out the relevant analyzer, but have not unlearned the original response attachments, whereas overtrained animals have not necessarily switched-out the original analyzer but have unlearned its response attachments. Sutherland performed this experiment with stat-rats using the standard parameter settings from Table 13.III and found that, at the end of extinction, the value of $V_W$ (response strength of selecting the stimulus that was positive in initial training) was .70 for overtrained stat-rats and .83 for nonovertrained. Although the difference was in the right direction, it appears to be too small to account for the whole effect. Stat-rats did, however, take about the right number of trials to extinction, 25 if not overtrained, 33 after overtraining; the comparable figures for real rats are 27 and 36 (Mackintosh, 1965b), although the experiment with real rats involved an easier discrimination, which was learned in 37 trials, whereas the stat-rats were run with standard parameter settings and learned in 70 trials.

It seems unlikely that Lovejoy's model could account for the progressive improvement in serial reversal learning that occurs when rats are trained to criterion at each reversal (E-44). In order to obtain improvement over serial reversals, criterion must be reached with the $A$ value of the relevant analyzer a little higher on each successive reversal. In fact if we examine the average $A$ values after training and first reversals in Sutherland's simulation (Experiment 4 in Table 13.IV), they are almost identical (.48 after initial learning, .49 after the first reversal). Moreover, no path-independent model such as those propounded by ourselves and Lovejoy can account for the fact that even-numbered reversals often tend initially to be easier than odd-numbered reversals (letting initial training be Reversal 0 and the first reversal, Reversal 1, etc.). Successive reversal training will also only retard the subsequent learning of a nonreversal shift if it results in progressively higher $A$ values at the end of each reversal (E-45).

### 5. *Different Schedules*

Lovejoy found that stat-rats trained with two relevant cues learned more about the weaker cue when trained under a probabilistic schedule than when trained under consistent reinforcement (E-45).

### 6. *Conclusions*

Before leaving our discussion of Lovejoy's model, it is worth summarizing our findings. It must be remembered that simulation has only been attempted for a very small number of parameter settings. The settings chosen were those that came nearest to reproducing the results of real rats in overtraining and reversal (see Experiment 4 in Table 13.IV), but there may be many other settings that would reproduce this phenomenon equally well, and reproduce some of the others rather better, particularly if the constants $x$ and $y$ are varied. Considering the limited variation in parameter settings that was explored initially, the success of the model on other phenomena is remarkable. However, it did have certain failures and, for improving the model, these may be more instructive than its successes.

The main points at which Lovejoy's model appears to fail concern position habits, the standard deviations of trials to criterion, consecutive errors at the start of reversal, and serial reversal learning. There are four points at which the model fails to deal adequately with position habits.

(1) Too many errors are made between a position habit being abandoned and criterion being met.
(2) Too few of these errors are to the nonpreferred side.
(3) In nonovertrained stat-rats, there is a correlation between the side to which position habits develop in reversal and the side to which they develop in training.
(4) There are too many sporadic jumps to the nonpreferred side in the middle of a position habit.

It would be possible to meet the failures in Lovejoy's model in dealing with position habits if selective attention were a more active process than we have represented it to be. It will be remembered that rats develop differential latencies to the positive and negative stimuli at a time when choice responses are still controlled by position. The rat in some sense "knows" what is good and what is bad, but its choice behavior is not controlled by this knowledge. This suggests the following anthropomorphic account: when at the beginning of training, the stimuli are first introduced and the rat experiences nonreward, it is confused and remembers very little of what it does over the first few trials; hence the lack of correlation between early position preferences and later position habits. Sooner or later it actively decides to store information about the most dominant cue (position) and this results in the formation of a position habit; knowledge about the reward contingencies associated with the values of the relevant cue is gradually built up by a passive process, and when it is sufficiently strong, the rat actively decides to select the positive on the nonpreferred side. The rat may now either abandon its previous hypothesis that the preferred side is better than the nonpreferred or even substitute the opposite hypothesis which would result in errors on the nonpreferred side. This will decrease the number of errors made after a position habit breaks and result in a lack of correlation between side of position habit in training and reversal.

The above account involves two kinds of learning that cut across the distinction between response learning and analyzer learning. It involves active hypothesis formation resulting in sudden changes in the selection of responses and analyzers, and a slower, more passive, process by which even when behavior is controlled by one hypothesis something is learned about other possible hypotheses. The above paragraph is based on anthropomorphic ideas that would be difficult, although not impossible to incorporate into a formal model. This may seem a retrograde step, but the problem is that no model based on simpler ideas is able to account for the complexities of rat behavior even in such a simplified situation as the Lashley jumping stand.

A second main problem with Lovejoy's model is the great variance it produces in the scores of individual stat-rats; the problem is even more serious than it might appear at first sight, since, whereas stat-rats within a group are run with the same initial parameter settings, this is most unlikely to be true for real rats. If the passive learning about features of the situation about which the rat is not entertaining an active hypothesis is a less probabilistic process than Lovejoy makes out, then it would proceed at a more constant pace over rats, and the variance in scores would accordingly be reduced.

We shall not discuss the model's other failures here. More will be said below about why overtraining increases the number of successive incorrect responses at the start of reversal learning. For the reasons given above it is unlikely that any path-independent model can adequately account for the data on serial reversal learning.

## IV. A Further Model

Sutherland has attempted to simulate a further model of two-process learning. It differs from Lovejoy's model in two main respects. First, analyzer strength is altered not simply by reward and nonreward: analyzers increase in strength when an "expected" outcome occurs and decrease when an unexpected outcome occurs (for a justification for this see Chapters 2 and 3). Second, the model allows both for control of behavior by more than one analyzer at a time, and for learning about several analyzers within a trial.

### A. Specification of Model

#### 1. *Changes in Analyzer Strength*

The symbols used in specifying the model are described in Table 13.V. It is assumed, as in Lovejoy's model, that all analyzers have a starting value $B$, and a value $A$ that can be changed. $B$ values are read in and always sum to 1.0; initially $A$ values are the same as $B$ values. On the conclusion of each trial $A$ values are altered by applying the appropriate equation or equations of those shown in Table 13.VI. A flow graph of the procedure for determining

TABLE 13.V

**Symbols used in Sutherland's Model**

| Symbol | Definition |
|---|---|
| [a] $B_i$ | Base strength of $i$th analyzer |
| $A_i$ | Additional strength of $i$th analyzer |
| $C_i$ | $(A_i + B_i)/2$, used to determine rate of response learning |
| [a] $V_X$ | Response strength of selecting value X |
| $V_i(\mathrm{R})$ | Response strength of selecting value chosen on given trial on $i$th dimension |
| $V_i(\mathrm{L})$ | Response strength of going left attached to $i$th analyzer taking into account which value of $i$th dimension is on the left |
| $CV(\mathrm{L})$ | Total response strength of going left combining all controlling analyzers |
| $pV(\mathrm{L})$ | Probability of going left |
| max | Index of analyzer having highest $V(\mathrm{R})$ value |
| min | Index of analyzer having lowest $V(\mathrm{R})$ value |
| [a] $y$ | Constant used to determine $pV(\mathrm{L})$ from $CV(\mathrm{L})$ |
| [a] $z$ | Constant determining range of analyzers used to control performance |
| [a] $w$ | Constant used in selecting analyzer making a strong prediction |
| [ab] $t$ | Constant used to determine which equation to use in decrementing strong analyzer and response strengths |
| [a] $\theta_1$ | Reward response operator |
| [a] $\theta_2$ | Nonreward response operator |
| [a] $\theta_3$ | Reward analyzer operator [used when $V_{\max}(\mathrm{R}) > w$] |
| [a] $\theta_4$ | Nonreward analyzer operator [used when $V_{\min}(\mathrm{R}) < (1 - w)$] |
| [a] $\theta_5$ | Analyzer operator used when no analyzer makes a good prediction |
| [ab] $\theta_6$ | Operator applied when a high response strength is decremented |
| [ab] $\theta_7$ | Operator applied when a high analyzer strength is decremented |

[a] Parameter of model.
[b] Additional parameter used in revised version of model only.

TABLE 13.VI

**Rules Determining Changes in Analyzer Strengths on Sutherland's Model**

**A.** If reward and $V_{\max}(\mathrm{R}) > w$

$$A_{\max} = A_{\max} + (1 - A_{\max}) \cdot \theta_3 \tag{13.15}$$

$$A_i = A_i + A \cdot \theta_3, \quad \text{where } A_i \neq A_{\max} \tag{13.16}$$

**B.** If no reward and $V_{\min}(\mathrm{R}) < (1 - w)$

$$A_{\min} = A_{\min} + (1 - A_{\min}) \cdot \theta_4 \tag{13.17}$$

$$A_i = A_i - A_i \cdot \theta_4, \quad \text{where } A_i \neq A_{\max} \tag{13.18}$$

**C.** Else

$$A_i = A_i - A_i \cdot \theta_5 + B_i \cdot \theta_5 \tag{13.19}$$

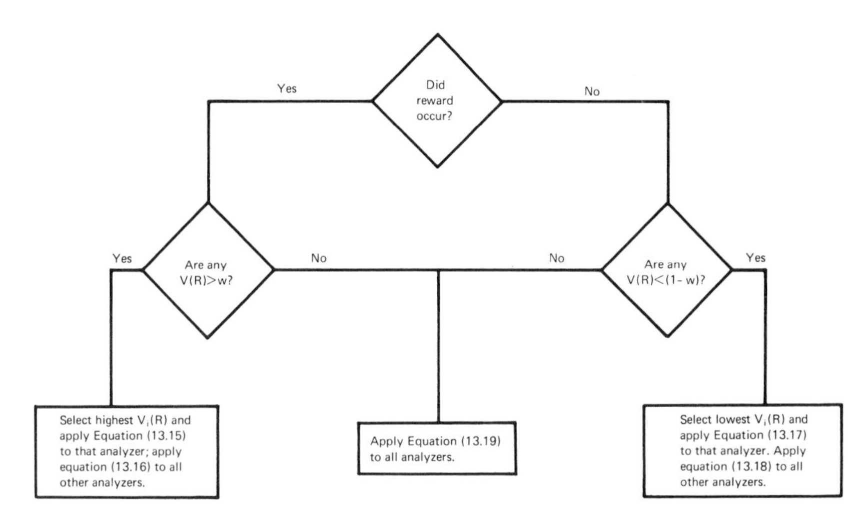


**Fig. 13.3.** Flow diagram of rules determining changes in analyzer strength in Sutherland's model.

changes in analyzer strength is presented in Figure 13.3. The $V$ strength of the response actually made on a given trial $V_i(\mathrm{R})$ is first examined for all analyzers; e.g., if the animal selected black, horizontal on the left, the $V$ strengths used in the following procedure would be $V_{\mathrm{B}}$, $V_{\mathrm{H}}$, and $V_{\mathrm{L}}$. If reward occurs on a trial and there is an analyzer that makes a strong correct prediction about the trial outcome [has a high $V(\mathrm{R})$ value], then the analyzer making the strongest prediction is increased, and others are decremented. If nonreward occurs on a trial then the analyzer making the best strong prediction [i.e., having the lowest $V(\mathrm{R})$ value] is increased and others decremented. If no analyzer makes a strong correct prediction, then all analyzers change in the direction of their base values. It will be noticed that analyzer changes no longer occur as a direct result of reward and punishment, but that analyzers making good predictions about trial outcomes are increased at the expense of others (as suggested in Chapter 2).

### 2. *Changes in Response Strength*

All response attachments are changed on each trial. The equations determining changes are identical to those used in Lovejoy's model (see Table 13.II) except that the amount learned is weighted by the factor $C_i$, where $C_i = (A_i + B_i)/2$. Formally, Equations 13.6–13.9 are altered by multiplying each $\theta$ by $C_i$. This device ensures that the rate at which response attachment learning proceeds is determined both by the base strengths $B$ of an analyzer and by its additional strength $A$. One effect of altering response attachments in this way is that there is less variance between individual animals than in Lovejoy's model.

3. *Performance Rules*

The analyzer with the highest $A$ value and all other analyzers with $A$ values within a constant $z$ of that analyzer are selected to control performance. Let the $V$ strength for going to the stimulus on the left be $V_i(\mathrm{L})$ for each analyzer: $V_{\mathrm{BR}}(\mathrm{L})$ will be equal to $V_{\mathrm{B}}$ where black is on the left, and to $V_{\mathrm{W}}$ where white is on the left. We now combine the $V_i(\mathrm{L})$ values of all analyzers selected to give a combined strength of going left $CV(\mathrm{L})$ according to the following equation:

$$CV(\mathrm{L}) = \frac{\prod V_i(\mathrm{L})}{\prod V_i(\mathrm{L}) + \prod [1 - V_i(\mathrm{L})]}. \qquad (13.20)$$

To find the probability of going left we now substitute $CV(\mathrm{L})$ for $V_{\mathrm{W}}$ in Lovejoy's equations (13.12)–(13.14). This simply scales the probability of going left so that it is 0 when $CV(\mathrm{L}) < y$, 1 when $CV(\mathrm{L}) > (1 - y)$ and increases linearly when $y > CV(\mathrm{L}) > (1 - y)$. It will be noticed that the effect of using the constant $z$ in the above procedure is to give multiple-cue control of behavior if several analyzers are high and close together; performance remains under single-cue control if one analyzer is higher than all the others by the quantity $z$ or more.

4. *Results Obtained*

It was found possible to obtain the ORE with this model. However, the model failed at several points. First, when stat-rats were reversed without overtraining, they reversed in about the same number of trials as it had taken them to learn the original discrimination. Second, as in Lovejoy's model, overtraining did not lead to more initial errors at the start of reversal. Third, although position habits were formed and broken in a way similar to the phenomena exhibited by real rats, the sides to which position habits developed in training and reversal were highly correlated; this defect also appeared in Lovejoy's model.

5. *Further Changes to Model*

It was decided to concentrate on amending the model in such a way that it would correctly reproduce the details of Mackintosh's (1962) experiment (see Experiment 0, Table 13.IV). Now, there is no empirical reason to suppose that internal variables such as analyzer strength and response attachment strength are altered simply by applying linear operators; this type of model was initially proposed solely because of its mathematical convenience. There is further very little reason to think that such variables change in exactly the same way regardless of their absolute values. It seems intuitively plausible to suppose that if either $A$ strength or $V$ strength is very high, they would be weakened more slowly than when they are at intermediate levels; if an animal has a very strong hypothesis, before weakening the hypothesis, it might require more evidence that it is no longer valid than when it holds a weak hypothesis. The following changes were therefore made in the rules governing changes in analyzer and response strength.

On trials on which no analyzer strongly predicts a correct trial outcome [i.e., on trials on which Equation (13.19) would be applied], if the highest analyzer has an $A$ strength of above $t$, instead of applying Equation (13.19) the following equation is applied to that analyzer (calling it the $j$th analyzer),

$$A_j = A_j - \theta_6 \cdot B_j. \tag{13.21}$$

The resulting decrement to the value of the strongest analyzer is distributed as an increment to other analyzers using the equation,

$$A_i = A_i + \theta_6 \cdot B_i/(1 - B_j). \tag{13.22}$$

These equations have the effect of slowing down the rate at which an analyzer with a high $A$ value is decremented when its response attachments make the wrong predictions until it drops below the value of $t$. When no analyzer has an $A$ value above $t$, the equations in Table 13.VI continue to be applied. A similar change was made to the rate at which very high $V$ strengths are changed. When a trial outcome results in a decrement to a given $V$ and that $V$ is greater than a constant $t$, instead of applying the equations in Table 13.II, the following equation was applied to the high $V$ value to be decremented:

$$V_i = V_i - C_i \cdot \theta_7. \tag{13.23}$$

Provided $\theta_6$ and $\theta_7$ have low values, the effects of these equations is to make high analyzer or response strengths "sticky": they can only be reduced slowly until their values fall below $t$.

### 6. *Results with Amended Model*

Using the parameter settings shown in Table 13.VII, it was found possible to replicate by simulation all Mackintosh's (1962) data on the ORE (see Experiment 6, Table 13.IV).

The model yielded the following effects that Lovejoy's model failed to produce:

TABLE 13.VII

**Standard Parameter Settings on Sutherland's Model**

| | | | | | | |
|---|---|---|---|---|---|---|
| $B_{BR} = .2$ | | $B_{POS} = .7$ | | $B_{BR} = .1$ | | |
| $V_W = .5$ | | $V_L = .5$ | | $V_H = .5$ | | |
| $y = .2$ | $z = .2$ | $w = .7$ | $t = .8$ | | | |
| $\theta_1 = .3$ | $\theta_2 = .01$ | $\theta_3 = .02$ | $\theta_4 = .1$ | $\theta_5 = .1$ | $\theta_6 = .02$ | $\theta_7 = .1$ |

(1) The number of successive errors at the start of reversal was greater for overtrained than for nonovertrained animals, and the absolute number of errors made in each case corresponded reasonably well to that made by real rats.

(2) The average number of errors made between breaking a position habit and reaching criterion was 4.2 which is about the same as the number made by real rats trained on a discrimination of comparable difficulty.

(3) The standard deviations of scores were lower than those of real rats; because real rats almost certainly start with different parameter settings, we regard this as a virtue.

However, in common with Lovejoy's model, Sutherland's model exhibited the following defects:

(1) When rats were not overtrained, the sides to which position habits developed in initial training and reversal were correlated.

(2) No errors were made after breaking a position habit when the positive stimulus was on the preferred side.

It is particularly regrettable that in none of the models discussed does the correlation between position habits in training and reversal disappear, since it was argued in Chapter 8 that the fact that real rats show no such correlation was a disproof of Spence's account of the ORE. It should be noted that on our own model the survival of position preferences into reversal does not itself cause the ORE; the effect is caused entirely by the greater $A$ strength of the relevant analyzer after overtraining. In Spence's account, on the other hand, the ORE can only occur if position preferences are stronger after criterion training than after overtraining, and, therefore, it is the very existence of position preferences after criterion training that causes the ORE. While the correlation between position habits in training and reversal in nonovertrained animals is, therefore, an incidental (and undesirable) product of our model, it is an intrinsic feature of Spence's model in so far as that model can be used to explain the ORE.

Because of the failures in our own model, it was decided that it was not worth simulating it further. The present simulation demonstrates that a model in which analyzers are altered according to whether their response attachments make good predictions can produce at least as good results over a narrow range of phenomena as a model in which analyzer changes depend upon the straightforward effects of reward and nonreward. However, these results have been achieved at the cost of considerably increased complexity and the addition of five extra parameters.

### 7. *The Problem of Complex Models*

There is, in fact, a stronger reason why it was decided not to continue with any further simulation. It will have been only too clear that many of the decisions taken by Lovejoy about how to set up his model and also those taken by Sutherland are extremely arbitrary. Such arbitrary decisions may have dramatic effects on the workings of a model. For example, when Sutherland first attempted to simulate Lovejoy's model, a programming mistake was made which resulted in Equation (13.10) being used to determine which analyzer was selected for response learning on trials on which the controlling analyzer was forgotten instead of Equation (13.5) as used by Lovejoy. Making this accidental and (apparently) small change in Lovejoy's model made it quite impossible to reproduce the ORE. The problem facing anyone who wishes to simulate a model of two-process learning is that there is an infinity of possible models and many of these are, on the surface, just as plausible as one another and just as elegant (or inelegant depending on the observer's standpoint). To demonstrate this we list below some of the decision points in constructing such models.

(1) Can more than one analyzer be increased on a given trial?

(2) Do response strengths play a role in determining which analyzers are increased or are they increased solely as a result of trial outcome?

(3) Do all *A* values change at the same rate or does the *B* value of an analyzer affect its rate of increase?

(4) What is the function determining how analyzers are changed? We have argued above that there is really no good empirical reason for assuming that they are changed by the application of linear operators.

(5) Are analyzers increased and decreased by applying the same function over the whole range of possible strengths?

(6) Does only one set of response attachments change on each trial or are there changes in the response attachments to many different analyzers within a trial?

(7) Is the rate of change (or probability of change) of response attachments determined only by *A* values or is it determined jointly by both *A* and *B* values and if so how are *A* and *B* values combined?

(8) What is the function determining the rate of change of $V$ values?

(9) Is it the same function no matter what the strength of the $V$ value?

(10) Is it adequate to work with a single value representing response strength or is it necessary to compute strengths separately for the tendency to choose each value and if so should inhibitory strengths be represented as well as excitatory?

(11) Is behavior controlled by one analyzer within a given trial or by many?

(12) Is control determined solely by $A$ values or does $B$ enter into it and if so how?

(13) Do response strengths enter into the determination of which analyzer controls behavior on a given trial?

(14) For each of the above alternatives there are many different forms that the rule determining control could take.

If each of these decisions were binary, there would be $2^{14}$ or about 16,000 possible models; the real situation is worse, since most of the decisions have many possible outcomes. Finally, with the number of parameters each model must have, it would be impossible to explore adequately the workings of even one model within a reasonable space of time.

It should, of course, ultimately be possible to collect evidence that allows us to decide at each decision point which way to go. Indeed, for the sake of making a point, we have been guilty of some exaggeration in the above account, since evidence already exists on several of the decision points. Thus, in Chapter 5, we produced evidence to show that analyzers cannot simply be incremented and decremented according to trial outcome (2). Zeaman and House have demonstrated that in retardates more than one response attachment changes within a trial and by using one-trial learning situations (e.g., the CER paradigm) it is possible to demonstrate the same effect in animals (6), see page 500. We have also argued that the rate of change of response attachments must to some extent be determined by the base value of the analyzer (7); unless this assumption is made it becomes very difficult to explain why position preferences are eliminated from one stage of training to another. Chapter 5 presents evidence favoring mlutiple-cue control (11), and Chapter 11 presents evidence suggesting that response strengths enter into the determination of which analyzer controls behavior on a given trial (13).

There is, therefore, evidence already available that rules out some possible answers to some of these questions, and it is reasonable to hope that future research will provide answers to others. Indeed, the present argument adds point to the cliché that theories are valuable because they guide research. The value of Lovejoy's third model and of our own attempts at model building is not so much that they make new, testable, predictions, as that it is only in the process of constructing them that areas of ignorance have been revealed where the theorist must make decisions in the absence of empirical data.

## V. Other Theoretical Analyses

Just as in the present chapter we have attempted to consider the advantages and disadvantages of different formulations of attentional theories, so throughout this book we have tried to take into account other analyses of the data that make no mention of the concept of selective attention. We do not, of course, delude ourselves that we have done so to the satisfaction of those who hold the theories in question, for we have rarely conceded the advantage to them. Some have been particular explanations restricted to a single experimental situation such as the attempt to explain transfer along a continuum in terms of generalization gradients, or the ORE in terms of a decline in the aversiveness of S− during overtraining; others have been general theories of discrimination learning that might encompass much the same range of data as we have attempted to explain; Spence's (1936) theory of discrimination learning, incorporating a system of orienting responses, is clearly applicable to many of these data. We have presented a number of arguments against this latter position; against the more restricted explanations, we have both advanced specific objections, and can also point to the much more extensive range of data that our own approach can (more or less satisfactorily) handle.

One particular set of data—that on blocking and overshadowing, reviewed in Chapters 4 and 5—seemed both particularly central to an attentional analysis, and, until very recently, unusually resistant to alternative explanations. Only last year, however, Rescorla and Wagner (1970) proposed a simple model that explains these phenomena elegantly and economically, and, although we have had little time to digest it fully, we feel bound to conclude this volume with some discussion of it.[1]

### A. The Rescorla–Wagner Model

Kamin (1968, 1969) found that if sufficient CER trials were given to condition suppression to a noise CS completely, then the further reinforced presentation of a noise plus light compound produced no conditioning to the light. Kamin suggested that this might be because an already fully expected US would not have any reinforcing function. In order for the US to reinforce suppression, the animal must be surprised when it occurs.

[1] We are indebted to the authors for allowing us to see a prepublication copy of their paper.

Rescorla (1969) and Wagner (1969b) both took up this implication, and Wagner pointed out that it could be simply formalized by saying that increments in associative strength between, say, the light and the shock were an inverse function of the current associative strength of the entire compound of which the light formed a part. It is this simple idea that has been elaborated by Rescorla and Wagner into a more substantial model.

The symbols used in describing the model are shown in Table 13.VIII; they are mercifully few. The model consists of one equation. Assume that two stimuli (A and X) are present on the *n*th trial and $US_1$ occurs, then the response strength $V$ to stimulus X changes according to the following equation:

$$V_{X(n+1)} = V_{X(n)} + \alpha_X \cdot \beta_1(\lambda_1 - \overline{V}_{AX}). \qquad (13.24)$$

TABLE 13.VIII
**Symbols Used in Rescorla–Wagner Model**

| Symbol | Definition |
|---|---|
| $V_A$ | Response strength to stimulus A |
| $\overline{V}$ | Total response strength to all stimuli present in situation |
| $\overline{V}_{AX}$ | Total response strength to A and X where it is assumed that these are the only stimuli in situation with strengths other than zero |
| $\alpha_A$ | Learning rate parameter for stimulus A |
| $\beta_1$ | Learning rate parameter for $US_1$ |
| $\lambda_1$ | Asymptote of $\overline{V}$ that can be supported by $US_1$ |

Substituting A for X and vice versa, we obtain the equation for the change in response strength to stimulus A. The symbol $\overline{V}$ represents the sum of the response strengths to all stimuli present. The parameter $\alpha$ corresponds to the conspicuousness of a given stimulus; the more salient the stimulus the higher the value of $\alpha$ and, hence, the more rapidly the response becomes attached to that stimulus. The parameter $\beta$ varies with the strength of the US; the stronger the US the more rapidly will learning proceed. The value of $\lambda$ also varies directly with the strength of the US, so that a higher asymptote of response strength will be reached with a strong US than with a weak one. When the US is omitted (for example in extinction) the value of $\lambda$ is zero, and the value of $\beta$ is assumed to be less than when the US is present.

## B. Evidence for Model

The model accounts very elegantly for the following phenomena:

(1) Blocking (E-9). If animals are trained with stimulus A present on its own, and are then trained on the compound AX, little learning about X can

occur since the value of $\overline{V}_{AX}$ will already be positive when X is introduced. In the case where animals have been trained to asymptote before X is introduced, $\overline{V}_{AX}$ will already equal $\lambda$ and hence no learning at all will occur about X.

(2) If (E-10) the strength of the US is increased at the time when X is introduced then even when animals are already trained to asymptote on A, some learning about X will occur since now the value of $\lambda$ will be greater than the value of $\overline{V}_{AX}$.

(3) Overshadowing (E-17). The explanation of overshadowing is very similar to that of blocking. If one animal is trained with the compound AX, where A is stronger than X, and another animal is trained with X alone, then after the first trial the value of $V_{AX}$ will be greater than the value of $V_X$, and increments to $V_X$ will be smaller in the former case then in the latter.

(4) The theory accounts for a number of experiments undertaken by Wagner showing that, if a stimulus X is imperfectly correlated with reinforcement conditions, that stimulus acquires lower asymptotic response strength if stimuli that are more correlated with reinforcement are included in the situation. We have already cited several of these studies [e.g., Wagner, 1969b, (see pages 118–121); Wagner *et al.*, 1968 (see page 152)]. In the latter experiment, for example, it was found that if a light (L) predicts the occurrence of the US on 50% of presentations, the light acquires much more control of behavior when it is paired with two tones ($A_1$ and $A_2$) that do not differentially predict the occurrence of the US than when it is paired with two tones one of which is always followed by the US, and the other of which is never followed by it. This result is predicted by the model since it is clear that when $A_1$ is always followed by the US, it will gain more strength than when it is sometimes accompanied by the US and sometimes not. The contribution of $A_1$ to the combined response strength $\overline{V}_{A_1L}$ will therefore be greater on reinforced trials than will be the contribution of $A_1$ and $A_2$ on reinforced trials in the case where $A_1$ and $A_2$ are not differentially paired with shock, and hence both the rate of gain of response strength to L and its asymptote will be lower in the former case than in the latter.

(5) The theory explains very successfully the conditions under which stimuli will become inhibitory. The model allows values of $V_X$ to become negative and this represents the stimulus X acquiring inhibitory power. This can clearly only happen if the expression, $\lambda_1 - \overline{V}_{AX}$, in Equation (13.24), is negative, i.e., if the value of $\lambda_1$ is less than the value of $\overline{V}_{AX}$. Since the minimum value of $\lambda$ is 0 (in the case where there is no US), this expression can only be negative where $\overline{V}_{AX}$ is positive, i.e., after some conditioning has occurred. It follows that X can only acquire inhibitory values if $V_A$ is positive and that the higher the value of $\overline{V}_A$ the more rapidly will a newly introduced stimulus X acquire inhibitory strength when not followed by the US. It should be noted that in most situations the background stimuli as well as a CS

explicitly paired with the US may acquire strength. It follows that if an animal is classically conditioned to stimulus A, and differentiation training is then given by introducing X without A and without US, X will acquire some inhibitory strength; although it is not paired with A, the extreme right-hand term should now be written as $\bar{V}_{BX}$, where B represents the background stimuli which may have acquired some positive response strength. All this fits in very neatly with what is known of Pavlovian conditioning. Moreover, Wagner has shown directly that the stronger the conditioning to A, the more rapidly X acquires inhibitory strength if it is introduced in the compound AX and the US is omitted.

The account has done scant justice to the range of experiments conducted by Rescorla and Wagner to test the implications of their theory, and the reader is referred to their article for a more comprehensive treatment.

## C. Further Discussion of the Model

We shall discuss the model under two headings. First, we shall consider how successfully it accounts for the phenomena of stimulus selection (e.g., blocking and overshadowing) that we deal with in attentional terms; and second, we shall consider how far a model of this type could be extended to account for the entire range of phenomena set out at the beginning of this chapter.

### 1. *Stimulus Selection*

(1) The model explains blocking by saying that, as a consequence of training on A alone, response strength to the AX compound is already very high, and little further learning can occur. This applies to simple cases of blocking but, as was pointed out in Chapter 4, not to those experiments where the response requirements are changed when compound training begins (E-11).

(2) As was also mentioned in Chapter 4, the model makes the surprising prediction that overtraining on cue A beyond asymptote (defined as the point that produces total blocking of X) should produce no detectable increase in response strength to A (as might be measured by increased resistance to extinction). If, on the other hand, blocking of X is due to attention to A, no such consequence follows. Rescorla and Wagner, in fact, report an experiment that indirectly suggests that their prediction might be correct. Animals were initially given 6 CER trials, 3 to a light (L) and 3 to a tone (T), sufficient to produce substantial suppression to the TL compound but less than complete suppression to either component presented alone. When one group was

given 10 further reinforced trials, all with the TL compound, and then extinguished to each component, their extinction performance did not differ from that of a control group given no further training. Thus further compound training did not add response strength to either component, although two other groups which received 10 additional training trials on one or other component in isolation did show significantly greater suppression to their trained component.

This result cannot readily be explained by our own theory and provides strong support for the Rescorla–Wagner model. Simultaneously, however, Mackintosh (unpublished) obtained results from a similarly conceived study which supported exactly the opposite conclusion. The first group in this experiment received 8 CER trials with a tone as CS, followed by 16 compound tone-plus-light trials. This pretraining on T completely blocked learning to L, since the group's suppression ratio to L (.48) did not differ from that of a group that received only tone training, but did differ from that of a group that simply received the 16 compound TL trials (.11). During their 16 TL trials, therefore, the first group learned nothing about L. That they did, however, learn something more about T was shown by the fact that over a series of 20 extinction trials to T alone they showed significantly more suppression than did a group given only 8 reinforced T trials before being extinguished. During compound training, therefore, animals pretrained on T learned more about T but learned nothing about L, a result which implies that failure of learning about L is at least partly a consequence of a failure of attention to L rather than of the impossibility of any further learning.

(3) The explanation of overshadowing derived from the Rescorla–Wagner model is an extension of their analysis of blocking. It rests upon the fact that after the first trial response strength to an AX compound will be greater than response strength to X alone, and therefore increments to X will be less in the former case than in the latter. From this it follows that no overshadowing should occur on the first training trial (since $\overline{V}_{AX}$, like $V_X$, will equal zero, provided that appropriate pretesting has been carried out). Mackintosh (unpublished) using conditioned suppression of water licking in rats (a situation in which substantial learning occurs in a single trial), gave a single conditioning trial in which for one group, a 63-dB tone was paired with shock, and for another group, the compound CS consisted of the tone plus a light. On subsequent test trials to the tone alone, animals given compound training showed significantly less suppression than those trained on tone alone. Overshadowing, therefore, appears to occur on the first trial of training, and in this instance cannot occur for the reasons implied by Rescorla and Wagner. Unfortunately for this argument, it is possible to explain many instances of overshadowing in terms of generalization decrement (see Chapter 5, page 148). The only way to control for the possibility that removal of the

light stimulus between training and testing disrupted performance in the compound group is to repeat the experiment using the same light but a very much stronger tone which would not be overshadowed by the light. With this procedure (one group given a single trial with an 83-dB tone, another group trained with the light and 83-dB tone), Mackintosh found no difference between groups on test trials to the tone: both groups showed significant suppression to the light, a result which (as mentioned on page 494) demonstrates that response attachments to more than one stimulus may be changed on a single trial.

We conclude that although the model has made a number of interesting predictions which have been confirmed, it is perhaps not completely successful in accounting for the phenomena of blocking and overshadowing. This is not to deny that it may contribute to these effects; and we have not discussed those phenomena of inhibitory learning with which it deals most successfully. In this area, we recognize that its contribution is extremely impressive.

### 2. *Extension of the Model*

We now consider how far the theory could be extended to account for the wider range of phenomena in which we believe attentional processes are implicated. The initial level of attention to a stimulus may be said to be represented in the model by the parameter $\alpha$ associated with each stimulus. This parameter does not change as a result of learning; if the model were to allow changes in $\alpha$ as a result of learning, then it would become an instance of a two-process model involving changes in attention. It would still differ from ours in that the strength of attention would change to a value of a stimulus on a dimension and not to the dimension as a whole. We shall not consider further the consequences of allowing changes in the parameter $\alpha$, since this simply turns the model into a special case of a two-process model.

It is clear that the Rescorla–Wagner model could explain many of the data on position habits if rules of a Spencian kind were incorporated relating response strengths to the probability of choice in a choice situation (E-1–E-8). Since we have already discussed in Chapter 4 how far a model not incorporating attention can deal with these phenomena, there is no need to pursue the point here.

The model clearly fails to explain why IDS learning should be easier than EDS learning in the case where the stimuli are changed from one stage of the problem to the next (E-23). It could also not explain why an IDS is learned more quickly than an EDS (E-22) if the response requirements are changed but the stimulus values remain the same (e.g., if the shift is from a simultaneous to a successive discrimination). The model, however, fares well on the phenomena associated with generalization (E-27–E-33), though because of the lack of dimensionality built into the model, it is not clear why training on the presence v. absence of a stimulus should lead to less steep generalization gradients than training on two stimuli differing along the dimension tested; it will be remembered that such training actually steepens the gradient for values along the dimension that are on the opposite side of S+ from S− (E-29).

The model cannot explain the ORE for the same reasons that Spence's model cannot explain this effect. On the model overtraining should increase the response strengths to the relevant stimuli. During reversal these response strengths must reverse and therefore pass through zero, the state at which training started. From that point onwards trials to relearn should be the same as in original learning, but overtraining should increase the number of trials to extinguish the original responses and therefore should slow down reversal learning. It is true that overtraining on the model might well decrease response strengths to irrelevant cues such as position, but we have already demonstrated in Chapter 9 (see page 300) that the ORE is due neither to a reduction in the difference between response strengths to irrelevant cues brought about by overtraining, nor solely to a reduction in control by position.

Without simulation, it is impossible to decide whether the model will predict the greater breadth of learning obtained under PR (E-46) nor whether it will predict the phenomena associated with probabilistic learning. It clearly predicts (falsely) that reversal learning will be faster after discrimination training on a probabilistic schedule than after training with CR (E-53).

Despite the model's success in predicting the results of training with multiple cues, we conclude that there are a great many other phenomena to which it cannot be successfully applied, but which can be accounted for in terms of a theory of selective attention. It seems likely that no model that incorporates only response learning will be able to account for such phenomena. Nevertheless the model is at its most successful in predicting the circumstances under which a stimulus will acquire inhibitory control and it might be profitable to develop a model incorporating the ideas of Rescorla and Wagner with our own, possibly by exploring the effects of allowing the parameter $\alpha$ to vary with learning and by allowing this parameter to affect the learning about dimensions rather than merely about a single value on a dimension. In fairness to Rescorla and Wagner, it should be added that a demonstration that their model cannot explain phenomena that it was never intended to explain, is in no sense a criticism of the model. We have discussed its application to such phenomena only in order to illustrate the more general point that an understanding of these phenomena requires the postulation of attentional learning as a separate process from response learning.

## VI. Summing Up

No lengthy summing up is needed here. The first section of this chapter gives a brief summary of the main experimental phenomena with which we have been concerned and the theoretical issues have reappeared in our discussion of the various models.

In theorizing about animal learning, workers are faced with a choice between putting up formalized models to explain a narrow range of data and working over a much broader range of phenomena with much looser models. The problem with the first approach is that it may leave out so much that the specific models put up to explain a narrow range of phenomena may be of little value when it comes to understanding a broader range of phenomena. The problem with the second approach is the uneasiness one feels at working with vague notions. Most workers have, in fact, taken the second approach, and it has led to the postulation of a large number of intervening factors such as selective attention, frustration, inhibition, and stimulus aftereffects in terms of which it seems possible to understand the results of existing experiments and to make predictions (albeit loose ones) about the outcome of further experiments.

There is no question but that our understanding of the factors determining the course of learning in animals has advanced enormously over the last 20 years. Most of this advance has come from working with rather loose concepts. One of the major problems at the moment is that to which we drew attention in Chapter 10. Although we are increasingly aware of many of the factors that determine animal learning, such as frustration, stimulus aftereffects, and the operation of selective attention, very few phenomena are determined by a single factor working in isolation, and it is extremely difficult to tease out the contribution made by each factor. Our own belief is that one of the most promising approaches to this problem lies in the still neglected area of the comparative psychology of the learning process. Working with different species in which the different factors contributing to performance have different weights should make it easier to isolate these factors and to specify the exact ways in which they influence behavior.

# References


ADAMS, D. K., & McCULLOCH, T. L. On the structure of acts. *Journal of Genetic Psychology*, 1934, **10**, 450–455.

ADELMAN, H. M., & MAATSCH, J. L. Learning and extinction based upon frustration, food reward, and exploratory tendency. *Journal of Experimental Psychology*, 1956, **52**, 311–315.

ADERMAN, M. The effect of differential training upon the relative strength of place vs. response habits. *Journal of Comparative & Physiological Psychology*, 1958, **51**, 372–375.

ADRIAN, E. D. Brain rhythms. *Nature*, 1944, **153**, 360–362.

ADRIAN, E. D., & MATTHEWS, B. H. C. The Berger rhythm: potential changes from the occipital poles in man. *Brain*, 1934, **57**, 355–385.

AIKEN, E. G. The effort variable in the acquisition, extinction and spontaneous recovery of an instrumental response. *Journal of Experimental Psychology*, 1957, **53**, 47–51.

ALLUISI, E. A., & SIDORSKY, R. C. The empirical validity of equal discriminability scaling. *Journal of Experimental Psychology*, 1958, **55**, 86–95.

AMSEL, A. Frustrative nonreward in partial reinforcement and discrimination learning. *Psychological Review*, 1962, **69**, 306–326.

AMSEL, A. Partial reinforcement effects on vigor and persistence. In K. W. Spence & J. T. Spence (Eds.), *The Psychology of learning and motivation.* New York: Academic Press, 1967, **1**, 1–65.

AMSEL, A., RASHOTTE, M. E., & MACKINNON, J. R. Partial reinforcement effects within subject and between subjects. *Psychological Monographs*, 1966, **80**, (20), 1–39.

AMSEL, A., & ROUSSEL, J. Motivational properties of frustration: I. Effect on a running response of the addition of frustration to the motivational complex. *Journal of Experimental Psychology*, 1952, **43**, 363–368.

AMSEL, A., & WARD, J. S. Frustration and persistence: resistance to discrimination following prior experience with the discriminanda. *Psychological Monographs*, 1965, **79**, (4), 1–41.

ANDELMAN, L., & SUTHERLAND, N. S. The effects of nondifferential reinforcement on two-cue discrimination learning. *Psychonomic Science*, 1970, **18**, 37–38.

APPELZWEIG, M. H. Response potential as a function of effort. *Journal of Comparative & Physiological Psychology*, 1951, **44**, 225–235.

ARMUS, H. L. Effect of magnitude of reinforcement on acquisition and extinction of a running response. *Journal of Experimental Psychology*, 1959, **58**, 61–63.

ATTNEAVE, F. Some informational aspects of visual perception. *Psychological Review*, 1954, **61**, 183–193.

BABB, H. Transfer from a stimulus complex to differentially discriminable components. *Journal of Comparative & Physiological Psychology*, 1957, **50**, 288–291.

BACON, W. E. Partial reinforcement extinction effect following different amounts of training. *Journal of Comparative & Physiological Psychology*, 1962, **55**, 998–1003.

BARON, M. R. The stimulus, stimulus control, and stimulus generalization. In D. I. Mostofsky (Ed.), *Stimulus generalization.* Stanford, California: Stanford University Press, 1965. Pp. 62–71.

BARRY, H., WAGNER, A. R., & MILLER, N. E. Effects of alcohol and amobarbital on performance inhibited by experimental extinction. *Journal of Comparative & Physiological Psychology*, 1962, **55**, 464–468.

BARLETT, F. C. *Remembering: A study in experimental and social psychology.* Cambridge: Cambridge University Press, 1932.

BEACH, F. A. The snark was a boojum. *American Psychologist*, 1950, **5**, 115–124.

BEACH, F. A., HEBB, D. O., MORGAN, C. T., & NISSEN, H. W. (Eds.), *The neuropsychology of Lashley.* New York: McGraw-Hill, 1960.

BECK, C. H., WARREN, J. M., & STERNER, R. Overtraining and reversal learning by cats and rhesus monkeys. *Journal of Comparative & Physiological Psychology*, 1966, **62**, 332–335.

BECK, E. C., DOTY, R. W., & KOOI, K. A. Electrocortical reactions associated with conditioned flexion reflexes. *Electroencephalography & Clinical Neurophysiology*, 1958, **10**, 279–289.

BECKER, P. W., & BRUNING, J. L. Goal gradient during acquisition, partial reinforcement and extinction of a five-part response chain. *Psychonomic Science*, 1966, **4**, 11–12.

BEHAR, I. Evaluation of the significance of positive and negative cue in discrimination learning. *Journal of Comparative & Physiological Psychology*, 1962, **55**, 502–504.

BEHREND, E. R., & BITTERMAN, M. E. Probability-matching in the fish. *American Journal of Psychology*, 1961, **74**, 542–551.

BEHREND, E. R., & BITTERMAN, M. E. Probability-matching in the goldfish. *Psychonomic Science*, 1966, **6**, 327–328.

BEHREND, E. R., & Bitterman, M. E. Further experiments on habit reversal in the fish. *Psychonomic Science*, 1967, **8**, 363–364.

BEHREND, E. R., DOMESICK, V. B., & BITTERMAN, M. E. Habit reversal in the fish. *Journal of Comparative & Physiological Psychology*, 1965, **60**, 407–411.

BERGER, B. D., YARCZOWER, M., & BITTERMAN, M. E. Effect of partial reinforcement on the extinction of a classically conditioned response in the goldfish. *Journal of Comparative & Physiological Psychology*, 1965, **59**, 399–405.

BERLYNE, D. E. *Conflict, arousal and curiosity.* New York: McGraw-Hill, 1960.

BERNSTEIN, B. B. Extinction as a function of frustration drive and frustration-drive stimulus. *Journal of Experimental Psychology*, 1957, **54**, 89–95.

BILLINGSLEY, B. A., FEDDERSEN, W. E., & BITTERMAN, M. E. Discrimination following nondifferential reinforcement with differential afferent consequences. *American Journal of Psychology*, 1954, **67**, 335–337.

BILLS, A. G. Blocking: a new principle of mental fatigue. *American Journal of Psychology*, 1931, **43**, 230–245.

BILLS, A. G. Blocking in mental fatigue and anoxemia compared. *Journal of Experimental Psychology*, 1937, **20**, 437–452.

BINDRA, D., & SEELY, J. F. Response decrement, produced by stimulus change, as a function of amount of training. *Journal of Experimental Psychology*, 1959, **57**, 317–322.

BIRCH, D. A motivational interpretation of extinction. *Nebraska symposium on motivation*, 1961, **9**, 179–202.

BIRCH, D., ISON, J. R., & SPERLING, S. E. Reversal learning under single stimulus presentation. *Journal of Experimental Psychology*, 1960, **60**, 36–40.

BIRNBAUM, I. M. Discrimination reversal, extinction, and acquisition after different amounts of overtraining. *American Journal of Psychology*, 1967, **80**, 363–369.

BITTERMAN, M. E. Towards a comparative psychology of learning. *American Psychologist*, 1960, **15**, 704–712.

BITTERMAN, M. E. Phyletic differences in learning. *American Psychologist*, 1965, **20**, 396–410.

BITTERMAN, M. E. Learning in animals. In H. Helson & W. Bevan (Eds.), *Contemporary approaches to psychology*. Princeton: Van Nostrand, 1966. Pp. 139–179.

BITTERMAN, M. E. Habit reversal and probability learning: rats, birds and fish. In R. Gilbert & N. S. Sutherland (Eds.), *Animal discrimination learning*. London: Academic Press, 1969. Pp. 163–175.

BITTERMAN, M. E., CALVIN, A. D., & ELAM, C. B. Perceptual differentiation in the course of non-differential reinforcement. *Journal of Comparative & Physiological Psychology*, 1953, **46**, 393–397.

BITTERMAN, M. E., & COATE, W. B. Some new experiments on the nature of discrimination learning in the rat. *Journal of Comparative & Physiological Psychology*, 1950, **43**, 198–210.

BITTERMAN, M. E., & ELAM, C. B. Discrimination following varying amounts of non-differential reinforcement. *American Journal of Psychology*, 1954, **67**, 133–137.

BITTERMAN, M. E., ELAM, C. B., & WORTZ, E. C. Perceptual differentiation as a function of non-differential reward and punishment. *Journal of Comparative & Physiological Psychology*, 1953, **46**, 475–478.

BITTERMAN, M. E., FEDDERSON, W. E., & TYLER, D. W. Secondary reinforcement and the discrimination hypothesis. *American Journal of Psychology*, 1953, **66**, 456–464.

BITTERMAN, M. E., & MCCONNELL, J. V. The role of set in successive discrimination. *American Journal of Psychology*, 1954, **67**, 129–132.

BITTERMAN, M. E., TYLER, D. W., & ELAM, C. B. Simultaneous and successive discrimination under identical stimulating conditions. *American Journal of Psychology*, 1955, **68**, 237–248.

BITTERMAN, M. E., WODINSKY, J., & CANDLAND, D. K. Some comparative psychology. *American Journal of Psychology*, 1958, **71**, 94–110.

BLACK, A. H., CARLSON, N. J., & SOLOMON, R. L. Exploratory studies of the conditioning of autonomic responses in curarized dogs. *Psychological Monographs*, 1962, **76** (29), 1–31.

BLACK, R. W. Differential conditioning, extinction, and secondary reinforcement. *Journal of Experimental Psychology*, 1965, **69**, 67–74.

BLACK, R. W. Shifts in magnitude of reward and contrast effects in instrumental and selective learning: a reinterpretation. *Psychological Review*, 1968, **75**, 114–126.

BLACK, R. W., & SPENCE, K. W. Effects of intertrial reinforcement on resistance to extinction following extended training. *Journal of Experimental Psychology*, 1965, **70**, 559–563.

BLODGETT, H. C., MCCUTCHAN, K., & MATHEWS, R. Spatial learning in the T-maze: the influence of direction, turn and food location. *Journal of Experimental Psychology*, 1949, **39**, 800–809.

BLOOM, J. M., & CAPALDI, E. J. The behavior of rats in relation to complex patterns of partial reinforcement. *Journal of Comparative & Physiological Psychology*, 1961, **54**, 261–265.

BLUM, R. A., & BLUM, J. S. Factual issues in the "continuity controversy." *Psychological Review*, 1949, **56**, 33–50.

BOLLINGER, P. An experimental study of the pre-solution period of discrimination learning in the white rat. Unpublished master's thesis, State University of Iowa, 1940.

BONEAU, C. A., & HONIG, W. K. Opposed generalization gradients based upon conditional discrimination training. *Journal of Experimental Psychology*, 1964, **66**, 89–93.

BORING, E. G. A new ambiguous figure. *American Journal of Psychology*, 1930, **42**, 444–445.

BOURNE, L. E., & RESTLE, F. Mathematical theory of concept identification. *Psychological Review*, 1959, **56**, 278–296.

BOWEN, J., & STRICKERT, D. Discrimination learning as a function of intertrial stimuli. *Psychonomic Science*, 1966, **5**, 297–298.

BOWER, G. H. A contrast effect in differential conditioning. *Journal of Experimental Psychology*, 1961, **62**, 196–199.

BOWER, G. H., & TRABASSO, T. Concept identification. In R. C. Atkinson (Ed.), *Studies in mathematical psychology*. Stanford: Stanford University Press, 1964. Pp. 32–94.

BOYCOTT, B. B., & YOUNG, J. Z. The comparative study of learning. *Symposium for the Society of Experimental Biology*, 1950, **4**, 432–453.

BOYER, W. N., & CROSS, H. A. Discrimination reversal learning in naive stump-tailed monkeys as a function of number of acquisition trials. *Psychonomic Science*, 1965, **2**, 139–140.

BRIMER, C. J., & DOCKRILL, F. J. Partial reinforcement and the CER. *Psychonomic Science*, 1966, **5**, 185–186.

BROADBENT, D. E. The role of auditory localization in attention and memory span. *Journal of Experimental Psychology*, 1954, **47**, 191–196.

BROADBENT, D. E. *Perception and communication*. London: Pergamon Press, 1958.

BROADBENT, D. E. Human perception and animal learning. In W. H. Thorpe & O. L. Zangwill (Eds.), *Current problems in animal behaviour*. Cambridge: Cambridge University Press, 1961. Pp. 248–272.

BROADHURST, P. L. Emotionality and the Yerkes–Dodson law. *Journal of Experimental Psychology*, 1957, **54**, 345–352.

BROGDEN, W. J. Sensory preconditioning. *Journal of Experimental Psychology*, 1939, **24**, 323–332.

BROOKSHIRE, K. H., WARREN, J. M., & BALL, G. G. Reversal and transfer learning following overtraining in rat and chicken. *Journal of Comparative & Physiological Psychology*, 1961, **54**, 98–102.

BROWN, J. S. The generalization of approach responses as a function of stimulus intensity and strength of motivation. *Journal of Comparative Psychology*, 1942, **33**, 209–226.

BROWN, J. S., & BASS, B. The acquisition and extinction of an instrumental response under constant and variable stimulus conditions. *Journal of Comparative & Physiological Psychology*, 1958, **51**, 499–504.

BROWN, R. T., & WAGNER, A. R. Resistance to punishment and extinction following training with shock or nonreinforcement. *Journal of Experimental Psychology*, 1964, **68**, 503–507.

BRUNER, J. S., MANDLER, J. M., O'DOWD, D., & WALLACH, M. S. The role of overlearning and drive level in reversal learning. *Journal of Comparative & Physiological Psychology*, 1958, **51**, 607–613.

BRUNER, J. S., MATTER, J., & PAPANEK, M. L. Breadth of learning as a function of drive level and mechanization. *Psychological Review*, 1955, **62**, 1–10.

BRUNSWIK, E. Probability as a determininer of rat behavior. *Journal of Experimental Psychology*, 1939, **25**, 175–197.

BRYANT, P. E. The role of the irrelevant stimulus dimension in transfer by young children. Paper presented at a Symposium on Discrimination Learning, University of Sussex, April, 1967.

BULLOCK, D. H., & BITTERMAN, M. E. Habit reversal in the pigeon. *Journal of Comparative & Physiological Psychology*, 1962, **55**, 958–962. (a)

BULLOCK, D. H., & BITTERMAN, M. E. Probability-matching in the pigeon. *American Journal of Psychology*, 1962, **75**, 634–639. (b)

BUSH, R. R., & MOSTELLER, F. A mathematical model for simple learning. *Psychological Review*, 1951, **58**, 313–323.

BUSS, A. H. Rigidity as a function of reversal and nonreversal shifts in the learning of successive discriminations. *Journal of Experimental Psychology*, 1953, **45**, 75–81.

BUSS, A. H. Reversal and nonreversal shifts in concept formation with partial reinforcement eliminated. *Journal of Experimental Psychology*, 1956, **52**, 152–166.

BUTTER, C. M., & GUTTMAN, N. Stimulus generalization and discrimination along the dimension of angular orientation. *American Psychologist*, 1957, **12**, 449.

BUYTENDIJK, F. J. J. Uber das Umlernen. *Archives neerlandaises de Physiologie*, 1930, **15**, 283–310.

CALFEE, R. C. Long-term behavior of rats under probabilistic reinforcement schedules. *Journal of Comparative & Physiological Psychology*, 1968, **65**, 232–237.

CALVIN, A. D. The growth of learning during nondifferential reinforcement, *Journal of Experimental Psychology*, 1953, **46**, 248–254.

CALVIN, A. D., & SEIBEL, L. J. A further investigation of response selection in simultaneous and successive discrimination. *Journal of Experimental Psychology*, 1954, **48**, 339–342.

CALVIN, A. D., & WILLIAMS, C. M. Simultaneous and successive discrimination in a single-unit, hollow-square maze. *Journal of Experimental Psychology*, 1956, **52**, 47–50.

CAMPIONE, J., HYMAN, L., & ZEAMAN, D. Dimensional shifts and reversals in retardate discrimination learning. *Journal of Experimental Child Psychology*, 1965, **2**, 255–263.

CAPALDI, E. J. The effect of different amounts of training on the resistance to extinction of different patterns of partially reinforced responses. *Journal of Comparative & Physiological Psychology*, 1958, **51**, 357–371.

CAPALDI, E. J. Overlearning reversal effect in a spatial discrimination task. *Perceptual and Motor Skills*, 1963, **16**, 335–336.

CAPALDI, E. J. Effects of N-length, number of different N-lengths, and number of reinforcements on resistance to extinction. *Journal of Experimental Psychology*, 1964, **68**, 230–239.

CAPALDI, E. J. Partial reinforcement: a hypothesis of sequential effects. *Psychological Review*, 1966, **73**, 459–477.

CAPALDI, E. J. A sequential hypothesis of instrumental learning. In K. W. Spence and J. T. Spence (Eds.), *Psychology of Learning and Motivation.* New York: Academic Press, 1967, **1**, 67-156.

CAPALDI, E. J., & DEUTSCH, E. A. Effects of severely limited acquisition training and pretraining on the partial reinforcement effect. *Psychonomic Science*, 1967, **9**, 171–172.

CAPALDI, E. J., & HART, D. Influence of a small number of partial reinforced training trials on resistance to extinction. *Journal of Experimental Psychology*, 1962, **64**, 166–171.

CAPALDI, E. J., HART, D., & STANLEY, L. R. Effect of intertrial reinforcement on the aftereffect of nonreinforcement and resistance to extinction. *Journal of Experimental Psychology*, 1963, **65**, 70–74.

CAPALDI, E. J., & LYNCH, D. Patterning at 24-hour ITI: Resolution of a discrepancy more apparent than real. *Psychonomic Science*, 1966, **6**, 229–230.

CAPALDI, E. J., & MINKOFF, R. Change in the stimulus produced by nonreward as a function of time. *Psychonomic Science*, 1966, **6**, 321–322.

CAPALDI, E. J., & SPIVEY, J. E. Effect of goal-box similarity on the aftereffect of nonreinforcement and resistance to extinction. *Journal of Experimental Psychology*, 1963, **66**, 461–465

CAPALDI, E. J., & SPIVEY, J. E. Intertrial reinforcement and aftereffects at 24-hour intervals. *Psychonomic Science*, 1964, **1**, 181–182.

CAPALDI, E. J., & SPIVEY, J. E. Schedule of partial delay of reinforcement and resistance to extinction. *Journal of Comparative & Physiological Psychology*, 1965, **60**, 274–276.

CAPLDI, E. J., & STANLEY, L. R. Temporal properties of reinforcement aftereffects. *Journal of Experimental Psychology*, 1963, **65**, 169–175.

CAPALDI, E. J., & STANLEY, L. R. Percentage of reward vs. *N*-length in the runway. *Psychonomic Science*, 1965, **3**, 263–264.

CAPALDI, E. J., & STEVENSON, H. W. Response reversal following different amounts of training. *Journal of Comparative & Physiological Psychology*, 1957, **50**, 195–198.

CAPALDI, E. J., VEATCH, R. L., & STEFANIAK, D. E. Stimulus control of patterning behavior. *Journal of Comparative & Physiological Psychology*, 1966, **61**, 161–164.

CARMICHAEL, L., HOGAN, H. P., & WALTER, A. A. An experimental study of the effect of language on the reproduction of visually perceived form. *Journal of Experimental Psychology*, 1932, **15**, 73–86.

CARR, H. A. Maze studies with the white rat. I. Normal Animals. *Journal of Animal Behavior*, 1917, **7**, 259–275.

CARR, H. A., & WATSON, J. B. Orientation in the white rat. *Journal of Comparative Neurology & Psychology*, 1908, **18**, 27–44.

CHAPMAN, D. W. Relative effects of determinate and indeterminate Aufgaben. *American Journal of Psychology*, 1932, **64**, 252–262.

CHERRY, E. C. Some experiments on the recognition of speech with one and with two ears. *Journal of the Acoustical Society of America*, 1953, **25**, 275–279.

CLAYTON, K. N. The relative effects of forced reward and forced non-reward during widely spaced successive discrimination reversal. *Journal of Comparative & Physiological Psychology*, 1962, **55**, 992–997.

CLAYTON, K. N. Overlearning and reversal of a spatial discrimination by rats. *Perceptual and Motor Skills*, 1963, **17**, 83–85. (a)

Clayton, K. N. Reversal performance by rats following overlearning with and without irrelevant stimuli. *Journal of Experimental Psychology*, 1963, **66**, 255–259. (b)

Clayton, K. N. The overlearning-reversal-effect: Dependent on a confound with drive? Paper presented at the Southeastern Psychological Association, Atlanta: 1965.

Clayton, K. N. T-maze acquisition and reversal as a function of intertrial interval. *Journal of Comparative & Physiological Psychology*, 1966, **62**, 409–414.

Clowes, M. B. Perception, picture-processing and computers. In N. L. Collins & D. Michie (Eds.), *Machine Intelligence*. Edinburgh: Oliver and Boyd, 1967, **1**, 181–197.

Cole, M., Belenky, G. L., Boucher, R. C., Fernandez, R. N., & Myers, D. L. Probability learning to escape from shock. *Psychonomic Science*, 1965, **3**, 127–128.

Cotton, J. W., Lewis, D. J., & Jensen, G. D. Partial reinforcement effects in a T-maze. *Journal of Comparative & Physiological Psychology*, 1959, **52**, 730–733.

Coutant, L. W., & Warren, J. M. Reversal and nonreversal shifts by cats and rhesus monkeys. *Journal of Comparative & Physiological Psychology*, 1966, **61**, 484–487.

Cowan, W. M., & Powell, T. P. S. Centrifugal fibres in the avian visual system. *Proceedings of the Royal Society* (*London*), *Ser. B*, 1963, **158**, 232–252.

Craik, K. J. W. Theory of the human operator in control systems. I. The operator as an engineering system. *British Journal of Psychology*, 1947, **38**, 56–61.

Crawford, F. T. Reversal learning to spatial cues by monkeys. *Journal of Comparative & Physiological Psychology*, 1962, **55**, 869–871.

Crespi, L. P. Quantitative variation in incentive and performance in the white rat. *American Journal of Psychology*, 1942, **5**, 467–517.

Cross, H. A., & Boyer, W. N. Influence of overtraining on single habit reversal in naive rhesus monkeys. *Psychonomic Science*, 1966, **4**, 245–246.

Cross, H. A., & Brown, L. T. Discrimination reversal learning in squirrel monkeys as a function of number of acquisition trials and prereversal experience. *Journal of Comparative & Physiological Psychology*, 1965, **59**, 429–431.

Cross, H. A., Fickling, R. M., Carpenter, J. B., & Brown, L. T. Discrimination reversal performance in squirrel monkeys as a function of prereversal experience and overlearning. *Psychonomic Science*, 1964, **1**, 353–354.

Dabrowska, J. An analysis of reversal learning in relation to the pattern of reversal in rats. *Acta Biologiae Experimentalis*, 1963, **23**, 11–24.

Dallenbach, K. M. Attention. *Psychological Bulletin*, 1926, **23**, 1–18.

Dallenbach, K. M. Attention. *Psychological Bulletin*, 1928, **25**, 493–512.

Dallenbach, K. M. Attention. *Psychological Bulletin*, 1930, **27**, 497–513.

D'Amato, M. R. The overlearning reversal effect in monkeys provided a salient irrelevant dimension. *Psychonomic Science*, 1965, **3**, 21–22.

D'Amato, M. R., & D'Amato, M. F. The partial reinforcement extinction effect following conventional and placed training trials. *Journal of Genetic Psychology*, 1962, **66**, 17–23.

D'Amato, M. R. & Fazzaro, J. Attention and cue-producing behavior in the monkey. *Journal of the Experimental Analysis of Behavior*, 1966, **9**, 469–473.

D'Amato, M. R., & Jagoda, H. Effects of extinction trials on discrimination reversal. *Journal of Experimental Psychology*, 1960, **59**, 254–260.

D'Amato, M. R., & Jagoda, H. Analysis of the role of overlearning in discrimination reversal. *Journal of Experimental Psychology*, 1961, **61**, 45–50.

D'Amato, M. R., & Jagoda, H. Overlearning and position reversal. *Journal of Experimental Psychology*, 1962, **64**, 117–122.

D'Amato, M. R., Lachman, R., & Kivy, P. Secondary reinforcement as affected by reward schedule and the testing situation. *Journal of Comparative & Physiological Psychology*, 1958, **51**, 737–741.

D'Amato, M. R., & Schiff, D. Further studies of overlearning and position reversal learning. *Psychological Reports*, 1964, **14**, 380–382.

D'Amato, M. R., & Schiff, D. Overlearning and brightness discrimination reversal. *Journal of Experimental Psychology*, 1965, **69**, 375–381.

Datta, L. G. Learning in the earthworm, *Lumbricus terrestris. American Journal of Psychology*, 1962, **75**, 531–553.

Davis, R. The limits of the psychological refractory period. *Quarterly Journal of Experimental Psychology*, 1956, **8**, 24–38.

Davis, R. The human operator as a single channel information system. *Quarterly Journal of Experimental Psychology*, 1957, **9**, 119–129.

Dawkins, R. A threshold model of choice behaviour. *Animal Behaviour*, 1969, **17**, 120–133. (a)

Dawkins, R. The attention threshold model. *Animal Behaviour*, 1969, **17**, 134–141. (b)

Denny, M. R. The role of secondary reinforcement in a partial reinforcement learning situation. *Journal of Experimental Psychology*, 1946, **36**, 373–389.

Denny, M. R. Learning through stimulus satiation. *Journal of Experimental Psychology*, 1957, **54**, 62–64.

Denny, M. R., Wells, R. H., & Maatsch, J. L. Resistance to extinction as a function of the discrimination habit established during fixed-ratio reinforcement. *Journal of Experimental Psychology*, 1957, **54**, 451–456.

Desmedt, J. E. Neurophysiological mechanisms controlling acoustic input. In G. L. Rasmussen & W. Windle (Eds.), *Neural mechanisms of the auditory and vestibular systems.* Springfield, Ill.: Charles Thomas, 1960. Pp. 152–164.

Desmedt, J. E. Auditory-evoked potentials from cochlea to cortex as influenced by activation of the efferent olivo-cochlear bundle. *Journal of the Acoustical Society of America*, 1962, **34**, 1478–1496.

Deutsch, J. A. A theory of shape recognition. *British Journal of Psychology*, 1955, **46**, 30–37.

Deutsch, J. A. *The structural basis of behaviour*. Cambridge: Cambridge University Press, 1960.

Deutsch, J. A., & Biederman, G. B. The monotonicity of the negative stimulus during learning. *Psychonomic Science*, 1965, **3**, 391–392.

Dewson, J. H., III. The olivocochlear bundle: effects upon noise-masked $N_1$ responses. *Journal of the Acoustical Society of America*, 1966, **40**, 1275.

Dewson, J. H., III. Efferent olivocochlear bundle: some relationships to noise masking and to stimulus attenuation. *Journal of Neurophysiology*, 1967, **30**, 817–832.

Dewson, J. H., III. Efferent olivocochlear bundle: some relationships to stimulus discrimination in noise. *Journal of Neurophysiology*, 1968, **31**, 122–130.

Dickerson, D. J. Performance of preschool children on three discrimination shifts. *Psychonomic Science*, 1966, **4**, 417–418.

DUFORT, R. H., GUTTMAN, N., & KIMBLE, G. A. One-trial discrimination reversal in the white rat. *Journal of Comparative & Physiological Psychology*, 1954, **47**, 248–249.

DUSEK, J. B., & GRICE, G. R. Visual discrimination learning following single-stimulus training. *Psychonomic Science*, 1968, **11**, 105–106.

ECKERMAN, D. A., Stimulus control by part of a complex $S^{\Delta}$. *Psychonomic Science*, 1967, **7**, 299–300.

EHRENFREUND, D. An experimental test of the continuity theory of discrimination learning with pattern vision. *Journal of Comparative & Physiological Psychology*, 1948, **41**, 408–422.

EHRENFREUND, D. A study of the transposition gradient. *Journal of Experimental Psychology*, 1952, **43**, 81–87.

EIMAS, P. D. Comment: Comparisons of reversal and nonreversal shifts. *Psychonomic Science*, 1965, **3**, 445–446.

EIMAS, P. D. Effects of overtraining and age on intradimensional and extradimensional shifts in children. *Journal of Experimental Child Psychology*, 1966, **3**, 348–355.

EIMAS, P. D. Overtraining and reversal discrimination learning in rats. *Psychological Record*, 1967, **17**, 239–248.

ELAM, C. B., & TYLER, D. W. Reversal-learning following partial reinforcement. *American Journal of Psychology*, 1958, **71**, 583–586.

ELAM, C. B., TYLER, D. W., & BITTERMAN, M. E. A further study of secondary reinforcement and the discrimination hypothesis. *Journal of Comparative & Physiological Psychology*, 1954, **47**, 381–384.

ELLISON, G. D. Differential salivary conditioning to traces. *Journal of Comparative & Physiological Psychology*, 1964, **57**, 373–380.

ENINGER, M. U. Habit summation in a selective learning problem. *Journal of Comparative & Physiological Psychology*, 1952, **45**, 604–608.

ENINGER, M. U. The role of generalised approach and avoidance tendencies in brightness discrimination. *Journal of Comparative & Physiological Psychology*, 1953, **46**, 398–402.

ERIKSEN, C. W., & HAKE, H. W. Absolute judgements as a function of the stimulus range and the number of stimulus and response categories. *Journal of Experimental Psychology*, 1955, **49**, 323–332. (a)

ERIKSEN, C. W., & HAKE, H. W. Multidimensional stimulus differences and accuracy of discrimination. *Journal of Experimental Psychology*, 1955, **50**, 153–160. (b)

ERLEBACHER, A. Reversal learning in rats as a function of percentage of reinforcement and degree of learning. *Journal of Experimental Psychology*, 1963, **66**, 84–90.

ESTES, W. K. Of models and men. *American Psychologist*, 1957, **12**, 609–617.

ESTES, W. K. The statistical approach to learning theory. In S. Koch (Ed.), *Psychology: A study of a science*. Vol. 2. New York: McGraw-Hill, 1959. Pp. 359–379.

ESTES, W. K. Learning theory. *Annual Review of Psychology*, 1962, **13**, 107–144.

ESTES, W. K., & LAUER, D. W. Conditions of invariance and modifiability in simple reversal learning. *Journal of Comparative & Physiological Psychology*, 1955, **50**, 199–206.

FANTZ, R. L. Form preferences in newly hatched chicks. *Journal of Comparative & Physiological Psychology*, 1957, **50**, 422–430.

FEHRER, E. Effects of amount of reinforcement and of pre- and post-reinforcement delays on learning and extinction. *Journal of Experimental Psychology*, 1956, **52**, 167–176.

FERNÁNDEZ-GUARDIOLA, A., ROLDÁN, R. E., FANJUL, M. L., & CASTELL, S. C. Role of the pupillary mechanism in the process of habituation of the visual pathways. *Electroencephalography & Clinical Neurophysiology*, 1961, **13**, 564–576.

Ferster, C. B. The effect on extinction responding of stimuli continuously present during conditioning. *Journal of Experimental Psychology*, 1951, **42**, 443–449.

Festinger, L. *A theory of cognitive dissonance.* Stanford: Stanford University Press, 1957.

Fidell, S., & Birch, D. The effect of overtraining on reversal learning under conditions of no reinforcement. *Psychonomic Science*, 1967, **8**, 27–28.

Fields, P. E. Studies in concept formation: I. The development of the concept of triangularity by the white rat. *Comparative Psychology Monographs*, 1932, **9**, 1–70.

Fields, P. E. Studies in concept formation: II. A new multiple stimulus jumping apparatus for visual figure discrimination. *Journal of Comparative Psychology*, 1935, **20**, 183–203.

Fitzgerald, R. D. Some effects of partial reinforcement with shock on classically conditioned heart-rate in dogs. *American Journal of Psychology*, 1966, **79**, 242–249.

Fitzgerald, R. D., Vardaris, R. M., & Teyler, T. J. Effects of partial reinforcement followed by continuous reinforcement on classically conditioned heart-rate in the dog. *Journal of Comparative & Physiological Psychology*, 1966, **62**, 483–486.

Franken, R. E. Stimulus change, attention, and brightness discrimination learning. *Journal of Comparative & Physiological Psychology*, 1967, **64**, 499–501.

Freeman, F. Unpublished M.A. Thesis. Kent, Ohio: Kent State University, 1967.

Freides, D. Goal-box cues and pattern of reinforcement. *Journal of Experimental Psychology*, 1957, **53**, 361–371.

Friedman, H., & Guttman, N. A further analysis of effects of discrimination training on stimulus generalization gradients. In D. I. Mostofsky, (Ed.), Stanford: Stanford University Press, 1965. Pp. 255–267.

Fritz, M. F. Long-time training of white rats on antagonistic visual habits. *Journal of Comparative Psychology*, 1930, **11**, 171–184.

Fuster, J. M. Effects of stimulation of brain-stem on tachistoscopic perception. *Science*, 1958, **127**, 150.

Galanter, E., & Bush, R. R. Some T-maze experiments. In R. R. Bush & W. K. Estes (Eds.), *Studies in mathematical learning theory.* Stanford: Stanford University Press, 1959. Pp. 265–289.

Galanter, E., & Shaw, W. A. "Cue" vs "reactive inhibition" in place and response learning. *Journal of Comparative & Physiological Psychology*, 1954, **47**, 395–398.

Ganz, L., & Riesen, A. H. Stimulus generalization to hue in the dark-reared Macaque. *Journal of Comparative & Physiological Psychology*, 1962, **55**, 92–99.

Garcia-Austt, E. Influence of the states of awareness upon sensory evoked potentials. *Electroencephalography & Clinical Neurophysiology*, 1963, Suppl. **24**, 76–89.

Gardner, R. A. On box score methodology as illustrated by three reviews of overtraining reversal effects. *Psychological Bulletin*, 1966, **66**, 416–418.

Gardner, R. A., & Coate, W. B. Reward versus nonreward in a simultaneous discrimination. *Journal of Experimental Psychology*, 1965, **69**, 579–582.

Garner, W. R. *Uncertainty and structure as psychological concepts.* New York: Wiley, 1962.

Gatling, F. P. A study of the continuity of the learning process as measured by habit reversal in the rat. *Journal of Comparative & Physiological Psychology*, 1951, **44**, 78–83.

GATLING, F. P. The effect of repeated stimulus reversals on learning in the rat. *Journal of Comparative & Physiological Psychology*, 1952, **45**, 347–351.

GELBER, B. Studies of the behavior of *Paramecium aurelia*. *Animal Behaviour Supplement*, 1965, **1**, 21–29.

GENTRY, G. V., OVERALL, J. E., & BROWN, W. L. Transpositional responses of rhesus monkeys to stimulus objects of intermediate size. *American Journal of Psychology*, 1959, **72**, 453–455.

GLANVILLE, A. D., & DALLENBACH, K. M. The range of attention. *American Journal of Psychology*, 1920, **41**, 207–236.

GONZALEZ, R. C., BEHREND, E. R., & BITTERMAN, M. E. Reversal learning and forgetting in bird and fish. *Science*, 1967, **158**, 519–521.

GONZALEZ, R. C., BERGER, B. D., & BITTERMAN, M. E. Improvement in habit-reversal as a function of amount of training per reversal and other variables. *American Journal of Psychology*, 1966, **79**, 517–530.

GONZALEZ, R. C., & BITTERMAN, M. E. Resistance to extinction in the rat as a function of percentage and distribution of reinforcement. *Journal of Comparative & Physiological Psychology*, 1964, **58**, 258–263.

GONZALEZ, R. C., & BITTERMAN, M. E. Partial reinforcement effect in the goldfish as a function of amount of reward. *Journal of Comparative & Physiological Psychology*, 1967, **64**, 163–167.

GONZALEZ, R. C., & BITTERMAN, M. E. Two-dimensional discriminative learning in the pigeon. *Journal of Comparative & Physiological Psychology*, 1968, **65**, 427–432.

GONZALEZ, R. C., ESKIN, R. M., & BITTERMAN, M. E. Further experiments on partial reinforcement in the fish. *American Journal of Psychology*, 1963, **76**, 366–375.

GONZALEZ, R. C., GENTRY, G. V., & BITTERMAN, M. E. Relational discrimination of intermediate size in the chimpanzee. *Journal of Comparative & Physiological Psychology*, 1954, **47**, 385–388.

GONZALEZ, R. C., GLEITMAN, H., & BITTERMAN, M. E. Some observations on the depression effect. *Journal of Comparative & Physiological Psychology*, 1962, **55**, 578–581.

GONZALEZ, R. C., LONGO, N., & BITTERMAN, M. E. Classical conditioning in the fish, exploratory studies of partial reinforcement. *Journal of Comparative & Physiological Psychology*, 1961, **54**, 452–456.

GONZALEZ, R. C., ROBERTS, W. A., & BITTERMAN, M. E. Learning in adult rats with extensive cortical lesions made in infancy. *American Journal of Psychology*, 1964, **77**, 547–562.

GOODRICH, K. P. Performance in different segments of an instrumental response chain as a function of reinforcement schedule. *Journal of Experimental Psychology*, 1959, **57**, 57–63.

GOODRICH, K. P., ROSS, L. E., & WAGNER, A. R. An examination of selected aspects of the continuity and noncontinuity positions in discrimination learning. *Psychological Record*, 1961, **11**, 105–117.

GOODWIN, W. R., & LAWRENCE, D. H. The functional independence of two discrimination habits associated with a constant stimulus situation. *Journal of Comparative & Physiological Psychology*, 1955, **48**, 437–443.

GOSSETTE, R. L. Successive discrimination reversal (SDR) performance of four avian species on a brightness discrimination task. *Psychonomic Science*, 1967, **8**, 17–18.

GOSSETTE, R. L., GOSSETTE, M. F., & INMAN, N. Successive discrimination reversal performance by the greater hill myna. *Animal Behaviour*, 1966, **14**, 50–53.

GOSSETTE, R. L., GOSETTE, M. F., & RIDDELL, W. Comparisons of successive discrimination reversal performances among closely and remotely related avian species. *Animal Behaviour*, 1966, **14**, 560–564.

GOTTSCHALDT, K. Uber den Einfluss der Erfahrung auf die Wahrnehmung von Figuren. II. *Psychologische Forschung*, 1929, **12**, 1–87.

GRAF, V., BULLOCK, D. H., & BITTERMAN, M. E. Further experiments on probability-matching in the pigeon. *Journal of the Experimental Analysis of Behavior*, 1964, **7**, 151–157.

GRAY, J. A. Sodium amobarbital and effects of frustrative nonreward. *Journal of Comparative & Physiological Psychology*, 1969, **69**, 55–64.

GRAY, J. A. Sodium amobarbital, the hippocampal theta rhythm, the partial reinforcement extinction effect. *Psychological Review*, 1970, **77**, 465–480.

GRICE, G. R. The acquisition of a visual discrimination habit following response to a single stimulus. *Journal of Experimental Psychology*, 1948, **38**, 633–642.

GRICE, G. R. Visual discrimination learning with simultaneous and successive presentation of stimuli. *Journal of Comparative & Physiological Psychology*, 1949, **42**, 365–373.

GRICE, G. R. The acquisition of a visual discrimination habit following extinction of response to one stimulus. *Journal of Comparative & Physiological Psychology*, 1951, **44**, 149–153.

GRICE, G. R., & SALZ, E. The generalization of an instrumental response to stimuli varying in the size dimension. *Journal of Experimental Psychology*, 1950, **40**, 702–708.

GRINDLEY, G. C. The formation of a simple habit in guinea-pigs. *British Journal of Psychology*, 1932, **23**, 127–147.

GROSSLIGHT, J. H., HALL, J. F., & SCOTT, W. Reinforcement schedules in habit reversal—a confirmation. *Journal of Experimental Psychology*, 1954, **48**, 173–174.

GROSSLIGHT, J. H., & RADLOW, R. Patterning effect of the nonreinforcement-reinforcement sequence in a discrimination situation. *Journal of Comparative & Physiological Psychology*, 1956, **49**, 542–546.

GROSSLIGHT, J. H., & RADLOW, R. Patterning effect of nonreinforcement-reinforcement sequence involving a single nonreinforced trial. *Journal of Comparative & Physiological Psychology*, 1957, **50**, 23–25.

GULLIKSEN, H., & WOLFE, H. L. A theory of learning and transfer: I. *Psychometrika*, 1938, **3**, 127–149.

GUTH, S. L. Patterning effects with compound stimuli. *Journal of Comparative & Physiological Psychology*, 1967, **63**, 480–485.

GUTTMAN, N. The pigeon and the spectrum and other perplexities. *Psychological Reports*, 1956, **2**, 449–460.

GUTTMAN, N., & KALISH, H. I. Discriminability and stimulus generalization. *Journal of Experimental Psychology*, 1956, **51**, 79–88.

HABER, A., & KALISH, H. I. Prediction of discrimination from generalization after variations in schedule of reinforcement. *Science*, 1963, **142**, 412–413.

HABERLANDT, K. F. Transfer along a continuum in classical conditioning. Unpublished Doctoral thesis. New Haven: Yale University, 1968.

HANSON, H. M. Effects of discrimination training on stimulus generalization. *Journal of Experimental Psychology*, 1959, **58**, 321–334.

HARA, K., & WARREN, J. M. Stimulus additivity and dominance in discrimination performance by cats. *Journal of Comparative & Physiological Psychology*, 1961, **54**, 86–90.

Harlow, H. F. Studies in discrimination learning by monkeys: VI. Discrimination between stimuli differing in both colour and form, only in color, and only in form. *Journal of General Psychology*, 1945, **33**, 225–235.

Harlow, H. F. Learning set and error factor theory. In S. Koch (Ed.) *Psychology: A study of a science*. Vol. II. New York: McGraw-Hill, 1959. Pp. 492–537.

Harlow, H. F., & Warren, J. M. Formation and transfer of discrimination learning sets. *Journal of Comparative & Physiological Psychology*, 1952, **45**, 482–489.

Harris, C. S., & Haber, R. N. Selective attention and coding in visual perception. *Journal of Experimental Psychology*, 1963, **65**, 328–333.

Harris, J. H., & Thomas, G. J. Learning single alternation of running speeds in a runway without handling between trials. *Psychonomic Science*, 1966, **6**, 329–330.

Harrow, M. Stimulus aspects responsible for the rapid acquisition of reversal shifts in concept formation. *Journal of Experimental Psychology*, 1964, **67**, 330–334.

Hearst, E. Simultaneous generalization gradients for appetitive and aversive behavior. *Science*, 1960, **132**, 1769–1770.

Hearst, E. Concurrent generalization gradients for food-controlled and shock-controlled behavior. *Journal of the Experimental Analysis of Behavior*, 1962, **5**, 19–31.

Hearst, E. Studies in stimulus generalization: Some behavioral, pharmacological and neuroanatomical factors. *Boletín del Instituto de Estudios Médicos y Biológicos, Universidad Nacional de Mexico*, 1963, **21**, 485–496.

Hearst, E. Approach, avoidance and stimulus generalization. In D. I. Mostofsky (Ed.), *Stimulus generalization*. Stanford: Stanford University Press, 1965. Pp. 331–355.

Hearst, E., & Koresko, M. B. Stimulus generalization and amount of prior training on variable-interval reinforcement. *Journal of Comparative & Psychological Psychology*, 1968, **66**, 133–138.

Hearst, E., Koresko, M. B., & Poppen, R. Stimulus generalization and the response-reinforcement contingency. *Journal of the Experimental Analysis of Behavior*, 1964, **7**, 369–380.

Hebb, D. O. *The organization of behavior*. New York: Wiley, 1949.

Heinemann, E. G., & Rudolph, R. L. The effect of discriminative training on the gradient of stimulus generalization. *American Journal of Psychology*, 1963, **76**, 653–658.

Hérnandéz-Péon, R., Jouvet, M., & Scherrer, H. Auditory potentials at the cochlear nucleus during acoustic habituation. *Acta Neurologica Latin America*, 1957, **3**, 114–116.

Hérnandéz-Péon, R., & Scherrer, H. "Habituation" to acoustic stimuli in cochlear nucleus. *Federation Proceedings*, 1955, **14**, 71.

Hérnandéz-Péon, R., Scherrer, H., & Jouvet, M. Modification of electrical activity in cochlear nucleus during "attention" in unanaesthetized cats. *Science*, 1956, **123**, 331–332.

Hick, W. E. On the rate of gain of information. *Quarterly Journal of Experimental Psychology*, 1952, **4**, 11–26.

Hicks, L. H. Effects of overtraining on acquisition and reversal of place and response learning. *Psychological Reports*, 1964, **15**, 459–462.

Hickson, R. H. Response probability in a two-choice learning situation with varying probability of reinforcement. *Journal of Experimental Psychology*, 1961, **62**, 138–144.

Hill, W. F., Cotton, J. W., & Clayton, K. N. Effect of reward magnitude, percentage of reinforcement, and training method on acquisition and reversal in a T-maze. *Journal of Experimental Psychology*, 1962, **64**, 81–86.

HILL, W. F., & SPEAR, N. E. A replication of overlearning and reversal in a T-maze. *Journal of Experimental Psychology*, 1963, **65**, 317. (a)

HILL, W. F., & SPEAR, N. E. Extinction in a runway as a function of acquisition level and reinforcement percentage. *Journal of Experimental Psychology*, 1963, **65**, 495–500. (b)

HILL, W. F., SPEAR, N. E., & CLAYTON, K. N. T-maze reversal after several different overtraining procedures. *Journal of Experimental Psychology*, 1962, **64**, 533–540.

HIRAYOSHI, I., & WARREN, J. M. Overtraining and reversal learning by experimentally naive kittens. *Journal of Comparative & Physiological Psychology*, 1967, **64**, 507–510.

HOFFELD, D. R. Primary stimulus generalisation and secondary extinction as a function of strength of conditioning. *Journal of Comparative & Physiological Psychology*, 1962, **55**, 27–31.

HOFFELD, D. R., KENDALL, S. B. THOMPSON, R. F., & BROGDEN, W. J. Effect of amount of preconditioning training upon the magnitude of sensory preconditioning. *Journal of Experimental Psychology*, 1960, **59**, 198–204.

HOFFELD, D. R., THOMPSON, R. F., & BROGDEN, W. J. Effect of stimuli-time relations during preconditioning training upon the magnitude of sensory preconditioning. *Journal of Experimental Psychology*, 1958, **56**, 437–442.

HOLDER, W. B., MARX, M. H., HOLDER, E. E., & COLLIER, G. Response strength as a function of delay of reward in a runway. *Journal of Experimental Psychology*, 1957, **53**, 316–323.

HONIG, W. K. Attentional factors governing the slope of the generalization gradient. In R. Gilbert & N. S. Sutherland (Eds.), *Animal discrimination learning*. London: Academic Press, 1969. Pp. 35–62.

HONIG, W. K. Attention and the modulation of stimulus control. In D. I. Mostofsky (Ed.), *Attention: Contemporary theory and analysis*. New York: Appleton-Century-Crofts, 1970. Pp. 193–238.

HONIG, W. K., BONEAU, C. A., BURSTEIN, K. R., & PENNYPACKER, A. S. Positive and negative generalization gradients obtained after equivalent training conditions. *Journal of Comparative & Physiological Psychology*, 1963, **56**, 111–116.

HONIG, W. K., THOMAS, D. R., & GUTTMAN, N. Differential effects of continuous extinction and discrimination training on the generalization gradient. *Journal of Experimental Psychology*, 1959, **58**, 145–152.

HOOPER, R. Variables controlling the overlearning reversal effect (ORE). *Journal of Experimental Psychology*, 1967, **73**, 612–619.

HORN, G. Electrical activity of the cerebral cortex of unanaesthetised cats during attentive behaviour. *Brain*, 1960, **83**, 57–76.

HORN, G. The response of single units in the striate cortex of unrestrained cats to photic and somaesthetic stimuli. *Journal of Physiology*, 1963, **165**, 80–81.

HORN, G. Physiological and psychological aspects of selective perception. In D. S. Lehrman, R. A. Hinde, & E. Shaw (Eds.), *Advances in the Study of Behavior*, New York: Academic Press, 1965, **1**, 155–215.

HORN, G., & HILL, R. M. Habituation of the response to sensory stimuli of neurons in the brain stem of rabbits. *Nature*, 1964, **202**, 296–298.

HORN, G., & HILL, R. M. Responsiveness to sensory stimulation of units in the superior colliculus and subjacent tectotegmental regions of the rabbit. *Experimental Neurology*, 1966, **14**, 199–223.

HOTHERSALL, D. Resistance to extinction when continuous reinforcement is followed by partial reinforcement. *Journal of Experimental Psychology*, 1966, **72**, 109–112.

House, B. J., & Zeaman, D. Miniature experiments in the discrimination learning of retardates. In L. P. Lipsitt & C. C. Spiker (Eds.), *Advances in Child Development and Behavior*. New York: Academic Press, 1963, **1**, 313–374.

Howarth, C. I., & Ellis, K. The relative intelligibility threshold for one's own name compared with other names. *Quarterly Journal of Experimental Psychology*, 1961, **13**, 236–239.

Hubel, D. H., Henson, C. O., Rupert, A., & Galambos, R. Attention units in the auditory cortex. *Science*, 1959, **129**, 1279–1280.

Hubel, D. H., & Wiesel, T. N. Receptive fields, binocular interaction and functional architecture in the cat's visual cortex. *Journal of Physiology*, 1962, **160**, 106–154.

Hubel, D. H., & Wiesel, T. N. Receptive fields of cells in striate cortex of very young, visually inexperienced kittens. *Journal of Neurophysiology*, 1963, **26**, 994–1002.

Hudson, B. B. One trial learning in the domestic rat. *Genetic Psychology Monographs*, 1950, **41**, 99–147.

Hugelin, A., Dumont, S., & Paillas N. Formation reticulaire et transmission des informations auditives au niveau de l'oreille moyenne et des voies acoustiques centrales. *Electroencephalogrophy & Clinical Neurophysiology*, 1960, **12**, 797–818.

Hughes, C. L., & North A. J. Effect of introducing a partial correlation between a critical cue and a previously irrelevant cue. *Journal of Comparative & Physiological Psychology*, 1959, **52**, 126–128.

Hulicka, I. M., Capehart, J., & Viney, W. The effect of stimulus variation on response probability during extinction. *Journal of Comparative & Physiological Psychology*, 1960, **53**, 79–82.

Hull, C. L. Quantitative aspects of the evolution of concepts. *Psychological Monographs*, 1920, **28**, No. 123.

Hull, C. L. The mechanism of the assembly of behavior segments in novel combinations suitable for problem solution. *Psychological Review*, 1935, **42**, 219–245.

Hull, C. L. The place of innate individual and species differences in a natural-science theory of behavior. *Psychological Review*, 1945, **52**, 55–60.

Hull, C. L. The problem of primary stimulus generalization. *Psychological Review*, 1947, **54**, 120–134.

Hull, C. L. Simple qualitative discrimination learning. *Psychological Review*, 1950, **57**, 303–313.

Hull, C. L. *A behavior system*. New Haven: Yale University Press, 1952.

Hulse, S. H., Jr. Amount and percentage of reinforcement and duration of goal confinement in conditioning and extinction. *Journal of Experimental Psychology*, 1958, **56**, 48–57.

Humphreys, L. G. The effect of random alternation of reinforcement on the acquisition and extinction of conditioned eyelid reactions. *Journal of Experimental Psychology*, 1939, **25**, 141–158.

Isaac, W. Arousal and reaction time in cats. *Journal of Comparative & Physiological Psychology*, 1960, **53**, 234–236.

Isaacs, I. D., & Duncan, C. P. Reversal and nonreversal shifts within and between dimensions in concept formation. *Journal of Experimental Psychology*, 1962, **64**, 580–585.

Ison, J. R. Resistance to extinction as a function of number of reinforcements. *Journal of Experimental Psychology*, 1962, **64**, 314–317.

Ison, J. R., & Birch, D. T-maze reversal following differential end box placement. *Journal of Experimental Psychology*, 1961, **62**, 200–202.

Ison, J. R., & Cook, P. E. Extinction performance as a function of incentive magnitude and number of acquisition trials. *Psychonomic Science*, 1964, **1**, 245–246.

Ison, J. R., Daly, H. B., & Glass, D. H. Amobarbital sodium and the effects of reward and nonreward in the Amsel double runway. *Psychological Reports*, 1967, **20**, 491–496.

Ison, J. R., & Pennes, E. S. Interaction of amobarbital sodium and reinforcement schedule in determining resistance to extinction of an instrumental running response. *Journal of Comparative & Physiological Psychology*, 1969, **68**, 215–219.

Ison, J. R., & Rosen, A. J. The effects of amobarbital sodium on differential instrumental conditioning and subsequent extinction. *Psychopharmacologia*, 1967, **10**, 417–425.

Jacobson, E. Electrophysiology of mental activities. *American Journal of Psychology*, 1932, **44**, 677–694.

James, W. *Principles of psychology*. New York: Holt, 1890.

James, W. *Psychology: Briefer course*. New York: Holt, 1892.

Jeeves, M. A., & North, A. J. Irrelevant or partially correlated stimuli in discrimination learning. *Journal of Experimental Psychology*, 1956, **52**, 90–94.

Jeffrey, W. E. Variables affecting reversal-shifts in young children. *American Journal of Psychology*, 1965, **78**, 589–595.

Jenkins, H. M. Resistance to extinction when partial reinforcement is followed by regular reinforcement. *Journal of Experimental Psychology*, 1962, **64**, 441–450.

Jenkins, H. M., & Harrison, R. H. Effect of discrimination training on auditory generalization. *Journal of Experimental Psychology*, 1960, **59**, 246–253.

Jenkins, H. M., & Harrison, R. H. Generalization gradients of inhibition following auditory discrimination learning. *Journal of the Experimental Analysis of Behavior*, 1962, **5**, 435–441.

Jensen, D. D. Paramecia, planaria and pseudo-learning. *Animal Behaviour Supplement*, 1965, **1**, 9–20.

Jensen, G. D. & Cotton, J. W. Running speed as a function of stimulus similarity and number of trials. *Journal of Comparative & Physiological Psychology*, 1961, **54**, 474–476.

Johnson, D. F. Determiners of selective discrimination stimulus control. Unpublished Doctoral dissertation. New York: Columbia University, 1966.

Johnson, D. F. Determiners of selective stimulus control in the pigeon. *Journal of Comparative & Physiological Psychology*, 1970, **70**, 298–307.

Johnson, D. F., & Cumming, W. W. Some determiners of attention. *Journal of the Experimental Analysis of Behavior*, 1968, **11**, 157–166.

Johnson, E. E. The role of motivational strength in latent learning. *Journal of Comparative & Physiological Psychology*, 1952, **45**, 526–530.

Johnson, P. J. Factors affecting transfer in concept-identification problems. *Journal of Experimental Psychology*, 1966, **72**, 655–660.

Jones, E. C. Latent learning and the partial reinforcement effect. *Psychonomic Science*, 1966, **6**, 119–120.

Jouvet, M., Schott, B., Courson, J., & Allegre, G. Documents neurophysiologiques relatifs aux mechanismes de l'attention chez l'homme. *Revue de Neurologie*, 1959, **100**, 437–450.

KAMIN, L. J. "Attention-like" processes in classical conditioning. In M. R. Jones (Ed.), *Miami symposium on the prediction of behavior: aversive stimulation.* Miami: University of Miami Press, 1968. Pp. 9–31.

KAMIN, L. J. Predictability, surprise, attention and conditioning. In R. Church & B. Campbell (Eds.), *Punishment and aversive behavior.* New York: Appleton-Century-Crofts, 1969. Pp. 279–296.

KATZ, S. Stimulus aftereffects and the partial reinforcement extinction effect. *Journal of Experimental Psychology*, 1957, **53**, 167–172.

KATZ, S., WOODS, G., & CARRITHERS, J. H. Reinforcement aftereffects and intertrial interval. *Journal of Experimental Psychology*, 1966, **72**, 624–626.

KELLEHER, R. T. Discrimination learning as a function of reversal and nonreversal shifts. *Journal of Experimental Psychology*, 1956, **51**, 379–384.

KENDLER, H. H., & D'AMATO, M. F. A comparison of reversal shifts and non-reversal shifts in human concept formation. *Journal of Experimental Psychology*, 1955, **49**, 165–174.

KENDLER, H. H., & GASSER, W. P. Variables in spatial learning. I: Number of reinforcements during training. *Journal of Comparative & Physiological Psychology*, 1948, **41**, 178–187.

KENDLER, H. H., & KENDLER, T. S. Vertical and horizontal processes in problem solving. *Psychological Review*, 1962, **69**, 1–16.

KENDLER, H. H., & KIMM, J. Reinforcement and cue factors in reversal learning. *Psychonomic Science*, 1964, **1**, 309–310.

KENDLER, H. H., & KIMM, J. Reversal learning as a function of the size of reward during acquisition and reversal. *Journal of Experimental Psychology*, 1967, **73**, 66–71.

KENDLER, H. H., & LACHMAN, R. Habit reversal as a function of schedule of reinforcement and drive strength. *Journal of Experimental Psychology*, 1958, **55**, 586–591.

KENDLER, T. S. An experimental investigation of transposition as a function of the difference between training and test stimuli. *Journal of Experimental Psychology*, 1950, **40**, 552–562.

KENDLER, T. S., KENDLER, H. H., & LEARNARD, B. Mediated responses to size and brightness as a function of age. *American Journal of Psychology*, 1962, **75**, 571–586.

KENDLER, T. S., KENDLER, H. H., & SILFAN, C. K. Optional shift behavior of albino rats. *Psychonomic Science*, 1964, **1**, 5–6.

KENDLER, T. S., KENDLER, H. H., & WELLS, D. Reversal and nonreversal shifts in nursery school children. *Journal of Comparative & Physiological Psychology*, 1960, **53**, 83–88.

KIMBALL, R. C., KIMBALL, L. T., & WEAVER, H. E. Latent learning as a function of the number of differential cues. *Journal of Comparative & Physiological Psychology*, 1953, **46**, 274–280.

KIMBLE, G. A. *Hilgard and Marquis' Conditioning and learning.* 2nd Ed. New York: Appleton-Century-Crofts, 1961.

KLEIN, R. M. Intermittent primary reinforcement as a parameter of secondary reinforcement. *Journal of Experiment Psychology*, 1959, **58**, 423–427.

KNAPP, R. K., KAUSE, R. H., & PERKINS, C. C. Jr. Immediate vs delayed shock in T-maze performance. *Journal of Experimental Psychology*, 1959, **58**, 357–362.

KOFFKA, K. *Principles of Gestalt Psychology.* London: Routledge & Kegan Paul, 1935.

KOMAKI, J. The facilitative effect of overlearning in discrimination learning by white rats. *Psychologia*, 1961, **4**, 28–35.

KOMAKI, J. Reversal retardation through introducing forced nonrewarding trials in overtraining by white rats. *Annual of Animal Psychology*, 1962, **12**, 1–10.

KOMISARUK, B. R. Basic synchrony of EEG theta rhythm, EKG and certain oscillatory patterns in awake rats. Proceedings of the International Union of Physiological Sciences, 24th International Congress, Vol. 7, 1968.

KRECHEVSKY, I. Hypotheses in rats. *Psychological Review*, 1932, **39**, 516–532. (a)

KRECHEVKSY, I. Antagonistic visual discrimination habits in the white rat. *Journal of Comparative Psychology*, 1932, **14**, 263–277. (b)

KRECHEVSKY, I. A study of the continuity of the problem-solving process. *Psychological Review*, 1938, **45**, 107–133.

KRECHEVSKY, I., & HONZIK, C. H. Fixation in the rat. *University of California Publications in Psychology*, 1932, **6**, 13–26.

KULPE, O. Versuche uber Abstraktion. *Berlin International Congress of Experimental Psychology*, 1904, 56–58.

LASHLEY, K. S. *Brain mechanisms and intelligence: A quantitative study of injuries to the brain.* Chicago: University of Chicago Press, 1929. Reprinted by Dover, New York, 1963.

LASHLEY, K. S. The mechanism of vision: XV. Preliminary studies of the rat's capacity for detail vision. *Journal of General Psychology*, 1938, **18**, 123–193.

LASHLEY, K. S. An examination of the continuity theory as applied to discriminative learning. *Journal of General Psychology*, 1942, **26**, 241–265.

LASHLEY, K. S., & BALL, J. Spinal conduction and kinesthetic sensitivity in the maze habit. *Journal of Comparative Psychology*, 1929, **9**, 71–105.

LASHLEY, K. S., & RUSSELL, J. T. The mechanism of vision. XI. A preliminary test of innate organization. *Journal of Genetic Psychology*, 1934, **45**, 136–144.

LASHLEY, K. S., & WADE, M. The Pavlovian theory of generalization. *Psychological Review*, 1946, **53**, 72–87.

LAWRENCE, D. H. Acquired distinctiveness of cues: I. Transfer between discriminations on the basis of familiarity with the stimulus. *Journal of Experimental Psychology*, 1949, **39**, 770–784.

LAWRENCE, D. H. Acquired distinctiveness of cues: II. Selective association in a constant stimulus situation. *Journal of Experimental Psychology*, 1950, **40**, 175–188.

LAWRENCE, D. H. The transfer of a discrimination along a continuum. *Journal of Comparative & Physiological Psychology*, 1952, **45**, 511–516.

LAWRENCE, D. H. The applicability of generalization gradients to the transfer of a discrimination along a continuum. *Journal of General Psychology*, 1955, **52**, 37–48.

LAWRENCE, D. H. Nature of a stimulus: some relationships between learning and perception. In S. Koch (Ed.), *Psychology: A study of a science* Vol. 5. New York: McGraw-Hill, 1959. Pp. 179–212.

LAWRENCE, D. H., & DE RIVERA, J. Evidence for relational transposition. *Journal of Comparative & Physiological Psychology*, 1954, **47**, 465–471.

LAWRENCE, D. H., & FESTINGER, L. *Deterrents and reinforcements.* Stanford: Stanford University Press, 1962.

LAWRENCE, D. H., & MASON, W. A. Systematic behavior during discrimination reversal and change of dimension. *Journal of Comparative & Physiological Psychology*, 1955, **48**, 1–7.

LEONARD, D. W. Amount and sequence of reward in partial and continuous reinforcement.

*Journal of Comparative & Physiological Psychology*, 1969, **67**, 204–211.

LEONARD, D. W., & THEIOS, J. Classical eyelid conditioning in rabbits under prolonged single alternation conditions of reinforcement. *Journal of Comparative & Physiological Psychology*, 1967, **64**, 273–276.

LETTVIN, J. Y., MATURANA, H. R., PITTS, W. H., & MCCULLOCH, W. S. Two remarks on the visual system of the frog. In W. A. Rosenblith (Ed.), *Sensory communication.* New York: Wiley, 1961. Pp. 757–776.

LEVINE, J. Studies in the interrelations of central nervous structures in binocular vision: II. The conditions under which interocular transfer of discriminative habits takes place in the pigeon. *Journal of Genetic Psychology*, 1945, **67**, 131–142.

LEVINE, M. Hypothesis Behavior. In A. M. Schrier, H. F. Harlow, & F. Stollnitz (Eds.), *Behavior of nonhuman primates: modern research trends.* Vol. 1. New York: Academic Press, 1965. Pp. 97–127.

LEWIS, D. J. Partial reinforcement: A selective review of the literature since 1950. *Psychological Bulletin*, 1960, **57**, 1–28.

LEWIS, D. J., & COTTON, J. W. Partial reinforcement and non-response acquisition. *Journal of Comparative & Physiological Psychology*, 1958, **51**, 251–254.

LEWIS, M. Some nondecremental effects of effort. *Journal of Comparative & Physiological Psychology*, 1964, **57**, 367–372.

LIKELY, D., & SCHNITZER, S. B. Dependence of the overtraining extinction effect on attention to runway cues. *Quarterly Journal of Experimental Psychology*, 1968, **20**, 193–196.

LINDSLEY, D. B. The reticular activating system and perceptual integration. In D. E. Sheer (Ed.), *Electrical stimulation of the brain.* Austin: University of Texas Press, 1961. Pp. 331–349.

LOCKARD, J. S. Choice of a warning signal or no warning signal in an unavoidable shock situation. *Journal of Comparative & Physiological Psychology*, 1963, **56**, 526–530.

LOGAN, F. A. *Incentive.* New Haven: Yale University Press, 1960.

LOGAN, F. A. Specificity of discrimination learning to the original context. *Science*, 1961, **133**, 1355–1356.

LOGAN, F. A. Transfer of discrimination. *Journal of Experimental Psychology*, 1966, **71**, 616–618.

LONGO, N., MILSTEIN, S., & BITTERMAN, M. E. Classical conditioning in the pigeon: Exploratory studies of partial reinforcement. *Journal of Comparative & Physiological Psychology*, 1962, **55**, 983–986.

LONGSTRETH, L. E. Partial reinforcement effect and extinction as a function of frustration and interfering responses. *Journal of Experimental Psychology*, 1964, **67**, 581–586.

LORENZ, K. The comparative method of studying innate behavior patterns. *Symposium of the Society for Experimental Biology*, 1950, **4**, 221–268.

LOVEJOY, E. An attention theory of discrimination learning. *Journal of Mathematical Psychology*, 1965, **2**, 342–362.

LOVEJOY, E. Analysis of the overlearning reversal effect. *Psychological Review*, 1966, **73**, 87–103.

LOVEJOY, E. *Attention in discrimination learning.* San Francisco: Holden-Day, 1968.

LOVEJOY, E., & RUSSELL, D. G. Suppression of learning about a hard cue by the presence of an easy cue. *Psychonomic Science*, 1967, **8**, 365–366.

LUBOW, R. E. Latent inhibition: effects of frequency of nonreinforced preexposure of the CS. *Journal of Comparative & Physiological Psychology*, 1965, **60**, 454–459.

LUBOW, R. E., & MOORE, A. U. Latent inhibition: the effect of nonreinforced preexposure to the conditioned stimulus. *Journal of Comparative & Physiological Psychology*, 1959, **52**, 415–419.

LUDVIGSON, H. W. A preliminary investigation of the effects of sodium amytal, prior reward in $G_1$, and activity level on the FE. *Psychonomic Science*, 1967, **8**, 115–116.

LUDVIGSON, H. W., & SYTSMA, D. The sweet smell of success: Apparent double alternation in the rat. *Psychonomic Science*, 1967, **9**, 283–284.

LUKASZEWSKA, I. Some further failures to find the visual overlearning reversal effect in rats. *Journal of Comparative & Physiological Psychology*, 1968, **65**, 359–361.

LYONS, J., & THOMAS, D. R. Effects of interdimensional training on stimulus generalization: II. Within subjects design. *Journal of Experimental Psychology*, 1967, **75**, 572–574.

MAATSCH, J. L., ADELMAN, H. M., & DENNY, M. R. Effort and resistance to extinction of the bar-pressing response. *Journal of Comparative & Physiological Psychology*, 1954, **47**, 47–50.

MCCAIN, G. Partial reinforcement effects following a small number of acquisition trials. *Psychonomic Monograph Supplements*, 1966, **1**, 251–270.

MCCAIN, G., & GARRETT, B. L. Generalization to stimuli of different brightness in three straight alley studies. *Psychological Reports*, 1964, **15**, 368–370.

MCCAIN, G., LEE, P., & POWELL, N. Extinction as a function of partial reinforcement and overtraining. *Journal of Comparative & Physiological Psychology*, 1962, **55**, 1004–1006.

MCCAIN, G., & POWER, R. Extinction as a function of reinforcement conditions in the start box. *Psychonomic Science*, 1966, **5**, 193–194.

MACCASLIN, E. F. Successive and simultaneous discrimination as a function of stimulus similarity. *American Journal of Psychology*, 1954, **67**, 308–314.

MACCASLIN, E. F., WODINSKY, J., & BITTERMAN, M. E. Stimulus-generalization as a function of prior training. *American Journal of Psychology*, 1952, **65**, 1–15.

MCCONNELL, J. V. Cannibals, chemicals and contiguity. *Animal Behaviour Supplement*, 1965, **1**, 61–68.

MACCORQUODALE, K., & MEEHL, P. E. Edward C. Tolman. In W. K. Estes, S. Koch, K. MacCorquodale, P. E. Meehl, C. G. Mueller, W. N. Schoenfeld, & W. S. Verplanck (Eds.), *Modern learning theory*. New York: Appleton-Centry-Crofts, 1954. Pp. 177–266.

MCCULLOCH, T. L., & PRATT, J. G. A study of the presolution period in weight discrimination by white rats. *Journal of Comparative & Physiological Psychology*, 1934, **18**, 271–290.

MCFARLAND, D. J. The role of attention in the disinhibition of displacement activities. *Quarterly Journal of Experimental Psychology*, 1966, **18**, 19–30.

MCFARLAND, D. J., & MCGONIGLE, B. Frustration tolerance and incidental learning as determinants of extinction. *Nature*, 1967, **215**, 786–787.

MACFARLANE, D. A. The role of kinesthesis in maze learning. *University of California Publications in Psychology*, 1930, **4**, 277–305.

MCGONIGLE, B. Stimulus additivity and dominance in visual discrimination performance by rats. *Journal of Comparative & Physiological Psychology*, 1967, **64**, 110–113.

MCGONIGLE, B., MACFARLAND, D. J., & COLLIER, P. Rapid extinction following drug-inhibited incidental learning. *Nature*, 1967, **214**, 531–532.

McHose, J. H. Patterned running as a function of the sequence of trial administration. *Psychonomic Science*, 1967, **9**, 281–282.

McHose, J. H., & Ludvigson, H. W. Differential conditioning with nondifferential reinforcement. *Psychonomic Science*, 1966, **6**, 485–486.

Mackintosh, J. An investigation of reversal learning in *Octopus Vulgaris* Lamarck. *Quarterly Journal of Experimental Psychology*, 1962, **14**, 15–22.

Mackintosh, N. J. The effect of overtraining on a reversal and a nonreversal shift. *Journal of Comparative & Physiological Psychology*, 1962, **55**, 555–559.

Mackintosh, N. J. The effect of irrelevant cues on reversal learning in the rat. *British Journal of Psychology*, 1963, **54**, 127–134. (a)

Mackintosh, N. J. Extinction of a discrimination habit as a function of overtraining. *Journal of Comparative & Physiological Psychology*, 1963, **56**, 842–847. (b)

Mackintosh, N. J. Direct transfer from a horizontal-vertical discrimination to a brightness discrimination in the rat. *Quarterly Journal of Experimental Psychology*, 1963, **15**, 212–213. (c)

Mackintosh, N. J. Overtraining and transfer within and between dimensions in the rat. *Quarterly Journal of Experimental Psychology*, 1964, **16**, 250–256.

Mackintosh, N. J. Selective attention in animal discrimination learning. *Psychological Bulletin*, 1965, **64**, 124–150. (a)

Mackintosh, N. J. Overtraining, extinction, and reversal in rats and chicks. *Journal of Comparative & Physiological Psychology*, 1965, **59**, 31–36. (b)

Mackintosh, N. J. Overtraining, transfer to proprioceptive control, and position reversal. *Quarterly Journal of Experimental Psychology*, 1965, **17**, 26–36. (c)

Mackintosh, N. J. Incidental cue learning in rats. *Quarterly Journal of Experimental Psychology*, 1965, **17**, 292–300. (d)

Mackintosh, N. J. The effect of attention on the slope of generalization gradients. *British Journal of Psychology*, 1965, **56**, 87–93. (e)

Mackintosh, N. J. Transposition after single-stimulus pretraining. *American Journal of Psychology*, 1965, **78**, 116–119. (f)

Mackintosh, N. J. Further analysis of the overtraining reversal effect. *Journal of Comparative & Physiological Psychology, Monograph Supplement*, 1969, **67**, 1–18. (a)

Mackintosh, N. J. Comparative psychology of serial reversal and probability learning: Rats, birds and fish. In R. Gilbert & N. S. Sutherland (Eds.), *Animal discrimination learning* London: Academic Press, 1969. Pp. 137–167. (b)

Mackintosh, N. J. Attention and probability learning. In D. I. Mostofsky (Ed.), *Attention: Contemporary theory and analysis*. New York: Appleton-Century-Crofts, 1970, pp. 173–191.

Mackintosh, N. J., & Holgate, V. Overtraining and the extinction of a discrimination in *Octopus*. *Journal of Comparative & Physiological Psychology*, 1965, **60**, 260–264.

Mackintosh, N. J., & Holgate, V. Effects of several pretraining procedures on brightness probability learning. *Perceptual and Motor Skills*, 1967, **25**, 629–637.

Mackintosh, N. J., & Holgate, V. Effects of inconsistent reinforcement on reversal and nonreversal shifts. *Journal of Experimental Psychology*, 1968, **76**, 154–159.

Mackintosh, N. J., & Holgate, V. Serial reversal training and nonreversal shift learning. *Journal of Comparative & Physiological Psychology*, 1969, **67**, 89–93.

Mackintosh, N. J., & Honig, W. K. Blocking and attentional enhancement in pigeons. *Journal of Comparative & Physiological Psychology*, 1970, **73**, 78–85.

Mackintosh, N. J., & Little, L. Intradimensional and extradimensional shift learning by pigeons. *Psychonomic Science*, 1969, **14**, 5–6.

Mackintosh, N. J., & Little, L. An analysis of transfer along a continuum. *Canadian Journal of Psychology*, 1970, **24**, 362–369.

Mackintosh, N. J., & Mackintosh, J. Reversal learning in *Octopus Vulgaris* Lamarck with and without irrelevant cues. *Quarterly Journal of Experimental Psychology*, 1963, **15**, 236–242.

Mackintosh, N. J., & Mackintosh, J. Performance of *Octopus* over a series of reversals of a simultaneous discrimination. *Animal Behaviour*, 1964, **12**, 321–324.

Mackintosh, N. J., Mackintosh, J., Safriel-Jorne, O., & Sutherland, N. S. Overtraining, reversal and extinction in the goldfish. *Animal Behaviour*, 1966, **14**, 314–318.

Mackintosh, N. J., McGonigle, B., Holgate, V., & Vanderver, V. Factors underlying improvement in serial reversal learning. *Canadian Journal of Psychology*, 1968, **22**, 85–95.

McNamara, H. J., & Wike, E. L. The effects of irregular learning conditions upon the rate and permanence of learning. *Journal of Comparative & Physiological Psychology*, 1958, **51**, 363–366.

Madison, H. L. Experimental extinction as a function of number of reinforcements. *Psychological Reports*, 1964, **14**, 647–650.

Madsen, M. C., & McGaugh, J. L. The effect of ECS on one-trial avoidance learning. *Journal of Comparative & Physiological Psychology*, 1961, **54**, 522–523.

Mahut, H. The effect of stimulus position on visual discrimination by the rat. *Canadian Journal of Psychology*, 1954, **8**, 130–138.

Maier, N. R. F. *Frustration: The study of behavior without a goal.* New York: McGraw-Hill, 1949.

Maier, N. R. F., & Schneirla T. C. *Principles of animal psychology.* New York: McGraw-Hill, 1935.

Maier, S. F., & Gleitman, H. Proactive interference in rats. *Psychonomic Science*, 1967, **7**, 25–26.

Malott, M. K. Stimulus control in stimulus-deprived chickens. *Journal of Comparative & Physiological Psychology*, 1968, **66**, 276–283.

Mandler, J. M. Behavior changes during overtraining and their effects on reversal and transfer. *Psychonomic Monograph Supplements*, 1966, **1**, 187–202.

Mandler, J. M. The effect of overtraining on the use of positive and negative stimuli in reversal and transfer. *Journal of Comparative & Physiological Psychology*, 1968, **66**, 110–115.

Mandler, J. M., & Hooper, W. R. Overtraining and goal approach strategies in discrimination reversal. *Quarterly Journal of Experimental Psychology*, 1967, **19**, 142–149.

Margolius, G. Stimulus generalization of an instrumental response as a function of the number of reinforced trials. *Journal of Experimental Psychology*, 1955, **49**, 105–111.

Margolius, G. Generalization and discrimination obtained within the same and different environments. *Conference on Stimulus Generalization.* Boston: Boston University, 1963.

Marrone, R., & Evans, S. Two-choice and three-choice probability learning in fish. *Psychonomic Science*, 1966, **5**, 327–328.

Marsh, G. Intradimensional transfer of discrimination along the hue continuum. *Psychonomic Science*, 1967, **8**, 411–412.

Marsh, G. An evaluation of three explanations for the transfer of discrimination effect. *Journal of Comparative & Physiological Psychology*, 1969, **68**, 268–275.

MARSH, G., & JOHNSON, R. The effect of irrelevant cues and overtraining on discrimination reversal in the pigeon. *Psychonomic Science*, 1968, **12**, 321–322.

MARSH, J. T., WORDEN, F. G., & HICKS, L. Some effects of room acoustics on evoked auditory potentials. *Science*, 1962, **137**, 280–282.

MARX, M. H. Resistance to extinction as a function of degree of reproduction of training conditions. *Journal of Experimental Psychology*, 1960, **59**, 337–344.

MARX, M. H., MCCOY, D. F., & TOMBAUGH, J. W. Resistance to extinction as a function of constant delay of reinforcement. *Psychonomic Science*, 1965, **2**, 333–334.

MASON, D. J. The relation of secondary reinforcement to partial reinforcement. *Journal of Comparative & Physiological Psychology*, 1957, **50**, 264–268.

MAX, L. W. An experimental study of the motor theory of consciousness: III. Action-current responses in the deaf during awaking, kinaesthetic imagery, and abstract thinking. *Journal of Comparative Psychology*, 1937, **21**, 301–344.

MEDIN, D. L., & DAVIS, R. Color discrimination by rhesus monkeys. *Psychonomic Science*, 1967, **7**, 33–34.

MELCHING, W. H. Reward value of an intermittent neutral stimulus. *Journal of Comparative & Physiological Psychology*, 1954, **47**, 370–374.

MELZACK, R. The genesis of emotional behavior: an experimental study of the dog. *Journal of Comparative & Physiological Psychology*, 1954, **47**, 160–168.

MIKULKA, P. J., LEHR, R., & PAVLIK, W. B. Effect of reinforcement schedules on reward shifts. *Journal of Experimental Psychology*, 1967, **74**, 57–61.

MILES, C. G. Acquisition of control by the features of a compound stimulus in discriminative operant conditioning. Unpublished Doctoral thesis, McMaster University, 1965.

MILES, C. G. Blocking the acquisition of control by an auditory stimulus with pretraining on brightness. *Psychonomic Science*, 1970, **19**, 133–134.

MILLER, G. A., GALANTER, E., & PRIBRAM, K. H. *Plans and the structure of behavior*. New York: Holt, 1960.

MILLER, G. A., & Isard, S. Some perceptual consequences of linguistic rules. *Journal of Verbal Learning & Verbal Behavior*, 1963, **2**, 217–228.

MILLER, N. E. Liberalization of basic S-R concepts: Extensions to conflict behavior, motivation, and social learning. In S. Koch (Ed.), *Psychology, a study of a science*. Vol. 2. New York: McGraw-Hill, 1959. Pp. 196–293.

MILLER, R. E., & MURPHY, J. V. Discrimination learning with vertical vs. horizontal stimulus relationships. *Journal of Comparative & Physiological Psychology*, 1956, **49**, 80–83.

MOORE, J. W., & GORMEZANO, I. Effects of omitted vs. delayed UCS on classical eyelid conditioning under partial reinforcement. *Journal of Experimental Psychology*, 1963, **65**, 248–257.

MOUNTJOY, P. T., & MALOTT, M. K. Wavelength generalization curves for chickens reared in restricted portions of the spectrum. *Psychological Records*, 1968, **18**, 575–583.

MOWBRAY, G. H. Simultaneous vision and audition: comprehension of prose passages with varying levels of difficulty. *Journal of Experimental Psychology*, 1953, **46**, 365–372.

MOWRER, O. H. Preparatory set (expectancy)—further evidence of its central locus. *Journal of Experimental Psychology*, 1941, **28**, 116–133.

MOWRER, O. H., & JONES, H. M. Extinction and behavior variability as functions of effortfulness of task. *Journal of Experimental Psychology*, 1943, **33**, 369–386.

MOWRER, O. H., & JONES, H. M. Habit strength as a function of the pattern of reinforcement. *Journal of Experimental Psychology*, 1945, **35**, 293–311.

MUENZINGER, K. F., & EVANS, W. O. Black preference or artifact? *Psychological Reports*, 1957, **3**, 493–495.

MUMMA, R., & WARREN, J. M. Two-cue discrimination learning by cats. *Journal of Comparative & Physiological Psychology*, 1968, **66**, 116–122.

MUNTZ, W. R. A. The development of phototaxis in the frog (*Rana temporaria*). *Journal of Experimental Biology*, 1963, **40**, 371–379. (a)

MUNTZ, W. R. A. Phototaxis and green rods in urodeles. *Nature*, 1963, **219**, 620. (b)

MUNTZ, W. R. A. Stimulus generalization following monocular training in the rat. *Journal of Comparative & Physiological Psychology*, 1963, **56**, 1003–1006. (c)

MYERS, R. E. Corpus callosum and visual gnosis. In J. F. Delafresnaye (Ed.), *Brain mechanisms and learning*. Oxford: Blackwell, 1961. Pp. 481–506.

NEEDHAM, J. G. Prior entry within a single sense department. *Journal of Experimental Psychology*, 1934, **17**, 400–411.

NEEDHAM, J. G. Some conditions of prior entry. *Journal of General Psychology*, 1936, **14**, 226–240.

NEFF, W. D. Neural mechanisms of auditory discrimination. In W. A. Rosenblith (Ed.), *Sensory Communication*. New York: Wiley, 1961. Pp. 259–278.

NEWELL, A., & SIMON, H. A. G.P.S. A program that simulates human thought. In E. A. Feigenbaum & J. Feldman (Eds.), *Computers and thought*. New York: McGraw-Hill, 1963. Pp. 279–293.

NEWMAN, F. L., & Baron, M. R. Stimulus generalization along the dimension of angularity. *Journal of Comparative & Physiological Psychology*, 1965, **60**, 59–63.

NEWMAN, F. L., & BENEFIELD, R. L. Stimulus control, cue utilization, and attention. Effects of discrimination training. *Journal of Comparative & Physiological Psychology*, 1968, **66**, 101–104.

NISSEN, H. W. Description of the learned response in discrimination behavior. *Psychological Review*, 1950, **59**, 121–137.

NISSEN, H. W., & JENKINS, W. O. Reduction and rivalry of cues in the discrimination behavior of chimpanzees. *Journal of Comparative Psychology*, 1943, **35**, 85–89.

NORTH, A. J. Performance during an extended series of discrimination reversals. *Journal of Comparative & Physiological Psychology*, 1950, **43**, 461–470.

NORTH, A. J. Acquired distinctiveness of form stimuli. *Journal of Comparative & Physiological Psychology*, 1959, **52**, 339–341.

NORTH, A. J. Discrimination reversal learning in a runway situation. *Journal of Comparative & Physiological Psychology*, 1962, **55**, 550–554.

NORTH, A. J., & CLAYTON, K. N. Irrelevant stimuli and degree of learning in discrimination learning and reversal. *Psychological Reports*, 1959, **5**, 405–408.

NORTH, A. J., & JEEVES, M. A. Interrelationships of successive and simultaneous discrimination. *Journal of Experimental Psychology*, 1956, **51**, 54–58.

NORTH, A. J., & LANG, P. Conditional discrimination in rats. *Journal of Genetic Psychology*, 1961, **98**, 113–118.

NORTH, A. J., & McDONALD, R. D. Discrimination learning as a function of the probability of reinforcement. *Journal of Comparative & Physiological Psychology*, 1959, **52**, 342–344.

North, A. J., Maller, O., & Hughes, C. L. Conditional discrimination and stimulus patterning. *Journal of Comparative & Physiological Psychology*, 1958, **51**, 711–715.

North, A. J., & Stimmel, D. T. Extinction of an instrumental response following a large number of reinforcements. *Psychological Reports*, 1960, **6**, 227–234.

Notterman, J. M. A study of some relations among aperiodic reinforcement, discrimination training, and secondary reinforcement. *Journal of Experimental Psychology*, 1951, **41**, 161–169.

O'Malley, J. J., Arnone, S. J., & Ziegenfus, F. Nondifferential reinforcement and discrimination learning. *Psychonomic Science*, 1969, **17**, 285–286.

Osgood, C. E. *Method and theory in experimental psychology*. New York: Oxford University Press, 1953.

Oswald, I. The human alpha rhythm and visual alertness. *Electroencephalography & Clinical Neurophysiology*, 1959, **11**, 601–602.

Paige, A. B., & McNamara, H. J. Secondary reinforcement and the discrimination hypothesis: the role of discrimination training. *Psychological Reports*, 1963, **13**, 679–686.

Parducci, A., & Polt, J. Correction vs noncorrection with changing reinforcement schedules. *Journal of Comparative & Physiological Psychology*, 1958, **51**, 492–495.

Paul, C. Effects of overlearning upon single habit reversal in rats. *Psychological Bulletin*, 1965, **63**, 65–72.

Paul, C. Effects of overtraining and two non-correction training procedures upon a brightness discrimination reversal. *Psychonomic Science*, 1966, **5**, 423–424.

Paul, C., & Havlena, J. Effects of overlearning and spatial delay of reinforcement upon a discrimination reversal. *Psychological Reports*, 1965, **16**, 79–83.

Paul, C., & Kesner, R. Effects of overlearning trials upon habit reversal under conditions of aversive stimulation. *Psychological Reports*, 1963, **13**, 361–363.

Pavlik, W. B., & Born, D. G. Partial reinforcement effects in selective learning. *Psychological Reports*, 1962, **11**, 575–590.

Pavlik, W. B., & Lehr, R. Strength of alternative responses and subsequent choices. *Journal of Experimental Psychology*, 1967, **74**, 562–573.

Pavlov, I. P. *Conditioned reflexes*. Oxford: Oxford University Press, 1927.

Perkins, C. C. Jr., Seymann, R. G., Levis, D. J., & Spencer, H. R., Jr. Factors affecting preference for signal-shock over shock-signal. *Journal of Experimental Psychology*, 1966, **72**, 190–196.

Perry, S. L., & Moore, J. W. The partial reinforcement effect sustained through blocks of continuous reinforcement in classical eyelid conditioning. *Journal of Experimental Psychology*, 1965, **69**, 158–161.

Peterson, L. R. Variable delayed reinforcement. *Journal of Comparative & Physiological Psychology*, 1956, **49**, 232–234.

Peterson, N. Effect of monochromatic rearing on the control of responding by wavelength. *Science*, 1962, **136**, 774–775.

Pillsbury, W. B. *Attention*. New York: Macmillan, 1908.

Pillsbury, W. B. "Fluctuations of Attention" and the Refractory Period. *Journal of Philosophy, Psychology and Scientific Method*, 1913, **10**, 181–185.

Pollack, I. The information of elementary auditory displays. *Journal of the Acoustical Society of America*, 1952, **24**, 745–749.

POLLACK, I. The information of elementary auditory displays, II. *Journal of the Acoustical Society of America*, 1953, **25**, 765–769.

POLLACK, I., & FICKS, L. Information of elementary multidimensional auditory displays. *Journal of the Acoustical Society of America*, 1954, **26**, 155–158.

POLYAK, S. L. *The retina*. Chicago: University of Chicago Press, 1941.

PREWITT, E. P. Number of preconditioning trials in sensory preconditioning using CER training. *Journal of Comparative & Physiological Psychology*, 1967, **64**, 360–362.

PROKASY, W. F. The acquisition of observing responses in the absence of differential external reinforcement. *Journal of Comparative & Physiological Psychology*, 1956, **49**, 131–134.

PROKASY, W. F., & HALL, J. F. Primary stimulus generalization. *Psychological Review*, 1963 **70**, 310–322.

PUBOLS, B. H., JR. The facilitation of visual and spatial discrimination reversal by over-learning. *Journal of Comparative & Physiological Psychology*, 1956, **49**, 243–248.

PUBOLS, B. H., JR. Serial reversal learning as a function of the number of trials per reversal. *Journal of Comparative & Physiological Psychology*, 1962, **55**, 66–68.

RASMUSSEN, G. L. The olivary peduncle and other fibre projections of the superior olivary complex. *Journal of Comparative Neurology*, 1946, **84**, 141–219.

RAZRAN, G. Stimulus generalization of conditioned responses. *Psychological Bulletin*, 1949, **46**, 337–365.

RAZRAN, G. Russian physiologists' psychology and American experimental psychology. *Psychological Bulletin*, 1965, **63**, 42–64.

REID, L. S. The development of noncontinuity behavior through continuity learning. *Journal of Experimental Psychology*, 1953, **46**, 107–112.

REID, R. L. Discrimination-reversal learning in pigeons. *Journal of Comparative & Physiological Psychology*, 1958, **51**, 716–720.

REINHOLD, D. B., & PERKINS, C. C., JR. Stimulus generalization following different methods of training. *Journal of Experimental Psychology*, 1955, **49**, 423–427.

RENNER, K. E. Delay of reinforcement: A historical review. *Psychological Bulletin*, 1964, **61**, 341–361.

RESCORLA, R. A. Probability of shock in the presence and absence of CS in fear conditioning. *Journal of Comparative & Physiological Psychology*, 1968, **66**, 1–5.

RESCORLA, R. A. Conditioned inhibition of fear. In N. J. Mackintosh & W. K. Honig (Eds.), *Fundamental issues in associative learning*. Halifax: Dalhousie University Press, 1969. Pp. 65–89.

RESCORLA, R. A., & Wagner, A. R. A theory of Pavlovian conditioning: variations in the effectiveness of reinforcement and nonreinforcement. In A. Black & W. F. Prokasy (Eds.), *Classical conditioning II*. New York: Appleton-Century-Crofts, 1970 (in press).

RESTLE, F. A theory of discrimination learning. *Psychological Review*, 1955, **62**, 11–19.

RESTLE, F. Discrimination of cues in mazes: a resolution of the "place-vs-response" question. *Psychological Review*, 1957, **64**, 217–228.

RESTLE, F. The selection of strategies in cue learning. *Psychological Review*, 1962, **69**, 329–343.

REYNOLDS, G. S. Attention in the pigeon. *Journal of the Experimental Analysis of Behavior*, 1961, **4**, 203–208.

RILEY, D. A., GOGGIN, J. P., & WRIGHT, D. C. Training level and cue separation as determiners of transposition and retention in rats. *Journal of Comparative & Physiological Psychology*, 1963, **56**, 1044–1049.

RILEY, D. A., RING, K., & THOMAS, J. The effect of stimulus comparison on discrimination learning and transposition. *Journal of Comparative & Physiological Psychology*, 1960, **53**, 415–421.

RIOPELLE, A. J., & CHINN, R. McC. Position habits and discrimination learning by monkeys. *Journal of Comparative & Physiological Psychology*, 1961, **54**, 178–180.

RIOPELLE, A. J., & COPELAN, E. L. Discrimination reversal to a sign. *Journal of Experimental Psychology*, 1954, **48**, 143–145.

RITCHIE, B. F., AESCHLIMAN, B., & PIERCE, P. Studies in spatial learning: VIII. Place performance and the acquisition of place dispositions. *Journal of Comparative & Physiological Psychology*, 1950, **43**, 73–85.

RITCHIE, B. F., EBELING, E., & ROTH, W. Evidence for continuity in the discrimination of vertical and horizontal patterns. *Journal of Comparative & Physiological Psychology*, 1950, **43**, 168–180.

ROBERTS, W. A. Learning and motivation in the immature rat. *American Journal of Psychology*, 1966, **79**, 3–23.

ROBBINS, D., & WEINSTOCK, S. The effects of nonreinforcement on subsequent running behavior under continuous reinforcement. *Psychonomic Science*, 1967, **9**, 7–8.

ROSENBLATT, F. Two theorems of statistical separability in perception. In *Proceedings of a symposium on the mechanization of thought processes.* London: Her Majesty's Stationery Office, 1959. Pp. 419–450.

ROSS, L. E. The effect of equal reinforcement of the positive and negative discriminanda of a learned discrimination. *Journal of Comparative & Physiological Psychology*, 1962, **55**, 260–266. (a)

ROSS, L. E. The response to previous discriminanda during the learning of a new problem. *Journal of Comparative & Physiological Psychology*, 1962, **55**, 944–946. (b)

ROSS, R. R. Positive and negative partial reinforcement extinction effects carried through continuous reinforcement, changed motivation and changed response. *Journal of Experimental Psychology*, 1964, **68**, 492–502.

ROSVOLD, H. E., & MISHKIN, M. Non-sensory effects of frontal lesions on discrimination learning and performance. In J. F. DelaFresnaye (Ed.), *Brain mechanisms and learning.* Oxford: Blackwells, 1961. Pp. 555–577.

RUDOLPH, R. L., HONIG, W. K., & GERRY, J. E. Effects of monochromatic rearing on the acquisition of stimulus control. *Journal of Comparative & Physiological Psychology*, 1969, **67**, 50–58.

SALTZMAN, I. J. Maze learning in the absence of primary reinforcement: A study of secondary reinforcement. *Journal of Comparative & Physiological Psychology*, 1949, **42**, 161–173.

SCHADE, A. F., & BITTERMAN, M. E. The relative difficulty of reversal and dimensional shifting as a function of overlearning. *Psychonomic Science*, 1965, **3**, 283–284.

SCHADE, A. F., & BITTERMAN, M. E. Improvement in habit reversal as related to dimensional set. *Journal of Comparative & Physiological Psychology*, 1966, **62**, 43–48.

SCHARLOCK, D. P. The role of extra maze cues in place and response learning. *Journal of Experimental Psychology*, 1955, **50**, 249–254.

SCHIPPER, L. An analysis of information transmitted to human observers with auditory signals as a function of number of stimuli and stimulus intensity interval size. Unpublished doctoral dissertation, University of Wisconsin, 1953.

SCHUSTERMAN, R. J. Transfer effects of successive discrimination reversal training in chimpanzees. *Science*, 1962, **137**, 422–423.

SCHUSTERMAN, R. J. Serial discrimination-reversal learning with and without errors by the California sea lion. *Journal of the Experimental Analysis of Behavior*, 1966, **9**, 593–600.

SECHENOV, I. M. Reflexes of the brain. *Meditsinskiy Vestnik*, 1863, **3**, 481–498.

SELIGMAN, M. E. P. Chronic fear produced by unpredictable electric shock. *Journal of Comparative & Physiological Psychology*, 1968, **66**, 402–411.

SETTERINGTON, R. G., & BISHOP, H. E. Habit reversal improvement in the fish. *Psychonomic Science*, 1967, **7**, 41–42.

SEWARD, J. P., PEREBOOM, A. C., BUTLER, B., & JONES, R. B. The role of prefeeding in an apparent frustration effect. *Journal of Experimental Psychology*, 1957, **54**, 445–450.

SGRO, J. A., DYAL, J. A., & ANASTASIO, E. J. Effects of constant delay of reinforcement on acquisition asymptote and resistance to extinction. *Journal of Experimental Psychology*, 1967, **73**, 634–636.

SGRO, J. A., & WEINSTOCK, S. Effects of delay on subsequent running under immediate reinforcement. *Journal of Experimental Psychology*, 1963, **66**, 260–263.

SHARPLESS, S., & JASPER, H. Habituation of the arousal reaction. *Brain*, 1956, **79**, 655–680.

SHEFFIELD, F. D. Relation between classical conditioning and instrumental learning. In W. F. Prokasy (Ed.), *Classical Conditioning*. New York: Appleton-Century-Crofts., 1965. Pp. 302–322.

SHEFFIELD, V. F. Extinction as a function of partial reinforcement and distribution of practice. *Journal of Experimental Psychology*, 1949, **39**, 511–526.

SHEFFIELD, V. F. Resistance to extinction as a function of the distribution of extinction trials. *Journal of Experimental Psychology*, 1950, **40**, 305–313.

SHELDON, M. H. Response learning. *Nature*, 1964, **202**, 1141–1142.

SHELDON, M. H. Some effects of discriminable goal-box conditions on the learning of a successive discrimination. *Quarterly Journal of Experimental Psychology*, 1967, **19**, 319–326.

SHEPP, B. E., & EIMAS, P. D. Intradimensional and extradimensional shifts in the rat. *Journal of Comparative & Physiological Psychology*, 1964, **57**, 357–361.

SHEPP, B. E., & SCHRIER, A. M. Consecutive intradimensional and extradimensional shifts in monkeys. *Journal of Comparative & Physiological Psychology*, 1969, **67**, 199–203.

SHIMP, C. P. Probabilistically reinforced choice behavior in pigeons. *Journal of the Experimental Analysis of Behavior*, 1966, **9**, 443–455.

SIDMAN, M. Two temporal parameters of the maintenace of avoidance behavior by the white rat. *Journal of Comparative & Physiological Psychology*, 1953, **46**, 253–261.

SIEGEL, P. S. The role of absolute initial response strength in simple trial-and-error learning. *Journal of Experimental Psychology*, 1945, **35**, 199–205.

SIEGEL, S. Overtraining and transfer processes. *Journal of Comparative & Physiological Psychology*, 1967, **64**, 471–477.

SIEGEL, S., & WAGNER, A. R. Extended acquisition training and resistance to extinction. *Journal of Experimental Psychology*, 1963, **66**, 308–310.

Silver, C. A., & Meyer, D. R. Temporal factors in sensory preconditioning. *Journal of Comparative & Physiological Psychology*, 1954, **47**, 51–59.

Simmons, F. B., & Beatty, D. L. A theory of middle ear muscle function at moderate sound levels. *Science*, 1962, **138**, 590–592.

Singer, B., Zentall, T., & Riley, D. A. Stimulus generalization and the easy-to-hard effect. *Journal of Comparative & Physiological Psychology*, 1969, **69**, 528–535.

Skinner, B. F. The extinction of chained reflexes. *Proceedings of the National Academy of Science*, 1934, **20**, 234–237.

Skinner, B. F. A case history in scientific method. In S. Koch (Ed.), *Psychology: A study of a science.* Vol. 2. New York: McGraw-Hill, 1959. Pp. 359–379.

Smith, S. M., Brown, H. P., Toman, J. E. P., & Goodman, L. S. The lack of cerebral effects of *d*-tubocurarine. *Anesthesiology*, 1947, **8**, 1–14.

Sokolov, E. N. Neuronal models and the orienting reflex. In M. A. B. Brazier, *The central nervous system and behavior.* New York: Josiah Macy, Jr. Foundation, 1960. Pp. 187–276.

Solomon, R. L., & Turner, L. H. Discriminative classical conditioning in dogs paralysed by curare can later control discriminative avoidance responses in the normal state. *Psychological Review*, 1962, **69**, 202–219.

Solomon, S. Effects of variations in rearing, drive level, and training procedure on performance in probability learning tasks. *Psychological Reports*, 1962, **10**, 679–689.

Spear, N. E., Hill, W. F., & O'Sullivan, D. J. Acquisition and extinction after initial trials without reward. *Journal of Experimental Psychology*, 1965, **69**, 25–29.

Spear, N. E., & Pavlik, W. B. Percentage of reinforcement and reward magnitude effects in a T-maze: between and within subjects. *Journal of Experimental Psychology*, 1966, **71**, 521–528.

Spear, N. E., & Spitzner, J. H. Simultaneous and successive contrast effects of reward magnitude in selective learning. *Psychological Monographs*, 1966, **80** (10), 1–31.

Spence, K. W. The nature of discrimination learning in animals. *Psychological Review*, 1936, **43**, 427–449.

Spence, K. W. The differential response in animals to stimuli varying within a single dimension. *Psychological Review*, 1937, **44**, 430–444.

Spence, K. W. Continuous versus non-continuous interpretations of discrimination learning. *Psychological Review*, 1940, **47**, 271–288.

Spence, K. W. An experimental test of continuity and non-continuity theories of discrimination learning. *Journal of Experimental Psychology*, 1945, **35**, 253–266.

Spence, K. W. The nature of the response in discrimination learning. *Psychological Review*, 1952, **59**, 89–93.

Spence, K. W. *Behavior theory and conditioning.* New Haven: Yale University Press, 1956.

Spence, K. W. *Behavior theory and learning.* Englewood Cliffs: Prentice-Hall, 1960.

Spence, K. W. Cognitive and drive factors in the extinction of the conditioned eye-blink in human subjects. *Psychological Review*, 1966, **73**, 445–458.

Spence, K. W., & Lippitt, R. Latent learning of a simple maze problem with relevant needs satiated. *Psychological Bulletin*, 1940, **37**, 429. (Abstract).

Spence, K. W., & Lippitt, R. An experimental test of the sign-gestalt theory of trial-and-error learning. *Journal of Experimental Psychology*, 1946, **36**, 491–502.

Spence, K. W., Platt, J. R., & Matsumoto, R. Intertrial reinforcement and the partial reinforcement effect as a function of number of training trials. *Psychonomic Science*, 1965, **3**, 205–206.

SPENCE, K. W., RUTLEDGE, E. F., & TALBOTT, J. H. Effect of number of acquisition trials and the presence or absence of the UCS on extinction of the eyelid CR. *Journal of Experimental Psychology*, 1963, **66**, 286–291.

SPERLING, S. E. Reversal learning and resistance to extinction: A review of the rat literature. *Psychological Bulletin*, 1965, **63**, 281–297.

SPERLING, S. E. (1967). Position responding and latency of choice in simultaneous discrimination learning. *Journal of Experimental Psychology*, 1967, **74**, 333–341.

SPERLING, S. E. The ORE in simultaneous and differential reversal: the acquisition task, the acquisition criterion and the reversal task. *Journal of Experimental Psychology*, 1970, **84**, 349–360.

SPIVEY, J. E. Resistance to extinction as a function of number of NR transitions and percentage of reinforcement. *Journal of Experimental Psychology*, 1967, **75**, 43–48.

SPIVEY, J. E., & HESS, D. T. Effect of partial reinforcement trial sequences on extinction performance. *Psychonomic Science*, 1968, **10**, 375–376.

SPIVEY, J. E., HESS, D. T., & APONTE, J. F. Modification of reinforcement aftereffects and level of training. *Psychological Reports*, 1968, **22**, 35–42.

SPIVEY, J. E., HESS, D. T., & Black, D. Influence of partial reinforcement pattern and intertrial reinforcement on extinction performance following abbreviated training. *Psychonomic Science*, 1968, **10**, 377–378.

SPIVEY, J. E., HESS, D. T., & KLEMIC, J. Extinction performance as a function of N–R transitions and intertrial reinforcement with extended partial reinforcement training. *Psychological Reports*, 1968, **22**, 765–771.

STANLEY, W. C., & AAMODT, M. S. Force of responding during extinction as a function of force requirement during conditioning. *Journal of Comparative & Physiological Psychology*, 1954, **47**, 462–464.

STEARNS, E. M., & BITTERMAN, M. E. A comparison of key-pecking with an ingestive technique for the study of discriminative learning in pigeons. *American Journal of Psychology*, 1965, **78**, 48–56.

STERNBERG, S. Stochastic learning theory. In R. D. Luce, R. R. Bush, & E. Galanter (Eds.), *Handbook of mathematical psychology*. Vol. 2. New York: Wiley, 1963. Pp. 1–120.

STETTNER, L. J. Effect of prior reversal and elimination of inhibition on the persistance of a discrimination despite subsequent equal reinforcement of the discriminanda. *Journal of Comparative & Physiological Psychology*, 1965, **60**, 262–264.

STETTNER, L. J., SCHULZ, W. J., & LEVY, A. Successive reversal learning in the bob-white quail (*Colinus virginianus*). *Animal Behaviour*, 1967, **15**, 1–5.

STRETCH, R. G., MCGONIGLE, B., & Morton, A. Serial position-reversal learning in the rat: trials/problem and the intertrial interval. *Journal of Comparative & Physiological Psychology*, 1964, **57**, 461–463.

STRETCH, R. G., MCGONIGLE, B., & RODGER, R. S. Serial position-reversal learning in the rat: a preliminary analysis of training criteria. *Journal of Comparative & Physiological Psychology*, 1963, **56**, 719–722.

SURRIDGE, C. T., & AMSEL, A. Performance under a single alternation schedule of reinforcement at 24-hour intertrial interval. *Psychonomic Science*, 1965, **3**, 131–132. (a)

SURRIDGE, C. T., & AMSEL, A. A patterning effect that seems unrelated to after-effects from reward and non-reward. *Psychonomic Science*, 1965, **3**, 373–374. (b)

Surridge, C. T., & Amsel, A. Confinement duration on rewarded and non-rewarded trials and patterning at 24-hour ITI. *Psychonomic Science*, 1968, **10**, 107–108.

Surridge, C. T., Boenhert, J., & Amsel, A. Effect of interpolated extinction on the re-acquisition of partially and continuously rewarded responses. *Journal of Experimental Psychology*, 1966, **72**, 564–570.

Sutherland, N. S. Visual discrimination of orientation by *Octopus*. *British Journal of Psychology*, 1957, **48**, 55–71. (a)

Sutherland, N. S. Visual discrimination of orientation and shape by *Octopus*. *Nature*, 1957, **179**, 11–13. (b)

Sutherland, N. S. Stimulus analysing mechanisms. In *Proceedings of a symposium on the mechanization of thought processes*. Vol. 2. London: Her Majesty's Stationery Office, 1959. Pp. 575–609.

Sutherland, N. S. The methods and findings of experiments on the visual discrimination of shape by animals. *Quarterly Journal of Experimental Psychology Monographs*, 1961, **1**, 1–68. (a)

Sutherland, N. S. Visual discrimination of horizontal and vertical rectangles by rats in a new discrimination training apparatus. *Quarterly Journal of Experimental Psychology*, 1961, **13**, 117–121. (b)

Sutherland, N. S. Cat's ability to discriminate oblique rectangles. *Science*, 1963, **139**, 209–210. (a)

Sutherland, N. S. Visual acuity and discrimination of stripe widths in *Octopus vulgaris* Lamarck. *Pubblicazione della Stazione Zoologica Napoli*, 1963, **33**, 92–109. (b)

Sutherland, N. S. The learning of discrimination by animals. *Endeavour*, 1964, **23**, 69–78.

Sutherland, N. S. Successive reversals involving two cues. *Quarterly Journal of Experimental Psychology*, 1966, **18**, 97–102. (a)

Sutherland, N. S. Partial reinforcement and breadth of learning. *Quarterly Journal of Experimental Psychology*, 1966, **18**, 289–302. (b)

Sutherland, N. S. Outlines of a theory of pattern recognition in animals and man. *Proceedings of the Royal Society* (*London*), *Ser. B*, 1968, **171**, 297–317.

Sutherland, N. S. Shape discrimination in rat, octopus and goldfish: a comparative study. *Journal of Comparative & Physiological Psychology*, 1969, **67**, 160–176.

Sutherland, N. S., & Andelman, L. Learning with one and two cues. *Psychonomic Science*, 1967, **7**, 107–108.

Sutherland, N. S., & Andelman, L. Effects of overtraining on intra- and extra-dimensional shifts. *Psychonomic Science*, 1969, **15**, 253–254.

Sutherland, N. S., Carr, A. E., & Mackintosh, J. A. Visual discrimination of open and closed shapes by rats. I. Training. *Quarterly Journal of Experimental Psychology*, 1962, **14**, 129–139.

Sutherland, N. S., & Holgate, V. Two-cue discrimination learning in rats. *Journal of Comparative & Physiological Psychology*, 1966, **61**, 198–207.

Sutherland, N. S., & Mackintosh, N. J. The learning of an optional extradimensional reversal shift problem by rats. *Psychonomic Science*, 1966, **5**, 343–344.

Sutherland, N. S. Mackintosh, N. J., & Mackintosh, J. Simultaneous discrimination training of *Octopus* and transfer of discrimination along a continuum. *Journal of Comparative & Physiological Psychology*, 1963, **56**, 150–156.

SUTHERLAND, N. S., MACKINTOSH, N. J., & MACKINTOSH, J. Shape and size discrimination in *Octopus*: the effects of pretraining along different dimensions. *Journal of Genetic Psychology*, 1965, **106**, 1–10.

SUTHERLAND, N. S., MACKINTOSH, N. J., & WOLFE, J. B. Extinction as a function of the order of partial and consistent reinforcement. *Journal of Experimental Psychology*, 1965, **69**, 56–59.

SWITALSKI, R. W., LYONS, J., & THOMAS, D. R. Effects of interdimensional training on stimulus generalization. *Journal of Experimental Psychology*, 1966, **72**, 661–666.

TAUB, E., & BERMAN, A. J. Movement and learning in the absence of sensory feedback. In S. J. Freedman (Ed.), *The neuropsychology of spatially oriented behavior*. Illinois: Dorsey Press, 1968. Pp. 172–193.

TAYLOR, R. W. The effect of certain stimuli upon the attention wave. *American Journal of Psychology*, 1901, **12**, 335–345.

TEAS, R. C., & BITTERMAN, M. E. Perceptual organization in the rat. *Psychological Review*, 1952, **59**, 130–140.

TERRACE, H. S. Discrimination learning with and without "errors". *Journal of the Experimental Analysis of Behavior*, 1963, **6**, 1–27. (a)

TERRACE, H. S. Errorless transfer of a discrimination across two continua. *Journal of the Experimental Analysis of Behavior*, 1963, **6**, 223–232. (b)

TERRACE, H. S. Behavioral contrast and the peak shift: effects of extended discrimination training. *Journal of the Experimental Analysis of Behavior*, 1966, **9**, 613–617.

THEIOS, J. The partial reinforcement effect sustained through blocks of continuous reinforcement. *Journal of Experimental Psychology*, 1962, **64**, 1–6.

THEIOS, J. The mathematical structure of reversal learning in a shock escape T-maze; overtraining and successive reversals. *Journal of Mathematical Psychology*, 1965, **2**, 26–52.

THEIOS, J., & BLOSSER, D. An incentive model for the overlearning reversal effect. *Psychonomic Science*, 1965, **2**, 37–38. (a)

THEIOS, J., & BLOSSER, D. The overlearning reversal effect and magnitude of reward. *Journal of Comparative & Physiological Psychology*, 1965, **59**, 252–256. (b)

THEIOS, J., & BRELSFORD, J. Overlearning-extinction effect as an incentive phenomenon. *Journal of Experimental Psychology*, 1964, **67**, 463–467.

THEIOS, J., & MCGINNIS, R. W. Partial reinforcement before and after continuous reinforcement. *Journal of Experimental Psychology*, 1967, **73**, 479–481.

THEIOS, J., & POLSON, P. Instrumental and goal responses in nonresponse partial reinforcement. *Journal of Comparative & Physiological Psychology*, 1962, **55**, 987–991.

THOMAS, D. R. The use of operant conditioning techniques to investigate perceptual processes in animals. In R. Gilbert & N. S. Sutherland (Eds.), *Animal discrimination learning*. London: Academic Press, 1969. Pp. 1–33.

THOMAS, D. R., & SWITALSKI, R. W. Comparison of stimulus generalization following variable-ratio and variable-interval training. *Journal of Experimental Psychology*, 1966, **71**, 236–240.

THOMAS, D. R., FREEMAN, F., SVINICKI, J. G., BURR, D. E. S., & LYONS, J. Effects of extradimensional training on stimulus generalization. *Journal of Experimental Psychology, Monograph*, 1970, **83**, 1–21.

THOMAS, E., & WAGNER, A. R. Partial reinforcement of the classically conditioned eyelid response in the rabbit. *Journal of Comparative & Physiological Psychology*, 1964, **58**, 157–158.

THOMPSON, C. P., & VAN HOESEN, G. W. Compound conditioning: effects of component intensity on acquisition and extinction. *Journal of Comparative & Physiological Psychology*, 1967, **64**, 128–132.

THOMPSON, M. E. Stimulus generalization of an instrumental response learned under distributed practice. *Psychological Reports*, 1962, **11**, 471–476.

THOMPSON, R. F. Primary stimulus generalization as a function of acquisition level in the cat. *Journal of Comparative & Physiological Psychology*, 1958, **51**, 601–606.

THOMPSON, R. F. Effect of acquisition level upon the magnitude of stimulus generalization across sensory modality. *Journal of Comparative & Physiological Psychology*, 1959, **52** 183–185.

TIGHE, L. S., & TIGHE, T. J. Overtraining and discrimination shift behavior in children. *Psychonomic Science*, 1965, **2**, 365–366.

TIGHE, T. J. Reversal and nonreversal shifts in monkeys. *Journal of Comparative & Physiological Psychology*, 1964, **58**, 324–326.

TIGHE, T. J. The effect of overtraining on reversal and extradimensional shifts. *Journal of Experimental Psychology*, 1965, **70**, 13–17.

TIGHE, T. J., BROWN, P. L., & YOUNGS, E. A. The effect of overtraining on the shift behavior of albino rats. *Psychonomic Science*, 1965, **2**, 141–142.

TIGHE, T. J., & TIGHE, L. S. Overtraining and optional shift behavior in rats and children. *Journal of Comparative & Physiological Psychology*, 1966, **62**, 49–54.

TITCHENER, E. B. *The psychology of feeling and attention.* New York: Macmillan, 1908.

TOLMAN, E. C. *Purposive behavior in animals and men.* New York: Appleton-Century-Crofts, 1932.

TOLMAN, E. C., RITCHIE, B. F., & KALISH, D. Studies in spatial learning. I. Orientation and the short-cut. *Journal of Experimental Psychology*, 1946, **36**, 13–24.

TRABASSO, T. R. Stimulus emphasis and all-or-none learning in concept identification. *Journal of Experimental Psychology*, 1963, **65**, 398–406.

TRABASSO, T. R., & BOWER, G. H. *Attention in learning: Theory and research.* New York: Wiley, 1968.

TRABASSO, T., DEUTSCH, J. A., & GELMAN, R. Attention and discrimination learning of young children. *Journal of Experimental Child Psychology*, 1966, **4**, 9–19.

TRACY, W. K. Wavelength generalization and preference in monochromatically reared ducklings. *Journal of the Experimental Analysis of Behavior*, 1970, **13**, 163–178.

TRAPOLD, M. A., & DOREN, D. G. Effect of noncontingent partial reinforcement on the resistance to extinction of a runway response. *Journal of Experimental Psychology*, 1966, **71**, 429–431.

TRAPOLD, M. A., & HOLDEN, D. Noncontingent partial reinforcement of running: a replication. *Psychonomic Science*, 1966, **5**, 449–450.

TREISMAN, A. M. Verbal cues, language, and meaning in selective attention. *American Journal of Psychology*, 1964, **77**, 206–219. (a)

TREISMAN, A. M. Selective attention in man. *British Medical Bulletin*, 1964, **20**, 12–16. (b)

TREISMAN, A., & GEFFEN, G. Selective attention: perception or response? *Quarterly Journal of Experimental Psychology*, 1967, **19**, 1–17.

TROTTER, J. R. The physical properties of bar-pressing behaviour and the problem of reactive inhibition. *Quarterly Journal of Experimental Psychology*, 1956, **8**, 97–106.

TSANG, Y-C. The functions of the visual area of the cerebral cortex of the rat in the learning and retention of the maze. *Comparative Psychology Monographs*, 1934, **10**, (4) 1–56.

TURNER, C. Models of Discrimination Learning. Unpublished doctoral thesis, Oxford University, 1968.

TURNER, C., & MACKINTOSH, N. J. Continuity theory revisited: comments on Wolford and Bower. *Psychological Review*, 1970, in press.

TURRISI, F. D., SHEPP, B. E., & EIMAS, P. D. Intra- and extra-dimensional shifts with constant-and variable-irrelevant dimensions in the rat. *Psychonomic Science*, 1969, **14**, 19–20.

TYLER, D. W., WORTZ, E. C., & BITTERMAN, M. E. The effect of random and alternating partial reinforcement on resistance to extinction in the rat. *American Journal of Psychology*, 1953, **66**, 37–65.

UHL, C. N. Two-choice probability learning in the rat as a function of incentive, probability of reinforcement, and training procedure. *Journal of Experimental Psychology*, 1963, **66**, 443–449.

UHL, C. N. Effects of overtraining on reversal and non-reversal discrimination shifts in a free operant situation. *Perceptual and Motor Skills*, 1964, **19**, 927–934.

UHL, C. N. Persistence in punishment and extinction testing as a function of percentages of punishment and reward in training. *Psychonomic Science*, 1967, **8**, 193–194.

UHL, C. N., PARKER, B. K., & WOOTON, P. B. Overtraining effects on reversal and nonreversal discrimination shifts with a continuously reinforced free operant. *Perceptual and Motor Skills*, 1967, **24**, 75–82.

UHL, C. N., & YOUNG, A. G. Resistance to extinction as a function of incentive, percentage of reinforcement, and number of nonreinforced trials. *Journal of Experimental Psychology*, 1967, **73**, 556–564.

UHL, N. P. Intradimensional and extradimensional shifts as a function of amount of training and similarity between training and shift stimuli. *Journal of Experimental Psychology*, 1966, **72**, 429–433.

UNDERWOOD, B. J. Interference and forgetting. *Psychological Review*, 1957, **64**, 49–60.

UTTLEY, A. M. Temporal and spatial patterns in a conditioned probability machine. In C. E. Shannon, & J. McCarthy (Eds.), *Automata studies*. Princeton: Princeton University Press, 1956. Pp. 277–285.

VARDARIS, R. M., & FITZGERALD, R. D. Effects of partial reinforcement on a classically conditioned eyeblink response in dogs. *Journal of Comparative & Physiological Psychology*, 1969, **67**, 531–534.

VINCENT, S. B. The white rat and the maze problem. II. The introduction of an olfactory control. *Journal of Animal Behavior*, 1915, **5**, 140–157.

VOM SAAL, W. Blocking the acquisition of stimulus control in operant discrimination learning. Unpublished Master's Thesis, McMaster University, 1967.

VOM SAAL, W., & JENKINS, H. M. Blocking the development of stimulus control. *Learning & Motivation*, 1970, **1**, 52–64.

WAGNER, A. R. The role of reinforcement and nonreinforcement in an "apparent frustration effect." *Journal of Experimental Psychology*, 1959, **57**, 130–136.

WAGNER, A. R. Effects of amount and percentage of reinforcement and number of acquisition trials on conditioning and extinction. *Journal of Experimental Psychology*, 1961, **62**, 234–242.

Wagner, A. R. Sodium amytal and partially reinforced runway performance. *Journal of Experimental Psychology*, 1963, **65**, 474–477. (a)

Wagner, A. R. Conditioned frustration as a learned drive. *Journal of Experimental Psychology*, 1963, **66**, 142–148. (b)

Wagner, A. R. Frustration and punishment. In R. N. Haber (Ed.), *Current research in motivation.* New York: Holt, 1966. Pp. 229–239.

Wagner, A. R. Incidental stimuli and discrimination learning. In R. Gilbert & N. S. Sutherland (Eds.), *Animal discrimination learning.* London: Academic Press, 1969. Pp. 83–111. (a)

Wagner, A. R. Stimulus validity and stimulus selection in associative learning. In N. J. Mackintosh & W. K. Honig (Eds.), *Fundamental issues in associative learning.* Halifax: Dalhousie University Press, 1969. Pp. 90–122. (b)

Wagner, A. R., Logan, F. A., Haberlandt, K., & Price, T. Stimulus selection in animal discrimination learning. *Journal of Experimental Psychology*, 1968, **76**, 171–180.

Wagner, A. R., Siegel, S., Thomas, E., & Ellison, G. D. Reinforcement history and the extinction of a conditioned salivary response. *Journal of Comparative & Physiological Psychology*, 1964, **58**, 354–358.

Walk, R. D., & Gibson, E. J. A comparative and analytical study of visual depth perception. *Psychological Monographs*, 1961, **75** (15), 1–44.

Wall, P. D., Freeman, J., & Major, D. Dorsal horn cells in spinal and in freely moving rats. *Experimental Neurology*, 1967, **19**, 519–529.

Waller, T. G. The effect of magnitude of reward on acquisition and extinction in spatial and brightness discrimination tasks. *Journal of Comparative & Physiological Psychology*, 1968, **66**, 122–127.

Warren, J. M. Additivity of cues in visual pattern discriminations by monkeys. *Journal of Comparative & Physiological Psychology*, 1953, **46**, 484–486.

Warren, J. M. Reversal learning by paradise fish (*Macropodus opercularis*). *Journal of Comparative & Physiological Psychology*, 1960, **53**, 376–378. (a)

Warren, J. M. Discrimination reversal learning by cats. *Journal of Genetic Psychology*, 1960, **97**, 317–327. (b)

Warren, J. M. Primate learning in comparative perspective. In A. M. Schrier, H. F. Harlow, & F. Stollnitz (Eds.), *Behavior of nonhuman primates: Modern research trends.* Vol. 1. New York: Academic Press, 1965. Pp. 249–281.

Warren, J. M. Reversal learning and the formation of learning sets by cats and rhesus monkeys. *Journal of Comparative & Physiological Psychology*, 1966, **61**, 421–428.

Warren, J. M., & Baron, A. The formation of learning sets by cats. *Journal of Comparative & Physiological Psychology*, 1956, **52**, 336–338.

Warren, J. M., & McGonigle, B. Attention theory and discrimination learning. In R. Gilbert & N. S. Sutherland (Eds.), *Animal discrimination learning.* London: Academic Press, 1969. Pp. 113–136.

Watson, A. J. Some questions concerning the explanation of learning in animals. In *Proceedings of a symposium on the mechanization of thought processes.* London: Her Majesty's Stationery Office, 1959, Pp. 691–720.

WATSON, J. B. Psychology as a behaviorist views it. *Psychological Review*, 1913, **20**, 158–177.

WEINSTOCK, S. Resistance to extinction of a running response following partial reinforcement under widely spaced trials. *Journal of Comparative & Physiological Psychology*, 1954, **47**, 318–322.

WEINSTOCK, S. Acquisition and extinction of partially reinforced running response at a 24-hour intertrial interval. *Journal of Experimental Psychology*, 1958, **56**, 151–158.

WEINSTOCK, S., NORTH, A. J., BRODY, A. L., & LO GUIDICE, J. Probability learning in a T-maze with noncorrection. *Journal of Comparative & Physiological Psychology*, 1965, **60**, 76–81.

WEISS, P., & BITTERMAN, M. E. Response selection in discrimination learning. *Psychological Review*, 1951, **58**, 185–195.

WEITZMAN, R. A. Positional matching in rats and fish. *Journal of Comparative & Physiological Psychology*, 1967, **63**, 54–59.

WELFORD, A. T. The psychological refractory period, and the timing of high speed performance—a review and a theory. *British Journal of Psychology*, 1952, **43**, 2–19.

WELLS, M. J. Factors affecting reactions to *Mysis* by newly hatched *Sepia*. *Behaviour*, 1958, **8**, 96–111.

WEYANT, R. G. Reversal learning in rats as a function of the type of discrimination and the criterion of learning. *Animal Behaviour*, 1966, **14**, 480–484.

WHITE, B. N., & SPIKER, C. C. The effect of stimulus similarity on amount of cue-position patterning in discrimination problems. *Journal of Experimental Psychology*, 1960, **59**, 131–136.

WICKELGREN, W. O. Effect of state of arousal on click-evoked responses in cats. *Journal of Neurophysiology*, 1968, **31**, 757–768. (a)

WICKELGREN, W. O. Effect of acoustic habituation on click-evoked responses in cats. *Journal of Neurophysiology*, 1968, **31**, 777–784. (b)

WIKE, E. L., KINTSCH, W., & GUTEKINST, R. The effects of variable drive, reward and response upon instrumental performance. *Journal of Comparative & Physiological Psychology*, 1959, **52**, 403–407.

WIKE, E. L., & MCNAMARA, H. J. The effects of percentage of partially delayed reinforcement in the acquisition and extinction of an instrumental response. *Journal of Comparative & Physiological Psychology*, 1957, **50**, 348–351.

WILCOXON, H. C. Abnormal fixation and learning. *Journal of Experimental Psychology*, 1952, **44**, 324–333.

WILLIAMS, D. I. Constant irrelevant cue learning in the pigeon. *Animal Behaviour*, 1967, **15**, 229–230. (a)

WILLIAMS, D. I. The overtraining reversal effect in the pigeon. *Psychonomic Science*, 1967, **7**, 261–262. (b)

WILSON, J. J. Level of training and goal-box movements as parameters of the partial reinforcement effect. *Journal of Comparative & Physiological Psychology*, 1964, **57**, 211–213.

WILSON, W., WEISS, E. J., & AMSEL, A. Two tests of the Sheffield hypothesis concerning resistance to extinction, partial reinforcement and distribution of practice. *Journal of Experimental Psychology*, 1955, **50**, 51–60.

WISE, L. M. Supplementary report: The Weinstock partial reinforcement effect and habit reversal. *Journal of Experimental Psychology*, 1962, **64**, 647–648.

WITTE, R. S. A stimulus-trace hypothesis for statistical learning theory. *Journal of Experimental Psychology*, 1959, **57**, 273–283.

WODINSKY, J., & BITTERMAN, M. E. Discrimination-reversal in the fish. *American Journal of Psychology*, 1957, **70**, 569–576.

WODINSKY, J., VARLEY, M. A., & BITTERMAN, M. E. Situational determinants of the relative difficulty of simultaneous and successive discrimination. *Journal of Comparative & Physiological Psychology*, 1954, **47**, 337–340.

WOLFORD, G., & BOWER, G. H. Continuity theory revisited: rejected for the wrong reasons? *Psychological Review*, 1969, **76**, 515–518.

WORDEN, F. G., & MARSH, J. T. Amplitude changes of auditory potentials evoked at cochlear nucleus during acoustic habituation. *Electroencephalography & Clinical Neurophysiology*, 1963, **15**, 866–881.

WORTZ, E. C., & BITTERMAN, M. E. On the effect of an irrelevant relation. *American Journal of Psychology*, 1953, **66**, 491–493.

WRIGHT, R. L. D. Motivational effects in probability learning. *Psychonomic Science*, 1967, **7**, 329–330.

YAMAGUCHI, H. G. The effect of continuous, partial and varied magnitude reinforcement on acquisition and extinction. *Journal of Experimental Psychology*, 1961, **61**, 319–321.

YERKES, R. M., & DODSON, J. D. The relation of strength of stimulus to rapidity of habit-formation. *Journal of Comparative Neurology & Psychology*, 1908, **18**, 458–482.

YOUNG, J. Z. The visual system of *Octopus*. (1) Regularities in the retina and optic lobe of *Octopus* in relation to form discrimination. *Nature*, 1960, **186**, 836–839.

YOUNG, J. Z. Repeated reversal of training in *Octopus*. *Quarterly Journal of Experimental Psychology*, 1962, **14**, 206–222.

ZANGWILL, O. L. A study of the significance of attitude in recognition. *British Journal of Psychology*, 1937, **28**, 12–17.

ZEAMAN, D. Response latency as a function of the amount of reinforcement. *Journal of Experimental Psychology*, 1949, **39**, 466–483.

ZEAMAN, D., and HOUSE, B. J. The role of attention in retardate discrimination learning. In N. R. Ellis (Ed.), *Handbook of mental deficiency: Psychological theory and research*. New York: McGraw-Hill, 1963. Pp. 159–223.

ZEIGLER, H. P. Learning-set formation in pigeons. *Journal of Comparative & Physiological Psychology*, 1961, **54**, 252–254.

ZEILER, M. D., & PAUL, B. J. Intra-pair similarity as a determinant of component and configuration discrimination. *American Journal of Psychology*, 1965, **78**, 476–480.

ZIMBARDO, P. G., & MONTGOMERY, K. C. The relative strength of consummatory responses in hunger, thirst and exploratory drive. *Journal of Comparative & Physiological Psychology*, 1957, **50**, 504–508.

# Author Index

Numbers in italics refer to the pages on which the complete references are listed.

## C



## D

N

O



P



R



S

## T

Y



Z

# Subject Index

## H



## I



## J



## K



## L

## Q



## R